ORGANIZED MULTIENZYME SYSTEMS:
Catalytic Properties

BIOTECHNOLOGY AND APPLIED BIOCHEMISTRY SERIES

A list of books in this series is available from the publisher on request.

ORGANIZED MULTIENZYME SYSTEMS:
Catalytic Properties

Edited by

G. Rickey Welch

Department of Biological Sciences
University of New Orleans
New Orleans, Louisiana

1985

ACADEMIC PRESS, INC.

(Harcourt Brace Jovanovich, Publishers)

Orlando   San Diego   New York   London
Toronto   Montreal   Sydney   Tokyo

ACADEMIC PRESS INC.
Orlando, Florida 32887

United Kingdom Edition published by
ACADEMIC PRESS INC. (LONDON) LTD.
24–28 Oval Road, London NW1 7DX

LIBRARY OF CONGRESS CATALOG CARD NUMBER: 85-11198

ISBN: 0-12-744040-2

PRINTED IN THE UNITED STATES OF AMERICA

85 86 87 88   9 8 7 6 5 4 3 2 1

Contents

85-5461

Chapter 6.   Models of Organized Multienzyme Systems: Use in Microenvironmental Characterization and in Practical Application

Nils Siegbahn, Klaus Mosbach, and G. Rickey Welch

Chapter 7.   Kinetic Analysis of Multienzyme Systems in Homogeneous Solution

Philip W. Kuchel

Chapter 8.   Theoretical and Experimental Studies on the Behavior of Immobilized Multienzyme Systems

J. F. Hervagault and D. Thomas

Chapter 9.   Long-Range Energy Continua and the Coordination of Multienzyme Sequences *in Vivo*

G. Rickey Welch and Michael N. Berry

Contributors

Numbers in parentheses indicate the pages on which the authors' contributions begin.

MICHAEL N. BERRY (419), Department of Clinical Biochemistry, School of Medicine, Flinders University of South Australia, Bedford Park, South Australia 5042, Australia

PETER FRIEDRICH (141), Institute of Enzymology, Biological Research Center, Hungarian Academy of Sciences, Budapest H-1502, Hungary

J. F. HERVAGAULT (381), Laboratoire de Technologie Enzymatique, E.R.A. n°338 du C.N.R.S., Université de Compiègne, 60206 Compiègne, France

DOUGLAS B. KELL (63), Department of Botany and Microbiology, University College of Wales, Aberystwyth SY23 3DA, Wales

PHILIP W. KUCHEL (303), Department of Biochemistry, University of Sydney, Sydney, New South Wales 2006, Australia

BORIS I. KURGANOV (241), The All-Union Vitamin Research Institute, Moscow Nauchny, proezd 14A, Moscow 117246, U.S.S.R.

KLAUS MOSBACH (271), Pure and Applied Biochemistry, Chemical Center, University of Lund, S-220 07 Lund, Sweden

JACQUES RICARD (177), Centre de Biochimie et de Biologie Moléculaire du C.N.R.S., 13402 Marseille, Cedex 9, France

NILS SIEGBAHN (271), Pure and Applied Biochemistry, Chemical Center, University of Lund, S-220 07 Lund, Sweden

PAUL A. SRERE (1), Pre-Clinical Science Unit, Veterans Administration Medical Center, and Department of Biochemistry, University of Texas Health Science Center at Dallas, Dallas, Texas 75216

D. THOMAS (381), Laboratoire de Technologie Enzymatique, E.R.A. n° 338 du C.N.R.S., Université de Compiègne, 60206 Compiègne, France

G. RICKEY WELCH (271, 419), Department of Biological Sciences, University of New Orleans, New Orleans, Louisiana 70148

HANS V. WESTERHOFF (63), National Institutes of Health, Bethesda, Maryland 20205

Preface

When we picture cell metabolism we usually think of *multienzyme sequences*, rather than of individual enzymes. Although the study of specific isolated enzyme activities has carried us far in elaborating the kinetic/catalytic basis of the metabolic machinery, to understand the "whole" demands that we integrate the individual enzyme with its metabolically sequential neighbors and with the cellular infrastructure. There is now compelling evidence, both empirical and theoretical, that most (if not all) of intermediary metabolism takes place in organized states. The mode of organization of the component enzymes may entail one or more of the following: (i) protein–protein complexes, (ii) association on (or in) a membrane, and (iii) attachment to fibrous cytoskeletal elements. In some cases the organization is strong, such that the system can be extracted from cells and studied by standard analytical–enzymological methods. For others, however, the interaction is weak (or transient) and readily disrupted by extraction, dilution, etc., in which case special analytical techniques and indirect methods must be used.

Importantly, these organized multienzyme systems have exhibited unique forms of kinetic/catalytic facilitation, unlike the counterpart systems free in bulk solution. Studies with such designs are pointing us to a more realistic appreciation of the nature (and role) of the *microenvironment* in enzyme action. Indeed, microenvironmental factors dictate the very kinetics of enzyme systems, as regards influence on coupled reaction–diffusion flows as well as effects (mediated by protein–protein or protein–matrix interaction) on the intrinsic catalytic properties of the individual enzymes. Continued exploration of structure/function relationships in these organized regimes will be vital to the gradual, dialectical course of constructing a "cellular biochemistry," as well as to applied uses in enzyme technology.

Contributions to our understanding of the kinetic/catalytic properties of organized enzyme systems have come from two sectors. First, we have studies on multienzyme clusters isolated from living cells (or reconstructed with individual components isolated therefrom). These are "Nature's own" designs, a knowledge of which is foremost to biochemists. Second, there is the work on artificially immobilized enzymes. From the biological perspective these systems have served a significant role, as stable "macroscopic" models wherein organizational and microenvironmental parameters can be defined and manipulated more discerningly. And, of course, the immobilized schemes have had an impact in areas of applied biochemistry.

Our purpose in this volume is to review current work in a number of areas concerning enzyme organization, largely from the biological angle. We include both immobilized and naturally occurring systems, although the heavier emphasis is on the latter. Chapter 1 provides an examination of the nature and function of enzyme organization in perhaps the most well-studied cellular organelle, the mitochondrion. Just as structuralization of metabolism permeates the cytoplasm, so is the case for the *milieu intèrieur* of organelles as well. Chapter 2 is a discourse on structural/functional coupling of the components in energy-transducing membrane systems. Exciting new advances are being made in the elucidation of these bioenergetic designs, giving us a more "local" view of protonmotivated ATP synthesis. Chapter 3 is a discussion of "dynamic compartmentation" in soluble multienzyme systems, with particular emphasis on glycolysis. Such transiently interacting designs may constitute the most prevalent form of organization for cytosolic metabolic processes. Chapter 4 is a detailed review of allosteric enzyme systems. The goal is to define the difference between the behavior of an isolated globular protein and that of a corresponding subunit in a polymeric system. Chapter 5 is a treatment of allosterism in reversibly adsorptive enzyme systems. There is increasing evidence that many metabolic enzymes can partition between particulate structures and the cytosol, depending on physiological conditions. Chapter 6 relates model studies with specific immobilized multienzyme sequences, as regards the analysis of microenvironmental effects. Both biological and applied implications are discussed. Chapter 7 gives a detailed mathematical exposition on the kinetic analysis of multienzyme systems in homogeneous solution. Some new methods are provided for solving the differential rate equations describing consecutive enzyme reactions, so as to allow parameter evaluation (for noninteracting as well as for interacting enzyme systems), optimization of coupled assays, *inter alia*. Chapter 8 includes some theoretical and experimental studies on the behavior of immobilized systems, wherein we see how standard reaction–diffusion equations must be altered for heterogeneous states, and how some interesting biological phenomena in such states can be modeled. As an indication of new directions in which "cellular biochemistry" may be going, Chapter 9 presents a somewhat speculative integrative view of the kind of functional coherence that may be operative in organized states *in vivo*. This view is based on a melding of the following features: (i) the reactivity of the very protein fabric of enzymes, (ii) the nature of the material substratum in which enzymes are organized *in vivo*, and (iii) the ambience of long-range "energy continua" (e.g., electrical fields, mobile protonic states) in cytological substructures.

The accumulation of knowledge on the behavior of enzymes in organized states is bound to yield fruits for basic and applied biochemistry alike. It is hoped that this volume will help foster continued dialogue and cooperation between these two sides of biochemistry. To this end, I gratefully acknowledge the editors of the "Biotechnology and Applied Biochemistry Series" for their receptiveness to the subject matter herein. In particular, I owe a large debt of gratitude to Dr. Lemuel B. Wingard, Jr., for his interest and encouragement and, also, for his assistance on numerous questions regarding organization of the book, style, etc. Also, I appreciate the assistance and patience of the Academic Press editorial staff in the construction of the text. Of course, my heartiest thanks go to the contributing authors, for their time and effort invested in producing such a fine collection of review chapters. It is, after all, *their* book.

G. RICKEY WELCH

1

Organization of Proteins within the Mitochondrion

Paul A. Srere

Pre-Clinical Science Unit
Veterans Administration Medical Center
and Department of Biochemistry
University of Texas Health Science Center at Dallas
Dallas, Texas

I. BIOLOGICAL ORGANIZATION

In no biological entity, be it a subcellular organelle, a bacterial cell, or a eukaryote, does it appear that structure depends only upon random juxtaposition of its components. Organized arrays of biological components permit effective use of the elements by the reduction in the numbers of molecules needed to achieve the same thermodynamic chemical potentials and separation of elements allows reactions with seemingly opposite objectives to be achieved within the same biospace.

The structural complexity of cells—even of the simplest micro-organisms—has been made clearly visible by electron microscopy. In spite of the apparent need for, usefulness of, and physical existence of compartments in cells, a major problem has existed in terms of the difficulty of obtaining experimental evidence showing the metabolic functioning of compartments *in vivo*. Further, what about the existence and function of unseen compartments, the microenvironments? Do they exist and, if so, what is their role?

There are many easily recognized forms of biological compartmentation. (1) In the biosphere each life form is a separate compartment and the interaction between them is referrred to as the study of ecology. (2) In each multicellular species the carrying out of special functions by differentiated cells is another example of compartmentation. (3) Within the individual cells compartmentation is achieved with organelles (nuclei, mitochondria, chloroplasts, lysosomes, peroxisomes, vacuoles) separated from each other by enclosing membranes. Even prokaryotic cells, at one time thought to be devoid of organelles, have minimally the separate regions of cytosol and periplasmic space. (4) In the absence of membrane-limited organelle compartmentation, functional separation exists by means of isolatable, stable complexes of enzymes and by skeletal networks of proteins found in various subcellular compartments. (5) Finally, it seems probable that compartmentation exists even in the absence of membranes and stable enzyme complexes. In the latter case the compartment may be the microenvironment in a region of weakly interacting proteins or the microenvironment near a surface due to unstirred water layers or due to a proposed water structure in the cell. Indeed certain catalytic systems have been shown to become an oscillating structure, which is indicative that inhomogeneties can be formed in initially homogeneous solutions.

Below the level of stable multienzyme complexes direct evidence for microenvironmental effects in cells is unavailable. If separate microenvironments exist within the same physical compartment of a cell, all the present experimental attempts to evaluate which metabolites are regulatory in the various pathways are far from the mark.

Many aspects of metabolic regulation can be attributed to the kinetic properties of enzymes and the amounts of the enzymes that are present in cells. The metabolic advantages of microenvironments and compartmentalization are manifold but the data to support their role in regulation or even their existence are at present little more than suggestive. A number of reviews have appeared that discuss various aspects of the subject of compartmentation (Srere and Mosbach, 1974; Srere and Estabrook, 1978; Welch, 1977; Mosbach and Mattiasson, 1978; Ottaway and Mowbray, 1977; Masters, 1977; Nover *et al.*, 1980).

This paper will present and examine the available evidence for the existence of quinary structures (interactions between proteins) (McConkey, 1982) of metabolically related enzymes within the matrix and membranes of mitochondria.

II. THE MITOCHONDRION

A. Introduction

A mitochondrion has four separate metabolic compartments: the outer membrane, the intermembrane space, the inner membrane, and the matrix space (Racker, 1970; Tandler and Hoppel, 1972, Lloyd, 1974; Munn, 1974; Fleischer et al., 1978; Tzagoloff, 1982) (Fig. 1).

For the purpose of this review I will adopt the accepted notion that the cristal membrane is identical to the inner boundary membrane, even though there does exist some contrary evidence (Brdiczka et al., 1974). The question that I will examine in this article is whether or not an organization of proteins exists in the inner membrane and in the matrix space of the mitochondrion.

Both the inner membrane and matrix compartments are characterized by their high protein concentration. The inner membrane is reported as being 72% protein and 28% lipid (see Capaldi, 1982, for review), while the matrix space has a varying protein concentration, depending on its physiological state, of 36 to 50% (Hackenbrock, 1968). One of the main metabolic activities of these two contiguous compartments is to oxidize reduced carbon of carbohydrates, fats, and amino acids to CO_2 and H_2O and efficiently trap the energy as ATP. This is accomplished by three sets of enzymes: one set (the converting enzymes) converts carbohydrates, lipids, and amino acids to tricarboxylic acid cycle intermediates; a second set is the tricarboxylic acid

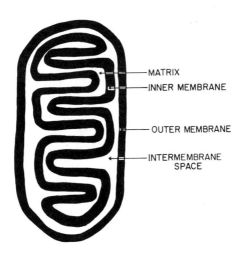

Fig. 1. The four mitochondrial compartments.

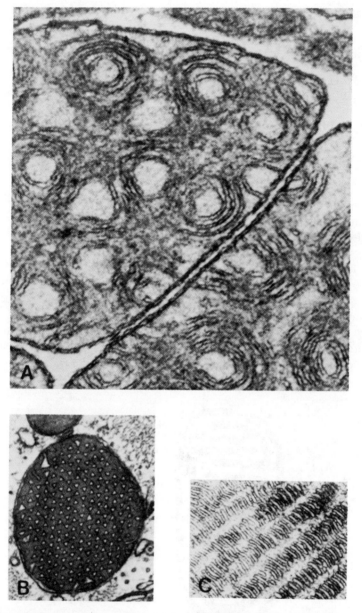

Fig. 2. Electron micrographs of some mitochondria showing a variety of cristal forms (A) Part of a mitochondrion from the ellipsoid of the retina of a tree shrew (Samorajski *et al.* 1966) with permission of the Rockefeller Univ. Press. (B) Mitochondrion in an astrocyte of the hamster brain

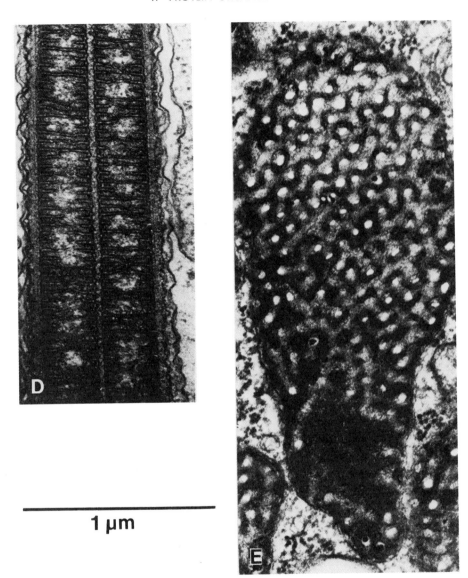

1 µm

(Blinzinger *et al.*, 1965) with permission of the Rockefeller Univ. Press. (C) Part of a mitochondrion of a housefly (*Musca domestica*) from Smith *et al.* (1970) with permission. (D) Dragonfly spermatozoan mitochondria with a regular arrangement of cristae (Kessel, 1966) · with permission. (E) A mitochondrion from slime mold (*Didymum*) (from Tandler and Hoppel, 1972, micrograph courtesy of F. Schuster).

(TCA)[1] cycle enzymes themselves, some of which remove hydrogen atoms from the intermediate compounds; and the third set of enzymes is in the inner membrane involved in the transfer of the electrons to oxygen and the conversion of ADP and P_i to ATP. An associated set of membrane proteins is involved in the transport of substrates and products across the inner membrane. Therefore, the question I will examine will be:

Are the Krebs tricarboxylic acid cycle enzymes and the converting enzymes in the matrix sequentially organized by quinary interactions?

I will also briefly discuss the questions:

Are the electron transport proteins within the inner membrane organized?

Are there specific interactions between the proteins of these two enzyme systems in the two separate compartments?

When we examine the Krebs TCA[1] cycle enzymes of the matrix, even though they constitute the components of the major metabolic activity ascribed to mitochondria, they do not constitute a majority of the matrix proteins (Henslee and Srere, 1979). The electron transport proteins and ATP synthase of the inner membrane, however, constitute a major fraction of the total inner membrane protein (Capaldi, 1982).

The complexity of biological systems has led us to expect that organization of related components is a necessity for all cells. This has been confirmed not only at the electron microscopic level but also at the molecular level. One must note again the elegant experiments of Zalokar (1960) and Kempner and Miller (1968), which showed that centrifugation of *Neurospora* hyphae and *Euglena* cells indicated that all proteins in those cells are bound to high-molecular-weight material. I will examine all the evidence for and against the notion of *in situ* interactions of sequential enzymes, recently termed quinary (or quintary) interactions (McConkey, 1982), of the metabolically related enzymes in the matrix and in the inner membrane of the mitochondrion.

Even though all mitochondria are topologically identical in their four-compartment structure, there are numerous mitochondrial forms that have quite a different appearance (Fig. 2) by electron microscopy. I will discuss in this article the relation of surface density of cristae to mitochondrial function, but there are no current theories to explain the differences in forms that are observed.

---

[1] Abbreviations: TCA, tricarboxylic acid; SDS, sodium dodecyl sulfate; CS, citrate synthase; mMDH, mitochondrial malate dehydrogenase; cMDH, cytosolic malate dehydrogenase; mAAT, mitochondrial aspartate amino transferase; LDH, lactate dehydrogenase; PDC, pyruvate dehydrogenase complex; FUM, fumarase; STK, succinate thiokinase; αKGDC, α-ketoglutarate dehydrogenase complex; OAA, oxaloacetate; ICDH (NADP), NADP-dependent isocitrate dehydrogenase; ICDH (NAD), NAD-dependent isocitrate dehydrogenase; GDH, glutamate dehydrogenase; PEG, polyethylene glycol; CPSI, carbamoyl phosphate synthetase I; M_r, molecular weight.

Functionally, the aerobic production of energy is the critical function of mitochondria since certain cells can exist without functional mitochondria only as long as energy can be supplied by an alternate route such as glycolysis. It is interesting to note that the products of mitochondrial DNA are proteins involved in the energy production pathway or in the biosynthesis of these proteins.

B. Mitochondrial Matrix

1. Proteins of the Matrix

a. Volume and protein distribution of mitochondria. The concentration of proteins in the matrix can vary since the volume of the matrix is not fixed and the volume varies by a factor of two (Hackenbrock, 1968) both *in vivo* and *in vitro*. The volume of the matrix of liver mitochondria (presumably in State II) has been measured carefully by electron microscopy by several groups. A compilation of data of the studies of four different groups (Reith *et al.*, 1976) can be summarized as follows for an average rat liver mitochondrion, volume 0.89 μm^3 (0.65–1.25). There about 290×10^9 mitochondria in 1 gm of rat liver.

Loud (1968) estimates that about 16% of this volume is outer membrane and intermembrane space, 27.5% is cristae, and 56% is matrix volume. Thus the matrix volume of an average liver mitochondrion in State IV is 0.50 μm^3. Brunner and Bucher (1970) have estimated that 66% of the total protein of the mitochondria is in the matrix with about 9.6% in the outer membranes and 24.4% in the inner membranes. Schnaitman and Greenawalt (1968) have estimated the protein distribution in mitochondria based on the specific activities of marker enzymes. Their results indicate a protein distribution of: outer membrane, 4.1%; intermembrane space, 6.4%; inner membrane, 21.6%; and matrix, 67.9%.

b. Number of polypeptides in the matrix. The polypeptides of the mitochondrial matrix have been examined by several laboratories. Using SDS polyacrylamide gels, Clarke (1976) and others (Melnick *et al.*, 1973) indicated about 40 separate components. When the more powerful technique of O'Farrell (1975), which uses both isoelectric focusing and SDS polyacrylamide gel electrophoresis, is applied approximately 140 separate polypeptides can be resolved (Henslee and Srere, 1979). This is a minimum number since I have calculated that matrix polypeptides with fewer than about 500 molecules per mitochondrion would not be detected by the staining procedure.

The two-dimensional gel patterns from a number of different rat tissues were compared and found to be quite similar. We have identified about 10 of these matrix proteins on the two-dimensional gel pattern (Henslee and Srere, 1979). A list of known matrix proteins includes ~ 60–70 enzymes (Altman and Katz, 1976; Henslee, 1978).

These matrix proteins exhibit no particularly unusual global characteristics. That is no say, the size distribution of the matrix polypeptides compared to size distribution of cytosolic polypeptides appears quite similar. The matrix polypeptides have apparent isoelectric points that cover the entire pH range. Rat liver mitochondrial matrix contains a very high proportion of carbamoyl phosphate synthetase, which is absent in the matrix fraction of mitochondria from other tissues.

There appear to be at least 90–100 polypeptides that are common to the mitochondria examined from a number of rat tissues as judged by the two-dimensional locations of the polypeptides. It is also clear that the mitochondrial matrix from each tissue contains polypeptides that are unique to each tissue.

c. The Krebs TCA Cycle Enzymes. We have examined citrate synthase (CS) from a number of rat tissues. The pure enzymes from brain, heart, kidney, and liver appear to be identical proteins based upon physicochemical, kinetic, and immunological data (Moriyama and Srere, 1971; Matsuoka and Srere, 1973).

It is tempting to conclude on the basis of the similar two-dimensional locations of polypeptides of the mitochondrial matrix of several tissues and from the data with citrate synthase from the several rat tissues that most of the proteins of the mitochondrial matrix are identical from tissue to tissue in an animal and that these identities are related to the need for protein–protein interactions as well as for the catalytic activities. McConkey (1982) has shown by high-resolution two-dimensional polyacrylamide gel electrophoresis that at least half of 370 denatured polypeptides from hamster cells and human cells are indistinguishable. He postulates that this observation is one of a conservative molecular evolutionary process and that it may be due to the many molecular interactions of proteins necessary in a cell; he has termed these as *quinary* interactions.

The enzymes of the Krebs TCA cycle carry out the following overall reaction:

$$CH_3COSCoA + 3NAD^+ + (FAD) + MgGDP^- + H_2PO_4^- + 2H_2O \longrightarrow$$
$$2CO_2 + CoASH + 3NADH + (FADH_2) + MgGTP^{2-} + 3H^+$$

The enzymes involved in the sequence of reactions responsible for this overall reaction are listed in Table I. There is still a question about whether ICDH (NAD) or (NADP) is involved in the cycle (Srere, 1972; Ottaway, 1983). The cycle can be sustained by the inflow of any of its intermediates: acetyl-

TABLE I
ENZYMES OF THE KREBS TCA CYCLE

Enzyme	EC number	$M_r (10^5)$	Subunit composition	Cytosolic isozyme
Citrate synthase	4.1.3.7	0.98	α_2	No
Aconitase	4.1.1.3	0.66	α	Yes
Isocitrate dehydrogenase (NAD)	1.1.1.41	1.6	$\alpha_2\beta\lambda$	No
Isocitrate dehydrogenase (NADP)	1.1.1.42	0.58	α	Yes
α-Ketoglutarate dehydrogenase	1.2.4.1			
complex	2.3.1.61	40 (?)	$(\alpha\beta\gamma)$	No
	1.6.4.3			
Succinate thiokinase	6.2.1.4	0.7	$\alpha\beta$	No
Succinate dehydrogenase	1.3.99.1	0.97	$\alpha\beta$	No
Fumarase	4.2.1.2	1.94	α_4	Yes?
Malate dehydrogenase	1.1.1.37	0.7	α_2	Yes

CoA, citrate, cis-aconitate, isocitrate, α-ketoglutarate, succinate, succinyl-CoA, fumarate, malate, or oxalacetate.

Two of the major sources of acetyl-CoA (AcCoA) are the pyruvate dehydrogenase complex and the β-ketoacyl-CoA thiolase of the fatty acid oxidizing system. Additionally, the transaminases will generate α-ketoglutarate, oxaloacetate, and pyruvate from amino acid precursors for oxidation in the cycle. Succinyl-CoA can be generated from methylmalonyl-CoA, which in turn is formed from propionyl-CoA and CO_2. Some auxiliary enzymes are listed in Table II.

The location of almost all the enzymes listed in Tables I and II is considered to be the matrix compartment. It is certain that succinate dehydrogenase is in

TABLE II
SOME AUXILIARY ENZYMES OF THE KREBS TCA CYCLE

Enzyme	EC number	Function
Pyruvate dehydrogenase complex	1.2.4.1	Generate acetyl-CoA from carbohydrate
	2.3.1.12	
	1.6.4.3	
Acetyl-CoA synthase	6.2.1.1	Generate acetyl-CoA
β-Ketoacyl CoA thiolase	2.3.1.16	Generate acetyl-CoA from fatty acids
Thiolase	2.3.1.9	Generate acetyl-CoA from acetoacetate
Aspartate aminotransferase	2.6.1.1	Generate oxaloacetate from aspartate
Glutamate dehydrogenase	1.4.1.3	Generate α-ketoglutarate from glutamate
Propionyl-CoA carboxylase	4.1.1.41	Generate succinyl-CoA from propionate
Methylmalonyl-CoA mutase	5.4.99.2	Generate succinyl-CoA from propionate

the inner membrane but it is considered at present to be an extrinsic protein. The two keto acid dehydrogenase complexes are attached to the membranes but are more easily removed than is succinate dehydrogenase. The other Krebs TCA cycle enzymes may or may not be attached to the membrane, and that point will be discussed in Section II,B,4.

It should be noted that many enzymes of the Krebs TCA cycle have cytosolic counterparts (isozymes). These include aconitase, isocitrate dehydrogenase (NADP), and malate dehydrogenase. Fumarase may have an isozyme as shown by genetic experiments but Halper and Srere (1979) showed that the cytosolic fumarase isozyme in pig heart could not be distinguished from its mitochondrial counterpart.

A number of interesting observations concerning the content of Krebs TCA cycle enzymes in mitochondria have been reported. First, there is a direct correlation between the oxidative capacity of a cell and the content of Krebs TCA enzymes. Although this point has not been documented for each of the enzymes, several examples can be given. Table III lists a series of different cell types, their Q_{O_2}, and their citrate synthase content (Srere, 1969). A second observation has been that of Pette et al. (1962), who have shown that the enzymes of the Krebs TCA cycle seem to be a constant proportion group of enzymes. Comparison of these activities in a series of different cells showed a remarkable constancy in their relative activities, using cytochrome c content of mitochondria as a reference point. It has been observed also that conditions that cause the induction of one Krebs TCA cycle enzyme also cause an induction of all of them. Thus, we have shown that vitamin B_{12} deficiency in rats leads to an increase in citrate synthase in rat liver (Frenkel et al., 1976). When all the enzymes of the cycle are measured an increase is found in all of the Krebs TCA cycle enzymes (Matlib et al., 1979). Similar results have been reported for exercised rats, in which an increase of citrate synthase and other Krebs TCA cycle enzymes has been reported (Holloszy et al., 1970). Injection of T_3 to rats causes an increase in several Krebs cycle enzymes in rat muscle

TABLE III

CITRATE SYNTHASE CONTENT AND THE RESPIRATION OF SOME CELLS

Cell	Q_{O_2}	Citrate synthase (μmol C_2 unit/min/g)	Ratio
Rat heart (active)	(5.6)	77	14
Rat liver	0.92	9	10
Yeast (on glucose)	0.96	2.1	2.2
Yeast	4.7	14	3.1
E. coli	10.2	45	4.5
Azotobacter	280	660	2.6

(Winder *et al.*, 1975). In a study of the citrate synthase in yeast it was shown that a direct linear relation existed between the Q_{O_2} of the yeast cell and the specific activity of citrate synthase (Nunez de Castro *et al.*, 1976).

The question of whether a stoichiometric relationship exists among the matrix Krebs TCA cycle enzymes has never been studied directly. In order to make such a calculation one must have an accurate measure of total mitochondrial enzyme activity, specific activity of the pure enzyme, and molecular weight of the pure enzyme. Unfortunately, these data are not available for a single tissue. I have calculated the stoichiometry of Krebs TCA cycle enzymes using the data from a number of different laboratories (Srere, 1972). A comparison of the molar ratios of active sites of these enzymes reveals that the amounts are within one order of magnitude. All these results would seem to indicate a careful control of the total amount of each enzyme and that no enzyme is rate limiting. Although for many years it has been assumed that the rate-limiting step of the Krebs TCA cycle was at the citrate synthase reaction, recent analysis of the control of the pathway using the method of Kacser and Burns (1973) would distribute control among several steps of the cycle (Ottaway, 1976).

There is little information concerning the turnover times of the enzymes of the Krebs TCA cycle in the same tissue. We have determined that the citrate synthase of rat liver has a half-life of about 94 h (Mukherjee *et al.*, 1976). No other data are available for other rat liver Krebs TCA cycle enzymes. One study showed that the turnover times for the mitochondrial matrix proteins

TABLE IV

LOCATION OF MITOCHONDRIAL KREBS TCA CYCLE AND AUXILIARY ENZYMES ON HUMAN CHROMOSOMES

Enzyme	EC number	Chromosome
Aconitase	4.2.1.3	22[a]
Citrate synthase	4.1.3.7	12
Diaphorase 1	1.6.4.3	22[a]
Diaphorase 2	1.6.4.3	7 (?)
Diaphorase 4	1.6.4.3	16
Fumarate hydratase	4.2.1.2	1[b]
Aspartate aminotransferase	2.6.1.1	16
Isocitrate dehydrogenase (NADP)	1.1.1.42	15
Malate dehydrogenase	1.1.1.37	7
Malic enzyme	1.1.1.40	11(?)
Succinate dehydrogenase	1.3.99.1	1[b]

[a] Not adjacent on chromosome 22.

[b] Not adjacent on chromosome 1.

varied widely in rat hepatoma cells but no identification of the individual proteins was made (Hare and Hodges, 1982).

The publication of the human gene map (McKusick, 1983) lists the chromosomal location for a number of mitochondrial Krebs TCA cycle enzymes (Table IV). It is clear that they are distributed on many different chromosomes and also that their isozymes are also widely separated. In *Escherichia coli* the genes for succinate dehydrogenase and two components for the α-ketoglutarate dehydrogenase (αKGD) cluster lie very close together at 17 min on the standard map; although citrate synthase is also at 17 min it is separated from these three. However, two components of pruvate dehydrogenase (3 min) and lipoamide dehydrogenase (3 min), isocitrate dehydrogenase (NADP) (25 min), malate dehydrogenase (70 min), and four fumarate reductase components (94) are quite widely distributed in the *E. coli* genome (Bachman, 1983). Since the genes for enzymes of the Krebs TCA cycle are not consecutive in either *E. coli* or in the human and since the amounts of the enzymes apparently are maintained in a constant ratio to each other, one must minimally propose a common control element for the genes (Table V).

There is little information concerning the control of biosynthesis of Krebs TCA cycle enzymes even in simple systems of either prokaryotes or eukaryotes. The isolation of the messenger RNA for citrate synthase has been reported for *Neurospora* (Harmey and Neupert, 1979) and for yeast (Alam *et al.*, 1982).

TABLE V

LOCATION OF KREBS TCA CYCLE ENZYMES ON THE *E. coli* GENOME[a]

Symbol	Enzyme	Minutes
ace E	Pyruvate dehydrogenase decarboxylating component (EC 1.2.4.1)	3
frd A	Fumarate reductase (EC 1.3.99.1) flavoprotein subunit	94
frd B	Fumarate reductase iron sulfur subunit	94
frd C	Fumarate reductase membrane anchor polypeptide	94
frd D	Fumarate reductase membrane anchor polypeptide	94
glt A	Citrate synthase (EC 4.1.3.7)	17
icd	Isocitrate dehydrogenase (NADP) (EC 1.1.1.42)	25
mdh	Malate dehydrogenase (EC 1.1.1.37)	70
lpd	Lipoamide dehydrogenase (NADH) (EC 1.6.4.3)	3
nuR	Regulatory gene for fumarate reductase	29
ndh	Respiratory NADH dehydrogenase	22
Sdh	Succinate dehydrogenase (EC 1.2.4.2)	17
suc A	αKGD decarboxylating component (EC 1.2.4.2)	17
suc B	αKGD dehydrolipoyl transuccinate component (EC 2.3.1.61)	17

[a] Bachman (1983).

2. Interactions between Matrix Proteins

a. *Physical Evidence.* The advantage of organized arrays of metabolically related enzymes over a random array of enzymes seems to be obvious (see
Welch, 1977, for review; Welch and Keleti, 1981). Briefly, one could imagine
that a fewer number of intermediate molecules confined to a microenvironment would have a higher chemical potential than the same number of
molecules in a disorganized system. (That this is indeed the case has been
demonstrated in a number of different systems and will be discussed more fully
in Section II,B,2,b).

Backman and Johansson (1976) showed that mitochondrial aspartate
transaminase and mitochondrial malate dehydrogenase specifically interacted
with each other. Their method of analysis consisted of determining the
countercurrent distribution of the enzymes in a biphasic mixture of H_2O,
dextran, and carboxymethyl polyethylene glycol when they were mixed and
when they were separate. Interaction was indicated by a change in the
distribution of the enzymes when they were mixed. If either enzyme was
replaced by its cytosolic isozyme, then no interaction was detected. Interestingly, the two cytosolic isozymes together also show an interaction by this
method.

Following these experiments Halper and Srere (1977) showed that citrate
synthase and mitochondrial malate dehydrogenase (mMDH) coprecipitated
in polyethylene glycol, as measured by an increase in optical density of the
solution. Analysis of the resulting precipitate showed it contained both
enzyme activities. No precipitate was formed when the mitochondrial malate
dehydrogenase was replaced with cytosolic malate dehydrogenase (cMDH),
bovine serum albumin, or a number of other proteins. It has been shown by
Miekka and Ingham (1980) that polyethylene glycol acts by reducing the
solubility of heterocomplexes of proteins. Its action cannot be attributed to
the induction of complex formation and therefore the interactions observed
using this method are probably functionally important. We have shown that
similar results can be obtained when the solvent is a concentrated ammonium
sulfate solution. Thus at ammonium sulfate concentrations at which no
precipitation of mMDH, CS alone, or a mixture of CS and cMDH occurs, one
finds a rapid precipitation of a mixture of CS and mMDH.

It has also been observed that it is possible to cross-link mMDH and CS
using glutaraldehyde (Koch-Schmidt *et al.*, 1977). These complexes show
enhanced catalytic activity and will be discussed in the next section (II,B,2,b).

Beeckmans and Kanarek (1981a,b) have studied the interactions between
the Krebs TCA cycle enzymes, MDH, CS, and fumarase (FUM), and also
interactions between these and aspartate aminotransferase (AAT). They

immobilized one of the enzymes on a gel and then studied the binding of the other enzymes to the immobilized enzyme. Using this technique they reported interactions between fumarase and malate dehydrogenase, malate dehydrogenase and citrate synthase, and fumarase and citrate synthase. In addition, they showed interactions between malate dehydrogenase and aspartate aminotransferase. In a slightly different method of immobilization with antibodies to fumarase or malate dehydrogenase that had been bound to Sepharose–Protein A they were able to show interaction between fumarase and malate dehydrogenase. Both pig heart and chicken heart enzymes were used in these studies. It was possible to calculate a stoichiometry of four molecules of malate dehydrogenase bound to each fumarase. The MDH–FUM complex can bind either CS or AAT but not both and it was proposed that alternate binding as possibly affected by substrates might constitute a switch mechanism between citric acid cycle activity and aspartate–malate shuttle activity. No specificity was seen for mitochondrial malate dehydrogenase over cytosolic malate dehydrogenase in their binding to citrate synthase or AAT. However, there was specificity in the binding of fumarase to mMDH but not cMDH. Aldolase and lysozyme proteins with pI values very close to that of mMDH did not show any of the interactions observed for MDH. The observed interactions between the proteins were sensitive to ionic strength but did not show any special sensitivity to metabolites.

Dulin and Harrison (1983) have reported that CS exhibits an apparent molecular weight by chromatography on Sephacryl S-300 of between 87,000 and 138,000 in the presence of mMDH. The interaction depends upon ionic strength, enzyme concentration, and the presence or absence of certain buffer anions. An apparent K_d for CS–mMDH of 0.2 μM was determined. Mullinax and Harrison (1983) using an immobilized enzyme technique similar to that of Beeckmans and Kanarek (1981a) showed that immobilized MDH bound CS and AAT and that the amount of CS bound could be increased by the addition of oxaloacetate.

Sumegi et al. (1980) reported an interaction between citrate synthase and the pyruvate dehydrogenase complex (PDC). This interaction was demonstrated using ultracentrifugation, in which the CS (M_r 97,900) is shown to sediment with the much larger PDC (M_r 1 × 10^6). In addition, CS would coprecipitate with PDC in 3.5% polyethylene glycol. Several chromatographic techniques, frontal analysis, the Hummel–Dreyer method, and active enzyme chromatography all showed that, in the presence of PDC, the elution position of CS changed in a way that indicated that it was bound in a high-molecular-weight species. Several controls were run with these experiments. First, other high-molecular-weight complexes, such as αKGDC and glutamate dehydrogenase, could not substitute for PDC in these experiments. Second, cMDH could not substitute for CS in the experiments. Sumegi et al. (1980) calculated that about

10 CS molecules bind with 1 PDC molecule with a K_d of about 5 μM. These workers showed that the citrate synthase interacts with the E_2 (trans-acetylase) portion of PDC with much greater affinity than with E_1 and E_3.

In a similar series of experiments Z. Porpaczy *et al.* (1983) have shown a specific interaction between the α-ketoglutarate dehydrogenase complex and succinate thiokinase (STK). All the physical characteristics of the PDC–CS interactions were shown to be true for the αKGDC–STK interaction.

Fahien and colleagues published a series of papers concerning the interaction between glutamate dehydrogenase and both mMDH and mAAT to form complexes (Fahien *et al.*, 1977, 1978, 1979; Fahien and Kmiotek, 1978, 1979; Fahien and Smith, 1974; Fahien and van Engelen, 1976). The evidence included gel filtration, fluorescence, kinetic, and polyethylene glycol precipitation data. Experiments with a divalent cross-linker also indicated that complex formation occurred with these enzymes (Fahien and Kmiotek, 1983). Additional studies from Fahien's group confirmed the interaction between AAT and CS. This latter complex is favored in the presence of NADH and malate. Palmitoyl-CoA is found to enhance the formation of complexes between the pairs of enzymes. These studies have been controlled for specificity by showing that cMDH, cAAT, and BSA do not show complex formation with the mitochondrial enzymes.

b. Kinetic Evidence.   It seems obvious that if soluble sequential metabolic enzymes form a complex or are immobilized in the vicinity of each other, and if forces (electrostatic, for example) or physical situations (such as an unstirred water layer or direct delivery of substrates to active sites) allow the product of the first enzyme reaction (and the substrate of the second enzyme reaction) to remain in the vicinity of the second active site longer than the turnover time of the second enzyme, then a kinetic advantage will be seen for this coupled situation over one in which the same quantities of enzyme are free in the same bulk solution.

The crux of there being a kinetic advantage depends mainly upon the diffusion times for the product–substrate compared to the turnover time of the second enzyme. Such calculations concerning diffusion times as they apply to the mitochondrial matrix will be made below (Section II,B,6,b). It can be said, however, that if the matrix compartment of the mitochondria has the characteristics of water, then the time for diffusion for small molecules between enzymes within a cell (10^{-8} s) is much faster than the turnover time of most enzymes (10^{-3} s). It would appear therefore that the kinetic advantages of the association of two sequential enzymes are probably due to the fact that diffusion does not occur as it does in a bulk solution and increased product–

substrate concentration in the microenvironment of the second enzyme can occur (Koch-Schmidt *et al.*, 1977). The fact that immobilized enzymes (sequential) had an advantage over the free enzymes was shown first by Mosbach and Mattiasson (1970) for hexokinase and the glucose-6-phosphate dehydrogenase system.

   i. Immobilized enzyme studies. For mitochondrial matrix enzymes the kinetic advantage of immobilized over free enzyme was first demonstrated by Srere *et al.* (1973b). These workers showed that when CS and MDH were immobilized together either by several different covalent coupling methods or entrapped in gels together the rate of the coupled reaction was greater than that of the free enzymes in solution. The problem arose when the rate of citrate synthase was considered in regard to the rate of the Krebs TCA cycle (Srere, 1972). A further increase in activity was seen when the third enzyme, lactate dehydrogenase (LDH), was activated. In this case LDH acts as an NADH acceptor mimicking the NADH acceptor system in the inner mitochondrial membrane. We have shown that no change in the K_m for oxaloacetate (OAA) occurs when citrate synthase is immobilized, so an explanation for this phenomenon is probably related to restricted diffusion and is not due to a change in the enzyme properties when it is immobilized (Mukherjee and Srere, 1978).

   The experiments of Koch-Schmidt *et al.* (1977) showed that soluble complexes of cross-linked CS and mMDH did not show a kinetic advantage over a soluble system of CS and mMDH. However, when these two enzymes either separately or in the cross-linked aggregated state were immobilized on Sepharose beads, a steady state was reached immediately, which contrasted to the nonimmobilized free or aggregated stage, in which a pronounced lag was observed. Since cross-linking the aggregated enzyme complex reduced the proximity of the enzymes from an average of about 2000 nm (in free solution) to 2 nm (cross-linked) with no change of kinetics between the two, it is clear that change in proximity does not confer a kinetic advantage on this system. Immobilization, however, creates a favorable condition in that diffusional hindrance by the gel phase of the oxaloacetate intermediate results in its higher microenvironmental concentration and elimination of the lag phase.

   In another elegant experiment Mansson *et al.* (1983) immobilized alcohol dehydrogenase and LDH to Sepharose in a manner such that their active sites were facing each other. In this case a kinetic enhancement was observed over the system in which the enzymes were immobilized to the same bead. Thus the lack of a proximity effect observed in the MDH and CS experiments by Koch-Schmidt *et al.* (1977) may have been due to the random orientation of the active sites of the two linked enzymes.

   ii. Enzyme kinetics. Sumegi *et al.* (1980) have reported that the K_m for CoA for the pyruvate dehydrogenase complex is decreased from 10 μM for the PDC alone to 1.5 μM for the PDC–CS complex. No change for the K_m for pyruvate or NAD, or for V_{max}, was observed between the PDC and PDC–CS complex. Furthermore, when other enzymes that could convert AcCoA to CoA, such as arylamine acetyltransferase, carnitine acetyltransferase, or phosphotransacetylase, were substituted for CS no change in the K_m for CoA was observed. These workers also showed the K_m for acetyl-CoA in the CS reaction was 3.1 μM for the PDC–CS complex compared to 12 μM for CS alone. This would indicate not only a proximity effect in the complex, but a possible alteration in the active sites of the two enzymes when the complex forms. A similar interpretation for other enzyme systems is treated more extensively by Welch (1977).

In the analogous interaction of αKGDC with STK, Z. Porpaczy et al. (1983) have shown that the k_m for CoA in the αKDGC reaction was 5.5 μM for this enzyme alone and 3.5 μM for the αKGDC–STK complex. The other kinetic constants for the αKGDC reaction did not change. A more dramatic change in K_m for succinyl-CoA in the STK reaction was observed. For STK alone, K_m was 65 μM and for the αKGDC–STK complex it was 1.5 μM. As in the previous case, the complex of the sequential metabolic enzymes shows a distinct kinetic advantage over the individual enzymes.

At about the same time that Backman and Johansson (1976) showed that complex formation occurred between mMDH and mAAT, Bryce et al. (1976) showed that a kinetic advantage existed for the coupled system in that no lag was observed. Indeed isotope experiments indicated that OAA produced as an intermediate in that reaction did not equilibrate with the bulk OAA of the solution, which indicated that a channeling of some sort was taking place. More recent evidence questions their results and at present there is doubt concerning the kinetic change (Manley et al., 1980). However, the evidence for the physical interaction between mMDH and mATT from other experiments remains quite good.

Cohen et al. (1982) have studied carbamoyl phosphate and citrulline synthesis in rat liver mitochondria. In earlier reports they had shown that the behavior of carbamoyl phosphate synthetase (CPS) in liver mitochondria is different from its behavior in solution in that ornithine stimulates the mitochondrial enzyme in situ but not the enzyme in solution. They showed further that the ornithine effect could be prevented when ornithine transcarbamylase was inhibited. The authors conclude that one explanation for these results is the existence of an interaction between carbamoyl phosphate synthetase and ornithine transcarbamylase within the mitochondrial matrix.

iii. Isotope studies. Several other experiments have indicated with kinetic and isotope data that a compartmentation of metabolites exists within the mitochondrial matrix. Schoolwerth and LaNoue (1980) studied the release of $^{14}CO_2$ from labeled glutamine in the presence of unlabeled glutamate in rat kidney mitochondria. Their results led them to conclude that matrix glutamate was channeled through the glutamate dehydrogenase reaction whereas external glutamate was metabolized through the glutamate–oxaloacetate transaminase (GOT) pathway. The existence of two different mitochondrial populations was tested for and found not to provide an explanation for the results.

These experiments are similar to those of Fritz (1968), who showed that unlabeled pyruvate decreased the $^{14}CO_2$ evolution from $[1-^{14}C]$palmitate but did not decrease the incorporation of $[1-^{14}C]$palmitate into $[^{14}C]$acetoacetate. Later, Fritz and Lee (1974) used $[1-^{14}C]$pyruvate and measured its conversion by rat liver mitochondria to $^{14}CO_2$ and $[^{14}C]$ketone bodies in the presence and absence of unlabeled octonate. In this case greater dilution of label into CO_2 was observed than for ketone bodies but calculations indicated that in the presence of fatty acyl carnitine the relative specific activities were lowered to the same extent.

McKinley and Trelease (1980) studied the incorporation of $[^{14}C]$acetate into mitochondria of the free-living nematode *Turbatrix aceti*. They examined the distribution of radioactivity into organic acids, amino acids, and sugars in the absence of and the presence of added unlabeled pyruvate and palmitate. They concluded that there were multiple separate pools of acetyl CoA within the mitochondrion.

A similar approach used by Lopes-Cardozo *et al.* (1978) yielded data consistent with the proposal that only a single pool of acetyl-CoA existed within rat liver mitochondia. Von Glutz and Walter (1975), however, using rat liver mitochondria oxidizing labeled palmitate in the absence and presence of unlabeled pyruvate, concluded that these organelles contained more than one compartment of acetyl-CoA.

There is therefore no clear consensus on these difficult-to-interpret isotope data. It can be said, however, that consideration of the high protein concentration in the matrix can lead to a physical representation of acetyl-CoA compartmentation, even allowing a normal diffusion coefficient of acetyl-CoA and the small size of the matrix. This hypothesis will be discussed later (Section II,6,b).

c. Studies in Situ

i. Studies with intact swollen mitochondria.   We have studied the kinetics of the Krebs TCA cycle *in situ* in two different ways to see if we could obtain evidence for a complex of these enzymes. In our first approach the rates of oxidation of various substrates were measured in intact and swollen rat liver mitochondria (Matlib and Srere, 1976). In swollen mitochondria one would expect a drastic reduction in the rate of oxidation of a substrate (if diffusion is limiting) since the mean free path between an intermediate and the next enzyme is increased. For an eight-step reaction, such as the Krebs TCA cyele, the decrease in rate might be large. It was not expected that swelling of the mitochondria would have a rate effect on electron transport since this is carried out by the inner membrane, which, even in the absence of organization of the electron-transport proteins, is a compartment that cannot be diluted by water. Our second approach used mitochondria made permeable with toluene and will be discussed later in this section (Matlib et al., 1977).

Extensive swelling of mitochondria accompanied by gross disorganization of matrix and at least threefold increase in mitochondrial volume resulted in the reduction of rates of respiration with NAD-linked Krebs TCA cycle intermediates. This reduction of rates of respiration was not due solely to the loss of intramitochondrial NAD, since added NAD did not restore respiration completely and could be explained by a disruption of the putative Krebs TCA cycle complex. Succinate- or β-hydroxybutyrate-supported respiration was not affected, probably because the enzymes are bound to the inner membrane; thus the electron-transport chain is not a rate-limiting step in the system.

When mitochondria were incubated in the presence of $MgCl_2$, the volume changes were small compared to incubation in the absence of $MgCl_2$, and no drastic matrix disruption was seen. The rate of respiration with malate and pyruvate by these mitochondria could be restored with added NAD.

At an intermediate stage of swelling (15 min incubation in the absence of $MgCl_2$) the loss of respiration with pyruvate and malate was small and could be recovered with the addition of NAD. It is therefore unlikely that the volume change or loss of any substrate other than NAD was responsible for the irreversible loss of respiration in the mitochondria swollen for 30 min in the absence of $MgCl_2$.

Most previous studies on respiration of swollen mitochondria were carried out using the substrates succinate and β-hydroxybutyrate, which are oxidized by membrane-bound dehydrogenases. In some cases both phosphorylation and respiration could be restored under certain conditions. The ability of NAD to restore respiration is apparently dependent on the degree of disorganization of the matrix.

Optical density changes and electron microscopic studies in that investigation indicated that after a 30-min incubation at 25°C in the absence of $MgCl_2$, the matrix was grossly disorganized, which may account for the inability of the mitochondria to regain their respiratory control and phosphorylative activities.

The inhibition of respiration at the late swelling stage may be due to a disorganization of a loosely associated putative complex of Krebs TCA cycle enzymes on the inner surface of the inner membrane. In swollen mitochondria the gellike matrix is disorganized and the so-called matrix enzymes are probably free within the inner membrane; respiration supported by these enzymes in this state is reduced and cannot be restored by the addition of NAD. In intact or swollen but not disrupted mitochondria, however, the Krebs TCA cycle enzymes may be organized in assemblies with a fixed relation to each other and the inner membrane. The statistical distance between the enzymes, their active site orientations, and their substrates in a complex are such that there may be a microenvironment of higher substrate concentration for each enzyme in the cycle. In such a situation the Krebs cycle enzymes would function more efficiently at very low overall apparent concentrations of substrates, since locally high concentrations could be maintained in the regions of each enzyme's active site.

Based on transit times for substrate and enzyme concentrations, an organization of glycolytic enzymes was proposed in rat liver cells (Hubscher et al., 1971) and ruled out in yeast cells (Hess and Boiteux, 1972) (for discussion see Section II,B,6,b). Nothing is known of the transit times for intermediates of Krebs TCA cycle enzymes in the mitochondrial matrix. Even if diffusion of intermediates is not a rate-limiting process in the mitochondrial matrix, there are advantages in an organized system of enzymes as described earlier (Srere, 1972; Srere et al., 1973b; Srere and Mosbach, 1974).

ii. Studies with mitochondria made permeable with toluene. In another approach to the question of matrix enzyme organization we have used toluene to make mitochondria permeable (Matlib et al., 1977) to the substrates of the Krebs TCA cycle. Toluene treatment of microorganisms to render them permeable to metabolites has been employed in *in vivo* studies of synthesis of DNA (Moses and Richardson, 1970) and RNA (Peterson et al., 1971), studies of the kinetic properties of phosphofructokinase (Reeves and Sols, 1973), and studies of the regulation of citrate synthase (Weitzman, 1973; Weitzman and Hewson, 1973; Swissa et al., 1976). Toluene-treated yeast cells have been used in the study of *in situ* kinetic properties of hexokinase and pyruvate kinase (Serrano et al., 1973). By some unknown mechanism, toluene treatment makes the membranes permeable to charged or large molecules normally excluded by the intact cells, but the cells do not become permeable to macromolecules.

Hilderman and Deutscher (1974) and Hilderman et al. (1975) observed maximal activities of aminoacyl-tRNA synthetases in toluene-treated rat liver hepatocytes with only a small portion of the enzyme activities and proteins released from inside the cells into the suspending medium. The use of this method on mitochondria had the obvious advantage of possibly producing a mitochondrial system that was a closer approximation of the enzymes' normal *in vivo* environment.

The results of these studies indicate that isolated rat liver mitochondria can be made permeable to substrates and cofactors by treatment with toluene. Enzyme loss from the treated mitochondria, which occurred at low concentrations of mitochondria, could be prevented by using polyethylene glycol. Activities of matrix enzymes can be measured *in situ* with exogenously added substrates and cofactors. The enzyme activities obtained by this procedure were similar to those of sonicated mitochondria. Low succinate–cytochrome c reductase activity in toluene-treated mitochondria indicates that the cristae have been "pinched off" in a way such that the cytochrome c_1, which had been on the outer surface of the inner membrane (Racker and Horstman, 1972), is now on the inside of the vesicles and probably not available to the added cytochrome c. Low recovery of α-ketoglutarate dehydrogenase activity after assay is probably due to the instability of the liver enzyme (Linn, 1974). Higher leakage of sulfite oxidase ($\sim 50\%$) may be due to its easy access to the external solution from its actual location in the intermembrane space. Comparatively higher losses occur of malate dehydrogenase and isocitrate dehydrogenase (NADP), among the Krebs cycle enzymes, but whether this is due to smaller size (M_r 65,000) or to their possible dual localization in mitochondria was not determined. Loss of respiratory control of intact mitochondria in $\sim 8.5\%$ polyethylene glycol solution may be due to its effect on NAD-linked flavoproteins of the respiratory chain. In toluene-treated mitochondria, loss of respiratory control may be due to either leaky membranes or activation of mitochondrial ATPase or both. There was little damage caused to the mitochondrial structure as judged by electron microscopic studies (Shannon *et al.*, 1977). A major fraction of the mitochondrial constituents including outer and inner membrane remained intact in the treated mitochondria. It is possible that the toluene extraction of some cholesterol and phospholipids from the membranes makes them less rigid and allows certain membrane proteins to aggregate in a way such that a channel is formed, allowing nucleotides and metabolite to penetrate the mitochondrial membranes.

Toluene treatment of isolated mitochondria appears to be a gentle method of unmasking mitochondrial enzymes, particularly those enzymes that are located in the matrix compartment. Sonication totally disrupts the mitochondria and releases most of the enzymes into the medium and is therefore suitable only for studies on the total activities of enzymes. Among detergents, digitonin has been widely used to unmask matrix enzymes (Schnaitman and Greenawalt, 1968; Wojtczak and Zaluska, 1969; Addink *et al.*, 1972; Matlib and O'Brien, 1975). Loss of matrix structure occurs with digitonin, particularly at concentrations required to completely unmask all the matrix enzymes.

Other detergents, like Nonidet P-40, Triton, and deoxycholate, completely solubilized mitochondria even at very low concentrations. Another disadvantage of detergent treatment is the difficulty of subsequent detergent removal without loss of enzyme activity. One advantage of the use of toluene is that it does not remain in the mitochondrial pellet.

The mechanism of action of polyethylene glycol in stabilizing toluene-treated mitochondria is not known. It has been used as a protecting agent for ultrastructure of cellular organelles as well as an embedding medium in histochemistry in localization of proteins (Sandstrom and Westman, 1969; Mazurkiewicz and Nakane, 1972).

Studies have shown that toluene-treated mitochondria resemble intact mitochondria with regard to their enzyme content. Since most enzymes are studied at concentrations much lower than their physiological concentration (Srere, 1967, 1968, 1970), the system should be useful in studies of kinetic and regulatory properties of enzymes situated inside the mitochondrion at their normal *in vivo* protein concentration and protein microenvironment. For example, Weitzman and Hewson (1973), using toluene-treated yeast cells, have shown that citrate synthase *in situ* is insensitive to ATP inhibition. This observation contradicts the hypothesis of citrate synthase regulation based upon *in vitro* inhibition of citrate synthase by ATP (Hathaway and Atkinson, 1965). Reeves and Sols (1973), using yeast cells made permeable with toluene, discovered kinetic and regulatory properties of phosphofructokinase that were distinct from its properties *in vitro*.

Studies on the *in situ* kinetic and regulatory properties of citrate synthase in toluene-treated mitochondria indicated no change in the V_{max} or K_m for oxaloacetate in the toluene-treated mitochondria, in which the enzyme concentration is unchanged from its *in vivo* concentration, compared to the enzyme in dilute solutions (Matlib *et al.*, 1978). The K_m for acetyl-CoA was increased, contrary to what one would expect based upon relative rates of citrate synthase and Krebs TCA cycle activities. This higher K_m for acetyl-CoA did not appear to be due to a diffusion barrier since increased stirring of the mixture did not effect a decrease in the K_m. There was no problem for CoA penetration as a substrate for succinate thiokinase since the K_m values for the enzyme within and outside the mitochondrion are identical. Thus, the hypothesis that the environment *in situ* would change the kinetic characteristics of citrate synthase so that faster rates could be achieved at low substrate but high enzyme concentration was not upheld by these results.

In agreement with the result of Weitzman and Hewson (1973), who used toluene-treated yeast cells, ATP, a putative regulator of citrate synthase, had little effect on the activity of this enzyme *in situ*. The possibility of a diffusion barrier against ATP is ruled out by the results, which showed that the inner membrane ATPase behaved identically *in situ* and in the supernatant following sonication. Evidence against the regulation of citrate synthase by ATP in isolated rat heart mitochondria has also been obtained (Williamson and Olson, 1968). Two other known citrate synthase inhibitors, NADPH and tricarballylate (Srere *et al.*, 1973a), also showed reduced inhibitory action on the *in situ* enzyme as compared to the free enzyme. These observations of effects of inhibitors may be due to the high enzyme concentration in toluene-treated mitochondria. Lack of pyruvate inhibition of lactate dehydrogenase at high enzyme concentration has been observed (Wuntch *et al.*, 1970).

It is clear therefore that good physical and chemical evidence exists that specific complexes of metabolically sequential enzymes of the Krebs TCA cycle and TCA cycle-associated enzymes are formed. The results of these studies on protein interactions are summarized in Fig. 3. Not all of the possible relevant interactions have been studied as yet but the data thus far presented I think argue strongly for the existence of such complexes.

3. Matrix Protein Interactions as Studied by Cross-Linking Techniques

Cross-linking techniques have been used to stabilize the weak interactions of proteins *in situ* with the aim of studying the possible subcompartmentation of different proteins in cells and cellular organelles (Peters and Richards, 1977). Cross-linking techniques have been used to study the association of proteins in intact erythrocytes (Wang and Richards, 1975). Boulikas *et al.*

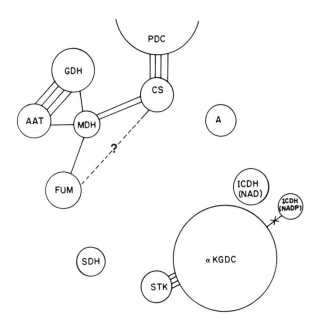

Fig. 3. Interactions between Krebs TCA cycle enzymes and some auxilliary enzymes. A line connecting two enzymes indicates one kind of evidence for interaction (see text). The line with a cross indicates no interaction seen. A dotted line is weak evidence. No line indicates not tested.

(1980) have studied the interactions of histones in nuclei using the cross-linking methods. Using glutaraldehyde as a cross-linker, Keokitichai and Wrigglesworth (1979) have confirmed glyceraldehyde 3-PO$_4$ dehydrogenase is a membrane-bound enzyme in intact erythrocytes. Glutaraldehyde cross-linking has also indicated that compartmentation of specific glycolytic enzymes occurs on the membranes of intact nerve endings (Knull, 1980).

Although several studies (Tinberg et al., 1975, 1976; Rendon et al., 1980) have demonstrated that the cross-linking of either intact mitochondria or the inner mitochondrial membrane can modify certain membrane properties and functions, no information is available on the effect of cross-linking on the matrix proteins.

We have examined various methods to cross-link mitochondrial matrix proteins in situ to determine if some organization of these proteins exists within the matrix space. Although the cross-linking technique itself presents many problems, the ability to use cross-linkers of various group specificities, of various arm lengths, with the ability for short cross-linking time, and selective site specificities with photoactivatable cross-linkers has enabled investigators to gather important nearest neighbor data for protein–protein, protein–nucleic acid, and protein–lipid interactions (Peters and Richards, 1977). Cross-linking reagents have been used in identifying nearest neighbor relationships of subunits in polymeric proteins and in the structure of ribosomes.

Unlike their effect on soluble enzymes or on isolated membranes, one of the limitations of the cross-linking technique in the study of whole organelles is the problem of permeability of the mitochondrial membrane to the cross-linker. Even though techniques using toluene are available to alter the permeability of mitochondria such treatments may bring about changes in the gross organization of the mitochondrial matrix (Matlib et al., 1977). In order to study the infrastructure of the mitochondrial matrix using a cross-linker, it should be freely permeable and able to cross-link matrix proteins at a physiological pH (7.4) in a short time so that the gross infrastructure of the mitochondria is not altered.

The earlier studies on cross-linking of mitochondria involved reactions with relatively long time intervals and a pH greater than 8, since alkaline pH is favorable for cross-linking with imidates (Tinberg et al., 1975, 1976). Such high pH conditions may be detrimental to the preservation of mitochondria in the coupled state. Recently, Rendon et al. (1980) have indicated that mitochondria could be cross-linked at pH 7.5 using dimethyl suberimidate but not much information has been available on the effect of other cross-linkers. The studies of D'Souza and Srere (1983a) indicate that most of the cross-linkers used could cross-link mitochondrial proteins at pH 7.4 even at times as short as 2 min when a 10 mM concentration of the cross-linker was used. A number of different cross-linkers were tried and the overall resuls indicated no differences in the pattern of cross-linking of matrix proteins. Glutaraldehyde was chosen as the cross-linker to use routinely.

Glutaraldehyde cross-links mitochondrial matrix proteins easily and differentiates organizational changes that could be induced by different substrates in the mitochondrial matrix by the mitochondrion's altered sensitivity to cross-linking. This indicates that glutaraldehyde does not randomly cross-link mitochondrial proteins.

A decrease in the rate of glutaraldehyde-induced cross-linking was detected when mitochondrial matrix proteins were disorganized by swelling. Earlier studies have indicated that in swollen mitochondria succinate-supported respiration was unchanged but the respiration of NAD-linked substrates was reduced (Matlib and Srere, 1976). This has been attributed to the disorganization of the putative complex of Krebs cycle enzymes on the inner surface of the inner membrane (Matlib and Srere, 1976). The complete destruction of the infrastructure of the mitochondria by sonication drastically reduced the rate of cross-linking. It is not clear at this stage whether or not these effects are due to a dilution effect or due to the disruption of the multienzyme system. The studies on cross-linking of the matrix fraction indicate that membranes are essential for optimal cross-linking of mitochondrial proteins.

Although the composition of the high-molecular-weight cross-linked polypeptide complexes is not known, it is apparent that all the cross-linkers used caused extensive cross-linking of the mitochondrial materials. The nature of cross-linking is complex since amino groups on both the lipids and proteins can cross-link resulting in protein–protein, protein–lipid, or lipid–lipid conjugates. However, cross-linking involving SH groups may give rise to predominantly protein–protein interaction. Furthermore, the cross-linking may involve both intramolecular and intermolecular cross-links.

The most surprising result of these studies is the appearance of very-high-molecular-weight complexes without the formation of large amounts of intermediate-molecular-weight products. Although we have used relatively short time intervals for cross-linking compared to other studies, these time intervals are very long compared to the theoretical time of diffusion of one protein through the length of a mitochondrion. These diffusion calculations, however, may well be incorrect considering the probable close packing of protein molecules within the matrix (Srere, 1981). It is not possible to conclude from these results whether or not the proteins in the matrix are present in a random array, but one is surprised that an entire spectrum of different molecular weight complexes does not appear as a "smear" of staining material on top of the gel. A possible explanation of our results would be (1) the immediate cross-linking of inner membrane proteins to give a very-high-molecular-weight material followed by (2) the cross-linking of matrix proteins to the cross-linked membrane. One would not expect to observe the formation of intermediate complexes with this sequence.

A number of other experiments support this hypothesized sequence of events. The swelling of mitochondria by osmotic means causes a drastic reduction in the amount of cross-linking. The cross-linking of isolated fractions, inner membrane, or matrix proteins does not lead to the same pattern of cross-linking. The use of cross-linkers with different group specificities leads to similar patterns of cross-linking. It is apparent that longer cross-linking times change the pattern of cross-linking only in its degree. The observation that cross-linkers of different group specificities yield similar patterns rules out an important role for lipid cross-linking in the results presented here.

The results can best be interpreted using a model of the inner membrane–matrix compartment as being a system in which all the matrix proteins are close to the inner membrane. Calculations using the morphometric measurements of rat liver and rat heart mitochondria show this to be a distinct possibility (see below, Section II,B,6,c).

a. *Cross-linking Krebs TCA Cycle Enzymes.* In order to extend these general studies on cross-linking mitochondrial matrix proteins we have looked at three specific enzymes of the TCA cycle after the glutaraldehyde cross-linking process. The enzymes citrate synthase, malate dehydrogenase, and fumarase were selected in our studies as these are known to be present mainly in mitochondrial matrix and catalyze a connected sequence of the Krebs TCA cycle reaction. Moreover, citrate synthase has been shown to interact with mitochondrial malate dehydrogenase *in vitro* (Halper and Srere, 1977). The kinetic advantage of simultaneous binding of these two enzymes on an insoluble support is also known (Srere *et al.*, 1973a). Fumarase was included in our experiments as it could be the anchor: it links the next enzyme of the Krebs cycle, succinate dehydrogenase, an enzyme that is firmly bound to the mitochondrial inner membrane. Interaction of fumarase with malate dehydrogenase and citrate synthase has also been observed (Beeckmans and Kanarek, 1981a,b).

The results indicated that a large number of matrix proteins, including the Krebs cycle enzymes CS, MDH, and fumarase, are cross-linked into high-molecular-weight complexes under the cross-linking conditions employed.

An enhancement in the specific activity of the three Krebs cycle enzymes was observed in liver mitochondria and it was suggested that these enzymes are relatively closer to the membranes, thus being cross-linked more specifically than other matrix proteins. The fact that no dramatic changes were observed of enhancement in specific activities in cross-linked heart mitochondria may indicate that most of the matrix proteins in heart mitochondria are close to the membrane. Thus even though more protein was cross-linked when heart mitochondria was cross-linked, not much enhancement in the specific activities of Krebs cycle enzymes was seen. Calculations based on stereomorphometry of heart mitochondria indicate that proteins in the heart mitochondrial matrix are, on the average, closer to the membranes than those in liver mitochondria (Section II,B,6,c).

Comparatively higher specific activity of fumarase and its ability to be retained completely by the cross-linked mitochondria on toluene treatment as compared to citrate synthase and malate dehydrogenase suggests that fumarase is cross-linked more efficiently than other enzymes. This might suggest closer proximity or association of fumarase with the inner membrane or better proximity and orientation of glutaraldehyde reactive groups. The earlier studies of Matlib and O'Brien (1975) have indicated that fumarase activity is unmasked at low concentrations of digitonin as compared to the concentrations necessary for the unmasking of citrate synthase and malate dehydrogenase. These authors concluded that fumarase is located closer to the inner membrane than citrate synthase and malate dehydrogenase.

Initial studies on the analysis of the soluble fraction obtained from cross-linked mitochondria indicate the presence of nonsedimentable cross-linked proteins, thus indicating that glutaraldehyde treatment causes extensive cross-linking of the matrix proteins.

Devaney and Powers-Lee (1983) reported that they have isolated a cross-linked complex of GDH and carbamoyl phosphate synthetase I (CPSI) from treatment of rat liver mitochondria with low levels of several bisimido esters. Analysis of the products in the matrix fraction by SDS polyacrylamide gel electrophoresis indicated a component the M_r of which indicated the possibility that it was a GDH–CPSI complex. This putative complex reacted with both GDH and CPSI antisera. This complex could be a metabolically significant one, with the NH_3 formed by the GDH reaction being used by CPSI in urea formation. It should be remembered, however, that CPSI is the most abundant liver mitochondrial matrix protein and that our studies (D'Souza and Srere, 1983b) indicated that it is an easily cross-linked protein. Thus cross-linking to another protein must be shown to have some degree of specificity, which has not been demonstrated in this case.

The cross-linking method is a potentially powerful one for indicating whether or not specific protein–protein interactions exist within the mitochondrial matrix. There are not yet sufficient data to allow firm conclusions to be drawn on the basis of these studies alone.

4. Binding of Matrix Proteins to the Inner Membrane

There are a number of reports concerning the binding of matrix enzymes to the inner membrane of mitochondria. There are some disagreements in these results but when one considers the large variations in the conditions (such as ionic composition and ionic strength) used in the various experiments then the differences reported are not surprising.

Succinate dehydrogenase is often used as a marker enzyme for the inner membrane compartment of mitochondria. It has been considered an intrinsic protein of the inner membrane since most isolation media for the membrane did not remove it. Electron microscopy using histochemical methods located it as inner membrane enzyme (see below, Section II,B,5). Capaldi (1982) has depicted this enzyme's main subunits (M_r 73,000 and M_r 25,000) as being on the inner surface of the membrane and complexed with three other proteins, which are buried in the membrane.

The isolated PDC and αKGDC are large protein complexes ($M_r = 7$–8.5×10^6 and 4.5×10^6, respectively) and are known to be difficult to remove from the mitochondrial membranes. Sonication, detergent treatment, freeze–thawing, and special ionic conditions have all been used to render these enzymes membrane-free. Certain purification procedures make use of detergents to remove these enzymes from the membrane (Stanley and Perham, 1980; Matlib and O'Brien, 1975; T. C. Linn, personal communication, 1980). In spite of these indications of the tightness of binding of these enzymes to the membrane they are considered primarily matrix enzymes.

A number of years ago Matlib and O'Brien (1975) used digitonin to examine the latency and releasability of matrix enzymes. Digitonin binds specifically to 3β-hydroxysterols and thus will complex and precipitate cholesterol. Membranes that contain cholesterol will therefore be disrupted or made "leaky" by this reagent. The outer mitochondrial membrane has six times the cholesterol content of the inner membrane and thus digitonin has proved to be a good reagent to use to separate the outer membranes from the mitoplast (inner membrane–matrix complex).

In the experiments of Matlib and O'Brien (1975) rat liver mitochondria were titrated with digitonin. The α-keto acid dehydrogenase complexes, aconitase, ICDH (NAD), and fumarase were unmasked at 0.3 mg digitonin/mg mitochondrial protein. Much higher concentrations were needed to unmask completely the MDH, CS, and ICDH (NADP). The interpretation given to these data was that the first set of enzymes was membrane bound while the second set of enzymes was "deeper" in the matrix. This interpretation is partially in error as I will show later. What is actually being measured in this experiment is the permeability of the substrates and products of each of the various reactions. When the release of fumarase and MDH were compared, although MDH was released faster than FUM, the difference was smaller than indicated by the latency experiment.

Other reports on the location of the Krebs TCA cycle and other matrix enzymes show contradictory results. Thus, some mMDH of pig heart (Comte and Gautheron, 1978) and of chicken heart (Elduque *et al.*, 1982) is reported to be bound to the inner mitochondrial membrane when this fraction is isolated. Elduque *et al.* (1982) also reported binding to a lesser extent than MDH for FUM, GDH (glutamate dehydrogenase), and AAT. A number of workers have reported that AAT is firmly bound to guinea pig heart inner mitochondrial membranes (Wit-Peeters *et al.*, 1971; Scholte, 1969; Wit-Peeters, 1969).

Wit-Peeters *et al.* (1971) concluded that probably all matrix enzymes bind to the inner membrane and differ only in the degree of the tightness of this binding. Contrariwise Landriscina *et al.* (1970) have reported that AAT, ICDH (NAD), GDH, and ICDH (NADP) are isolated in the matrix fractions, which contain no membranes. When one considers the variation in the methods used in obtaining these results it is not surprising that no clear agreement could be reached concerning the internal location of the TCA enzymes.

A number of experiments in our laboratory as well as certain calculations (see below, Section II,B,6,c) led us to reinvestigate the problem of the binding of Krebs TCA cycle enzymes to the inner mitochondrial membrane. We have shown thus far that CS, mMDH, and FUM bind to inner membrane in a saturable process (D'Souza and Srere, 1983a). The binding is fairly tight in 2 m*M* Hepes pH 7.0 but it is quite sensitive to ionic strength. The binding is specific for matrix isozymes and for the inner mitochondrial membrane. Neither mitochondria (outside of outer membrane) nor mitoplasts (outside of inner membrane) could bind these enzymes. In other experiments we found that membranes isolated from red blood cells, the endoplasmic reticulum, or artificial phospholipid liposomes could not bind these enzymes. It is possible to isolate a substance from inner membranes that is sensitive to trypsin and sulfhydryl reagents and that when incorporated into phospholipid vesicles can bind CS and mMDH. The binding capacity of inner mitochondrial membranes can be calculated from the saturation curves for CS binding. We found that based on the normal concentration of CS in mitochondria about one-half the citrate synthase may be bound.

Rendon and Waksman (1973) have shown that MDH and AAT could be bound by inner mitochondrial membranes and various metabolites could trigger specific release of these enzymes. Masters (1981) and others (see Coleman, 1973, for review) have reviewed thoroughly the metabolic advantages of enzyme binding to membranes. The specificity for ligand (matrix enzymes only) and for membrane (inner surface of inner membrane) indicates that this interaction is physiologically important. The extreme sensitivity of this interaction to ionic strength detracts, however, from its biological importance. In this connection two points are pertinent: first, the ionic strength of the matrix compartment is unknown, and second, the conditions within the matrix, which has a high total protein concentration, make the *in situ* conditions much different from the conditions used *in vitro* in the membrane-binding experiments. In a series of experiments designed to mimic the high protein concentration of the matrix milieu we have found that binding of CS to inner membrane vesicles is much better if 14% polyethylene glycol (PEG) is used as solvent and that under these conditions, the binding is less sensitive to the ionic strength of the solution. Since PEG has some properties in common with proteins, e.g., polymer exclusion, it is possible that if we were able to mimic matrix conditions then the sensitivity to ionic strength of enzyme binding would disappear. It is worth noting that the binding of glyceraldehyde-3-phosphate dehydrogenase to red cell membranes is also sensitive to ionic strength but that Kliman and Steck (1980) have shown that a consideration of mass action makes it likely that the poor binding of the enzyme at isotonic ionic strength is an artifact of high dilution rather than due to a low affinity.

If we now consider the results presented in this section, another diagram (Fig. 4) can be drawn to indicate Krebs TCA cycle enzyme–membrane interactions that have some experimental support.

5. Electron Microscopy of Mitochondria

a. General. Mitochondria contain four compartments: the outer membrane, the intermembrane space, the inner membrane, and the matrix (Fig. 1). Large variations can be seen in the structure of the inner membrane matrix compartments (Fig. 2) and in their volume (Hackenbrock, 1968). The intermembrane space tends to be featureless and usually quite small (see below for exceptions). The inner membrane has been considered by some to consist of two regions. One region is that membrane that runs parallel to the outer membrane, referred to as the inner membrane. The other region, which is contiguous with the inner membrane but invaginates into the matrix, is

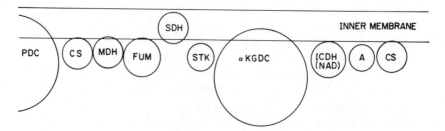

Fig. 4. A schematic representation of tightness of binding of Krebs TCA cycle enzymes (and PDC) to the inner mitochondrial membrane. The depth of protein "burying" in the membrane is proportional to the presumed strength of binding. Binding may be to binding proteins located within the membrane but is not depicted in this scheme.

referred to as the cristae. There have been some reports that these two regions of the inner membrane differ in composition (Brdiczka et al., 1974). I will make the assumption, however, that the inner membrane is continuous and uniform.

The outer membrane as viewed by electron microscopy is featureless. Using a freeze-fracture technique intermembranous particles can be observed and it has been reported that the convex face (inner half of membrane) has four times as many particles as the concave face (half of membrane adjacent to cytosol). On the other hand, the inner membrane has more particles in the concave face (half of membrane facing intermembrane space) than in the convex face (Wrigglesworth et al., 1970). Protuberances can be seen using negative staining on the inner membrane but none are visible on the outer membrane using this technique. The protuberances on the inner membrane have been identified as part of the ATPase (see Section II,C).

Rat liver mitochondria exhibit several distinct morphologies under the electron microscope. The actively respiring mitochondria (State III respiring) have a condensed matrix with a volume only approximately one-half that of the matrix of resting, nonrespiring mitochondria (State IV resting). Hackenbrock (1968), using several electron microscopic techniques including freeze-etching of physically and chemically fixed specimens, has shown that the matrix of rat liver mitochondria in the condensed respiring form appears as a homogeneous electron-dense compartment. In the resting state, by contrast, Hackenbrock showed the existence within the matrix of an electron-dense reticular network (with electron-lucent interstices) physically continuous with the inner membrane. Freeze-etching results supported this observation since etched cavities appeared within the membrane during sublimation. The nonsublimable network (the protein) and the inner membrane also appeared continuous. Pihl and Bahr (1970) later described a skeleton in the matrix of rat liver mitochondria in unfixed samples that had been dried at the critical point. This skeleton was described as "... fairly uniform, solid, tortuous, and branched strands of about 200 Å diameter..." that accounted for about one-third of the total mitochondrial protein. Thus both these studies support the concept of the matrix being a compartment containing closely packed protein molecules, which do not become dispersed even when the volume of the matrix is increased.

b. Succinate Dehydrogenase.   It is well known that succinate dehydrogenase is associated with the inner membrane. The electron acceptors used have been ferricyanide and tetrazolium salts and the electron-dense accumulations have been on the membrane or in the intercristal space. With short-term incubation, accumulations are first seen on the matrix side of the membrane (see Harmon et al., 1974, for review).

c. *Citrate Synthase.* We have used a histochemical method for the ultrastructural localization of citrate synthase (Matlib *et al.*, 1976). Partially fixed (paraformaldehyde) rat heart tissue was incubated with acetyl-CoA, oxaloacetate, $CuSO_4$, and $K_3Fe(CN)_6$. The CoASH released in the reaction would react with the $Fe(CN)_6^{-3}$ and Cu to give an electron-dense precipitate. Initial deposits were seen on the inner surface of inner membrane. Longer incubations showed greater deposits and these seemed to have "moved" to the intercristal space. No deposits were observed in the locations in the absence of oxaloacetate, but deposits were observed in the sarcoplasmic reticulum, probably due to an acetyl-CoA deacylase. The location of CS thus seems to be similar to that of succinate dehydrogenase.

d. *Glutamate Oxaloacetate Transaminase (Asparate Amino Trans-ferase).* Lee and Toback (1968) have used a histochemical method to locate glutamate–oxaloacetate transaminase (GOT) (aspartate aminotransferase, AAT) in rat liver mitochondria. Previous studies using differential centrifugation had indicated that AAT was firmly bound to the inner mitochondrial membrane. The enzymatic reaction product could be demonstrated in the cristae of mitochondria only when the mitochondria had been fixed *in situ*. When mitochondria were isolated and fixed, then no GOT activity could be detected histochemically or biochemically.

e. *Carbamoyl Phosphate Synthethase.* The first published report using labeled antibodies for the localization of a mitochondrial matrix enzyme was for carbamoyl phosphate synthetase. Knecht *et al.* (1979) labeled CPS antibodies with ferritin and then incubated them with tissue that had been treated with (postembedding) sodium methoxide. This treatment was necessary to render the cells and organelles permeable to the large ferritin–antibody molecule. Individual ferritin molecules could be observed with the electron microscope. The ferritin molecules were distributed homogeneously throughout the mitochondrial matrix. By counting the number of ferritin molecules per square micrometer of mitochondrial surface an independent measure of carbamoyl phosphate synthetase in rat liver could be made. This method gave a concentration of 0.12 mM for CPS, which can be compared with the concentration of 0.5 mM based upon activity measurements (Raijman and Jones, 1976; Cohen *et al.*, 1982).

Bendayan and Shore (1982) have used a protein A–gold labeling technique to localize CPS in rat hepatocytes. This technique also showed the CPS to be uniformly distributed in the mitochondrial matrix. They observed 84 particles of gold/μm^2, which is considerably less than the 6,000 particles/μm^2 observed by Knecht *et al.* (1979).

f. Relation of Amount of Inner Membrane to Oxidative Activity. I have pointed out earlier that there is a good correlation between the amount of inner membrane as measured by electron microscopy in a mitochondrion and the amount of Krebs cycle enzymes (Srere, 1972). In this section I will present some calculations that relate to this. I cite some studies that support this relationship. Jakovic *et al.* (1978) measured cytochrome *a* concentrations and inner membrane area in rat liver mitochondria from normal and thyroidectomized animals before and after thyroxine administration. A direct correlation could be seen, with the cytochrome *a* rising from 0.19 nmol/mg protein in mitochondria from thyroidectomized animals to 0.33 nmol/mg protein in mitochondria from thyroxine-treated normal animals. The inner membrane area also changed from 19,000 μm^2/cell to about 40,000 μm^2/cell in the hypothyroid versus hyperthyroid livers.

In similar studies McCallister and Page (1973) and Page and McCallister (1973) showed that thyroxin treatment of rats resulted in an increase in mitochondrial number and volume and an increase in the inner membrane area per unit cell volume in rat myocardial cells. A series of other studies showed that whenever the oxidative activity of the tissue under study increases, enzymes of the inner membrane and the total surface area of the inner membrane increase concomitantly. Plattner and co-workers (Plattner *et al.*, 1969; Plattner, 1968, 1973) have shown this correlation for cytochrome *c* oxidase and surface density when comparing mouse liver (high respiration) to beef liver (low respiration). In addition he compared mouse, frog, pigeon, and rat livers as to respiration rate, cytochrome *c* oxidase, succinate dehydrogenase, and inner membrane surface density and found a positive correlation among these. Reith and Fuchs (1973) have shown an increase in rat heart cristae membrane area when riboflavin and T_3 were administered to riboflavin-deficient animals.

We have found that in vitamin B_{12} deficiency rat liver shows an increased content of citrate synthase (Mukherjee *et al.*, 1976). Morphometric examination of the liver mitochondria showed a concomitant increase in the inner membrane surface in B_{12}-deficient animals (Frenkel *et al.*, 1976). Further studies indicated that all the Krebs TCA cycle enzymes were increased in the liver from B_{12}-deficient animals (Matlib *et al.*, 1979). Studies by Holloszy and colleagues have shown that TCA cycle enzymes are increased in the muscles of exercised rats as well as in T_3-treated animals (Winder *et al.*, 1979).

In addition to these electron microscopic studies on the mitochondrion and the matrix there is a very large literature on matrix granules and paracrystalline arrays within the matrix of many different mitochondria (see Munn, 1974, for review; also Wakabayashi *et al.*, 1971). It is apparent that the nature of these granules and paracrystalline arrays is not well understood. There is some evidence that some of the granules are calcium deposits, whereas certain others are apparently lipid and protein in nature. Certain mitochondria seem to contain glycogen granules but no firm evidence is available as to how the glycogen gets into the mitochondrion. Even less is known concerning the paracrystalline arrays that are seen. If our speculations are correct about a TCA cycle complex, then at the high protein concentration of the matrix one might imagine an orderly arrangement of the proteins to be the origin of such arrays.

It is clear that electron microscopic studies, especially with the use of labeled antibodies, will be helpful in answering some of these questions.

6. Theoretical Aspects

a. The Protein Content of the Matrix. Hackenbrock (1968) noted that in the matrix of rat liver mitochondria, presumably in the resting state (state IV), there was 1 g of mitochondrial protein/ml and that matrix may contain as much as 67% of the total mitochondrial protein (Schnaitman and Greenawalt, 1968); thus the protein concentration in the matrix would be 560 mg/ml water (i.e., 56%). Hackenbrock pointed out that "it is likely that matrix protein is not in true solution." Another determination of matrix protein in rat liver mitochondria was reported to be 47% (Hoppel, 1972) and a value of 27% by weight (Landriscina *et al.*, 1970) was reported for pigeon liver mitochondria.

Garlid (1979) and Gamble and Garlid (1970) reported a value of 1 g H_2O/g dry weight of mitochondria while Pfaff *et al.* (1968) reported values of 0.8–1.8 ml H_2O/g protein for sucrose-inaccessible space. These values are presumably for State II mitochondria. Tarr and Gamble (1966) report values of 1.43–0.53 g H_2O/g dry weight rabbit liver mitochondria. Similar values were reported by Soboll *et al.* (1976).

Hackenbrock's calculations do not take into account the space occupied by the protein, which must be added to the water space, and when this is done, one gets a figure of about 40% protein instead of 56%. This still represents a very high protein concentration. It is clear that the protein concentration depends completely on the respiratory state of the mitochondrion, so in the completely condensed state the concentration may be higher than this figure.

The mitochondrial matrix is therefore an unusual biological compartment in which the protein concentration is very high. Two important consequences of this fact are (1) the protein molecules are probably in very close proximity to each other, which may be relevant to the existence of multienzyme complexes in the matrix, and (2) the protein at these concentrations occupies about half of the total volume of the matrix compartment. Changes in the matrix volume of about twofold that occur during metabolism must represent the exit of almost all of the matrix bulk water, which would cause large changes in the concentrations of metabolites. These large changes in metabolite concentration may be an important regulatory feature of the metabolic pathways within the matrix.

To obtain a physical model of a high concentration of protein in a cellular space, I have made the following calculation (Srere, 1980). Assume spherical nonpenetrable protein molecules of the same diameter d. Assume that another molecule cannot enter the imaginary cube (side d) of water surrounding the protein molecule: cube volume $= d^3$, protein volume $= \pi d^3/6$, water volume $= d^3 - \pi d^3/6$, water weight $= (d^3 - \pi d^3/6) p_{H_2O} (p_{H_2O} = 1 \text{ g/ml})$, protein weight $= \pi d^3/6 \, p_{Prot}, p_{Prot} = 1/V \, (V = 0.74 \text{ ml/g})$, then,

$$\text{protein weight per volume } \% = \frac{100 \, \pi d^3}{6V} \left/ \frac{\pi d^3}{6V} + \left(d^3 - \frac{\pi d^3}{6} \right) \right.$$

$$= 100 \, \pi/[6V + \pi(1 - V)] = 59.76\%.$$

Thus a space in which protein molecules are stacked in cubical arrays with each molecule touching six other protein molecules contains about 60% protein by weight. This figure is close to the value calculated by Hackenbrock (1968) for matrix protein concentration and this calculated packing of protein molecules agrees with the electron microscopic evidence.

i. Metabolic consequences of high protein content. Let us consider three questions that arise from the fact that the matrix has a high protein concentration.

1. Are these closely packed protein molecules free to move in relation to each other?

2. Are they arranged in multienzyme complexes? With the high protein concentration even small associative forces between protein molecules would be favored so that weak interactions not apparent in dilute solution might become important.

3. What effect do changes in matrix volume have on the concentration of matrix metabolites? Since the volume of the protein molecule is $\pi d^3/6$ and the volume of the cube containing it is d^3, the percentage of the total volume occupied by the protein is $(100 \, \pi d^3/6)/d^3 = (100 \, \pi)/6 = 52.36\%$. With a closer packing arrangement and an assortment of sphere sizes the protein weight percentage might go as high as 70%. The conversion of the matrix from the resting ("inactive") to the actively respiring condensed state conformation ("active") involves a twofold change of its total volume of the matrix. Since the protein occupies about half of the total volume, the change in matrix volume that occurs must involve virtually all the *free water* (the water not bound to protein) in the matrix. Consequently, the concentration of metabolites in that free water space will vary by a factor of much greater than two and may even

increase by as much as close to two orders of magnitude. If we assume that the metabolites are not simultaneously removed with the water, then a great increase in metabolite concentration and subsequent increase in the rates of the individual enzyme steps would occur without an increase in the total number of metabolite molecules. Thus the rate of the Krebs cycle could be controlled by changing the water content of the mitochondrial matrix. In the condensed actively respiring State III, concentrations of ADP and P_i are increased and, more important, the concentration of free water is greatly decreased. Since the formation of ATP essentially involves the dehydration of ADP and P_i, these conditions would favor ATP synthesis. The movement of water out of the matrix is a potential metabolic control system that is not usually considered.

We can also recalculate the apparent concentrations of the enzymes within the matrix space. A number of years ago I pointed out that the apparent concentrations of many enzymes of the major metabolic pathways were in the range of 10^{-6} to 10^{-5} M (Srere, 1967, 1968, 1970). At these concentrations, which are much higher than those usually used in studies of the enzymes *in vitro*, the total free metabolite concentrations measured in cells may be overestimated because a substantial portion may be bound to the enzymes. The calculated apparent concentration of the enzymes were minimal figures and we can revise these now to obtain a better estimate. Thus, using the whole mitochondrial volume, I originally calculated that citrate synthase (M_r 97,938, two active sites/mol) had an apparent concentration in rat liver mitochondria of 10^{-6} M. Since in the resting state, the matrix space occupies only one-half of the total mitochondrial volume and one-half of the matrix is water space, then the new enzyme concentration is probably four times that reported earlier. Moreover, the apparent concentration of the enzymes will increase further when the mitochondria begin to respire and the matrix condenses.

From experiments in which the matrix volume was held constant and the concentration of solutes in the medium was varied, Garlid (1979) concluded that the mitochondrial matrix had at least two different water compartments. This could reflect the existence of a layer of water two or three molecules deep around each protein molecule (Peschel and Belouschek, 1979) in addition to the bulk water. If this is true, the apparent concentration of limiting metabolites might be much higher than the calculated values.

The metabolic consequences of the data presented earlier by Hackenbrock (1968) and others have been largely ignored. Since the bulk of the proteins of the matrix are easily soluble in water, this compartment is usually thought of as the soluble compartment. However, the calculations presented here emphasize the unusual characteristics of the mitochondrial matrix that indicate that the compartment and the metabolism therein must be considered in special ways.

The kinetic behavior of enzyme systems in the matrix might be expected to be somewhere between that of enzymes in solution and that of enzymes in an ordered complex, as for example in a membrane. Citrate synthase and malate dehydrogenase when immobilized as a multienzyme system have a significant kinetic advantage over enzymes in free solutions. This advantage may exist *in vivo* for these mitochondrial matrix enzymes. Clearly, the very high concentration of protein that exists in the matrix might well have important implications not only for metabolic regulation, but also for ion transport and for mitochondrial biogenesis as well.

b. Protein Crystals as a Model for Mitochondrial Matrix. The model of matrix proteins consisting of spherical protein molecules packed in a cubical array (six protein–protein contact points per molecule) would represent a protein concentration of 59.76% (w/w) (0.674 g H_2O/g protein). The protein in this model would occupy 52.36% of the volume of the array as calculated above. I now wish to extend these ideas by noting the similarity of such an array to the situation that occurs in protein crystals. The information available for the structure of protein crystals and the activity of crystalline enzymes may then be used to help understand the structure of the mitochondrial matrix and the activity of the proteins therein.

Protein crystals are known to contain $\sim 50\%$ water (40–60%) (Matthews, 1968; Crick and Kendrew, 1957) and, for some proteins studied, contact points with other protein molecules in the crystal have been reported to number between 5 and 10 (Rupley, 1969). Since these figures agree with those of our model of the matrix outlined above, protein crystals and their properties may represent an acceptable paradigm for the proteins of the mitochondria matrix.

i. Structure of water in concentrated protein solutions. An important question to be answered in our matrix model (and protein crystals) concerns the state of the water in the presence of such a high protein concentration. Researchers in this area agree that the first layer of water around each protein molecule (and perhaps part of the second layer) is firmly bound to ionic groups on the protein and that the properties of this monomolecular water layer differ from the properties of bulk water. This first layer of water is essential for the structure and activity of the protein. Careri *et al.* (1980) correlated infrared spectroscopic, heat capacity, diamagnetic susceptibility, and enzymatic measurements on lysozyme as it was hydrated from 0 g of H_2O per g of protein $(h = 0)$ to $h = 0.38$, a point at which the protein had only a monomolecular layer of water. These properties change during hydration from $h = 0$ to $h = 0.38$ to give a protein with essentially the same properties, except for full enzyme activity, as it has in bulk water. Thus, beyond a single layer of water, they conclude that additional layers of water in the sense of the properties they studied had the same structure as bulk water. Cooke and

Kuntz (1974) have postulated the existence of three types of water in biological systems. By NMR relaxation measurements they estimated that for a 20% protein solution there would be 90% Type I water (bulk water), 10% Type II water ("bound water") and 0.1% Type III water (irrotationally bound water within the protein molecule). For crystals (50% protein) the percentage of Type II water would be about 15–30%. We can calculate that in our model a monomolecular water layer (0.96 nm) around each protein molecule would give ~35% of the total water as bound water. (It can be shown that if the protein has a radius of 3.2 nm, then the water in our model is sufficient for approximately three water layers on the average for each protein molecule). Richards (1977) pointed out that in crystals the roughness of protein surfaces would theoretically increase the surface area by a factor of 1.7, but he points out that since all techniques detect only about half this amount there must be slippage between parts of the monolayer and the protein surface. These calculations would hold true for the mitochondrial matrix proteins as well.

Even if the proteins in the matrix are larger than 3.2 nm in radius, the calculations are not altered significantly. Carbamoyl phosphate synthetase in rat liver matrix constitutes 20% of the total protein mass and its M_r is 130,000 (Clarke, 1976; Raijman and Jones, 1976). Assuming a spherical shape, its radius can be estimated as 4.0 nm and the hydration shell 0.96 nm or 3.3 molecules of H_2O. For protein molecules smaller than 68,000 Da, hydration layers would be less than 2.8 molecules of H_2O. Since a sphere has the smallest surface for a given mass, any deviation from the spherical shape would tend to increase the surface area for a given protein mass and thus diminish the thickness of the water layer. Therefore, the estimate of a water layer of about three molecule thickness for each protein molecule in the matrix is probably not an overestimate.

ii. Biological effects of limited amounts of water. One must now ask whether or not this surprisingly small number of water layers on each protein molecule has any biological consequences, that is, does the small number of water layers affect the activity of an enzyme in a crystal or *in situ* or an enzyme in the mitochondrial matrix, or are the properties of H_2O in these situations altered?

In interesting experiments related to this problem, Clegg (1979) studied the metabolic changes that occur when anhydrous cysts of *Artemia salina* are hydrated. No metabolism could be detected between 0 g of H_2O and 0.3 g of water/g dried cysts. Krebs TCA cycle activity occurred at a slightly higher water content and conventional total metabolism occurred at 0.6 g of H_2O/g dried cysts. Assuming that protein represents 70% of the dry weight of the cysts, these values correspond to $h = 0.42$ and $h = 0.85$. These experiments are difficult to interpret specifically for mitochondrial metabolism because we are dealing with a mixture of biological compartments in which proteins may exist with different h values. It is interesting to note that in our theoretical construct the h value is 0.68, a value between the two values of beginning and full metabolism reported by Clegg (1979).

There is disagreement concerning the properties of the water in cells that lie beyond the monomolecular layer. Rupley (1969), Richards (1977), and Cooke and Kuntz (1974) maintain that the water (beyond the monomolecular layer) can be considered as bulk water since a variety of physical measurements (nuclear magnetic resonance spectroscopy, electron spin resonance spectroscopy, freezing point data) and diffusion of small molecules in crystals indicate that the water acts as bulk water. A number of workers (see Drost-Hansen and Clegg, 1979; Keith, 1979; Ling, 1972) maintain, however, that structured (altered) water exists three layers or more from the surface of macromolecules and that its properties are not the same as bulk water and this is an important factor to consider in most biological control mechanisms.

iii. Solvent pores in crystals and matrix. In the model of a cubical array of protein molecules water channels must exist in the interstices of the protein array that will allow passage of small solutes as it does in protein crystals (Rupley, 1969). Although it has been shown that while equilibrium properties of the reaction of an enzyme in crystals and enzymes in solution are the same, the kinetic properties of enzymes in crystals are usually altered; thus there may be within the crystal changes in the diffusion coefficients for small molecules (Rupley, 1969; Makinen and Fink, 1977). For larger molecules such as the nucleotides (1–1.6 nm) there may be inaccessible bulk water regions near the protein contact points as well.

Bishop and Richards (1968) have shown that the pore size in cross-linked β-lactoglobulin crystals was 0.8–1.3 nm. Differences in diffusion rates were seen for small molecules between crystals and free solution but are explained by physical pore restrictions. Outside the bound water layer the remainder of the water in crystals behaves like bulk water. That high concentrations of sucrose cause a much larger effect on the diffusion of a small molecule (KBr) in the protein crystal than it does in free solution is explained by a decrease in pore radius caused by the hydrated sucrose molecules.

Pore sizes can be calculated for our model of a cubical array of protein molecules (Srere, 1981). We calculate that the maximum pore size can be represented by a sphere with a radius of $0.41r$, where r is the radius of the protein molecules. For a spherical protein of M_r 50,000 then $r = 2.45$ nm and $0.41r = 1.00$ nm. This means that particles whose hydrated dimensions are greater than 2 nm would have difficulty moving through the pores. (Assuming β-lactoglobulin, M_r 65,000, to be spherical and in a cubical array a similar calculation yields a pore size of 1.2 nm). A space-filling model of CoA can be put in a volume of about $1.7 \times 1.2 \times 1.1$ nm and would barely fit through the pores of our model. Diffusion of other mitochondrial nucleotides, which may be smaller, may be less affected by pores of this dimension.

A great many mitochondrial enzymes have a nucleotide as a cosubstrate. The total concentration of the nucleotides in the rat liver mitochondrial matrix is in the range 10–15 mM (Altman and Katz, 1976). It can be calculated that in our model the concentration of the spherical protein molecules with $M\hat{r}$ 50,000 is 14 mM. Since the total concentration of nucleotides and nucleotide-binding proteins are of the same order of magnitude, it is likely that much of the total nucleotide pool is bound to protein. It is difficult to determine the free concentration of these nucleotides since neither the number of free nucleotide molecules nor the volume of water they are in is known. It is possible that nucleotide molecules shuttle between adjacent "coupled" enzymes in a manner similar to the immobilized coenzyme system constructed by Mansson et al. (1978). It should be noted that closer packing of the protein molecules would reduce the pore sizes so that the potential difficulty for the movement of large coenzyme molecules would be exacerbated. Changes in the molecular weight of protein molecules must be very large before large differences in their radii or in the pore sizes occur. For spherical protein molecules in cubical array the pore sizes for M_r 50,000, 10,000, and 1,000 proteins are 1.0, 0.59, and 0.27 nm, respectively.

   iv. Consequences of acceptance of model.   The protein structure in crystals and the structure postulated herein in mitochondrial matrix put an important restriction on the system. In the model of proteins in a cubical array (and in some crystals) or the mitochondrial matrix each protein contacts on the order of six other protein molecules. Therefore the protein molecules must be relatively fixed in position to one another. It is easy to see how crystals of a single protein can have interacting surfaces that are repeating and hold the protein molecules in fixed positions within the crystal lattice. However, when the six contact points occur between different proteins, as might occur in the mitochondrial matrix, then each protein must present to the others contact points that chemically interact with others in a way to attain the most attraction to (or least repulsion from) its neighbors. The moving around to find other points of favorable contacts would be difficult because of the extremely high viscosity for protein molecules in this space. If the protein molecules are fixed in relation to each other then the enzymatic active sites would be best oriented toward channels or a good portion of their activity would be lost. An alternate and simplifying situation would be if all the protein molecules of a single protein species would gather together to form a crystallike lattice in the matrix. It is possible that during evolution proteins in a single metabolic pathway developed structures that would have points of specific but weak interactions with each other (McConkey, 1982). Such an arrangement would have kinetic advantages, as we have shown previously (Srere *et al.*, 1973a; Srere and Mosbach, 1974). This factor might overcome the kinetic disadvantages that may arise at the high protein concentrations described above.

v. The problem of diffusion in the matrix. The physicochemical problem involved in this system rests mainly on the issues of diffusion coefficients, enzyme turnover times, and the volume of cells. The rates of all chemical reactions are bounded at an upper limit by the time it takes for the two reactants to encounter each other. This time is dependent upon the size and shape of the reactants, the viscosity of the medium, the distance between the two particles, and the dimensionality of the system (Weisz, 1962; Eigen, 1973; Hardt, 1981). When one considers the small distances involved in living cells, assumes that water is the medium, and measures the diffusion coefficients of both large and small biological molecules, one finds that calculated encounter times are very fast compared to the turnover times of all but the most rapidly catalyzed bioreactions. It is generally assumed that diffusion of small molecules in the matrix would be sufficiently rapid so that complete mixing would occur faster than reaction times. These calculations on diffusion rates, however, are dependent on a number of assumptions. Using the simple Einstein mean square equation,

$$\Delta^2 \text{ (distance)} = 2D \text{ (diffusion coefficient) } \Delta t \text{ (time)},$$

$$\Delta^2 = \text{mean square of } \Delta X \text{ (distance)}.$$

In a dimensional analysis this would be $cm^2 = cm^2 \, s^{-1} \, s$. D for small molecules in dilute solution of pure water is about $10^{-5} \, cm^2 \, s^{-1}$, so the time to diffuse 1 μm (1×10^{-4} cm) (the length of a mitochondrion) would be given by $(1 \times 10^{-4} \, cm)^2 = 2(10^{-5} \, cm^2 \, s^{-1}) \, \Delta t$; thus $\Delta t = 0.5 \times 10^{-3}$ s.

The turnover number of enzymes varies from 10^2 to $10^6 \, s^{-1}$. Citrate synthase is $10^4 \, min^{-1}$ or about $250 \, s^{-1}$. So that in this case it would appear at first that diffusion time would be meaningful. The assumptions were one substrate molecule, one enzyme molecule, pure water, and an unhindered path for mean square calculation, and it was also assumed that the molecules will meet in a productive encounter. None of these assumptions is true for the situation in a mitochondrial matrix. Thus, in most cases we are dealing with many substrate molecules and many enzyme molecules factors that would make diffusion encounters more rapid. The high concentration of proteins and other substances in the matrix would, however, contribute to the tortuosity of the system and relatively fixed active sites would all tend to increase the time necessary for a successful encounter. Even if the diffusion coefficient for small molecules is changed by only a factor of three, the rotational diffusion times for the closely packed proteins would be much larger so that productive encounters would take longer.

Another approach, which in a way takes this into account, is derived from encounter theory (Metzler, 1977), in which the volume that is swept by a particle in 1 s is calculated. This volume is 1.4×10^{-11} cm^3 (a cube 2.4 μm on a side or a sphere with a radius of 1.6×10^{-4} cm or 1.6 μm). This means that to make sure in a sphere of 1.6 μm radius that one molecule will encounter another stationary molecule 1 s is required.

Hess' calculations in relation to the diffusion time from one glycolytic enzyme to the next in yeast are correct if you assume that a substrate molecule leaves an enzyme heading in a mean square path directly for an active site 50 Å away. Assuming viscosity of water, then this time is given by (5 $\times$ 10^{-7} cm)$^2 = 2(10^{-5}) \Delta t$; thus $\Delta t = 1.25 \times 10^{-8}$ s (Hess and Boiteux, 1972). This time is insignificant compared to turnover times of enzymes (10^{-3} s). However, the same oversimplifying assumptions had to be made, but, as indicated, the known deviations from ideality that exist in cells and the fact that the diffusion coefficients are not known under cellular conditions negate any conclusions that can be drawn from the calculations.

As an example of how simple diffusion is modified in a biological environment it is valuable to consider the diffusion of ions in the extracellular microenvironment of the rat cerebellum (Nicholson and Phillips, 1981). In that elegant study they point out that for the diffusion of ions in a complex medium Fick's law must be modified by the two constants of volume fraction (α) and tortuosity (λ). α is the fraction volume available to the ion compared to the whole system. The tortuosity "is related to the increase in path length of a diffusing particle in a complex medium compared to that in a simple one" (Nicholson and Phillips, 1981). The effect of diffusion is by the factor λ^2/α and in the system they studied they found $\lambda = 1.55$ and $\alpha = 0.21$, thus λ^2/α approached 12 and modified the time course of diffusion. In control experiments using dilute agar solutions both α and λ were measured as 1. If we extend these results to the microenvironment of the matrix it seems quite likely that diffusion would be slowed by these factors.

Another problem arises if one is to depend solely on diffusion of substrates between randomly distributed enzymes. It is generally assumed that the concentrations of most substrates in the cell are near their Michaelis constant for the reaction. Atkinson (1969) has noted that unless special strategies were employed the volume occupied by the total metabolites would be so large that the solvent capacity of the cell would be a limiting factor. Among the strategies the cell has used, noted by Atkinson, are activation of metabolic intermediates, coordinate derepression or induction, and enzyme modulation. Another strategy that could be added to this list would be the creation of microenvironments. The arrangement of enzymes of a single pathway next to each other would lead to intracellular microenvironments whereby locally high concentrations of metabolites could be maintained with a low total cellular metabolite concentration (Srere and Estabrook, 1978; Srere and Mosbach, 1974). Thus, a high rate of an enzyme activity in that region could be maintained with the locally high metabolite concentration whose total average cellular concentration and therefore volume occupancy would be low. Several demonstrations of increased rates of reaction when immobilized multistep enzyme systems are compared to their soluble counterparts have been reported. Presumably a high local concentration exists as the substrate is produced (Srere *et al.*, 1973b) in the immobilized system. The mechanism to explain these observations is not known precisely since diffusion times are presumably much smaller than turnover times and some energy input would seem to be necessary to maintain this apparently nonequilibrium situation. However, one might imagine non-membrane-trapped metabolites restricted by the ionic or chemical nature of the support or enzymes or trapping due to close packing of molecules to produce small pores, which may restrict diffusion (see above). Another explanation, offered by Welch *et al.* (1983), is related to viscosity effects in the microenvironment of active sites.

One reason for the acceptance of random diffusion-controlled metabolic events may stem from the fact that electron microscopy, though showing a rich complex cellular structure, also reveals areas that are apparently randomly organized. These areas would include the cytosol, membrane proteins, and the mitochondrial matrix proteins. However, investigations on cytosolic structural elements have revealed a complex trabecular structure that exists there as well as in the nucleus (Schliwa *et al.*, 1981). This has stimulated a renewed interest in the structure of the cytoplasm (*Cold Spring Harbor Symp.*, 1982). In addition, experiments showing interactions between soluble enzymes and subcellular structures have given rise to more intense considerations of metabolic control without the intervention of diffusional processes (Masters, 1977, 1981). Welch has presented forcefully the theoretical basis that supports the role of organized enzyme systems in metabolism (Welch, 1977).

I wish to point out that the available data indicate that the inner membrane matrix compartment of many mitochondria may have a structure in which most, if not all, of the matrix proteins are adjacent (perhaps bound) to the inner membrane. This proposal may have important consequences concerning the control (diffusional or structural) of the chief energy producing pathway of most aerobic cells and in the biogenesis of mitochondria.

c. *The Infrastructure of the Mitochondrial Matrix.*

i. *Theoretical considerations of membrane surface area and the volume it encloses.* The literature contains abundant data on the stereomorphology of mitochondria. One piece of evidence available is the surface density of the inner membrane, which is the ratio of inner membrane (including cristae) surface area to mitochondrial volume ($\mu m^2/\mu m^3$). The theoretical limits for this value can be arrived at by calculating this ratio for a spherical inner membrane matrix particle, a configuration with the smallest surface area per volume, and for sheets of closely packed inner membrane (no matrix) a configuration with the largest surface area per volume (Srere, 1982).

For a smooth 60 Å lipid bilayer enclosing a spherical volume of 1 μm^3, the value 4.74 $\mu m^2/\mu m^3$ represents the minimum surface area–volume ratio for a theoretical spherical inner membrane matrix particle.

Consider now a situation in which our hypothetical smooth 60 Å bilayer is as densely packed as possible. The ratio of the area of the total inner surface, that is, the facing inner surfaces, to the volume would be 166.7 $\mu m^2/\mu m^3$. A similar calculation has been presented by Reith et al. (1976). This value then defines the maximum surface area of inner membrane to volume ratio of an inner membrane–matrix particle.

ii. Experimental measurements of mitochondrial inner membrane surface and the volume of the mitochondrion.   The stereomorphology of mitochondria, especially that of rat liver (Loud, 1968), and to a lesser extent, rat heart (Page, 1973; Smith and Page, 1976), has been well studied for years. For many years there appeared to be good agreement between various laboratories with a value for rat liver mitochondria of about 35 μm^2 inner membrane (cristae) surface area per μm^3 of mitochondrial volume (Loud, 1968; Weibel et al., 1969). More recently, however, a value of 16.5 $\mu m^2/\mu m^3$ (Reith, 1973; Reith and Barnard, 1976) has been reported by several laboratories. Reith and Mayhew (1980) have pointed to a possible systematic error in the previous estimates based upon an incorrect use of a correction factor employed by the earlier workers. They reason that if the high values reported for liver (35 $\mu m^2/\mu m^3$) were correct, then the volume percentage of total membranes in the mitochondria would be about 50%. This value, they argue, would be reasonable for heart or brown fat mitochondria but not for liver mitochondria. This is based to a large part upon visual comparison of the cristal content of the mitochondria of liver, heart, and brown adipose tissue (Reith and Barnard, 1976). They therefore feel that a systematic error has been made for the surface area–volume value for liver mitochondria and that the values of Loud (1968) and Weibel et al. (1969) of about 35 $\mu m^2/\mu m^3$ (35 cm^2/cm^3) are too high and that the values of 16.5 $\mu m^2/\mu m^3$ are more acceptable. Rat papillary muscle mitochondria have been reported to have a surface density of 37 $\mu m^2/\mu m^3$ (Anversa et al., 1980). Stereomorphometric measurements of rat heart mitochondria have yielded surface density data in the range of 61 $\mu m^2/\mu m^3$ (Page, 1973). This value together with the "appearance" of rat liver and rat heart mitochondria lead one to believe that the surface density of rat liver mitochondria would be closer to 16.5 $\mu m^2/\mu m^3$. The difference in the cristal content of rat liver, kidney, and heart is apparent to the eye (Fig. 5).

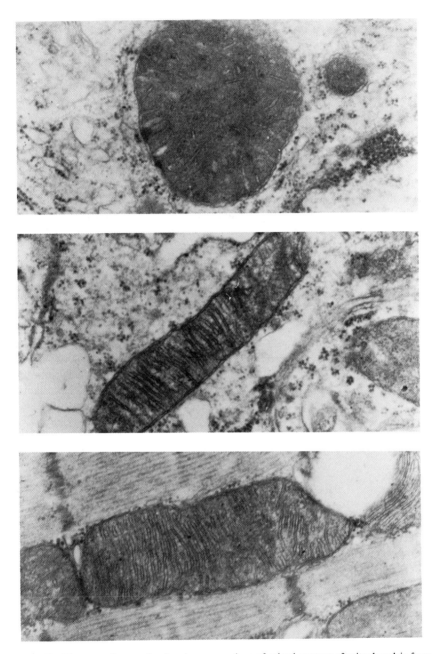

Fig. 5. Electron micrographs showing comparison of cristal content of mitochondria from rat liver (top), rat kidney (center), and rat heart (bottom).

iii. Model of inner membrane–matrix configuration.   These surface area–volume ratios can be used to calculate an average spacing between inner surfaces of cristal membranes in the matrix of mitochondria of rat liver and rat heart. This is another way of considering the packing density discussed by Reith and Barnard (1976) but I believe the calculation presented here gives a more graphic description of the compartment (Srere, 1982).

It was shown above, and also according to Reith *et al.* (1976), with 60-Å (6.0-nm) membranes that the maximum surface density of cristal membranes could be 166.7 $\mu m^2/\mu m^3$, that is, no matrix space. It should be remembered also that these calculations are based upon the volume of inner membrane and matrix, whereas the experimental data are based on the volume of the entire mitochondrion. Since the volume of inner membrane plus matrix is about 80% of the volume of the whole mitochondrion then our final figures are subject to this correction. If one assumes an average spherical protein molecule of the mitochondrial matrix to be 60 Å, then it is possible to calculate the average distance between cristal surfaces in terms of surface area–volume ratios and protein molecules. This can then be compared to the measured surface area–volume ratio for mitochondria and a model of the inner membrane–matrix compartment can be visualized. I have made these calculations (Table VI). The low experimental value for rat liver mitochondria, 16.5 $\mu m^2/\mu m^3$, would correspond to a distance equivalent to about 15 protein molecules (assuming 6.0-nm diameter proteins) between the inner membrane inner surfaces. For heart, however, the value of 61 $\mu m^2/\mu m^3$ is equivalent to about 3.5 molecules between the surfaces. A value of two molecules distance, which would

TABLE VI

THE MATRIX DISTANCE BETWEEN INNER MEMBRANES AT
VARIOUS INNER SURFACE MEMBRANE DENSITIES

Inner membrane inner surface area per volume mitochondria[a] ($\mu m^2/\mu m^3$)	Average distance between inner surfaces (Å)
132.8	0
88.8	60
66.4	120
53.6	180
44.8	240
38.4	300
22.4	600
16.0	900

[a] Assume 80% of the mitochondrial volume is inner membrane–matrix volume.

correspond to a morphometric measurement of 83 $\mu m^2/\mu m^3$, would indicate that every protein molecule in the matrix contacted an inner membrane surface of the mitochondrion. It is not difficult to imagine that such a situation might exist in mitochondria such as those from brown fat, which Reith and Barnard (1976) have identified by eye as having a higher membrane percentage in the inner membrane–matrix compartment than even heart mitochondria. The mitochondria of insect flight muscle also have a very high volume percentage of cristae and presumably a similar situation exists there.

iv. A partial model of the rat heart mitochondrial inner membrane–matrix compartment. Several assumptions were made in order to generate the model and these would tend to minimize the number of matrix proteins adjacent to the inner surface of the inner membrane. In this section I would like to examine the effect of using values that better approximate reality.

1. It is known that ATPase, an intrinsic protein of the inner membrane, extends into the matrix space about 90 Å. Since the membranes cannot be closer than 90 Å (unless these proteins are buried *in situ*), then the newly corrected figure (see above) of 132.8 $\mu m^2/\mu m^3$ would be altered to about 76.2 μm^2 inner membrane surface area to μm^3 mitochondrial volume.

2. I have assumed the diameter of a typical matrix protein subunit to be 60 Å, which corresponds to an M_r of about 100,000. Our SDS gel analysis of mitochondrial matrix proteins would indicate that about 60% of the polypeptides have a subunit $M_r > 40,000$. However, some Krebs TCA cycle enzymes exist as dimers with larger sizes. Thus, a dimer of citrate synthase (M_r 97,900) has the dimensions of $70 \times 60 \times 90$ Å (in the crystal) (Wiegand *et al.*, 1979) and the size of α-ketoglutarate dehydrogenase is estimated to be 250 Å. A number of years ago I estimated the cross-sectional area of a hypothetical Krebs TCA cycle complex to be about 2.3×10^{-4} μm^2 (Srere, 1972). This did not include a pyruvate dehydrogenase complex or the associated ATPases. If one assumes one ATPase (90 Å each) and one α-keto acid dehydrogenase complex (250 Å diameter) then the estimated overall complex is about 5×10^{-4} μm^2. If a rat heart mitochondrion has about 45 μm^2 of inner membrane surface and about 20,000 Krebs TCA cycle assemblies this would indicate that as much as one-fifth its inner membrane surface could be occupied by Krebs TCA cycle assemblies.

The earlier constructs define limits for theoretical lipid bilayer configurations. The structure of the inner mitochondrial membrane differs significantly from these hypothetical constructs. The inner mitochondrial membrane, far from being a lipid bilayer, is one of the most proteinaceous membranes known, consisting of 75% protein and 25% lipid (Munn, 1974). If all the protein were within the 60 Å lipid bilayer, then this would constitute almost a solid protein layer. It is known that the proteins of the inner membrane are both of the intrinsic and extrinsic types so that the membrane bilayer itself does not appear "crowded" with proteins when viewed by electron microscopy (Hackenbrock, 1981). The protruding intrinsic proteins and the extrinsic proteins may be crowded on the inner surface of the inner membrane. One can compare, for instance, the acetylcholine receptor proteins in their membrane, where the protein stalks do not appear crowded in the membrane but the larger diameter head pieces are crowded on the surface of the membrane (Anholt *et al.*, 1981). Proteins attached to the inner surface of the inner mitochondrial membrane cannot contribute to the osmotic behavior of matrix proteins because of their attachment and must be considered as membrane proteins.

A more realistic appearance of heart mitochondrial inner membranes and most of the Krebs TCA cycle and electron-transport proteins would be as sketched in Fig. 6. It should be remembered that in heart mitochondria the total proteins of the Krebs cycle probably represent only 10–20% of the total matrix protein, so the apparently unoccupied volume is densely packed with other proteins.

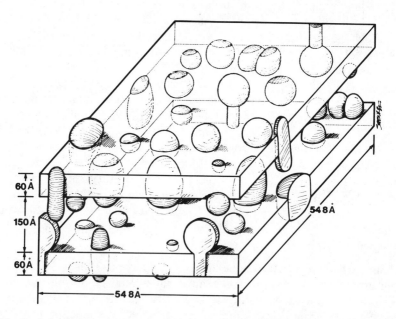

Fig. 6. A theoretical visualization of a portion of a heart mitochondrion showing only TCA cycle enzymes and enzymes of electron transport. The proteins and membrane are drawn at a scale to approximate their relative concentrations and size in these mitochondria. Their distribution is depicted as random in this drawing. Other mitochondrial proteins are not shown.

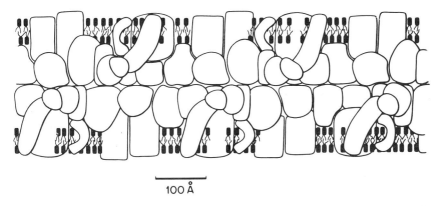

100 Å

Fig. 7. A schematic picture of a portion of the inner membrane–matrix compartments of a mitochondrion (from Williams, 1983, with permission).

Recently, Williams (1983) in commenting on this model presented a sketch of what the packed inner membrane–matrix might look like (Fig. 7). In this fully packed model he notes that the "chemi-osmotic descriptions of proton energies and movements are unlikely to be related to the mechanism of energy transduction."

It is apparent that use of experimental data on matrix proteins does not weaken the model of a rat heart mitochondrion in which the bulk of matrix protein ($>50\%$) is adjacent to the inner membrane surface but rather strengthens the conclusions. The corrections do not lead to a model that is physically naive (i.e., more protein than surface).

I have pointed out above that the matrix proteins in the mitochondria were very densely packed. Two consequences of this arrangement were the formation of protein-bound pores that were capable of limiting free diffusion of large metabolites (nucleotides) and the possible change in local water pool sizes, which could result in large changes in local microenvironmental concentrations of metabolites. When protein molecules in solution are in "crowded" conditions, Minton (1981) and Wilf and Minton (1981) have shown that there is a significant excluded volume effect on the activity coefficients of macromolecules. Some of these effects can be shown to occur on the self-association of the macromolecules.

The molecular arrangement of Krebs TCA cycle enzymes and electron-transport proteins on or in the membrane is shown as a random one (Fig. 6). As pointed out in this article accumulating evidence shows specific interactions between Krebs cycle enzymes (Section II.B.2) as well as between certain electron-transport proteins (Azzi *et al.*, 1981), so one must seriously consider the possibility that a complex of enzymes for this metabolic pathway exists, rather than this depicted random arrangement.

If the protein structure of the inner membrane matrix compartment is as I have postulated in this paper, then it would appear that the two elements form what may be a single continuous compartment with extensive interactions possible between the proteins of the lipid and aqueous phases. It has been noted by others that the matrix proteins of red blood cell may control protein diffusion in the red blood cell membrane (Koppel et al., 1981). Similar interactions, if they occur in the inner membrane–matrix fraction of the mitochondria, may have important consequences in our understanding of the control of aerobic oxidations and of mitochondrial biogenesis.

C. Inner Membrane

Two excellent reports have reviewed the composition and possible arrangements of proteins within the inner membrane (Capaldi, 1982; Harmon et al., 1974). I will not repeat all the information contained in those articles, but rather I would like to discuss just two points: first, the possibility of enzyme complexes within the membrane, and second, the possibility of interactions between matrix proteins and membrane proteins.

The inner membrane of mitochondria has a number of unusual features. It has one of the highest protein contents of any biological membrane and its lipid composition includes a high percentage of cardiolipin and a very low percentage of cholesterol. The phospholipids contain a high percentage of unsaturated fatty acids, which results in a highly fluid membrane. The phospholipids are asymmetrically distributed with phosphatidylethanolamine and cardiolipin (75%) mainly on the matrix side and phosphatidylcholine mainly on the intermembrane space side (Krebs et al., 1979).

With the high protein content, estimated by some to be 74%, it would be difficult for all the protein molecules to be solvated by the lipid molecules present. Thus, if one assumes the proteins are all M_r 50,000 spheres, one could calculate that there are 50 phospholipid molecules per protein molecule. This would result in a spacing of 7Å between each phospholipid molecule. Thus, the model of the protein completely solvated by phospholipid is not likely to occur, and the actual saturation must be that much of the so-called inner membrane protein is external to the lipid bilayer and the external protein is solvated by water. This scenario is confirmed by the fact that approximately 45% of the protein can be removed from inner membrane vesicles by dilute acid or urea washing (Fleischer et al., 1971). An external protein must therefore have two kinds of surface: hydrophobic and hydrophilic. It seems likely that when such proteins are removed from the membrane by aqueous solvents their normal membrane conformation will change such that their hydrophobic surfaces will be minimized by interactions between molecules.

The average concentration of eight inner membrane proteins is (nmol/mg total membrane protein) Complex I (NADH–ubiquinone reductase) 0.1, complex II (succinate–ubiquinone reductase) 0.2, Complex III (ubiquinol–cytochrome c oxidoreductase) 0.4, complex IV (cytochrome c oxidase) 0.8, cytochrome c 0.9, ATP synthase 0.5, ADP–ATP translocase 4.0, and transhydrogenase 0.05 (Capaldi, 1982). Since the molecular weights of these components are known, it is possible to calculate the weight of each protein present per mg total membrane protein. With the numbers presented above the eight proteins account for 750 μg of the 1 mg of mitochondrial protein. In addition to the proteins listed there must exist all the proteins necessary for the transport of metabolites in and out of the matrix compartment. There are no data available to estimate the quantities of the latter class of proteins that are present.

It is clear that the high protein content of the inner membrane makes it a special compartment, but although there is some indication that these crowded conditions have special effects on protein interactions in aqueous medium, there has been no work to indicate that this can be extended to a lipid medium. If one looks at the intermembrane particles by freeze etch electron microscopic methods the appearance is not a packed one. Hackenbrock (1981) has shown that the particles occupy about one-half of the area of the inner membrane. Wrigglesworth et al. (1970) has counted the particles in the freeze-fracture faces and found a total of 4525 particles/μm^2. Similar results were reported by Packer et al. (1974). The predominant size of particle was 85 Å. One can calculate that only 33% of the area would be occupied by these particles in a cubical packing pattern. With either estimate this still represents a very concentrated solution.

Even though the bulk of the protein of the inner membrane resides in just eight enzyme activities, SDS polyacrylamide gel analysis of the inner membrane protein is quite complex. This is true because each of the complexes is composed of several different polypeptide chains. Complex I is estimated to contain 26 different polypeptides, Complex II 5, Complex III 10, and Complex IV 8, or a total of 49 different polypeptides just for these four complexes (Capaldi, 1982).

It would appear to be very efficient if the complexes in the membrane responsible for electron transport and ATP synthesis formed a megacomplex to accomplish this task. Such a proposal has been made by Racker (1970) showing interaction between SDH–cyt b–cyt c_1–cyt c–cyt a–cyt a_3 and the ATPase vectorially arranged across the membrane. Several factors should be considered before making such proposals. First, one must consider the stoichiometry of the various components, and second, one must consider the necessity for the superstructure. In relation to the stoichiometry, the ratio of Complex I:Complex II:Complex III:Complex IV:ATP synthase is 1:2:4:8:5 for the monomers. Most of these are present in the dimeric state, but on the basis of present data it is not possible to propose a simple scheme for the possible association into functional complexes.

Hackenbrock and colleagues have presented a variety of evidence to indicate that the lateral diffusion rates of the various complexes are much more rapid than the turnover times of the individual reactions (Sowers and Hackenbrock, 1981; Schneider *et al.*, 1980a,b; Hochli and Hackenbrock, 1969; Hackenbrock, 1981). Cherry and his colleagues (Kawato *et al.*, 1980; Muller *et al.*, 1982) have indicated, however, that the ADP–ATP translocase and cytochrome c oxidase are relatively immobile in the inner membrane.

The experiments of Hackenbrock's group use inner membrane vesicles so that the membrane proteins are not in contact with the matrix compartment components, which are themselves highly viscous. Koppel *et al.* (1981) have shown that the effect of matrix viscosity on lateral movement of membrane proteins is considerable so in intact mitochondria it is quite likely that the membrane proteins are not diffusing at rates comparable to that measured by Hackenbrock. Ubiquinone, however, which is quite likely to be entirely within the membrane, can diffuse freely, as can cytochrome c, which is on the outer surface of the inner membrane. However, even cytochrome c may be slowed by the proteins of the intermembrane space.

Recently Hochman *et al.* (1983) showed that in giant mitochondria the diffusion of cytochrome c and cytochrome $a–a_3$ is not sufficiently rapid to account for the rate of succinate oxidation. These authors suggest that some mechanism other than random diffusion must be operating in order to account for the activity observed. Mitchell (1979, 1981) has also considered the organization membrane proteins to be essential for the vectorial theory of ATP generation to operate adequately.

III. CONCLUDING REMARKS

The following questions should be considered. How important a biological concept is the organization of proteins within cellular compartments? Is the evidence on hand sufficiently strong for or against the concept to justify continued work in the area? What experimental approaches might be considered to augment and/or confirm the available data concerning the organization of metabolic pathways?

One may say that the early studies of Zalokar (1960) and Kempner and Miller (1968) established unequivocally that no enzymes were free in cells. One sees, however, a continued implicit use of the notion of free enzymes when all aspects of cellular activities are discussed. In any case those experiments cannot be used to argue for an organization of mitochondrial matrix proteins.

It has been amply demonstrated that the kinetic behavior of enzymes free in solution, upon which many of our concepts of metabolic regulation are based, is quite different from their behavior when they are bound. That vectorial arrangements of proteins are important has been well demonstrated for energy generation (Mitchell, 1981) and such arrangements probably exist also for the biosynthesis of all macromolecules.

Let us consider the notion that metabolic pathways are brought about by randomly oriented enzymes that coexist in a single compartment. This would imply that apart from determining its cellular location as either an aqueous compartment protein or a lipid compartment protein that there was no other information content of a protein's surface. In terms of the structure of enzymes the only important aspect would be that of its active site. The rest of the protein structure of the enzyme in a certain sense would be superfluous. One has always wondered why such a large macromolecule was necessary to carry a relatively small active grouping of amino acid residues. It certainly is partially a consequence of having to bring together these residues in a fairly precise geometrical relation to each other, but recent experiments with large rate enhancements by chiral catalysts (see Maugh, 1983, for review) would lead one to believe that active sites could have been formed with a smaller number of linear residues. Recent work in protein chemistry has also made it quite clear that what is evolutionarily conserved is tertiary structure (the protein domains) and not the absolute primary sequence. This leads me to believe that the surface structure of a protein also must serve other functions. One such function may be to determine its location as membranal or not. But surface structure may also determine interactions with other specific protein surfaces. Such a function has been largely overlooked, since studies of these proteins have been usually under conditions that were physiologically accurate, except with respect to the total protein concentration. As the proteins evolved, interactions would develop that were sufficiently strong only under their normal microenvironmental conditions of high protein concentrations. The stimulating paper by McConkey (1982) has made these quinary interactions an important real possibility.

Another function would be the binding of substrates at the surface away from an active site and the subsequent diffusion of the substrate to the active site at an increased rate due to the decrease in dimensionality of the system (Richter and Eigen, 1974; Berg *et al.*, 1982).

Not only does the concept of biological organization have important repercussions in the consideration of the protein structure and function but it presents an important insight into the physical nature of the cell's interior. We have seen a gradual refinement of the theoretical basis of considering the nature of the cellular environment. Physical models now include the notions of tortuosity, crowdedness, and interactions between lipid-solvated and water-solvated proteins.

I feel that the most important novel aspect concerning mitochondrial structure that is emerging is the relatively close association of all matrix proteins with the membrane and the fact that the membrane and matrix protein populations represent a continuum of proteins as illustrated in Fig. 8 from partially H_2O- and lipid-solvated molecules to lipid-solvated molecules to partially lipid–H_2O-solvated molecules to H_2O-solvated molecules. It seems highly likely both from a theoretical standpoint and from the accumulating experimental data that some degree of organization exists and that the organization itself and its possible reversible dissociation and formation may be an important aspect of metabolic control. The recent data of McConkey (1982) indicate that the quinary interactions necessary for such organizations must have been highly conserved during evolution, thus indicating in another way their importance to the living condition.

The available evidence indicates that the mitochondrial matrix and inner membrane compartments are a semicontinuous compartment crowded with proteins that are interacting with each other in each semiautonomous phase and with proteins in the adjacent phase. This solid-state organization of proteins, solvated by lipids in the membrane phase and water in the matrix phase, has clusters of interacting enzymes of specific metabolic sequences of both inter- and intraphase types. In such an array it is not possible at present to assess physical properties usually assigned to solutes in solution such as diffusion coefficients, ionic strength, free concentrations, and osmotic pressure.

Also, the concentrations of substrates measured in whole cells and tissues are misleading since not only do they not reflect compartmentation at the

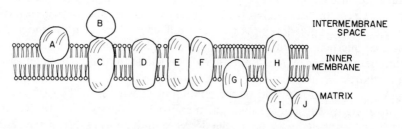

Fig. 8.   Some possible interactions between proteins and a membrane. A, extrinsic membrane protein; B, C, intermembrane space protein with intrinsic membrane protein; D, intrinsic membrane protein; E, F, interaction between two intrinsic membrane proteins; G, extrinsic membrane protein; H, I, intrinsic membrane protein with matrix protein, I, J, interaction between matrix proteins.

macro level of organelles, but in all probability do not represent their microenvironmental concentrations due to a considerable fraction being protein bound and in some compartments being excluded from the considerable quantities of bound water.

One is faced with a number of topological problems in the construction of an ordered arrangement of the Krebs TCA cycle complex of enzymes. For one thing, one must imagine that the large pyruvate and α-ketoglutarate dehydrogenase complexes do not exist as the multisubunit complexes that are usually isolated and studied. In the mitochondrion it is simpler to consider them as simple complexes of a small number of each subunit (i.e., $1E_1 : 1E_2 : 1E_3$). Also, for a complete complex one cannot consider directly apposite active sites since each enzyme must communicate with more than one enzyme. Thus, for citrate synthase one can consider an input of acetyl-CoA from PDC or β-ketoacyl-CoA thiolase, input of oxaloacetate from MDH, and delivery of citrate to aconitase. Aconitase in turn must communicate with isocitrate dehydrogenase, and MDH must receive malate from fumarase, and so forth around the cycle. One could imagine a three-dimensional arrangement of these enzymes partially attached to the membrane with the substrates in a central open "pore space", or a linear juxtaposition of enzymes might occur with the substrates flowing in an "unstirred" layer, kept there by charge effects. Another possibility would be that only portions of the cycle are organized, such as FUM–MDH–CS–PDC, which might be a citrate-forming portion of the cycle, or the dehydrogenases might be bound to Complex I in the membrane. One could propose a variety of such organizational variants. I believe such variations may exist, be dependent on cell type, and be related to the metabolic differences and different structures observed in various mitochondria.

What remains to be done seems quite clear. I believe an extension of the work of McConkey (1982) is necessary. Although we have shown that mitochondrial matrix proteins from tissues of single animals behave similarly in terms of size and charge distribution it would be interesting to compare the mitochondrial matrix proteins from different animals to see the degree of similarity. Interactions between other metabolic pathways such as fatty acid metabolism, urea cycle, and Krebs TCA cycle enzymes should also be explored. Specific interactions between transport proteins and these enzymes should also be investigated.

Furthermore, every effort should be made to obtain mutant cells (probably yeast) that lack the binding proteins or have mutant enzymes with normal active sites but altered quinary binding sites. Only with such mutants will we be able to assess the biological importance of the putative organization of cellular proteins.

ACKNOWLEDGMENTS

I am indebted to all my colleagues who, over the years, have contributed to the work and ideas discussed in this chapter. I am especially indebted to Drs. D'Souza, Matlib, Robinson, and Henslee.

I am grateful to Ms. Penny Perkins for her excellent secretarial assistance.

This work was supported by research from the Veterans Administration, NIH, and NSF.

REFERENCES

Addink, A. D. F., Boer, P., Wakabayashi, T., and Green, D. E. (1972). *J. Biochem.* **29,** 47–59.

Alam, T., Finkelstein, D., and Srere, P. A. (1982) **257,** *J. Biol. Chem.* 11181–11185.

Altman, P. L., and Katz, D. D., eds. (1976). "Cell Biology" pp. 160–179. Fed. Am. Soc. Exp. Biol., Bethesda, Maryland.

Anholt, R., Lindstrom, J., and Montal, M. (1981). *J. Biol. Chem.* **256,** 4377–4387.

Anversa, P., Olivetti, G., Melissari, M., and Loud, A. V. (1980). *J. Mol. Cell. Cardiol.* **12,** 781–795.

Atkinson, D. E. (1969). *Curr. Top. Cell. Regul.* pp. 29–42.

Azzi, A., Bill, K., and Broger, C. (1981). *Proc. Natl. Acad. Sci. U.S.A.* **79,** 2447–2450.

Bachman, B. J. (1983). *Microbiol. Rev.* **47,** 180–230.

Backman, L., and Johansson, G. (1976). *FEBS Lett.* **65,** 39–42.

Beeckmans, S., and Kanarek, L. (1981a). *Eur. J. Biochem.* **117,** 527–535.

Beeckmans, S., and Kanarek, L. (1981b). *Arch. Int. Physiol. Biochem.* **89,** B1–B46.

Bendayan, M., and Shore, G. C. (1982). *J. Histochem. Cytochem.* **30,** 139–147.

Berg, O. G., Winter, R. B., and von Hippel, P. H. (1982). *Trends Biochem. Sci.* **7,** 52–55.

Bishop, W. H., and Richards, F. M. (1968). *J. Mol. Biol.* **38,** 315–328.

Blinzinger, K., Rewcastle, N. B., and Hager, H. (1965). *J. Cell. Biol.* **25,** 293–303.

Boulikas, T., Wiseman, J. M., and Garrard, W. T. (1980). *Proc. Natl. Acad. Sci. U.S.A.* **77,** 127–131.

Brdiczka, D., Dolken, G., Krebs, W., and Hofmann, D. (1974). *Z. Physiol. Chem.* **355,** 731–743.

Brunner, G., and Bucher, Th. (1970). *FEBS Lett.* **6,** 105–108.

Bryce, C. F. A., Williams, D. C., John, R. A., and Fasella, P. (1976). *Biochem. J.* **153,** 571–577.

Capaldi, R. A. (1982). *Biochim. Biophys. Acta* **694,** 291–306.

Careri, G., Gratton, E., Yang, P.-H., and Rupley, J. A. (1980). *Nature (London)* **284,** 572–573.

Clarke, S. (1976). *J. Biol. Chem.* **251,** 950–959.

Clegg, J. S. (1979). *In* "Cell-Associated Water" (W. Drost-Hansen and J. Clegg, eds.), pp. 363–413. Academic Press, New York.

Cohen, N.S., Cheung, C. W., Kyan, F. S., Jones, E. E., and Raijman, L. (1982). *J. Biol. Chem.* **257,** 6898–6907.

Cold Spring Harbor Symposium on Quantitative Biology (1982). **46.**

Coleman, R. (1973). *Biochim. Biophys. Acta* **300,** 1–30.

Comte, J., and Gautheron, D. C. (1978). *Biochimie* **60,** 1299–1305.

Cooke, R., and Kuntz, I. D. (1974). *Annu. Rev. Biophys. Bioeng.* **3,** 95–125.

Crick, F. H. C., and Kendrew, J. C. (1957). *Adv. Protein Chem.* **12,** 134.

DeVaney, M., and Powers-Lee, S. (1983). *Fed. Proc., Fed. Am. Soc. Exp. Biol.* **42,** 2106.

Drost-Hansen, W., and Clegg, J., eds. (1979). "Cell-Associated Water". Academic Press, New York.

D'Souza, S. F., and Srere, P. A. (1983a). *J. Biol. Chem.* **258,** 4706–4709.

D'Souza, S. F., and Srere, P. A. (1983b). *Biochim. Biophys. Acta,* **724,** 40–51.

Dulin, D. R., and Harrison, J. H. (1983). *Fed. Proc., Fed. Am. Soc. Exp. Biol.* **42,** 2106.

Eigen, M. (1973). *In* "Quantum Statistical Mechanics in the Natural Sciences (S. L. Mintz and S. M. Widmayer, eds.), pg. 37–61. Plenum, New York.

Elduque, A., Casado, F., Cortes, A., and Bozal, J. (1982). *Int. J. Biochem.* **14**, 221–229.

Fahien, L. A., and Kmiotek, E. (1978). "Symposium on Microenvironments and Cellular Compartmentation" (P. A. Srere and R. W. Estabrook, eds.), pp. 355–370. Academic Press, New York.

Fahien, L. A., and Kmiotek, E. (1979). *J. Biol. Chem.* **254**, 5983–5990.

Fahien, L. A., and Kmiotek, E. (1983). *Arch Biochem. Biophys.* **220**, 386–397.

Fahien, L. A., and Smith, S. E. (1974). *J. Biol. Chem.* **249**, 2696–2703.

Fahien, L. A., and van Engelen, D. L. (1976). *Arch. Biochem. Biophys.* **176**, 298–305.

Fahien, L. A., Hsu, S. L., and Kmiotek, E. (1977). *J. Biol. Chem.* **252**, 1250–1256.

Fahien, L. A., Ruoho, A. E., and Kmiotek, E. (1978). *J. Biol. Chem.* **253**, 5745–5751.

Fahien, L. A., Kmiotek, E., and Smith, L. (1979). *Arch. Biochem. Biophys.* **192**, 33–46.

Fleischer, S., Zahler, W. L., and Ozawa, H. (1971). *Biomembranes* **2**, 105–119.

Fleischer, S., Hatefi, Y., MacLennan, D. H., and Tzagoloff, A. eds. (1978). "The Molecular Biology of Membranes." Plenum, New York.

Frenkel, E. P., Mukherjee, A., Hackenbrock, C. R., and Srere, P. A. (1976). *J. Biol. Chem.* **251**, 2147–2154.

Fritz, I. B. (1968). *In* "Cellular Compartmentalization and Control of Fatty Acid Metabolism" (F. C. Gran, ed.), *FEBS Proc. 4th Meet.*, Vol. 4, pp. 39–63. Academic Press, New York.

Fritz, I. B., and Lee, L. (1974). *In* "Regulation of Hepatic Metabolism" (F. Luncquist and N. Tigstrup, eds.), Benzon Symp 6, pp. 224–234. Munksgaard, Copenhagen.

Gamble, J. L., Jr., and Garlid, K. D. (1970). *Biochim. Biophys. Acta* **211**, 223–232.

Garlid, K. D. (1979). *In* "Cell Associated Water" W. Drost-Hansen and J. Clegg, eds.), pp. 293–362. Academic Press, New York.

Hackenbrock, C. R. (1968). *Proc. Natl. Acad. Sci. U.S.A.* **61**, 598–605.

Hackenbrock, C. R. (1981). *Trends Biochem. Sci.* **6**, 151–154.

Halper, L. A., and Srere, P. A. (1977). *Arch. Biochem. Biophys.* **184**, 529–534.

Halper, L. A., and Srere, P. A. (1979). *J. Solid Phase Biochem.* **4**, 1–13.

Hardt, S. L. (1981). *Bull. Math. Biol.* **43**, 89–99.

Hare, J. F., and Hodges, R. (1982). *J. Biol. Chem.* **257**, 12950–12953.

Harmey, M. A., and Neupert, W. (1979). *FEBS Lett.* **108**, 385–389.

Harmon, H. J., Hall, J. D., and Crane, F. L. (1974). *Biochim. Biophys. Acta* **344**, 119–155.

Hathaway, J. A., and Atkinson, D. E. (1965). *Biochem, Biophys. Res. Commun.* **20**, 661–665.

Henslee, J. G. (1978). Ph. D. Dissertation. Univ. Texas Heath Science Center, Dallas.

Henslee, J. G., and Srere, P. A. (1979). *J. Biol. Chem.* **254**, 5488–5497.

Hess, B., and Boiteux, A. (1972). "Protein-Protein Interactions" (R., Jaenicke and E. Helmreich, eds.), pp. 271–297. Springer-Verlag, Berlin and New York.

Hilderman, R. H., and Deutscher, M. P. (1974). *J. Biol. Chem.* **249**, 5346–5348.

Hilderman, R. H., Goldblatt, P. J., and Deutscher, M. P. (1975). *J. Biol. Chem.* **250**, 4796–4801.

Hochli, M., and Hackenbrock, C. R. (1979). *Proc. Natl. Acad. Sci. U.S.A.* **76**, 1236–1240.

Hochman, J. H., Schindler, M., and Ferguson-Miller, S. (1983). *Fed. Proc., Fed. Am. Soc. Exp. Biol.* **42**, 2061.

Holloszy, J. O., Oscai, L. B., Don, I. J., and Mole, P. A. (1970). *Biochem. Biophys. Res. Commun.* **40**, 1368–1373.

Hoppel, C. L. (1972). *J. Biol. Chem.* **247**, 832–841.

Hubscher, G., Mayer, R. J., and Hansen, H. J. M. (1971). *Bioenergetics* **2**, 215–218.

Jakovcic, S., Swift, H. H., Gross, N. J., and Rabinowitz, M. (1978). *J. Cell Biol.* **77**, 887–901.

Kacser, H., and Burns, J. (1973). *Symp. Soc. Exp. Biol.* **32**, 65–104.

Kawato, S., Sigel, E., Carafoli, E., and Cherry, R. J. (1980). *J. Biol. Chem.* **255**, 5508–5510.

Keith, A. D., ed. (1979). "The Aqueous Cytoplasm." Dekker, New York.

58 PAUL A. SRERE

Kempner, E. S., and Miller, J. H. (1968). *Exp. Cell Res.* **51**, 141–149.
Keokitichai, S., and Wrigglesworth, J. (1979). *Int. Congr. Biochem.*, *11th* p. 267.
Kessel, R. G. (1966). *J. Ultrastruct. Res.* **16**, 293–304.
Kliman, H. J., and Steck, T. L. (1980). *J. Biol. Chem.* **255**, 6314–6321.
Knecht, E., Hernandez, J., Wallace, R., and Grisolia, S. (1979). *J. Histochem. Cytochem.* **27**, 975–981.
Knull, H. R. (1980). *J. Biol. Chem.* **255**, 6439–6444.
Koch-Schmidt, A., Mattiasson, B., and Mosbach, K. (1977). *Eur. J. Biochem.* **81**, 71–78.
Koppel, D. E., Sheetz, M. P., and Schindler, M. (1981). *Proc. Natl. Acad. Sci. U.S.A.* **78**, 3576–3580.
Krebs, J. R., Hauser, H., and Carafoli, E. (1979). *J. Biol. Chem.* **254**, 5308–5316.
Landriscina, C., Papa, S., Coratelli, P., Mazzarella, L., and Quagliariello, E. (1970). *Biochim. Biophys. Acta* **205**, 136–147.
Lee, S. H., and Toback, R. M. (1968). *J. Cell Biol.* **39**, 725–732.
Ling, G. (1972). "Water and Aqueous Solutions" (R. A. Horne, ed.), pp. 663–700. Wiley (Interscience), New York.
Linn, T. C. (1974). *Arch. Biochem. Biophys.* **161**, 505–514.
Lloyd, D. (1974). "The Mitochondria of Microorganisms." Academic Press, New York.
Lopes-Cardozo, M., Klazinga, W., and van den Bergh, S. G. (1978). *Eur. J. Biochem.* **83**, 635–640.
Loud, A. V. (1968). *J. Cell Biol.* **37**, 27–46.
McCallister, L. P., and Page, E. (1973). *J. Ultrastruct. Res.* **42**, 136–155.
McConkey, E. H. (1982). *Proc. Natl. Acad. Sci. U.S.A.* **79**, 3236–3240.
McKinley, M. P., and Trelease, R. N. (1980). *Comp. Biochem. Physiol.* **67B**, 27–32.
McKusick, V. A. (1983). *Hum. Gene Map.*
Makinen, M. W., and Fink, A. L. (1977). *Annu. Rev. Biophys. Bioeng.* **6**, 301–343.
Manley, E. R., Webster, T. A., and Spivey, H. O. (1980). *Arch. Biochem. Biophys.* **205**, 380–387.
Mansson, M.-O., Larsson, P.-O., and Mosbach, K. (1978). *Eur. J. Biochem.* **86**, 455–463.
Mansson, M.-O., Siegbahn, N., and Mosbach, K. (1983). *Proc. Natl. Acad. Sci. U.S.A.* **80**, 1487–1491.
Masters, C. J. (1977). *Curr. Top. Cell. Regul.* **12**, 75–105.
Masters, C. J. (1981). *CRC Crit. Rev. Biochem.* pp. 105–143.
Matlib, M. A., and O'Brien, P. (1975). *Arch. Biochem. Biophys.* **167**, 193–202.
Matlib, M. A., and Srere, P.A. (1976). *Arch. Biochem. Biophys.* **174**, 705–712.
Matlib, M. A., Shannon, W. A., Jr., and Srere, P. A. (1976). *Annu. Proc. Electron Microsc. Soc. Am.*, *34th*, Miami.
Matlib, M. A., Shannon, W. A., Jr., and Srere, P. A. (1977). *Arch. Biochem. Biophys.* **178**, 396–407.
Matlib, M. A., Boesman-Finkelstein, M., and Srere, P. A. (1978). *Arch. Biochem. Biophys.* **191**, 426–430.
Matlib, M. A., Frenkel, E. P., Mukherjee, A., Henslee, J., and Srere, P. A. (1979). *Arch. Biochem. Biophys.* **197**, 388–395.
Matsuoka, Y., and Srere, P. A. (1973). *J. Biol. Chem.* **248**, 8022–8030.
Matthews, B. W. (1968). *J. Mol. Biol.* **33**, 491–497.
Maugh, T. J., III. (1983). *Science* **221**, 351–354.
Mazurkiewicz, J. E., and Nakane, P. K. (1972). *J. Histochem. Cytochem.* **20**, 969–974.
Melnick, R. L., Tinberg, H. M., Magiure, J., and Packer, L. (1973). *Biochim. Biophys. Acta* **311**, 230–241.
Metzler, D. E. (1977). "Biochemistry." Academic Press, New York.
Miekka, S. I., and Ingham, K. C. (1980). *Arch. Biochem. Biophys.* **203**, 630–641.
Minton, A. P. (1981). *Biopolymers* **20**, 2093–2120.
Minton, A. P., and Wilf, J. (1981). *Biochemistry* **20**, 4821–4826.
Mitchell, P. (1979). *Eur. J. Biochem.* **95**, 1–20.

Mitchell, P. (1981). *In* "Bioenergetic Aspects of Unity in Biochemistry: Evolution of the Concept of Ligand Conduction in Chemical, Osmotic, and Chemiosmotic Reaction Mechanisms" (G. Semenza, ed.). Wiley, New York.

Moriyama, T., and Srere, P. A. (1971). *J. Biol. Chem.* **246**, 3217–3223.

Mosbach, K., and Mattiasson, B. (1970). *Acta Chem. Scand.* **24**, 2093–2100.

Mosbach, K., and Mattiasson, B. (1978). *Curr. Top. Cell. Regul.* **14**, 197–241.

Moses, R. E., and Richardson, C. C. (1970). *Proc. Natl. Acad. Sci. U.S.A.* **67**, 674–681.

Mukherjee, A., and Srere, P. A. (1978). *J. Solid Phase Biochem.* **3**, 85–94.

Mukherjee, A., Srere, P. A., and Frenkel, E. P. (1976). *J. Biol. Chem.*, **251**, 2155–2160.

Muller, M., Krebs, J. J. R., Cherry, R. J., and Kawato, S. (1982). *J. Biol. Chem.* **257**, 117–120.

Mullinax, T. R., and Harrison, J. H. (1983). *Fed. Proc., Fed. Am. Soc. Exp. Biol.* p. 2106.

Munn, E. A. (1974). "The Structure of Mitochondria." Academic Press, New York.

Nicholson, C., and Phillips, J. M. (1981). *J. Physiol. (London)* **321**, 225–257.

Nover, L., Lynen, F., and Mothes, K. (1980). "Cell Compartmentation and Metabolic Channeling." Elsevier, Amsterdam.

Nunez de Castro, I., Arias de Saavedra, J. M., Machado, A., and Mayor, F. (1976). *Mol. Cell. Biochem.* **12**, 161–169.

O'Farrell, P. (1975). *J. Biol. Chem.* **250**, 4007–4021.

Ottaway, J. H. (1976). *Biochem. Soc. Trans.* **4**, 371–376.

Ottaway, J. H. (1983). *Biochem. Soc. Trans.* **11**, 47–52.

Ottaway, J. H., and Mowbray, J. (1977). *Curr. Top. Cell. Regul.* **12**, 75–105.

Packer, L., Mehard, C. W., Meissner, G., Warren, L., and Fleischer, S. (1974). *Biochim. Biophys. Acta* **363**, 159–181.

Page, E. (1973). *J. Cell Biol.* **59**, 513.

Page, E., and McCallister, L. P. (1973). *Am. J. Cardiol.* **31**, 172–181.

Peschel, E., and Belouschek, P. (1979). *In* "Cell Associated Water" (W. Drost-Hansen and J. Clegg, eds.), pp. 3–52. Academic Press, New York.

Peters, K., and Richards, F. M. (1977). *Ann-. Rev. Biochem.* **46**, 523–551.

Peterson, R. E., Radcliff, C. W., and Pace, N. R. (1971). *J. Bacteriol.* **107**, 585–588.

Pette, D. G., Klingenberg, G., and Bucher, T. (1962). *Biochem. Biophys. Res. Commun.* **7**, 425–429.

Pfaff, E., Klingenberg, M., Ritt, E., and Vogell, W. (1968). *Eur. J. Biochem.* **5**, 222–232.

Pihl, E., and Bahr, G. F. (1970). *Exp. Cell Res.* **63**, 391–403.

Plattner, H. (1968). *Eur. Reg. Conf. Electron Microsc. 4th, Rome.*

Plattner, H. (1973). "Quantitative Correlation of Structure and Function on Biomembranes. Fischer, Stuttgart.

Plattner, H., Pfaller, W., and Tiefenbrunner, F. J. (1969). *Comp. Morphol. Resp. Act. Mitochondria Several Animal Cells* **43**, 258.

Porpaczy, Z., Sumegi, B., and Alkonyi, I. (1983). *Biochim. Biophys. Acta* **749**, 172–179.

Racker, E. (1970). *Am. Chem. Soc.* No. 165.

Racker, E., and Horstman, L. L. (1972). *In* "Energy Metabolism and the Regulation of Metabolic Processes in Mitochondria" (M. A. Mehlman and R. W. Hanson, eds.), pp. 1–25. Academic Press, New York.

Raijman, L., and Jones, M. E. (1976). *Arch. Biochem. Biophys.* **175**, 260–278.

Reeves, R. E., and Sols, A. (1973). *Biochem. Biophys. Res. Commun.* **50**, 459–466.

Reith, A. (1973). *Lab Invest.* **29**, 216–228.

Reith, A., and Barnard, T. (1976). *Proc. Int. Congr. Stereol., 4th* 427–428.

Reith, A., and Fuchs, S. (1973). *Lab. Invest.* **29**, 229–235.

Reith, A., and Mayhew, T. M. (1980). *Gegenbaurs Morphol. Jahrb.* **126**, 206–215.

Reith, A., Barnard, T., and Rohr, H.-P. (1976). *CRC Crit. Rev. Toxicol.* **4**, 219–269.

Rendon, A., and Waksman, A. (1973). *Biochem. Biophys. Res. Commun.* **50**, 814–819.

Rendon, A., Rott, R., and Avi-Dor, Y. (1980). *Biochim. Biophys. Acta* **590**, 290–299.

Richards, F. M. (1977). *Annu. Rev. Biophys. Bioeng.* **6**, 151–176.

Richter, P. H., and Eigen, M. (1974). *Biophys. Chem.* **2**, 255–263.

Rupley, J. A. (1969). *In* "Structure and Stability of Biological Macromolecules" (S. N. Timasheff and D. G. Fasman, eds.), Ch. 4. pp. 291–352. Dekker, New York.

Samorajski, T., Ordy, J. M., and Keefe, J. R. (1966). *J. Cell Biol.* **28**, 489–504.

Sandstrom, B., and Westman, J. (1969). *Histochemie* **19**, 181–183.

Schliwa, M., van Blerkom, J., and Porter, K. R. (1981). *Proc. Natl. Acad. Sci. U.S.A.* **78**, 4329–4333.

Schnaitman, C., and Greenawalt, J. W. (1968). *J. Cell Biol.* **38**, 158–175.

Schneider, H., Hochli, M., and Hackenbrock, C. R. (1980a). *Proc. Natl. Acad. Sci. U.S.A.* **77**, 387–393.

Schneider, H., Lemasters, J. J., Hochli, M., and Hackenbrock, C. R. (1980b). *J. Biol. Chem.* **255**, 3748–3756.

Scholte, H. R. (1969). *Biochim. Biophys. Acta* **176**, 137–144.

Schoolwerth, A. C., and LaNoue, K. F. (1980). *J. Biol. Chem.* **255**, 3403–3411.

Serrano, R., Gancedo, J. M., and Gancedo, C. (1973). *Eur. J. Biochem.* **34**, 479–482.

Shannon, W. A., Jr., Matlib, M. A., Robinson, J. B., Hochli, M., and Srere, Paul A. (1977). *Ann. EMSA Meet., 35th* 428–429.

Smith, H. E., and Page, E. (1976). *J. Ultrastruct. Res.* **55**, 31–41.

Smith, U., Smith, D. S., and Yunis, A. A. (1970). *J. Cell Sci.* **7**, 501–521.

Soboll, S., Scholz, R., Freisl, M., Elbers, R., and Heldt, H. W. (1976). *In* "Use of Isolated Liver Cells and Kidney Tubules in Metabolic Studies" (J. M. Tager, H. D. Soling, and J. R. Williamson, eds.), pp. 29–39. North-Holland, Publ., Amsterdam.

Sowers, A. E., and Hackenbrock, C. R. (1981). *Proc. Natl. Acad. Sci. U.S.A.* **78**, 6246–6250.

Srere, P. A. (1967). *Science* **158**, 936–937.

Srere, P. A. (1968). *In* "The Metabolic Roles of Citrate" (T. W. Goodwin, ed.), pp. 11–21. Academic Press, New York.

Srere, P. A. (1969). *Biochem. Med.* **3**, 61–72.

Srere, P. A. (1970). *Biochem. Med.* **4**, 43–46.

Srere, P. A. (1972). *In* Energy Metabolism and the Regulation of Metabolic Processes in Mitochondria" (R. W. Hanson and W. A. Mehlman, eds.), pp. 79–91. Wiley, New York.

Srere, P. A. (1976). *In* "Gluconeogenesis: Its Regulation in Mammalian Species" (R. W. Hanson and W. A. Mehlman, eds.), pp. 153–161. Wiley, New York.

Srere, P. A. (1980). *Trends Biochem Sci.* **5**, 120–121.

Srere, P. A. (1981). *Trends Biochem. Sci.* **6**, 4–6.

Srere, P. A. (1982). *Trends Biochem. Sci.* **7**, 375–378.

Srere, P. A., and Estabrook, R. W., eds. (1978). "Microenvironments and Metabolic Compartmentation." Academic Press, New York.

Srere, P. A., and Mosbach, K. (1974). *Annu. Rev. Microbiol.* **28**, 61–83.

Srere, P. A., Matsuoka, Y., and Mukherjee, A. (1973a). *J. Biol. Chem.* **248**, 8031–8035.

Srere, P. A., Mattiasson, B., and Mosbach, K. (1973b). *Proc. Natl. Acad. Sci. U.S.A.* **70**, 2534–2538.

Stanley, C. J., and Perham, R. N. (1980). *Biochem. J.* **191**, 147–154.

Sumegi, B., Gyocsi, L., and Alkonyi, I. (1980). *Biochim. Biophys. Acta* **616**, 158–166.

Swissa, M., Weinhouse, H., and Benziman, M. (1976). Biochem. J. **153**, 499–501.

Tandler, B., and Hoppel, C. L. (1972). "Mitochondria." Academic Press, New York.

Tarr, J. S., Jr., and Gamble, James, L., Jr. (1966). *Am. J. Physiol.* **211**, 1187–1191.

Tinberg, H. M., Lee, C., and Packer, L. (1975). *J. Supramol. Struct.* **3**, 275–283.

Tinberg, H. M., Nayudu, P. R. V., and Packer, L. (1976). *Arch. Biochem. Biophys.* **172**, 734–740.

Tzagoloff, A. (1982). "Mitochondria." Plenum, New York.

Von Glutz, G., and Walter, P. (1975). *Eur. J. Biochem.* **60**, 147–152.

Wakabayashi, T., Smoly, J. M., Hatase, O., and Green, D. E. (1971). *Bioenergetics* **2**, 167–182.
Wang, K., and Richards, F. M. (1975). *J. Biol. Chem.* **250**, 6622–6626.
Weibel, E. R., Staubli, W., Gnagi, H. R., and Hess, F. A. (1969). *J. Cell Biol.* **42**, 68–91.
Weisz, P. B. (1962). *Nature (London)* **195**, 772–774.
Weitzman, P. D. J. (1973). *FEBS Lett.* **32**, 247–250.
Weitzman, P. D. J., and Hewson, J. K. (1973). *FEBS Lett.* **36**, 227–231.
Welch, G. R. (1977). *Prog. Biophys. Mol. Biol.* **32**, 103–191.
Welch, G. R., and Keleti, T. (1981). *J. Theor. Biol.* **93**, 701–735.
Welch, G. R., Somogyi, B., Matko, J., and Papp, S. (1983). *J. Theor. Biol.* **100**, 211–238.
Wiegand, G., Kukla, D., Scholze, H., Jones, T. A., and Huber, R. (1979). *Eur. J. Biochem.* **93**, 41–50.
Wilf, J., and Minton, A. P. (1981). *Biochim. Biophys. Acta* **670**, 316–322.
Williams, R. J. P. (1983). *Trends Biochem. Sci.* **8**, 48.
Williamson, J. R., and Olson, M. S. (1968). *Biochem. Biophys. Res. Commun.* **32**, 794.
Winder, W. W., Baldwin, K. M., Terjung, R. L., and Holloszy, J. O. (1975). *Am. J. Physiol.* **228**, 1341–1345.
Wit-Peeters, E. M. (1969). *Biochim. Biophys. Acta* **178**, 137–144.
Wit-Peeters, M., Scholte, H. R., Van Den Akker, F., and De Nie, L. (1971). *Biochim. Biophys. Acta* **231**, 23–31.
Wojtczak, L., and Zaluska, H. (1969). *Biochim. Biophys. Acta* **193**, 64–72.
Wrigglesworth, J. M., Packer, L., and Branton, D. (1970). *Biochim. Biophys. Acta* **205**, 125–135.
Wuntch, T., Chen, R. F., and Vessell, E. S. (1970). *Science* **167**, 63–65.
Zalokar, M. (1960). *Exp. Cell Res.* **19**, 114–132.

2

Catalytic Facilitation and Membrane Bioenergetics

Douglas B. Kell

Department of Botany and Microbiology
University College of Wales
Aberystwyth, Wales

Hans V. Westerhoff

National Institutes of Health
Bethesda, Maryland

ORGANIZED
MULTIENZYME SYSTEMS

It seems to me that there is no single idea in biology which is hard to understand, in the way that ideas in physics can be hard. If biology is difficult, it is because of the bewildering number and variety of things that one must hold in one's head.

(Maynard-Smith, 1977)

I. INTRODUCTION

What is a cell like? The simplest attempt at a conceptual subdivision contrasts those (soluble) enzymes and molecules which, upon cell disruption, are released into the supernatant of a high-speed centrifugation step (105,000 × g; 1 h) with those (membrane bound) which are not. Now, while the evidence for the organization, aggregation, and indeed membrane association *in vivo* of many enzymes, such as those of glycolysis (Gorringe and Moses, 1980; Masters, 1981), is now extensive (e.g., Welch 1977; Welch and Keleti, 1981; Clegg, 1983a,b), this topic has been widely reviewed and is also excellently covered by other contributors to this volume.

Equally, just as a prime purpose of glycolysis is to synthesize ATP by substrate-level phosphorylation, a prime purpose of many biological membranes (*energy coupling membranes*) is also to catalyze ATP synthesis. These coupling membranes are exemplified by the inner mitochondrial membrane, the chloroplast thylakoid membrane, and the plasma (cytoplasmic) membrane of respiratory and photosynthetic bacteria. Our main purpose in this chapter will thus be to describe the nature, role, and organization of the free energy-transducing devices that these energy-coupling membranes contain, placing special emphasis upon the process of electron-transport phosphorylation. The literature is covered comprehensively through 1983.

Since many membranes as isolated catalyze electron-transport phosphorylation at rates, and with efficiencies, similar to those implicated *in vivo*, we regard it as permissible to consider such membranes in isolation from their normal "cytosociological" (Welch and Keleti, 1981) habitat. To this end, we next consider the gross structural features of typical membranes catalyzing electron-transport phosphorylation. We shall concentrate in particular upon the inner mitochondrial membrane, although when necessary or appropriate we shall address ourselves to other systems. We shall see that in electron-transport phosphorylation as well, enzymes that were previously thought to operate independently turn out to function in rather tightly linked units.

II. STRUCTURAL DYNAMICS OF ENERGY-COUPLING MEMBRANES

A. The Occurrence and Nature of Oligomolecular Complexes in Energy-Coupling Membranes

1. Introduction

As is now well known (e.g., Stryer, 1981; Lehninger, 1982), the best *general* biomembrane model (Singer and Nicolson, 1972; Finean *et al.*, 1978; Houslay and Stanley, 1982) visualizes biological membranes as consisting of a fluid phospholipid bilayer in, on, and through which are dispersed protein molecules, either as monomers or, more commonly in energy-coupling membranes, as oligomeric, polytopic (Blobel, 1980; Brock and Tanner, 1982) complexes. Thermodynamic considerations dictate (e.g., Tanford, 1978, 1980; Jähnig, 1983) that the major organizing force determining the degree of penetration of such complexes into or through the bilayer is the favorable free energy of transfer of the hydrophobic areas of proteins from an aqueous to a lipidic environment. There are also reasons to suppose that "the hydrophobic domain of a transmembrane protein is vertically delimited at both its upper and lower ends by two collars of charged amino acid residues, which interact with the polar head groups of lipids" (Montecucco *et al.*, 1982). Such interactions can be of great importance to the catalytic activity of such proteins (Johansson *et al.*, 1981a,b).

The likely arrangement of the polypeptide chain in, say, bacteriorhodospin (e.g., Engelman *et al.*, 1980) is consistent with the probable generality of this view, and, since this is not, perhaps, to be regarded as unexpected, we do not pursue it here. We may add, at this stage, that the flip-flop motion of even phospholipids (Kornberg and McConnell, 1971), let alone proteins, is negligible on a time scale relevant to electron-transport phosphorylation. Finally, there is abundant calorimetric and other evidence that the gel–liquid phase transition of membrane phospholipids usually occurs well below physiological temperatures, as documented, for instance, for the membranes of thylakoids (Murato and Fork, 1975), mammalian mitochondria (Hackenbrock *et al.*, 1976), and respiratory (McElhaney, 1974; Mechler and Stein, 1976) and photosynthetic (Fraley *et al.*, 1978; Kenyon, 1978; Kaiser and Oelze, 1980) bacteria (for possible exceptions see Raison, 1973). The gel–liquid phase transition does, however, occur over a range of temperatures, which tends to indicate heterogeneity in the fluidity of biological membranes (Vaz *et al.*, 1982).

We may therefore proceed from our general fluid mosaic picture to ask some slightly more detailed questions concerning the structural dynamics of (generally unenergized) energy-coupling protein complexes in biological membranes. The insertion and assembly of these complexes in the membrane, though a fascinating area of study, is beyond the scope of this chapter (see e.g., Brock and Tanner, 1982). Because of its special properties, we shall in general treat the quasi-crystalline "membrane" bacteriorhodospin as a separate complex.

2. How Much Phospholipid Bilayer Is There in Energy-Coupling Membranes?

Energy-coupling membranes are among the richest in protein. Thus, the inner mitochondrial membrane contains approximately 75–80% protein and 20–25% phospholipid (e.g., Tzagoloff, 1982; Kröger and Klingenberg, 1970), and similar ratios are found in other energy-coupling membranes (e.g., John and Whatley, 1977). It is sometimes remarked (e.g., Ling, 1981) that this low percentage of phospholipid cannot be sufficient to form a continuous bilayer. Now, many proteinaceous complexes pass right through the plane of the bilayer, for example, bacteriorhodopsin (Henderson and Unwin, 1975; Ovchinnikov *et al.*, 1979; Engelman *et al.*, 1980) and cytochrome oxidase (e.g., Henderson *et al.*, 1977; Fuller *et al.*, 1979; Azzi, 1980; Brunori and Wilson, 1982; Capaldi, 1982a,b), and while the former is fairly flush with the plane of the membrane, the latter extends well beyond it, as do many other complexes that we shall discuss, such as the H^+-ATP synthase (Soper *et al.*, 1979) and ubiquinol–cytochrome c reductase (EC 1.10.2.2) (e.g., von Jagow and Engel, 1980). Thus, a figure of, say, 20% for the percentage of phospholipid in an energy-coupling membrane provides a minimum for the percentage of surface area that may be in a bilayer configuration, since energy-coupling membranes are not thought to contain large areas of protein–phospholipid monolayer. In any event, NMR work (e.g., Arvidson *et al.*, 1975; Cullis *et al.*, 1980; de Kruijff *et al.*, 1982) shows very clearly that the great majority of phospholipids of the inner mitochondrial membrane are in a bilayer configuration, and we do not find much experimental evidence (but cf. Sjöstrand, 1978) to dissuade us from this interpretation. Whereas it is of interest that rather extensive delipidation of mitochondria causes no gross structural changes in the inner membrane (Fleischer *et al.*, 1967), as judged by electron microscopy, our own present interest must be centered primarily upon systems that are capable of carrying out free energy transduction, and we shall not consider these observations further, save to mention that they simply indicate that lipid–protein interactions may be substituted for by hydrophobic protein–protein interactions under appropriate conditions.

Studies using differential scanning calorimetry and freeze-fracture electron microscopy by Hackenbrock, Höchli, and colleagues (Hackenbrock *et al.*, 1976; Höchli and Hackenbrock 1976, 1977; Hackenbrock, 1976, 1981) show rather clearly that the total protein mass of the native inner mitochondrial membrane occupies less than one-half of the total lateral area of the membrane, and freeze-fracture work by others indicates a similar picture for the thylakoid membrane (e.g., Anderson, 1975; Arntzen, 1975, 1978). Most importantly, however, quite small changes in physical parameters have been shown to exert a marked effect upon the state of distribution of the intrinsic protein complexes of the inner mitochondrial membrane (Hackenbrock, 1976), and this leads us nicely to two important ideas that form the basis of the next two sections: proteins exist, and can move, as oligometric complexes in the plane of energy-coupling membranes. Topics related to this question, such as phospholipid organization (e.g., Cullis and de Kruiff, 1979; Israelachvili *et al.*, 1980; Blaurock, 1982; Davis, 1983) and asymmetry (e.g., Etemadi, 1980), are largely beyond our present scope.

3. Complexes of the Inner Mitochondrial Membrane

The inherent complexity of living systems in general, and of the inner mitochondrial and other energy-coupling membranes in particular, has led many workers to seek to isolate enzymes and/or complexes that carry out defined, partial reactions of processes such as oxidative phosphorylation. As is well known, pioneering work in the laboratories of Green, Hatefi, and Racker led to the isolation and characterization of proteinaceous complexes of the mitochondrial electron-transport chain (complexes I to IV) and the H^+-ATPase (complex V) components that could catalyze defined electron transfer or ATPase–ATP-P_i exchangease reactions, respectively. Not surprisingly, each complex has provided a rich area of study in its own right, and this approach has been extended by a great many workers to broadly comparable complexes in thylakoid and bacterial energy coupling membranes. The reviews by De Pierre and Ernster (1977), by Hatefi (1976), and by Capaldi (1982b) (see Table I) and the books by Tedeschi (1975), by Tzagoloff (1982), and by Nicholls (1982) provide useful introductions to the voluminous mitochondrial literature.

As pointed out by Saraste (1983), the criteria usually used in assessing the polypeptide composition of a particular complex often leave much to be desired. Fortunately, we shall largely be able, for our present purposes, to treat the complexes, especially complexes I to IV, more or less as black boxes, and will tend to assume that they contain those polypeptides ascribed to them as a consensus by the authorities in this field. The "complexes" of beef heart mitochondria are given in Table I.

The crucial, if well-known, message for our present purposes is this: oligometric protein complexes, which can be visualized by electron microscopy, exist as defined entities in energy-coupling membranes and catalyze reactions coupled to macroscopically observable chemical changes such as electron transport or ATP hydrolysis. Since, to preempt some of our later discussion, we shall wish to consider in detail the degree of localization of free energy transfer between different electron-transport and ATPase complexes, it

TABLE I

SOME GENERALLY RECOGNIZED COMPONENTS OF THE BEEF HEART MITOCHONDRIAL INNER MEMBRANE[a]

Components	Concentration range (nmol/mg membrane protein)	Approximate molecular weight of monomer (kD)
Complex I (NADH–ubiquinone oxidoreductase)	0.06–0.13	700
Complex II (succinate–ubiquinone oxidoreductase)	0.19	200
Complex III (bc₁ complex) (ubiquinol–cytochrome c oxidoreductase)	0.25–0.53	300
Complex IV (cytochrome c oxidase)	0.6–1.0	160
Complex V (ATP synthase)	0.52–0.54	500
Cytochrome c	0.8–1.0	12
ADP–ATP translocase	3.4–4.6	30
Transhydrogenase	0.05	120
Ubiquinone	6–8	—
Phospholipid	440–590	—

[a] Data after Capaldi (1982b).

is of interest to know the rate at which such complexes may collide with each other, a topic to which we now turn.

4. Lateral Diffusion of Protein Complexes in Energy-Coupling Membranes

That oligomeric protein complexes can, in many cases, diffuse laterally at rather rapid rates in the plane of energy-coupling and other biological membranes, although an important implicit corollary of the concept of the fluid mosaic membrane has now been demonstrated directly by a variety of experimental techniques (e.g., Edidin, 1974; Poo and Cone, 1973; Jaffe, 1977; Poo and Robinson, 1977; Webb, 1977; Evans and Hochmuth, 1978; Finean et al., 1978; Cherry, 1979; Barber, 1980; Sowers and Hackenbrock, 1981; Anderson and Anderson, 1982; Houslay and Stanley, 1982; Vaz et al., 1982; Webb et al., 1982; Zimmerman, 1982; Zimmerman and Vienken, 1982; Axelrod, 1983; Barber, 1983; Kell, 1983; Robertson, 1983). The different apparent mobilities of complexes I–IV in the inner mitochondrial membrane have been tabulated by Hackenbrock (1981) (Table II). Thus, even the least mobile complexes can move nearly 0.05 μm during the passage of a single electron through them. The distance diffused by ubiquinone molecules per turnover is some 40-fold greater, and this phenomenon can evidently account for the "Q-pool" behavior established by Kröger and Klingenberg (1973) in mitochondria, and the well-established and broadly comparable function of plastoquinone in thylakoids as a mobile electron-transfer agent between the spatially separate (Anderson, 1981; Andersson and Haehnel, 1982), proteinaceous light-harvesting complexes containing photosystems I and II. It is worth stressing that one may make these conclusions based on experimental data such as the rates of electron transfer in mitochondria (Hackenbrock, 1981) and the amount of spillover between the two photosystems in thylakoids (Barber 1980, 1982b; Anderson and Andersson, 1982) that have negligible interpretational complexities relative to those of certain more physical, as opposed to functional, approaches. However, we should mention that the exact status of the "Q-pool" concept (Heron et al., 1978; Gutman, 1980; Yu and Yu, 1981; Haehnel, 1982; Rich, 1982; Trumpower, 1982; Hauska et al., 1983) represents an area of active debate and great uncertainty.

For our discussions below it is important that Hackenbrock (1981) claims that the different protein complexes (i.e., complex I, II, III, and IV (Tables I and II) diffuse independently of one another. This might be extrapolated to a suggestion that also complex V (i.e., the H^+-ATPase) diffuses independently. We wish to point at a number of uncertainties in the basis of such a claim. First, we are not aware of any experimental evidence indicating diffusion of H^+-ATPases independent from electron-transfer chain components. Secondly,

TABLE II

EXPECTED LATERAL DISTANCE DIFFUSED PER UNIT TIME BY REDOX COMPONENTS OF THE INNER MITOCHONDRIAL MEMBRANE[a]

Redox component	Approximate stoichiometry	Lateral diffusion coefficient in cm^2/s	Electrons transferred in one turnover	One turnover in state 3 in ms	Distance diffused during one turnover in nm^b	Distance diffused during 20 ms in nm
Complex I	1	8.3×10^{-10}	2	5.5	43	81
Complex II	2	8.3×10^{-10}	2	11.0	60	81
Complex III	3	8.3×10^{-10}	1	8.6	53	81
Complex IV	7	8.3×10^{-10}	1	20.0	81	81
Cytochrome c	9	1.0×10^{-8}	1	28.0	335	280
Ubiquinone	63	1.0×10^{-8}	2	780.0	1760	280

[a] Data after Hackenbrock (1981).

[b] Distances are based on the lateral diffusion coefficients given and a state 3 respiratory rate for succinate oxidation of 50 electrons s^{-1} heme^{-1} aa$_3$.

Hackenbrock bases his claim on the effect of fusion of liposomes to mitoplasts (i.e., mitochondria depleted of their outer membrane), which is reduction of the rates of redox reactions that involve two or more of the complexes. It should be noted that the explanation alternative to Hackenbrock's (who proposes that the rates are reduced because the proteins have to diffuse over longer distances), that is, that low-molecular-weight (lipid-soluble) ecofactors such as ubiquinone are diluted out, has not been fully ruled out. Moreover, mitoplasts fused to liposomes are immensely different from intact mitochondria with their much more folded inner membrane (cf. Sjöstrand, 1978); the former have been shown not (Hackenbrock, 1981; Westerhoff, unpublished observations) to catalyze oxidative phosphorylation. Thirdly, the diffusion coefficients measured by Sowers and Hackenbrock (1981) may be representative of a case where the existing protein–protein interactions have been disrupted (see below).

For our present purposes, however, we may state three facts: (1) electron-transport events may take place in isolated, reconstituted systems *in vitro*, containing only the recognized complexes of the appropriate region(s) of the electron transport chain(s) of interest; (2) such complexes do not need to be membrane incorporated to express such activity (e.g., Lam and Malkin, 1982); and (3) *electron-transport* events can be reconstituted using protein complexes from different types of energy-coupling membrane (e.g., Packham *et al.*, 1980). Such findings give weight to the widely held belief in the close functional relationships between electron-transfer components in different types of membrane (see e.g., Hauska *et al.*, 1983), as well as to the permissibility of treating these complexes as black boxes.

5. Rotational, Bending, and Stretching Behavior of the Components of Energy-Transducing Membranes

It is now well known that the rotation of individual complexes is generally rather fast (e.g., Cherry, 1979; Muller *et al.*, 1982) and occurs on a time scale similar to that of the exchange of so-called boundary lipid (e.g., Vanderkooi, 1978; Jardetzky and Roberts, 1981; Gennis and Jonas, 1977; Chapman *et al.*, 1982; Marsh, 1983) with that of the bulk lipid of the bilayer. These rates are far more rapid than the turnover of the apparatus of electron-transport phosphorylation, but we will not pursue this issue here.

In modelling a biomembrane, it is common to draw it as a flat sheet, as though there were no mobile deformations in a plane perpendicular to the plane of the bilayer. This is certainly an unjustified and erroneous oversimplification (e.g., Evans and Hochmuth, 1978; Haines, 1979; Miller, 1981; Crilly and Earnshaw, 1983; Brown *et al.*, 1983; Robertson, 1983); however, the extent to which such motions are of significance in energy coupling remains, regrettably, as yet unknown, although Haines (1979) proposes a causal role of lipid-mediated potential compaction waves in oxidative phosphorylation (see also Haines, 1982).

6. Fluctuational Behavior of Energy-Transducing Proteins

It is now widely recognized that proteins exhibit thermally activated motions of varying degrees of cooperativity, on a time scale from picoseconds upwards (e.g., Gurd and Rothgeb, 1979; Clementi and Sarma, 1983; McCammon and Karplus, 1983; Welch, 1985). There is also evidence of very slow conformational transitions in energy transducing ATP synthases, on a time scale of minutes (e.g., Slooten and Branders, 1979). However, since our main interest lies in the organization of these membranes during the time necessary to synthesize a single ATP molecule, we shall confine our thoughts to the time scale below 100 ms. Some of the typical motions that have been characterized for globular proteins are noted in Fig. 1. Our main purpose here is (1) to indicate the extensive nature of these fluctuations, (2) to remind readers of the probability that these fluctuations play an *essential* role in enzyme catalytic processes (Welch *et al.*, 1982; Careri *et al.*, 1979; Welch, 1985) and (3) to state that a complete description of electron-transport phosphorylation should preferably take such fluctuational behavior into account. The philosophical problems inherent in the description of free energy-transducing devices of molecular size, which are subject to such thermal fluctuations yet work under macroscopically isothermal conditions, are explored elsewhere (Welch and Kell, 1985; Somogyi *et al.*, 1984).

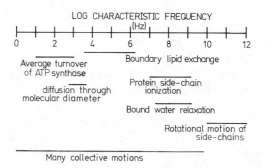

Fig. 1. Some time events of biomembrane proteins of relevance to electron transport phosphorylation. For many further details, and the evidence for such motions, see Gurd and Rothgeb (1979), Careri *et al.* (1979), and Welch *et al.* (1982). It is evident that the overall process of electron transport phosphorylation occurs on a very slow time scale relative to that of other molecular motions.

7. Disposition of Protein Complexes under Energized Conditions

In the absence of electron transport- or ATP-derived free energy, it is to be assumed that the disposition of protein complexes in the fluid mosaic membrane follows a more or less random distribution, that is, that their topological relationship to each other constitutes an equilibrium, or quasi-equilibrium process. The question arises, therefore, as to what infuence, if any, the input of free energy (whether from electron transport or from ATP) exerts upon the disposition of the energy-transducing protein complexes themselves.

Remarkably enough, there is relatively little direct evidence to indicate, whether such free energy inputs can promote the (transient) association of such protein complexes that may be of relevance to the free energy-transducing processes themselves. However, several lines of circumstantial evidence indicate that this indeed may be the case. Energization was shown to affect the distribution of complexes in the inner mitochondrial membrane (Hackenbrock, 1972), and Goodchild et al. (1983) have shown that electron transport does indeed result in a somewhat closer association of intramembranous particles, including the light-harvesting and the cytochrome b/f complexes, in chloroplast thylakoids.

A number of recent detailed and elegant studies have shown that appropriately oriented electrical fields can drive the migration of intramembrane protein complexes (see e.g., Jaffe, 1977; Poo and Robinson, 1977; Sowers and Hackenbrock, 1981; Poo, 1981; Zimmermann, 1982; see also Kell, 1983). This phenomenon has been termed *lateral electrophoresis*, and it has been shown that the clustering of certain membrane components can have significant effects upon their catalytic properties (Young and Poo, 1983). In particular, since, as we shall see shortly, electron-transport and ATP hydrolase reactions are accompanied by the generation of (initially) localized electrical fields, it does not seem unreasonable that some type of topological arrangement of electron transport and ATP synthase complexes, *inter alia*, is of significance to the energy-coupling process.

We may consider the recent elegant work of Zimmermann and colleagues (see Zimmermann et al., 1981; Zimmermann, 1982; Zimmermann and Vienken, 1982, for reviews) on electrically mediated cell–cell fusion. In this process, cells are apposed by dielectrophoresis in a medium of low electrical conductivity, and then fused by the application of a short (microsecond) electrical pulse of high field strength. Molecularly, it is thought that the area of cell–cell contact (strictly, protoplast–protoplast contact) involves solely phospholipid bilayer areas of the membrane, and that, *ergo*, part of the role of dielectrophoretic induction involves the lateral electrophoresis of membrane proteins (and/or charged phospholipids (Zimmermann and Vienken 1982; Kell, 1983)). Most significantly, it was shown using *Avena sativa* protoplasts that while the adenylate energy charge of the cells was irrelevant to fusion, the time taken for the fused double cell to round up was essentially inversely proportional to the ATP/ADP ratio (Verhoek-Köhler et al., 1983). Although, as the authors mention, this may be due to the requirement for phosphorylation of one or more membrane proteins, it seems possible that ATP energization of the plasma membrane may be required for the normal effective disposition of proteins in these membranes (as well as the rearrangement of the cytoskeleton).

We may also consider the isolation by simple mechanical disruption of chromatophores from the plasma membrane of photosynthetic bacteria. As is well known, these chromatophores contain all the apparatus of electron-transport phosphorylation, yet possess a composition quite distinct from that of the other parts of the cytoplasmic membrane (Garcia et al., 1981; Dierstein et al., 1981; Kaufmann et al., 1982; Oelze and Drews, 1981). Given the apparent absence of any anchorage between membrane proteins and cytoplasmic structures in these prokaryotes (Drews, 1982), one is virtually bound to conclude either that the complexes bind to each other or to other membrane proteins as an equilibrium process and/or that some factors, requiring the input of free energy, act to array the complexes of electron-transport phosphorylation in chromatophores in a reasonably close spatial relationship to each other in situ in the intact cell. This conclusion is very strongly reinforced by the elegant observations of a highly nonrandom segregation of membrane components between daughter cells in a variety of dividing microorganisms (Kepes and Autissier, 1972; Poole, 1981; Edwards, 1981; Lloyd et al., 1982).

To put this point on a quantitative footing, let us take a fairly typical protein complex diffusion coefficient (D) of 10^{-9} cm^2/s (Hackenbrock, 1981). The relaxation time (τ) for the randomization of such a complex in a spherical shell membrane of radius R is given by $\tau = R^2/2D$ (Huang, 1973; Benz and Zimmermann, 1981; Sowers and Hackenbrock, 1981; Zimmermann, 1982). As remarked elsewhere (Kell, 1983), this equation assumes that the complexes take up a negligible volume fraction of the bilayer; for a given τ this will overestimate D (or underestimate R), but we will ignore this. Thus the relaxation time for randomization of the disposition of a protein complex in a spherical bacterial cell of radius 0.5 μm ($= 5 \times 10^{-5}$ cm) is $25 \times 10^{-10}/2 \times 10^{-9}$ s, that is, 1.25 s. Since the fastest doubling time for a microorganism, ~20 min, is some 1000-fold greater, there would seem to be an extremely serious discrepancy between (1) the more biophysical studies indicating rapid and random diffusion of electron-transport complexes and (2) the more biochemical studies demonstrating highly nonrandom segregation of membrane components. Obviously, such calculations depend greatly upon the value chosen for D. However, the lowest value we have found for an energy-coupling membrane ($D = 10^{-11}$ cm^2/s) is that given by Barber (1982b) in thylakoids, based on chlorophyll fluorescence induction. We can perceive one possible resolution of this problem which seems consistent with the available data. Using the same equation as Sowers and Hackenbrock (1981), Kell (1983) obtained, from dielectric measurements in chromatophores, a diffusion coefficient of 1.35×10^{-7} cm^2/s, on the assumption that individual complexes could diffuse freely within the entire chromatophore membrane; this value is evidently far too large and forces one to the conclusion that the radius one should construe is not the actual geometric radius of the chromatophore but a more restricted domain. [Actually, as stressed by Peter Rich (personal communication), the dielectric method cannot alone distinguish lateral and rotational movements of charged or dipolar groups; however, the relative effective dipole moments involved will mean that lateral, rather than rotational, motions should, if present, dominate the dielectric response.] In these dielectric measurements the applied field was approximately 50 mV/cm (Kell, 1983), while in the experiments of Sowers and Hackenbrock (1981) the applied field was approximately 650 V/cm, a value large enough actually to drive ATP synthesis in particles of this size (e.g., Vinkler and Korenstein, 1982; Hamamoto *et al.*, 1982; Schlodder *et al.*, 1982). Thus the energy in the field induced across the membranes in the experiments of Sowers and Hackenbrock (1981) could easily have been enough to rupture noncovalent, intercomplex interactions, while that induced in the dielectric measurements (which will be very substantially below kT) was not. Obviously the free energy available in energized membranes could promote *different* intercomplex associations from those existing under nonenergized conditions. Dielectric measurements at widely varying field strengths (see Delalic *et al.*, 1983) might shed light on this possibility.

8. Summary

The foregoing sections may be summarized as follows:

1. Electron transport and ATP synthase components exist as discrete complexes in energy-coupling membranes.

2. These complexes, in common with other globular proteins, exhibit a variety of intramolecular fluctuations that may be cooperative and may be of crucial significance to their catalytic activity.

3. The complexes can apparantly diffuse at rates sufficient to account for the role of bimolecular collisions in controlling the rate of electron transport.

4. Under energized conditions there is a significant amount of evidence that the disposition of these complexes in energy-coupling membranes is not random; this may be important in regulating the transfer of free energy between them.

5. There are some serious discrepancies between many *biophysical* measurements of membrane protein diffusion coefficients and those obtained from more biochemical approaches (Kell, 1984).

III. PROTONMOTIVE FUNCTIONS OF ELECTRON TRANSPORT AND ATP SYNTHASE COMPLEXES

A. Introduction

As is now well known, P. Mitchell (see e.g., Mitchell, 1966, 1968, 1979, 1981; Nicholls, 1982), in his chemiosmotic coupling hypothesis, proposed that, *inter alia*, the function of electron transport and ATP synthase complexes was to catalyze the vectorial translocation of protons across the coupling membrane in which the complexes are embeded. We take as axiomatic the fact that such activity is now proven. However, there are two areas of outstanding controversy in relation to the general problem of so-called protonmotive systems.

1. What *type* of general mechanism is exploited by these complexes in catalyzing protonmotive activity?;
2. What pathway is taken by energized protons translocated across the coupling membrane by the electron-transport-linked proton pumps *en route* to sinks such as the H^+-ATP synthase?

To a certain extent, these two questions may be dealt with separately. However, as mentioned by others (e.g., Wikström and Krab, 1980; Wikström *et al.*, 1981; Ferguson and Sorgato, 1982), it is not easy to see how a direct type of protonmotive mechanism might exhibit other than an orthodox, "delocalized" chemiosmotic coupling activity. What does this mean? The original version of the chemiosmotic coupling hypothesis posited direct mechanisms by which the reactions of electron transport might be coupled to proton motive activity. The simplest version, the redox loop concept, rests on the appropriate spatial relationship between electron-transport carriers and the alternation of hydrogen and electron carriers (Fig. 2a). Such mechanisms, including the so-called Q-cycle (Mitchell, 1976), have been referred to as *direct* mechanisms (Mitchell, 1977a), a semantic convention which we shall adopt. By contrast, the possibility exists (see e.g., Papa, 1976; Wikström and Krab, 1980; Fillingame, 1980; von Jagow and Engel, 1980; Wikström *et al.*, 1981; Kell *et al.*, 1981a; Vignais *et al.*, 1981; Nicholls, 1982) that the protonmotive and electron-transport–ATP synthase–hydrolase activities are coupled by a purely indirect, conformational type of mechanism (Fig. 2b). Although there are other differences between the various types of model, such as the protonmotive stoichiometries with which they are consistent, in

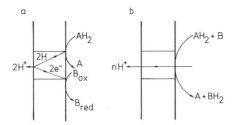

Fig. 2. Direct and indirect mechanisms of redox-linked protonmotive activity: (a) The classical direct method: the simplest possible redox loop depends upon the alternation of hydrogen (H) and electron (e^-) carriers at opposite faces of the coupling membrane. This type of model has, *for a given pathway of electron transfer*, an invariant $\rightarrow H^+/2e^-$ stoichiometry of, in this case, 2. (b) An indirect, conformationally coupled proton pump. In this case the redox reactions are functionally linked, through conformational changes of the protein complex, to transmembrane proton translocation. The $\rightarrow H^+/2e^-$ stoichiometry may in this case be variable due to incomplete coupling (slip) at the molecular level. This type of model is compatible both with delocalized chemiosmosis (as drawn) or with more localized energy-coupling theories. This type of device requires some kind of molecularly ratcheted proton channel within the protein complex.

our view a crucial difference between them is as follows: in the case of redox-linked protonmotive systems the direct mechanism, for a given pathway of electron transport between redox centers, requires an invariant ratio of protons translocated per electron passing through the region of interest, i.e., a constant $\rightarrow H^+/e^-$ ratio. By contrast, redox-linked proton pumps *sensu stricto* (i.e., those operating via an indirect mechanism) are more likely to exhibit what has been termed *slip* (Rottenberg, 1973; Baccarini-Melandri et al., 1977; Hill, 1977; Pietrobon et al., 1981, 1982; Kell and Morris, 1981; Stucki, 1982; Walz, 1983). That is, the electron- and proton-transport reactions themselves may be incompletely coupled at the molecular level. This concept is illustrated in a simplified fashion, using a so-called Hill diagram (Hill, 1977), in Fig. 3.

By and large, we shall be able to avoid the controversy concerning the direct/indirect type of mechanism of protonmotive activity. Nevertheless, a few comments are in order.

1. Endogenous slip in redox-linked (Pietrobon et al., 1981) and ATP synthase-linked (Pietrobon et al., 1983) proton pumps has been noted in rat liver mitchondria (see, however, Westerhoff et al., 1983c, 1984b).

2. Slip in redox-linked proton pumps has been demonstrated directly in a number of systems (e.g., Casey et al., 1980; Anderson et al., 1981; Phelps and Hatefi, 1981; Tu et al., 1981; Wikström and Penttilä, 1982; Price and Brand, 1983; Walz, 1983, and see later.)

3. There has been a historical tendency in some quarters to interpret experimental data solely in terms of direct mechanisms; readers should not

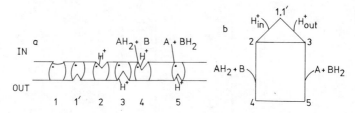

Fig. 3. Coupled and uncoupled (slip) protonmotive cycles in a redox-linked proton pump. (a) Diagrammatic representation of a redox-linked proton pump in an energy-coupling membrane. The protein contains one (shown) or more key proton-binding amino acid side chains on each face of the membrane and can adopt various types of conformational state (1 to 5) depending upon the redox level and degree of energization. The bonds between H^+ and the protein indicate localized coupling; for diagramming delocalized chemiosmosis the bonds are relaxed and the protons are free to come into electrochemical equilibrium with those in the bulk aqueous phases. (b) Hill (1977) diagram for this type of redox-linked proton pump. For a fully coupled cycle (an $\rightarrow H^+/2e^-$ cycle of only 1 is shown in the diagram), the cycle of the Hill diagram is 1–2–4–5–3–1'. There are two other, uncoupled slip cycles. 1–2–3–1' is a proton leak (proton slip: charges separated during the protonmotive activity relax without concomitant reversal of electron transfer). *Redox slip*, cycle 2–4–5–3, occurs when the redox reactions occur without concomitant protonmotive activity. Other reactions, such as the interaction of such a device with membrane-permeant ions, are not considered here.

think that this reflects the current status of this controversy, which possibly favors the universality of the indirect pump type of mechanism (Kell *et al.*, 1981a; Nicholls, 1982).

4. Slip is often referred to loosely as a variable stoichiometry in these proton pumps; we shall use the terms interchangeably.

Given our intention to treat these protonmotive complexes more or less as black boxes (i.e., we take a coarse-grained approach), we may turn to the problem of the localization of the free-energy transfer and proton current pathway during electron-transport phosphorylation.

Independently of Mitchell, R. J. P. Williams (see e.g., Williams, 1961, 1978a,b, 1982) proposed a more general view on protonic coupling in electron-transport phosphorylation. Our purpose here, however, is not to distinguish the views of Mitchell and Williams (see e.g., Mitchell, 1977b; Williams, 1978a) in explicit detail nor historical context. It is because Mitchell's ideas generated a much greater number of testable predictions that most workers in the field of membrane bioenergetics have come to regard them as the more useful framework for experimental activity. For our part, we will take it that chemiosmosis (or delocalized chemiosmosis) constitutes a special case of Williams' more general ideas (Williams, 1978a).

What then is the crucial feature of chemiosmotic coupling? It is usually taken as the central dogma of the chemiosmotic coupling model (Nicholls, 1982) that under stationary-state conditions the proton electrochemical potential difference $\Delta\tilde{\mu}_{H^+}$ or protonmotive force ($\Delta p = \Delta\tilde{\mu}_{H^+}/F$) possesses the properties appropriate to its being the high-energy intermediate of electron-transport phosphorylation (and other processes). This concept is illustrated in Fig. 4a. Quantitative calculations concerning proton diffusion rates indicate that in the time range of redox or ATP synthetic reactions, no significant protonic potential differences can be maintained between points in space (in the same aqueous compartment) at a distance corresponding to the size of a

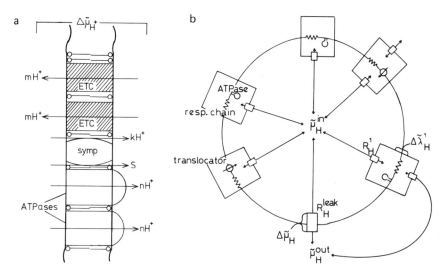

Fig. 4. (a) The principle of delocalized chemiosmotic coupling. The diagram shows a phospholipid bilayer energy-coupling membrane incorporating two redox-linked proton pumps (ETC), an electrogenic substrate (S) carrier (symp), and two ATP synthases (ATPases). The proton motive activities of the ETC set up a proton electrochemical potential difference ($\Delta\tilde{\mu}_{H^+}$) across the coupling membrane, and this proton electrochemical potential difference acts to drive H^+ back through the ATP synthases and to carry out active transport by various means, etc. Since these membranes are isolated as topologically closed vesicular structures, the proton electrochemical potential difference is homogeneous, in the sense that free energy released by a particular electron-transport chain passes via $\Delta\tilde{\mu}_{H^+}$ and may therefore be utilized by any of the ATP synthases embedded in the same coupling membrane. Membrane leakiness or the presence of ionophoric compounds leads to a decrease in $\Delta\tilde{\mu}_{H^+}$, and hence to a decrease in the rate of phosphorylation (or active substrate transport). The membrane itself is thus viewed, in the chemiosmotic model, simply as an inert, insulating phase that prevents uncoupled proton backflow and does not itself become energized during electron-transport phosphorylation. (b) The principle of mosaic protonic coupling. Free-energy coupling is visualized as occurring in n independently operating coupling units. Each of these units is organized in a fashion broadly comparable to that in Fig. 4a, except that the number of proton pumps per unit is confined to two. Because the combination of proton pumps may be different for each coupling unit, and because there are also leak units (the lower unit in Fig. 4b), energy coupling is essentially proposed to be mosaic in nature. The nature of the functional connections between the different proton pumps, and the relative proton current carried via the separate pathways, remain as yet undefined.

bacterium or an intracytoplasmic organelle (Mitchell, 1981; Melandri *et al.*, 1983; contrast Hong and Junge, 1983; but see Westerhoff, 1983), and experimental measurements (Gutman *et al.*, 1981, 1982, 1983) confirm this. Further, as pointed out previously (Kell and Hitchens, 1983; Hitchens and Kell, 1982a, 1983b), the fact that the $P/2e^-$ ratio is for practical purposes independent of electron transport over a wide range (e.g., van Dam and Tsou, 1970; Ernster and Nordenbrand, 1974; Ferguson *et al.*, 1976; Gräber and Witt, 1976; Graan *et al.*, 1981; Jackson *et al.*, 1981) indicates that any distinction between delocalized chemiosmosis and alternative, more localized energy-coupling schemes may be regarded as clear-cut. Thus the proton electrochemical potential difference constitutes a macroscopic, delocalized, ensemble property of the two aqueous phases that coupling membranes serve to separate (Walz, 1983).

In common with Ort and Melandri (1982), it is this minimal model (Fig. 4a) which we use as a "straw man" to assess the agreement between experimental results and the predictions of chemiosmosis in the following sections. As we shall see, the agreement is not good when the experimental tests which are applied are stringent, and it will be necessary to enquire as to what minimal changes we may make to the chemiosmotic model in order to make it accord more fully with experimental findings.

We begin by considering the protonmotive properties of systems catalyzing electron-transport phosphorylation as observable in (phases in equilibrium with) bulk aqueous phases.

B. Measurement of Proton Translocation by Systems Catalyzing Electron-Transport Phosphorylation

1. Respiratory Systems

Following pioneering experiments with mitochondria (see Mitchell & Moyle, 1967), the oxygen pulse method was applied to suspensions of the respiratory microorganism *Micrococcus* (now *Paracoccus*) *denitrificans* by Scholes and Mitchell (1970). In this method (see e.g., Mitchell *et al.*, 1979; Reynafarje *et al.*, 1979; Wilkstöm and Krab, 1980; Kell and Hitchens, 1982; Nicholls, 1982), a pulse of O_2, as air-saturated saline, is added to a well-stirred, weakly buffered, anaerobic suspension of the membrane vesicles of interest, and the resultant pH changes in the external aqueous phase monitored with a sensitive glass electrode. The ratio of the measured number of H^+ translocated across the membrane to the number of O atoms reduced, the $\rightarrow H^+/O$ ratio, is greatly increased in the presence of compounds such as SCN^- or K^+/valinomycin that are believed to cross biological membranes rapidly in their charged forms, as are the rates of H^+ transfer to and from the bulk aqueous phase in equilibrium with the measuring electrode.

According to the conventional chemiosmotic explanation of this behavior (see e.g., Mitchell, 1967, 1968; Scholes and Mitchell, 1970; Kell, 1979; Archbold *et al.*, 1979; Conover and Azzone, 1981; Kell and Morris, 1981; Heinz *et al.*, 1981; Kell and Hitchens, 1982, 1983; Hitchens and Kell, 1984), the relatively low static electrical capacitance of the membrane means that the transmembrane transfer of a rather small number of electrically uncompensated H^+ ions leads to the formation of a large bulk–bulk phase membrane potential, which either inhibits further pumping or drives H^+ ions back across the membrane before they can be detected. SCN^- or K^+/valinomycin acts to dissipate this membrane potential and thus allows all the protons translocated into (or from) the bulk phase external to the microorganisms to be detected and to remain there sufficiently long to be measured as a true, limiting stoichiometric $\rightarrow H^+/O$ ratio.

A simple prediction that should be fulfilled if this explanation is correct is as follows: in the absence of added "permeant" ions the measured $\rightarrow H^+/O$ ratio should be a monotonically decreasing function of the size of the O_2 pulse, provided that the smallest size does not induce the maximum membrane potential (Sorgato and Ferguson, 1979; Heinz *et al.*, 1981). This is because, if the supposed membrane potential that has been set up by the translocation of a small fraction of the pumped H^+ is acting to inhibit further proton motive activity (and thus keep the $\rightarrow H^+/O$ ratio submaximal) as described above, this membrane potential should stop the further translocation of any more H^+ if the size of the O_2 pulse is increased, and thus lower the $\rightarrow H^+/O$ ratio. However, oxygen pulse experiments in mitochondria (Archbold *et al.*, 1974, 1979; Conover and Azzone, 1981), in *Escherichia coli* (Gould and Cramer, 1977a; Gould, 1979), and in *Paracoccus denitrificans* (Hitchens and Kell, 1982b, 1984; Kell and Hitchens, 1982) demonstrate that this prediction is not fulfilled; the $\rightarrow H^+/O$ ratio in the absence of added "permeant" ions and at low oxygen concentrations is independent of the size of the O_2 pulse. This type of observation has been interpreted by the above authors to indicate that there must be at least two types of proton circuits in these membranes, only one of which, not seemingly coupled to phosphorylation [but presumably concerned with pH regulation (Padan *et al.*, 1981; Booth and Kroll, 1983)], enters the bulk aqueous phase external to the membrane vesicles. Further, the half-time for H^+ ejection under many of these circumstances greatly exceeds the half-time of O_2 reduction, obviating arguments based on non-ohmic links or slips.

A converse experiment to the foregoing studies the $\rightarrow H^+/O$ ratio under conditions in which the number of O atoms added per cell (vesicle) present is so small that the maximum membrane potential that could be set up if chemiosmotic mechanisms are operative is so small as to be energetically insignificant. Under these conditions, the $\rightarrow H^+/O$ ratio should be the same whether or not "permeant" ions are present. Again, however, experiment shows that in the absence of added "permeant" ions the $\rightarrow H^+/O$ ratio remains submaximal (Gould and Cramer, 1977a,b; Hitchens and Kell, 1982b; Kell and Hitchens, 1982) consistent with the findings described above and with the view that the H^+ that *are* seen are not responsible for inhibiting the transfer of further protons under the typical conditions used.

2. Photosynthetic Systems

Photosynthetic electron transport in both thylakoid and bacterial membranes is also more or less tightly coupled to the reversible transmembrane translocation of protons. We will confine our discussion in this section to studies of chromatophores (inverted cytoplasmic membrane vesicles) from photosynthetic bacteria, although our remarks are essentially equally valid for thylakoids, with the exception that thylakoids as isolated generally have a much greater passive ion permeability than chromatophores. H^+ uptake by thylakoids has been reviewed, for example, by Murakami *et al.* (1975).

The relationship between photosynthetic electron flow and concomitant reversible proton uptake was first reported, in chromatophores for *Rhodospirillum rubrum* (von Stedingk and Baltscheffsky, 1966; von Stedingk, 1967). The extent of uptake was highly pH dependent, being negligible at the alkaline pH values characteristic of the intact cell interior. This is not inconsistent with a teleological view that such proton movements are normally concerned with the regulation of internal pH. The cyclic nature of chromatophore electron transport has in general (but cf. Jackson *et al.*, 1981) precluded the measurement of steady-state rates of electron transport, which might be correlated with the (quasi-)steady-state H^+ uptake which may be observed under conditions of continuous illumination so as to establish more clearly the molecular events underlying the macroscopically observable H^+ uptake. However, it is now understood (Cogdell & Crofts, 1974) hat these H^+ movements are elicited by the action of one or more electron-transport-linked H^+ pumps that act to move H^+ across the chromatophore membrane against their electrochemical potential.

This cyclic nature of chromatophore electron transport has promoted the use of light flashes of a duration sufficiently short to induce single turnovers of the chromatophore electron-transport chain and/or its associated phosphorylation apparatus (e.g., Crofts and Wood, 1978; Dutton and Prince, 1978; Wraight *et al.*, 1978; Baccarini-Melandri *et al.*, 1981; Junge and Jackson, 1982; Ort and Melandri, 1982; Prince *et al.*, 1982). However, despite the extensive and elegant use of this technique, there is currently a great deal of uncertainty concerning the exact nature and extent of the relationship between electron-transfer events and the observable H^+ movements in (phases in equilibrium with) the bulk aqueous phase external to illuminated chromatophore suspensions.

When chromatophores are illuminated with trains of single-turnover flashes of different frequencies, one can be sure that one is detecting all the H^+ pumped from the external aqueous phase, using the pH electrode technique, if this number is frequency independent. This was done by Kell and Hitchens (1983; see also Cogdell and Crofts, 1974), with results comparable to those described above for the measurement of respiration-driven H^+ translocation: for small numbers of flashes, the (submaximal) $\rightarrow H^+$/flash ratio observed in the absence of added ionophores was largely independent of the number of flashes (and their frequency) applied. Thus the conclusion again is that there must be at least two types of proton circuit in this type of system, only one of which enters the bulk phase external to the membrane vesicle of interest. Since chromatophores are inside out with respect to whole cells, the conclusion may be drawn that it is permissible to treat the two bulk phases that the coupling membrane separates as more or less equivalent. The protons which were *not* seen must still have been pumped *across* the membrane [as judged, e.g., by electrochromic measurements of membrane-located carotenoids (see e.g., Wraight et al., 1978)]; some, however, must eventually have returned across the membrane without driving concomitant phosphorylation, presumably via a slip pathway within the primary (electron-transport-linked) proton pump, since similiar behavior is observed with isolated electron-transport complexes reconstituted into liposomes.

Thus the experiments discussed in this and the preceding section show that one must be very careful to distinguish bulk-phase from non-bulk-phase proton movements in discussions of energy coupling, since *both* are coupled to electron transport. What we wish to know is which of them act in the entire energy-coupling process of electron-transport phosphorylation. It is, of course, a logical absurdity to ascribe a causal status as an intermediate between two processes to a process that is merely concomitant with them.

Lastly, it is worth reminding readers, in this section, that concentrations of ionophores sufficient to inhibit phosphorylation completely have relatively little effect upon the extent of light-induced H^+ uptake in bacterial chromatophores (Nishimura and Pressman, 1969; Pressman, 1972).

3. Summary

The discussion in the preceding two sections may be summarized as follows:

1. Electron-transport reactions are coupled to H^+ movements to and from bulk aqueous phases in contact with the coupling membranes in which the electron-transport reactions take place.

2. The measured kinetics and stoichiometeries of these H^+ movements, in the absence of specially added "permeant" ions, and under conditions in which, were ADP and phosphate present, they would have sufficient free energy to make ATP, are not easily consistent with the view that the *observable* H^+ movements might serve as an intermediate in electron-transport phosphorylation.

3. The addition of certain compounds ("permeant ions") indicates that "extra" H^+ were pumped but were not seen in the absence of such compounds; it is these protons that must be presumed to be a possible intermediate in electron-transport phosphorylation.

Similar statements to the foregoing may be made concerning ATP-linked transmembrane H^+ translocation, although all the pertinent experiments do not appear to have been done explicitly. In addition to the slip concept discussed earlier, then, one should thus be aware that variable stoichiometry may be construed to have different properties depending upon whether one is measuring redox-linked H^+ movements in bulk phases or more functional processes such as phosphorylation itself (for excellent examples of the latter, see, e.g., Heber and Kirk, 1975; Stucki, 1982).

Such studies on electron-transport-linked H^+ movements lead naturally to studies which have attempted to measure the thermodynamic magnitude of the proton motive force itself. This type of study forms the subject of the following section.

IV. THE PROTONMOTIVE FORCE AS AN INTERMEDIATE IN ELECTRON-TRANSPORT PHOSPHORYLATION?

"Reality" is what we take to be true. What we take to be true is what we believe. What we believe is based upon our perceptions. What we perceive depends upon what we look for. What we look for depends upon what we think. What we think depends upon what we perceive. What we perceive determines what we believe. What we believe determines what we take to be true. What we take to be true is our reality.

Zukav (1980)

A. Introduction

As indicated earlier, it is usually taken as the central dogma of the chemiosmotic coupling model that the proton motive force or proton electrochemical potential difference between two bulk aqueous phases is "the" (or at the very least "a") high-energy intermediate in processes such as electron-transport phosphorylation. Thus a great many studies have been devoted to an assessment of the role of this variable. However, the first problem arises in the *measurement* of the protonmotive force (pmf) ($\Delta\tilde{\mu}_{H^+}$).

As shown, the reactions of electron transport (and ATP hydrolysis) *are* linked to the vectorial, transmembrane translocation of protons between the two bulk aqueous phases that the coupling membrane serves to separate, it is evident that bulk phase pH gradients *are* produced by such reactions. (This does not, of course, tell us whether such gradients can serve as an intermediate between the two processes.) Methods for measuring such pH gradients are now well established, and the coincidence between different methods in the same system (e.g., Nicolay *et al.*, 1981) lends confidence to the view that certain methods such as weak acid/base distribution can measure the transmembrane pH difference quantitatively. Such methods have been reviewed by a number of workers (e.g., Kell, 1979; Rottenberg, 1979; Ferguson and Sorgato, 1982; Nicholls, 1982; Azzone *et al.*, 1984), and are not here considered further.

A bigger problem arises with the magnitude of the bulk-to-bulk phase membrane potential which is set up, in the chemiosmotic model, as an accompaniment to proton motive activity. The only absolute method, in principle, is measurement with microelectrodes. These very difficult measurements have indeed been reported for a number of systems. Tedeschi and colleagues have, for a number of years, in experiments of increasing refinement, reported that the metabolically induced membrane potential in mitochondria is energetically insignificant (see Tedeschi, 1980, 1981; Ling 1981) to account for ATP synthesis observed in the same mitochondrion (Maloff *et al.*, 1978). These important measurements represent the only direct methods used in mitochondria to date.

Vredenberg, Bulychev, and colleagues have used microelectrodes in giant chloroplasts (see, e.g., Vredenberg, 1976), and found again that the potential did not exceed a few tens of millivolts in the light, even under transient conditions. They also have given arguments indicating that the process of inserting microelectrodes did not damage the membrane significantly. Thus in this system, too, even in the first few turnovers following the initiation of illumination, the bulk-to-bulk phase membrane potential was far too small, assuming an $\rightarrow H^+/ATP$ ratio of 3, to permit ATP synthesis.

Only in giant *Escherichia coli* cells did Slayman and colleagues (Felle *et al.*, 1978, 1980) establish some degree of consonance between the membrane potential measured with microelectrodes and that calculated from the distribution of membrane-permeating lipophilic cations. However, it is by no means clear that the membrane potentials measured with the microelectrodes in these studies resulted from metabolically dependent processes, since (1) the effect of uncouplers or other inhibitors was not tested, (2) the dependence of the measured membrane potential on the external sodium concentration was semi-Nernstian (Felle *et al.*, 1978) (consistent with the view that it was at least partly a simple diffusion potential), and (3) the coincidence between the measurements with the microelectrodes and the permeant ion uptake methods was obtained with organisms suspended in *wholly distinct*, and thus non-comparable, media. Thus the status of these otherwise crucial experiments, which regrettably, have not yet been reproduced, is somewhat uncertain.

Partly on the basis of calibration with induced diffusion potentials, a great many workers (e.g., Westerhoff *et al.*, 1981) have assumed that, under common experimental conditions, and with appropriate corrections, ion-distribution methods are successfully measuring the transmembrane electrical potential difference. Although it is easily possible to envisage mechanisms of ion uptake that are independent of a bulk-to-bulk phase electrical potential difference, we may proceed, initially, by assuming that the generally used methods are measuring the bulk-to-bulk phase pH gradient and membrane potential components of the $\Delta \tilde{\mu}_{H^+}$. Even with this assumption, we find two major sets of anomalies between the data obtained and those expected upon a commonsense reading of the chemiosmotic model, incorporating appropriate additions such as a slightly variable $\rightarrow H^+/ATP$ stoichiometry and/or a non-ohmic leak of energy-coupling membranes to protons. Since we have co-authored a number of articles on this topic (e.g., Kell and Hitchens, 1983; Melandri *et al.*, 1983; Westerhoff *et al.*, 1983a, 1984b, we confine ourselves to some summary statements.

B. Limited Correlation between the Apparent $\Delta \tilde{\mu}_{H^+}$ and Rates of Electron Transfer or of ATP Synthesis

It was observed by many workers, under a variety of conditions, that rates of electron transport or of phosphorylation are not uniquely dependent upon the value of the apparent protonmotive force, but in many cases depend more upon the number of active electron-transport chains or ATP synthases (e.g., Padan and Rottenberg, 1973; Baccarini-Melandri *et al.*, 1977, 1981; Casadio *et al.*, 1978; Sorgato and Ferguson 1979; Kell *et al.*, 1978a; Melandri *et al.*, 1980; de Kouchovsky and Haraux, 1981; de Kouchovsky *et al.*, 1982, 1983; Zoratti *et al.*, 1982; Wilson and Forman, 1982; Mandolino *et al.*, 1983). Some of these observations have been taken to indicate that there may exist kinetic interactions between redox- and ATP synthase-linked proton pumps that are not free energy transducing. However, there is another type of otherwise comparable anomaly which is not subject to this type of interpretation.

C. The Force Ratio under Static Head Conditions Is Not Constant at Different Values of the Apparent $\Delta \tilde{\mu}_{H^+}$

If an energy-transducing system is allowed to synthesize ATP until no *net* ATP formation is observed, the maximal free-energy change for ATP synthesis may be evaluated and compared with the magnitude of the simultaneously measured $\Delta\tilde{\mu}_{H^+}$. Such stationary-state measurements are said to take place at *static head*. It is generally assumed that the ratio of the two free-energy terms may be used, in the chemiosmotic model, to give a minimal measure of the $\rightarrow H^+/ATP$ ratio. It is found, however, in a variety of experiments, that this ratio is heavily dependent upon the magnitude of the apparent pmf, and achieves wholly unrealistic values when the pmf is severely decreased (e.g., Wiechmann *et al.*, 1975; Kell *et al.*, 1978b,c; Azzone *et al.*, 1978; Guffanti *et al.*, 1978, 1981; Decker and Lang, 1977, 1978; Westerhoff *et al.*, 1981; Baccarini-Melandri *et al.*, 1977; Mandolino *et al.*, 1983). Since, at least in some of these experiments, the reaction catalyzed by the ATP synthase was apparently in *equilibrium* with the measured pmf, no kinetic arguments may be invoked (Westerhoff *et al.*, 1981).

These two main types of anomaly, which have been widely observed, have led many workers to assess the competence of the $\Delta\tilde{\mu}_{H^+}$ in energy coupling using methods that do not depend on the accuracy of methods that purport to be measuring the $\Delta\tilde{\mu}_{H^+}$ generated by electron transport.

D. Indirect Means Used to Assess the Competence of the Protonmotive Force in Energy Coupling

1. "Acid Bath" Experiments

Following the pioneering experiments of Jagendorf and Uribe (1966), a number of workers, in a variety of systems, have demonstrated that the imposition of an *artificial* $\Delta\tilde{\mu}_{H^+}$ (acid bath), sometimes partly constituted by a diffusion potential, can drive the synthesis of ATP (see e.g., Thayer and Hinkle, 1975; Schuldiner, 1977; Gräber, 1981; Schlodder *et al.*, 1982; Maloney, 1982; and contrast Malenkova *et al.*, 1982). Most workers find a sharp value of the applied pmf, typically 150 mV, below which the rate of ATP synthesis is negligible. When the value of the applied $\Delta\tilde{\mu}_{H^+}$ is very great, the rates of phosphorylation induced are as great as those driven by electron flow (e.g., Thayer and Hinkle, 1975; Smith *et al.*, 1976), indicating that were the electron-transport-induced $\Delta\tilde{\mu}_{H^+}$ as great as the artificially applied one, $\Delta\tilde{\mu}_{H^+}$ could serve as a kinetically competent intermediate in electron-transport phosphorylation. Unfortunately, the coincidence of $\Delta\tilde{\mu}_{H^+}$ values under the comparable conditions has not yet been tested adequately (see Schlodder *et al.*, 1982) to arrive at a conclusion on this point. It is obviously this comparison, of rates of phosphorylation with electron flow causing a properly measured $\Delta\tilde{\mu}_{H^+}$ equal to that applied in acid-bath experiments, which is the only meaningful one.

Parenthetically, it is worth mentioning that D_2O substitution for H_2O does not significantly affect acid bath-induced phosphorylation in *Streptococcus*

V4051 (Khan and Berg, 1983), while, at least in thylakoids, such a substitution markedly inhibits electron-transport phosphorylation (de Kouchovsky and Haraux, 1981). This would seem to argue against a pmf being the sole intermediate in the latter process.

It is usually assumed that the decay of the pmf applied in acid-bath experiments is concomitant with the phosphorylation, as expected in the chemiosmotic model. However, in an important series of experiments, Hangarter and Good (1982, 1984) have strongly called this assumption into question. They observed that the ability of a preilluminated thylakoid suspension to phosphorylate ADP decays exponentially, as does the light-induced H^+ uptake, and was not, in contrast to acid bath-induced phosphorylation, increased by the imposition of a potassium diffusion potential. The exponential dependence of the decay is actually wholly unexpected given the highly pH-dependent buffering power of the inner thylakoid space and the threshold pmf demonstrated in the acid-bath experiments. Thus the "energized state" set up by preillumination of thylakoids is not the same as that caused by the imposition of an acid bath. It is the presumed equality of these states that has provided for many a convincing demonstration of the accuracy of the chemiosmotic model in describing photophosphorylation. It can only be concluded that the *actual* pmf caused by electron transport is lower than the threshold observed in acid bath experiments when it is created in response to electron flow.

Blumenfeld (1983) also discusses many of the points raised in Section III in his outstanding monograph, which readers are urged to consult for fuller details.

2. "Antacid Bath" Experiments

If a pH gradient in thylakoids were to be driving ATP synthesis *in vivo*, then decreasing it, under conditions of a constant (or, better, negligible) membrane potential should affect the ability of such thylakoids to phosphorylate. The ways in which such experiments have been performed is to study the yield of ATP as a function of the number of single turnover flashes. Consistent with chemiosmotic expectations (Mitchell, 1968), given the reasonably significant ion permeability of thylakoids and the low electrical capacitance relative to the differential buffering power, it is found that the addition of valinomycin leads to a lag in the capacity for ATP formation (Ort *et al.*, 1976; Davenport and McCarty, 1980; Vinkler *et al.*, 1980; Graan *et al.*, 1981; Ort and Melandri, 1982). The lag is independent of the concentration of K^+ between 20 and 40 mM, indicating that under such conditions the pH gradient should be the sole contributor to $\Delta\tilde{\mu}_{H^+}$ (Graan *et al.*, 1981; but contrast Boork and Baltscheffsky, 1982, for chromatophore experiments of this type). However, it is found that if the thylakoids are preloaded with hydrophilic, non-uncoupling buffers, the lag remains essentially unaffected (Graan *et al.*, 1981). The pH gradient that should have been produced in the presence of the buffer can be calculated from H^+ uptake measurements and may be arranged to be significantly less than the threshold required in acid bath experiments. The conclusion to be drawn from this type of "antacid" bath experiment is that the protons used for making ATP did not pass through a space in equilibrium with the internal thylakoid space accessible to the buffer molecules. Thus this type of observation complements the findings discussed under the acid bath experimental section and indicates that the localization of the high-energy intermediate is more microscopic than implied in the delocalized chemiosmotic model.

A number of other workers have also sought a functional approach that considers whether the free energy released by electron transport is freely available to every enzyme molecule of a given type in a given vesicle, as assumed in the chemiosmotic model (Fig. 4a), approaches which form the subject of the next section.

E. The Intermediate Is Not a Pool: Dual Inhibitor Titrations

The principle underlying the general methods of double inhibitor titrations may be expounded with reference to Fig. 5a. This figure models the process of electron transport phosphorylation as four redox-linked H^+ pumps (ETC) plus four H^+-ATP synthases (written ATP in the figure) that are coupled via a high-energy intermediate ($\sim$). The box indicates a model in which the free energy released by a particular electron-transport complex may be used by any ATP synthase in the membrane vesicle preparation, as in the delocalized

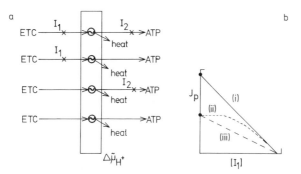

Fig. 5. Principle of dual inhibitor titrations using inhibitors of electron transport and ATP synthase enzymes. For discussion and explanation, see text.

chemiosmotic model. I_1 and I_2 represent tight-binding and specific inhibitors of electron transport and the ATP synthase, respectively. "Heat" represents all free-energy-dissipating processes.

As first expounded by Baum and colleagues (1971; Baum, 1978), a simple analysis (Fig. 5b) of this situation would suggest that partial inhibition of the overall flux of the system (i.e., the rate of ATP synthesis, J_p) using an I_2-type inhibitor [Fig. 5b(i), (ii)] should increase the titer of an I_1-type inhibitor initially necessary to inhibit phosphorylation relative to that in the absence of the I_2 inhibitor if energy coupling is delocalized (ii) (and see Section IX). Conversely, if energy coupling is fully localized, the titers will be unchanged (iii). As mentioned earlier, the very common finding of an independence of the $P/2e^-$ ratio on the rate of electron transport over a wide range in well-coupled systems indicates that energy leaks are not especially relevant to this type of analysis. Such a finding also militates against the thoughtful suggestions of Parsonage and Ferguson (1982) concerning putative $\Delta\bar{\mu}_{H^+}$ values during such titrations. Experimentally, it was found by several groups, in a number of systems, that the behavior observed corresponded to that expected for a fully localized system (Baum et al., 1971; Lee et al., 1969; Baum, 1978; Venturoli and Melandri, 1982; Hitchens and Kell, 1982a,c; Kell and Hitchens, 1983; Westerhoff et al., 1982, 1983a,c; Westerhoff, 1983; Welch and Kell, 1985). It is also worth remarking that the selection for unc phenotypes in aerobically grown E. coli, based on aminoglycoside resistance, follows from the assumption that their cytoplasmic membranes are less energized than those of unc^+ strains. In the case of the uncA phenotype, which should exert a lower drain on the energized state than the wild type, this observation is not easily reconciled with delocalized coupling, but is to be expected from considerations of the results of I_1/I_2 titrations discussed above. The finding (Schreiber and Del Valle-Tascon, 1982; Blumenfeld, 1983) that there is no nonzero threshold rate of electron transport necessary to drive ATP synthesis under level flow conditions has also been taken to indicate a localized energy-coupling interaction between redox chains and ATP synthase enzymes.

A variety of experiments following, in some aspects, this broad strategy has also been performed by Dilley and his colleagues (see Dilley et al., 1982), with results incompatible with a delocalized energized state caused by electron transport (cf. Haraux et al., 1983). It is also germane to draw attention to some beautifully done expositions of this type of approach in relation to the interaction between electron-transport chains (Heron et al., 1978; Haehnel, 1982; Mar et al., 1982; Packham and Barber, 1983).

Other dual inhibitor titrations using combinations of uncoupler and I_2-type inhibitors may be found in work by the present authors and others (e.g., Hitchens and Kell, 1982c, 1983a,b; Kell and Hitchens, 1983; Westerhoff et al., 1982, 1983a,b; Westerhoff, 1983; Herweijer et al., 1984). The crucial observation in these experiments was that inhibition of the output (secondary) proton pump decreased the titer of uncoupler required to uncouple the energy-linked reactions completely. A similar observation, interpreted somewhat differently, was earlier made in the case of the uncoupling antibiotic leucinostatin plus the F_o inhibitor venturicidin by Lardy et al. (1975) and by Reed and Lardy (1975).

Cotton et al. (1981) made a similar observation concerning the release of respiratory control in intact cells of Rps. capsulata; less uncoupler was required to induce loss of respiratory control in cells which had been treated with venturicidin, an H^+-ATP synthase inhibitor, than in cells which had not been so treated. These authors argued that the uncoupler might have decreased the pmf in a fashion which would have led to an inhibition of substrate uptake sufficient to explain the result in terms of delocalized coupling. Unfortunately, they did not seek to test this hypothesis in cells respiring on endogenous substrate.

The same group (Cotton and Jackson, 1983) has criticized some of the experiments of Hitchens and Kell (1982c, 1983a,b) in that, especially in the presence of energy-transfer inhibitors, they were unable to obtain linear rates of phosphorylation. They argued (Cotton and Jackson, 1983) that such effects might serve to explain the observations of Hitchens and Kell (1982c, 1983a,b). With regard to these criticisms, the following remarks, inter alia, may be made:

1.   In contrast to the results of Cotton and Jackson (1983), linear rates of photophosphorylation could easily be obtained by Hitchens and Kell; these were also observed, with the same resultant uncoupler titers, when uncoupler titrations were carried out on a succession of samples, with the uncoupler being added at the same time after illumination (Hitchens and Kell, unpublished).

2.   Figure 1 of Cotton and Jackson (1983) indicates that even in the absence of oligomycin the apparent uncoupler potencies measured by these authors could vary by a factor of approximately two in the same chromatophores; according to their Fig. 1, the uncoupler is less potent the longer it is present.

3.   The rapid initial proton uptake cannot be due to light-induced vectorial proton translocation since it is fully sensitive to energy transfer inhibitors (Hitchens and Kell, 1982c).

4.   The decrease in H^+ uptake after 8 min illumination cannot be due to the buildup of a high phosphorylation potential since the decrease in rate under these (latter) circumstances (using ADP, but not UDP) can only be observed very shortly before the attainment of static head (Hitchens & Kell, unpublished).

5.   Similar trends in uncoupler titrations are also observed in sub-mitochondrial particles using an entirely different assay (Westerhoff et al 1982, 1983a,b; Westerhoff, 1983; Herweijer et al., 1984).

Thus these purely functional approaches seem to indicate that electron-transport phosphorylation cannot be proceeding via a delocalized intermediate, and that some kind of catalytic facilitation in some kind of functionally organized multienzyme complex is an integral part of these processes. We shall return to this concept in more general terms at the end of this review (Sections IX and X). Our next task is to look at studies that have aimed to take apart and reconstitute the enzymes catalyzing electron-transport phosphorylation.

V. THE EFFECTIVENESS OF RECONSTITUTED SYSTEMS IN CATALYZING ATP SYNTHESIS

We find that the theories of physicists constantly undergo modification, so that no prudent man of science would expect any physical theory to be quite unchanged a hundred years hence. But when theories change, the alteration usually has only a small effect so far as observable phenomena are concerned.

(Russell, 1966)

A. Mitchondrial Oxidative Phosphorylation

It is widely believed (see, e.g., Racker, 1976; Tzagoloff, 1982) that the successful co-reconstitution of purified electron-transport complexes (or bacteriorhodospin) plus a *purified* ATPase complex in a liposomal system capable of catalyzing electron-transport phosphorylation at rates (and with efficiencies) characteristic of those *in vivo* has been amply demonstrated. In particular, the experiments of Racker and Stoeckenius (1974) on co-reconstitution of bacteriorhodopsin plus an oligomycin-sensitive ATPase preparation are regularly cited as the classic demonstration that energy coupling occurs via a purely (i.e., delocalized) chemiosmotic mechanism. This notion, which we shall here refer to as the simple interpretation, is illustrated in Fig. 6. To what extent does this notion actually fit the available data?

Kagawa and Racker (1971) isolated amorphous membrane fragments ("hydrophobic protein fraction"), depleted in phospholipids and cytochrome oxidase, from beef heart mitochondria, which could be reconstituted with phospholipids to catalyze ATP–P_i exchange activity. In the same year, Racker and Kandrach (1971) added cytochrome c and cytochrome c oxidase during the reconstitution of this hydrophobic protein fraction preparation and found that the new preparation would catalyze oxidative phosphorylation with a P:O ratio of 0.13. In his 1972 review, Kagawa (1972, p. 330) cautioned, "it should be stressed that the intrinsic membrane proteins used in the reconstitution are still crude preparations. Further purification and character-

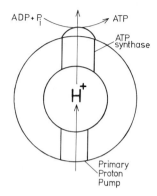

Fig. 6. The simple interpretation of reconstitution experiments involving proton-motivated ATP synthesis. In this interpretation, it is assumed that purified ATP synthases plus a purified primary proton pump (bacteriorhodopsin or a redox-linked proton pump) have been co-reconstituted into relatively ion-impermeable liposomes with the appropriate mutual polarity. According to a delocalized chemiosmotic viewpoint, the primary proton pump sets up a proton-motive force, which is then used to drive ATP synthesis at a rate characteristic of those *in situ*. Although no special spatial interaction between the complexes is required, whether one in fact occurs is not usually clarified. The extent to which the available data indicate that such a preparation catalyzes phosphorylation with turnover numbers akin to that *in vivo* is discussed in the text.

ization of the proteins essential to phosphorylation or respiration is [sic] necessary before we can discuss the molecular architecture of this biomembrane and the mechanism of oxidative phosphorylation." Regrettably, this advice has not always been followed.

Meanwhile, Racker and Kandrach were optimizing their technique, and by 1973 (Racker and Kandrach, 1973) had achieved a P:O ratio of 0.5 with ascorbate *plus* phenazine methosulfate (PMS) as electron donor to cytochrome oxidase, and Ragan and Racker (1973) had achieved a similar P:$2e^-$ ratio using electron flow through the NADH–ubiquinone–oxidoreductase segment of the respiratory chain. Such yields, rather close to those *in vivo*, might be (and indeed are widely) thought to have proven the simple interpretation. However, *every one* of these studies as well as that of Racker and Stoeckenius (1974) included the hydrophobic protein fraction; these pioneering reconstituted particles were *not purified* (i.e., pure) *preparations*.

What does the hydrophobic protein fraction contain? We know of only one attempt properly to characterize this material, that of Capaldi *et al.* (1973), who found that even a one-dimensional SDS–polyacrylamide gel separated a massive number of components; their Fig. 1 (Capaldi *et al.*, 1973) would seem to indicate, assuming a Gaussian peak shape with the width of the best resolved peak, 20 at the very least. How many of these might be ascribed to the F_oF_1-ATPase (complex V)? There seems to be a general consensus (e.g., Baird and Hammes, 1979; Galante et al., 1979; Kagawa, 1979; Senior, 1979; Berden and Hennecke, 1981; Pedersen et al., 1981; Pedersen, 1982; Senior and Wise, 1983) that one-dimensional SDS gels can separate approximately 11–13 components in *purified* beef heart mitochondrial complex V. [Up to 18 may be seen in two-dimensional gels (Ludwig et al., 1980; Capaldi, 1982b), although the role of many of these remains unknown.]

It may be concluded that the hydrophobic protein fraction may contain many polypeptide components that are not part of those of purified complex V.

B. Reconstitutions Using *Purified* Systems

How efficient are reconstitutions of phosphorylation performed without a hydrophobic protein fraction or something analogous from another energy-coupling membrane? We can give the answer at once: not at all. A few quite typical examples serve to illustrate this.

Winget et al. (1977) reconstituted bacteriorhodopsin and the chloroplast CF_oF_1-ATPase. Their best (highest) rates of photophosphorylation were approximately 1 nmol ATP min^{-1} mg^{-1} CF_oF_1 (their Fig. 6). This is already very low compared with that in thylakoids *in vivo*, but, as remarked by Hauska et al. (1980) and by Kell and Morris (1981), what we need to know is the turnover number of the ATP synthase, since these preparations are enriched in the sole proteins supposedly required (in the simple interpretation) for free-energy transduction. If we assume a fairly rough molecular weight for the CF_oF_1 of 500,000 daltons (Pick and Racker, 1979), 1 mg $CF_oF_1 = 2$ nmol CF_oF_1, so that the turnover number of photophosphorylation in the reconstituted system is approximately 0.5/min, i.e., about 0.01/s. Since the turnover number of the CF_oF_1 during photophosphorylation by thylakoids *in vivo* is approximately 100/s (Hauska et al., 1980), and this figure is essentially accurate for energy-coupling membranes generally (Kell and Morris, 1981), the reconstituted system has an efficiency of 0.01%, hardly consistent with the simple interpretation discussed above!

Oren et al. (1980) co-reconstituted bacteriorhodopsin and a purified F_oF_1 preparation from *Rhodospirillum rubrum*; their highest rates of photophosphorylation (their Table I) were approximately 8 nmol min^{-1} mg^{-1} F_oF_1, similar to those found by Ryrie et al. (1978) using bacteriorhodopsin plus a purified yeast F_oF_1 preparation and those of Berden et al. (1981) using bacteriorhodopsin plus beef heart complex V. Using the assumed molecular weight for the ATPase complexes given in the previous paragraph, we obtain a turnover number of about 0.08/s, again less than *0.1%* of the *in vivo* rate. It seems that van de Bend and colleagues may now have somewhat increased this value (van de Bend *et al.*, 1984).

Finally, we may consider the detailed experiments of Hauska *et al.* (1980), who carefully studied the ability of liposomes containing a purified photosystem I preparation and a purified CF_oF_1-ATPase preparation to catalyze photophosphorylation in the presence of phenazine methosulphate. Their best turnover number was 0.4% of that in the thylakoids from which the preparations were obtained, and, as far as we are aware, this is the highest published rate obtained in any system reconstituted with "purified" components to date. It may be mentioned that the proton tightness of these proteoliposomes was just as great as that of the thylakoids, as judged, insofar as is possible, by 9-aminoacridine fluorescence measurements (Hauska *et al.*, 1980).

We are, of course, aware that unidentified features such as a lack of proper orientation of some components may be responsible for the poor rates observed in these purified reconstituted systems. Nevertheless, it may be concluded that the simple interpretation of these admittedly difficult and elegant reconstitution experiments is, on the basis of presently available data, wholly incorrect and misleading. Further, the success of reconstitutions using the hydrophobic protein fraction may well indicate, in our view, that "for rapid and efficient protonmotivated energy coupling, something else is required in addition to the protonmotive sources and sinks [i.e., primary and secondary proton pumps] and an intact insulating coupling membrane" (Kell and Morris, 1981). We shall return to this shortly, but digress briefly to comment upon the role of so-called structural protein in mitochondria.

C. Structural Protein

Many early studies (e.g., Criddle *et al.*, 1962; Richardson *et al.*, 1964) indicated the presence of a so-called structural protein, which could be isolated from mitochondria, which might have a crucial role in oxidative phosphorylation, and which might yet be distinct from the (now) generally recognized protonmotive protein complexes. This concept was effectively nullified by the work of Schatz and Saltzgaber (1970), who showed that these preparations of structural protein contained, *inter alia*, large amounts of denatured F_1 protein. Since the publication of this article (Schatz and Saltzgaber, 1970), the literature has been essentially devoid of discussion of the role of some kind of (quasi-) structural protein (but cf. Sjöstrand, 1978). We suspect here that the baby may have been thrown out with the bathwater, since the studies summarized in the two previous sections, which are *wholly self-consistent*, indicate that some uncharacterized proteins important to energy coupling are now routinely being neglected. Let us look at some further studies which bolster this conclusion still further and indeed may shed important light upon why energy coupling may *not* be easily reconstituted in liposomes, using purified proton pumps alone.

D. Phospholipid-Enriched Mitochondria and Chromatophores

The fusion of liposomes with the inner mitochondrial membrane (as mitoplasts) to produce lipomitochondria, with the same protein composition but a greatly enhanced lipid: protein ratio, was reported in two elegant articles by Schneider *et al.* (1980a,b). The activities of individual electron-transport and ATPase complexes remained unchanged (or were slightly increased), while the rate of electron-transport events between the substrate-linked dehydrogenases and cytochrome-containing complexes was decreased by a factor in rough proportion to the diffusional mean free path of the complexes. This was taken to indicate (see also Hackenbrock, 1981; Schneider *et al.*, 1982; but cf. our discussion in Section I,4) that there was (at least in lipomito-chondria) a diffusion-limited step in intercomplex electron transfer. This limitation could be relieved by the incorporation of lipophilic ubiquinone homologs (Schneider *et al.*, 1982).

However, although the ATP hydrolase activity of these lipomitochondria was unimpaired or slightly increased, and remained oligomycin sensitive, the activity was insensitive to uncouplers, and the membranes were incapable of oxidative phosphorylation (Schneider *et al.*, 1980b; Westerhoff, unpublished). These workers indicated (Schneider *et al.*, 1980a,b) that the lipomitochondria were rather nonspecifically permeable to charged molecules. We find it a little difficult to concur with this conclusion for two reasons: (a) liposomes are routinely highly ion impermeable; indeed, they are significantly more so than typical energy coupling membranes; (b) lipochromatophores produced from *Rhodopseudomonas sphaeroides* by the incorporation of extra phospholipids (Casadio *et al.*, 1982) are just as ion impermeable as their parent chromato-phores. The finding (Casadio *et al.*, 1982) that lipid enrichment did not affect the extent of the spectral carotenoid signal induced by K^+/valinomycin diffusion potentials in the dark, but greatly decreased that of the light-dependent signal, adds weight to the finding (Ferguson *et al.*, 1979) that ion distribution and the carotenoid signal monitor *different* aspects of the energized state.

Although there was evidently a certain amount of scrambling (inversion) of protein complexes in the lipomitochondria, in that their ATP hydrolase activity was only partially sensitive to atractylate (Schneider *et al.*, 1980b), we wish to mention an alternative explanation for the incapability of this system to catalyze oxidative phosphorylation: that the normal functioning of the apparatus of oxidative phosphorylation requires some kind of special, supramolecular, topological relationship between the energy-transducing membrane components additional to an appropriate polarity.

In these sections we have been adducing circumstantial evidence that energy-coupling membranes require (a) some kind of special lateral arrangement of the protein complexes *known* to be involved in electron-transport phosphorylation and (b) one or more proteins distinct from those in the generally recognised quasi-reversible proton pumps. We will now turn to further evidence implicating such additional proteins, which have been termed *protoneural proteins* (Kell et al., 1981b; Kell and Morris 1981).

VI. HOW THEN MIGHT ENERGY COUPLING PROCEED?

A. Introduction

In the preceding sections we have given a picture of the properties of energy-coupling membranes. In Section IV we reviewed the evidence showing that the electrochemical potential difference for protons between two supposedly homogeneous, aqueous phases bordering the energy-coupling membrane is not competent to act as the free-energy intermediate in free-energy transduction. Thus, the (delocalized) chemiosmotic coupling hypothesis put forward by Mitchell (1961), which had this implication, would not seem to be a sufficient account of how free-energy transduction takes placce. A number of authors have proposed modifications of the chemiosmotic coupling hypothesis. In recent reviews (Melandri et al., 1983; Westerhoff, 1983; Westerhoff et al., 1983a–c, 1984b) we have discussed these and have distilled a minimum hypothesis (called *mosaic protonic coupling*) for the mechanism of membrane-linked free-energy transduction. This hypothesis gives a functional definition of a device that should be present among the membrane properties reviewed above. In Section VII we shall discuss a possible structural identity of this device. Here we shall summarize the minimum hypothesis and discuss the consequences that are most relevant in the context of this volume.

B. Mosaic Protonic Coupling: A Feasible Minimum Hypothesis for Energy Coupling

To be compatible with the results of numerous experiments designed to prove the delocalized chemiosmotic coupling hypothesis (see above), it is necessary to retain most elements of the latter in any alternative scheme. Consequently, (see Fig. 4b) the mosaic protonic coupling hypothesis assumes that free-energy-coupling membranes contain proton pumps that are driven by chemical reactions. Important ones are electron-transfer-driven proton pumps and F_oF_1-type proton-translocating ATP synthases. Just as in the delocalized chemiosmotic coupling hypothesis, free-energy transduction is proposed to materialize as a proton current: a primary proton pump (e.g., an electron-transfer chain) pumps protons across a membrane. As a consequence there is an increase in proton concentration (or rather occurrence) and in electric potential difference across that membrane itself. However, in mosaic protonic coupling, free energy remains stored in the membrane, as well as in the proton gradient. Thus the average tendency of the pumped protons will be to return across the membrane. Through this tendency the protons can reverse the ATP hydrolysis-driven proton pump (the secondary proton pump). This then leads to the synthesis of ATP.

The difference between the delocalized and the mosaic chemiosmotic coupling hypotheses is that the former proposes that there is one single compartment (domain) into which all individual proton pumps pump (the high-free-energy protons form a pool), whereas the latter proposes that, functionally, and on a given turnover, every primary proton pump has its own compartment, which it shares with only one (or a very limited number of) secondary proton pump(s). For different proton domains to be as functionally independent as proposed, there should be an isolation between these domains and the bulk aqueous phase bordering the membrane. This barrier should be such that it prevents proton equilibration between the local domain and the bulk aqueous phase for at least the turnover time of free-energy transduction (typically some 1–100 ms, but in certain cases for several seconds). Moreover, there should be a significant proton leakage, draining the proton electrochemical potential difference between the two aqueous bulk phases (i.e., $\tilde{\mu}_{H^+}$ "in" minus "out"). Figure 4b gives a scheme for this mosaic protonic coupling hypothesis: the coupling membrane essentially contains a multitude of independently operating and possibly heterogeneous (i.e., a mosaic of) coupling units.

C. Implications of Mosaic Protonic Coupling

1. Explanation of the Incompetence of the PMF ($\Delta \tilde{\mu}_{H^+}$)

In terms of the mosaic protonic coupling hypothesis it is readily understood why observed magnitudes of the proton electrochemical potential difference between the bulk aqueous phases bordering the membrane ($\Delta \tilde{\mu}_{H^+}$, or pmf) have often been too low to account for the free energy necessary for (the observed) synthesis of ATP: due to leakage of protons through R_H^{leak} in Fig. 4b, there is a continuous dissipation of $\Delta \tilde{\mu}_{H^+}$ such that the latter will always be lower than the proton gradient at the "site" of the proton pumps, which is denoted $\Delta \tilde{\lambda}_{H^+}$ (see Fig. 4b). It is this $\Delta \tilde{\lambda}_{H^+}$, and not $\Delta \tilde{\mu}_{H^+}$, that should be competent for the observed ATP synthesis.

2. Some Statistical Aspects of Energy Coupling

It can be estimated (Westerhoff, 1983) that if indeed the proton pumped by the electron transfer-linked proton pump remains localized to the coupling unit consisting of that proton pump plus an H^+-ATPase (i.e., the pumping of on the order of two protons) would create a local proton potential (or rather, see below, the energetic equivalent thereof) that would be energetically sufficient to drive ATP synthesis. The fact that this number is so small has important effects on the characteristics of energy coupling. First, a proton domain is either in the energized state (i.e., contains one or two protons), or in the unenergized state (i.e., contains no protons). If, say, 50 proton domains are energized and 50 others are not, then this cannot be approximately described by saying that 100 domains are half energized. The system is not ergodic (Welch and Kell, 1985).

To illustrate the relevance of this phenomenon, we shall give a brief description (for a more refined treatment see Westerhoff et al., 1984b) of the time dependence of energization in a system where the electron-transfer chains are regularly, and simultaneously, pumping protons at a frequency of $1/\tau$. If the number of coupling units is denoted by n and the probability of escape of the energized proton is represented by the rate constant κ, then the number of energized proton domains n^* as a function of time is given by Eq. (1).

$$n^*(t) = n^*(0)\,e^{-\kappa t} \qquad \text{(for } t < \tau) \tag{1}$$

At $t = \tau$ this number would again jump back to n. The probability that the H^+-ATPase present in an energized coupling unit would make ATP may be described by (cf. Van Dam et al., 1983; Westerhoff and Van Dam, 1985) the rate constant $K_p[1 - f(\Delta G_p)]$, which indicates that at increasing magnitudes of the phosphate potential (ΔG_p) this probability should be reduced. The amount of ATP synthesized per coupling unit can be calculated:

$$(\delta \text{ATP}/n) = (n/\kappa)(1 - e^{-\kappa \tau})K_p[1 - f(\Delta G_p)] \tag{2}$$

Equation (2) shows that at low turnover rates [compared to the rate at which the energized proton leaks out of the proton domain, i.e., $(1/\tau) \ll \kappa$], the ATP yield per coupling unit per turnover of the electron-transfer chain tends to become independent of the turnover rate, or, in other words, the $P/2e^-$ ratio tends to become independent of the rate of electron transfer. In many free-energy transducing systems this is indeed observed (e.g., Van Dam and Tsou, 1970; Ernster and Nordenbrand, 1974; Ferguson et al., 1976; Venturoli and Melandri, 1982; and see earlier).

We may also consider what would happen if a fraction α_p of the H^+-ATPase or a fraction α_e of the electron-transfer chains would be eliminated at random, e.g., through the addition of irreversible inhibitors. The ATP yield per turnover of the electron-transfer chain would then become:

$$(\delta ATP/n) = (n/k)(1 - e^{\kappa\tau})K_p[1 - f(\Delta G_p)](1 - \alpha_e)(1 - \alpha_p). \tag{3}$$

Importantly, both types of inhibition would have the same effect, simply reducing the number of coupling units (n). Consequently, inhibition of electron-transfer chains is not expected to lead to a reduction in the titer of the inhibitor of the H^+-ATPase (nor vice versa). The delocalized chemiosmotic coupling hypothesis would generally predict such a reduction in titer. The experimental results seem to plead in favor of mosaic protonic coupling (see Section IV,E).

If the effect of an uncoupler would be to increase the probability of escape of a "local proton" (i.e., increase κ), then inhibition of the H^+-ATPases would be expected to have no effect on the titer of the uncoupler in its inhibitory action on ATP synthesis. If some (F_1-type) ATPase activity would be present in the system, then the titer would even be reduced. This is in contrast to the predictions of the hypothesis of delocalized chemiosmotic coupling, but in keeping with the experimental results (cf. Section IV,E).

Thus, it seems that precisely the statistical (or quantal; see Welch and Kell, 1985) properties entailed by the mosaic chemiosmotic coupling hypothesis (Westerhoff et al., 1983a,c, 1984b) can account for most observations that were at odds with the delocalized chemiosmotic theory. Consequently, mosaic protonic coupling is a feasible minimum hypothesis. However, it lacks a definitive postulate concerning the structural basis for the extra device(s) that keep(s) the energized proton from equilibrating with the bulk aqueous phase adjacent to the membrane, and thus from forming a pool with its fellow energized protons. In the next section we will discuss the indications for the existence of proteinaceous devices of that kind that are distinct from the proteinaceous proton pumps themselves.

VII. PROTONEURAL PROTEINS?

A. Introduction

How odd it is that anyone should not see that all observation must be for or against some view if it is to be of any service.

(Darwin, 1903, cited by Howard, 1981)

A commonsense (and usual) reading of the delocalized chemiosmotic coupling hypothesis indicates that all that is required for electron-transport phosphorylation is a primary (redox-linked) H^+ pump and a secondary (ATP synthase-linked) H^+ pump embedded with the appropriate polarity in a suitably ion-impermeable membrane, as indicated in the simple interpretation of reconstitution experiments (Fig. 6). In the previous sections, we concluded that more localized coupling theories permit (and, according to the calculations of H^+ transfer rates given in Section I, probably *mandate*) the existence of additional, presumably proteinaceous, devices that are required for energy coupling in electron-transport phosphorylation and related processes.

Hong and Junge (1983) have claimed that high local buffer capacities would produce localization of the high-free-energy proton. This proposal has been criticized (Westerhoff, 1983) as being insufficient to explain indications for mosaic behavior in steady-state experiments. Van Dam *et al.* (1983; cf. Sjöstrand, 1978) proposed that localization effects might be due to close apposition of the external surfaces of the inner mitochondrial (cristae) membrane. An uncertainty here is how this proposal could be extended to intact bacteria and submitochondrial particles. An alternative is the presence of special proteins in energy-coupling membranes constituting the so-called protoneural networks (Kell *et al.*, 1981b). Here we shall elaborate on this latter possibility, which was first explicitly stated, to our knowledge, by Ji (1976).

How complex really are these energy-coupling membranes, then? Have we already identified all the polypeptides which they contain? The answer again is: not at all. Lest any doubt remain concerning the very great, if often unrecognized, complexity of energy-coupling membranes, it may suffice to draw attention to the number of polypeptides resolvable in SDS gel electrophoretograms of some of these membranes that have been published to date. Thus, one-dimensional SDS gels of bacterial chromatophores contain at least 33 polypeptides (Gabellini *et al.*, 1982; Kaufmann *et al.*, 1982), few of which have as yet been identified, while two-dimensional gels of thylakoid membranes (Fig. 3 of Roscoe and Ellis, 1982) and the cytoplasmic membrane of *Escherichia coli* (Fig. 6 of Gibson, 1983) indicate numbers in excess of 50 [or 130 (Buetow and Gilbert, 1982)] and 100 polypeptides, respectively. Thus, in our search to identify the putative protoneural proteins, one can hardly claim that all the polypeptides of these energy-coupling membranes have already been identified and characterized!

To return to mitochondria, then, let us take a leap in the dark, and see whether even those few membrane proteins encoded by the mammalian mitochondrial genome have been identified.

B. Unidentified Reading Frames in the Mammalian Mitochondrial Genome

The entire genomes of a number of mammalian mitochondria, that is, those of human (Anderson *et al.*, 1981; see also Borst *et al.*, 1981), bovine (Anderson *et al.*, 1982), and murine (Bibb *et al.*, 1981) mitochondria, have been completely sequenced. The importance of these beautiful studies to our present considerations is that, in each case, the presence of eight unidentified reading frames (URFs), coding for rather hydrophobic proteins of unknown function, was demonstrated. As Davies *et al.* (1982) have put it, in our view prophetically, "the discovery of these putative genes in organisms as different as man and *A[spergillus] nidulans* suggests that they code for functional proteins that play a hitherto unsuspected role in mitochondria." While one may imagine many possible indentities for the proteins encoded by the URFs, such as translocases, or ribosomal proteins, or processing enzymes (Attardi *et al.*, 1982), we wish to stress a possibility that our molecular biological colleagues, who may not be fully aware of the bioenergetic literature, might otherwise neglect: that these URFs code for proteins with the role of the protoneural proteins alluded to above (Kell and Morris, 1981). It should be stressed that these URFs do code for functional mRNA molecules. Finally, we would mention that two-dimensional gel autoradiograms (which, in contrast to Coomassie Blue staining, do not discriminate against very hydrophobic proteins) indicate that the mitochondrial translation products of HeLa cells include as many as 26 polypeptides (Ching and Attardi, 1982).

It may be remarked, in conclusion, that the coincidence of the hydrophobic nature of the proteins coded for by the URFs and the importance of using Racker's hydrophobic protein fraction rather than more purified H^+-ATP synthase preparations in reconstituting oxidative phosphorylation seems beyond fortuity. We now turn to some studies of plant mitochondrial molecular biology which give us a more forthright clue as to the role of at least one unidentified mitochondrial protein in mitochondrial free-energy transduction.

C. The Interaction of *Helminthosporium* T Toxin with Corn Mitochondria

"Unexpected, shocking and true"

[P. P. Slonimski, May 17, 1981, describing the kind of results he hoped to see in mitochondrial research; quoted by Bendich (1982)].

Cytoplasmic male sterility (CMS) is a maternally inherited trait that prevents the production of functional pollen but does not affect female fertility in a number of plants of commercial significance (Edwardson, 1970). The genetic determinants which control CMS in maize (corn) are located on mitochondrial DNA (see e.g., Leaver *et al.*, 1982; Leaver and Gray, 1982). Since certain plant lines carry nuclear fertility-restoring genes, the CMS phenotype is now widely used in the commercial production of F_1 hybrid seed varieties to prevent the self-pollination of the parent.

Three types of CMS, namely CMS-T, CMS-S and CMS-C, have been described in the maize. Our attention is concentrated on the T (Texas) cytoplasm, which was used in over 85% of hybrid corn grown in the United States by 1970 (Leaver and Gray, 1982). In that year, an epidemic of southern corn leaf blight disease virtually wiped out the maize crop, causing losses in excess of $1 billion. The disease is caused by a polyketide toxin (T toxin) (for probable structure see Kono *et al.*, 1980) produced by the fungus *Helminthosporium maydis* Race T, which strongly affects corn with the CMS-T trait but has little effect on normal fertile (N) or the CMS-S and CMS-C types (Ullstrup, 1972; Gregory *et al.*, 1977). A wealth of physiological studies, summarized by Gregory and colleagues (1977, 1980) and by Leaver and colleagues (Leaver and Gray, 1982; Leaver *et al.*, 1982), leaves little room for doubt that the primary target for the T toxin is a mitochondrially encoded protein resident in the inner mitochondrial membrane and that the interaction of T toxin with this protein leads, *inter alia*, even at a T toxin concentration of 10 pmol/mg mitochondrial protein (Gregory *et al.*, 1980), to the "uncoupling" of sensitive mitochondria. It should be stressed that normal (N) mitochondria are quite resistant to doses of this toxin at concentrations as much as 1000-fold higher.

What is the nature of the target protein? Leaver and colleagues have studied the synthesis of proteins by mitochondria isolated from a number of strains of corn possessing N and the CMS traits (Forde *et al.*, 1978; Forde and Leaver, 1980; Leaver and Gray, 1982; Leaver *et al.*, 1982; Dixon *et al.*, 1982); analysis of the radiolabeled products of mitochondrial translation by one-dimensional SDS–polyacrylamide gel electrophoresis and autoradiography shows that at least 20 polypeptides in the range 8000–54,000 Da can be resolved, and, with the exception of a 44-kDa polypeptide, all are normally membrane bound. Most strikingly, a 21-kDa polypeptide synthesized by N mitochondria is not seen in T mitochondria, which instead possess, uniquely, a 13-kDa polypeptide. It seems, therefore, very likely that this 13-kDa polypeptide is the (or at least a) target of the T toxin. Since the result of the interaction of the toxin with this protein is some kind of uncoupling of the sensitive T mitochondria, it is evident that the role of the 13-kDa polypeptide normally lies in energy coupling in such mitochondria. Since this polypeptide is not thought to be part of complexes I to V of these mitochondria, an intriguing possibility, in harmony with the considerations in the rest of this section, is that this target protein possesses the properties of a protoneural protein, as proposed previously (Kell *et al.*, 1981b).

Although the T toxin is not proteinaceous, the physiological effects of its interaction with sensitive mitochondria bear many striking resemblances to those elicited by the interaction of a number of proteinaceous membrane-active bacteriocins with their sensitive target bacterial cells, a topic which forms the subject of the next section.

D. Effects of Membrane-Active Bacteriocins on Sensitive Cells

Since the BLM [black lipid membrane] is an exquisitely sensitive tool with which to study the interaction of proteins with bilayers, it is important to keep in mind the possible molecular mechanisms underlying the observed conductance effects. True channel formation in a BLM may reflect the actual physiologic role of a particular protein. If, on the other hand, a protein is observed to increase conductance in a BLM, but in a manner suggestive of lipid perturbation, it may be quite misleading to declare it a channel-former and to suppose its biological action or function is carried out via discrete channel formation.

(Blumenthal and Klausner, 1982)

The colicins are a heterogeneous group of proteinaceous bactericidal agents produced by a variety of bacteria and active against many strains of *Escherichia coli.* Their physiological effects have been described in a number of reviews (Holland, 1975; Konisky, 1978, 1982; Konisky and Tokuda, 1979; Kell *et al.,* 1981b; Cramer *et al.,* 1983). Many other bacteriocins have been described, such as butyricin 7423 (see Clarke *et al.,* 1982), which is active against the anaerobe *Clostridium pasteurianum,* that, like a number of the colicins, have the ability to deenergize the cytoplasmic membrane of sensitive cells. Since most of these membrane-active bacteriocins elicit broadly comparable effects upon sensitive cells, we shall confine our comments to those studies which have sought to infer the mechanism by which the target cell membrane is deenergized by membrane-active colicins.

Upon the addition to a sensitive *E. coli* suspension of an appropriate concentration of membrane-active colicin, the following effects, *inter alia,* may be observed: inhibition of certain respiration-linked active transport processes (Fields and Luria, 1969a), a lowering of cellular ATP levels (Fields and Luria, 1969b; Hirata *et al.,* 1969), a rapid efflux of actively accumulated intracellular K^+ (Luria, 1964; Nomura and Maeda, 1965; Dandeu *et al.,* 1969; Feingold, 1970; Wendt, 1970), and inhibition of the energy-linked transhydrogenase (Sabet, 1976). Respiration is unaffected.

From a chemiosmotic standpoint, it seemed reasonable to suppose that these colicins might be protonophorous, since the addition of a protonophorous uncoupler would be expected to inhibit all proton-motivated bioenergetic processes in such organisms. However, it was found that colicin E1, a typical membrane-active colicin, exhibited no protonophorous activity, and its action could not be mimicked by known protonophores (Feingold, 1970; Luria, 1973; Konisky *et al.,* 1975). Further, although uncoupling of proton-motivated energy coupling systems occurred, colicins did *not* act to dissipate the pH gradient across the cell membrane (Brewer, 1976; Weiss and Luria, 1978; Tokuda and Konisky, 1978). This is consistent with other data, discussed earlier, that may indicate that the role of the pH gradient lies (in bacteria and mitochondria) in pH homeostasis and not in energy coupling.

Rosen and Kashket (1978), whose thoughtful article was unfortunately not noted in an earlier discussion of this problem (Kell *et al.*, 1981b), concluded that the available data best fit a model in which the binding of colicin to a membrane component caused the formation of an anion channel, although in some of the experimental conditions used, one might then expect a substantial increase in the pH gradient, which is not observed. Further, the low-molecular-weight anion content of these cells is too low to account, in this model, for a steady-state dissipation of a membrane potential (Gould and Cramer, 1977a), in the absence of such a high ΔpH.

One of the widely discussed features of the action of membrane-active colicins is that they appear to exhibit "single-hit killing," that is, the titer of active colicin necessary to kill an entire cell is one molecule. Actually, the available data (see Holland, 1975; Konisky, 1978) would seem to be just as consistent with oligo-hit killing, but this particular point is not quantitatively crucial to the following discussion. In an influential paper, Schein *et al.* (1979) demonstrated that the addition of colicin K to a planar phospholipid bilayer membrane (BLM) could lead to the formation of a rather unselective, gated, ion-permeable channel. These workers (Schein *et al.*, 1979) further drew together a number of previous observations, including the rates of K^+ efflux observed by Wendt (1970), and suggested that the pathophysiology of colicin K action could be adequately accounted for by the view that it acts merely to create a poorly selective ion channel across the bacterial cytoplasmic membrane. In particular, in a delocalized chemiosmotic model, such a model might simply explain the single-hit killing property discussed above.

This view, which we will refer to as the "orthodox view," has gained a degree of currency such that two recent reviews on bacteriocins have been entitled "Colicins and other bacteriocins with established modes of action" (Konisky, 1982) and "The membrane channel-forming bacteriocidal protein, colicin E2" (Cramer *et al.*, 1983). It is our opinion (see also Kell *et al.*, 1981b) that the orthodox view is a highly premature baby, and our task here is to draw attention to some of the reasons that we think it is unlikely to survive into puberty. First, it is highly dependent upon findings in artificial liposomes and planar BLM. However, the data available even with these systems are not self-consistent, as noted by Cramer *et al.* (1983). In the original work of Schein *et al.* (1979) the BLM conductance was found to increase only if the applied voltage had a polarity that was positive in the compartment to which the bacteriocin was added. This was also observed with colicin Ib (Weaver *et al.*, 1981) and colicin A (Pattus *et al.*, 1983). However, neither the single-channel conductances measured (Weaver *et al.*, 1981; Cleveland *et al.*, 1983) in planar BLM, nor the efflux of radiolabeled substances from liposomes induced by a variety of colicins (Tokuda and Konisky, 1979; Kayalar and Luria, 1979; Uratani and Cramer, 1981; Weaver *et al.*, 1981), was significantly dependent on the polarity of any applied membrane potential. Uratani and Cramer (1981) had to *incorporate* colicin E1 into liposomes in order to see any effect upon their permeability to low-molecular-weight compounds, which is hardly consonant with the situation *in vivo*. The rates of Rb^+ efflux caused by the addition of an unspecified amount of colicin K to liposomes (Kayalar and Luria, 1979) were virtually identical to those of Na^+ efflux elicited by the addition of albumin to liposomes (Kimelberg and Papahadjopoulos, 1971); it is therefore important to relate the amount of bacteriocin used in these *in vitro* experiments to that serving as the minimum lethal dose in intact bacterial cells.

As indicated above, colicins are extremely potent; a typical titer giving $>99\%$ killing might be 100 ng colicin/mg dry weight cells (or per 5×10^8 cells, approximately). The surface area of such a cell mass is well in excess of $10 \, cm^2$ if we (conservatively) model the cells as spheres of diameter 1 μm. In a typical experiment with BLM (Schein *et al.*, 1979) the amount of bacteriocin added was 700 ng for a BLM of area $0.01 \, mm^2$. Thus, it is evident that these *in vitro* experiments have been using amounts of bacteriocin far in excess of those constituting a minimal lethal dose in bacteria. The excellent review by Blumenthal and Klausner (1982) lists the proper criteria for deciding whether a *bona fide* channel is formed in BLM or not. The available data using colicins do not suggest that these criteria are likely to be met with an orthodox result. Thus, given the self-inconsistency of the BLM and liposome data, and the inability of these supposedly poorly selective channel formers to collapse a pH gradient *in vivo*, one is forced to conclude that the primary act of membrane-active colicins *in vivo* cannot be to knock a rather nonselective hole in the lipid bilayer portion of the target cell membrane. This conclusion is greatly strengthened by the studies carried out by a number of groups on mutant strains which are nontrivially resistant to the membrane-active bacteriocins, and which for some reason seem not to be considered by the proponents of the orthodox view.

Parenthetically, some of the studies referred to above as using "colicin K" were actually performed with the comparable membrane-active bacteriocin colicin A (Luria, 1982).

If, as we have indicated here, a simple biophysical insertion of a hydrophobic channel-forming colicin into the target cell membrane does not explain the effects observed *in vivo*, what then might be the target membrane protein of these deenergizing colicins? Hong *et al.* (1977) isolated a mutant of *E. coli* which was resistant to colicin K (but still sensitive to colicin E1) and concluded from its properties and those of revertants that the basis for colicin sensitivity lay in the membrane-located product of a gene termed *ecf*, which they had previously studied (Lieberman and Hong, 1974, 1976; Hong, 1977;

Lieberman *et al.*, 1977; see also Tomochika and Hong, 1978). The properties of these mutants, when transferred to nonpermissive conditions, including a fall in intracellular ATP and efflux of intracellular metabolites, were closely similar to those induced by the addition of colicin K to the wild-type strain. Such mutants were pleiotropically defective in the coupling of respiration to a variety of proton-linked active transport systems and it was concluded that the defect lay in an inability of the mutant to couple respiration-derived energy to active transport and ATP synthesis. The mutation maps (Lieberman *et al.*, 1977) at minute 65 on the recalibrated *E. coli* linkage map (Bachmann, 1983) and is thus quite distinct from the operon coding for the *E. coli* ATP synthase (Gibson, 1983). It should be mentioned that apart from that of Tomochika and Hong (1978) these strains showed no increase in permeability to protons. Unfortunately, it seems that the original isolates have now been lost. Thus, these very elegant studies show quite clearly that for energy coupling to take place in electron-transport phosphorylation and respiration-linked active transport a factor distinct from the primary and secondary proton pumps and from proton leakiness is required.

Plate and his colleagues have described a number of other mutant strains of *E. coli* which, although selected for resistance to the aminoglycoside neomycin, are also rather resistant to colicin K, grow poorly on respiratory substrates, and are impaired in the operation of a number of respiration-linked active-transport systems (Plate, 1976, 1979; Plate and Suit, 1981). The properties of these strains are similar to those of Hong described in the previous paragraph (see also Hengge and Boos, 1983; Booth *et al.*, 1984), with the exception that their *eup* lesion maps at minute 87.5–88. It was concluded by these workers (Plate and Suit, 1981) that the *eup* locus codes for a protein normally required for the coupling of H^+ movements to solute symport. These mutants were also quite unimpaired in their ability to maintain a pH gradient (Plate and Suit, 1981). Physiological and genetic studies are consistent with the view that the *eup* gene is related to the independently studied genes *ssd* (Morris and Newman, 1980; Newman *et al.*, 1981, 1982) and especially *ecf*B (Thorbjarnardóttir *et al.*, 1978). Finally, it was shown (Hitchens *et al.*, 1982) that the ability of an *eup* mutant to exhibit respiration-driven H^+ translocation was unimpaired relative to that of its otherwise isogenic wild-type parent. This latter study confirmed (a) that the lesion lay in an inability to *use* energized protons and (b) that the proton (and other ion) permeability of the mutants, at least from the external bulk aqueous phase, was no different from that of their parents. Studies by C. A. Plate (personal communication) have added some complexity to the simple interpretation discussed herein, and suggest to us that, in the absence of any *eup* genetic locus, yet other proteins may fulfil the role normally played by the *eup* gene product (Kell *et al.*, 1981b). These studies tend to indicate rather suggestively that there is now fairly abundant, if widely scattered, evidence that "energy-transducing membranes normally contain a number of proteinaceous components whose role is to act co-operatively as conformationally switchable proton conductors, permitting fast, controlled, *lateral* proton transfer along the surfaces of such energy-transducing membranes, and acting as the major energetic links between the various protonmotive sources and proton-accepting sinks embedded in such membranes" (Kell et al., 1981b).

We have mentioned uncoupling several times in this section. The orthodox view would assume, since the membrane potential and pH gradients of delocalized chemiosmosis act all over the membrane surface, that the more classical, ionophorous types of uncoupler do not *necessarily* have to act at specific sites in such membranes. More localized views predict the existence of more or less specific sites of uncoupler action. It is therefore important to determine the extent to which such sites may have been demonstrated, a survey which forms the subject of the next section.

VIII. MECHANISMS OF UNCOUPLING

A. Protonophorous Uncouplers

1. Introduction

As is now well known, the addition of low concentrations of any of a number of certain nonphysiological low-molecular-weight chemicals to well-coupled mitochondria respiring under static head (state 4) conditions leads to a dramatic increase in respiratory rate, and to a more or less concomitant loss in the ability to couple the exergonic reactions of electron transport to the endergonic reactions of ATP synthesis. This *respiratory control* is only rarely observed in bacteria (e.g., Scholes and Mitchell, 1970; John and Whatley, 1977; McCarthy and Ferguson, 1982) or in submitochondrial particles (e.g., Hinkle *et al.*, 1975) and is thus probably not crucial to the interpretation of the compounds; what is most important is that they uncouple electron transport from ATP synthesis, and they are thus known as uncouplers. That concentrations of uncoupler necessary to inhibit oxidative phosphorylation in well-coupled mitochondria do not decrease respiration distinguishes them from inhibitors of electron flow or energy-transfer (ATP synthase) inhibitors. Typical phenomena correlative to uncoupling in mitochondria include the inhibition of various exchange reactions such as ATP–$^{32}P_i$ exchange and stimulation of the apparent ATP hydrolase activity (e.g., Heytler, 1979). The concentration of uncoupler required for a half-maximal effect in a given suspension depends upon the reaction being considered, which is consistent with the view that one effect of uncouplers is to cause slip in primary and/or secondary proton pumps and also not easily reconciled with the classical chemiosmotic explanation of respiratory control.

The classical uncoupler is 2,4-dinitrophenol (DNP) (e.g., Loomis and Lipmann, 1948), the curious and interesting history of which is discussed by Racker (1976). One of the most important and (in our view) persuasive predictions of the chemiosmotic hypothesis was that uncouplers such as DNP could catalyze the electrogenic transport of protons, to and from bulk aqueous phases, across biological, and particularly across energy-coupling, membranes, by virtue of the fact that they are lipophilic weak acids or bases. DNP and comparable uncouplers of this type, which are often referred to as classical or *bona fide* uncouplers, are therefore known as protonophores or protonophorous uncouplers. Typical protonophores include DNP, carbonyl cyanide p-trifluoromethyoxyphenylhydrazone (FCCP) (Heytler, 1963), 4,5,6,7-tetrachloro-2-trifluoromethylbenzimidazole (TTFB), pentachloro-phenol (PCP), 5-chloro-3-*tert*-butyl-2'-chloro-4'-nitrosalicylanilide (S-13), bis(hexafluoroacetonyl)acetone (1799), and, the most potent so far discovered (in published literature), 3,5-di-*tert*-butyl-4-hydroxy-benzylidene malono-nitrile (SF 6847) (Muraoka and Terada, 1972; Terada, 1981). Since most investigations on energy coupling use only these compounds (amongst the protonophores) we shall in general concentrate our attention on these compounds. For an overview of protonophores, the reader is referred to five excellent and comprehensive reviews (Hatefi, 1975; Hanstein, 1976a,b; McLaughlin and Dilger, 1980; Terada, 1981).

Given the well-known and abundant evidence that protonophorous uncouplers can indeed catalyze the electrogenic transfer of protons across natural and artificial bilayer membranes from one bulk aqueous phase to another, it might be thought that the mechanism of their action was well understood, and indicative that a delocalized chemiosmotic *coupling* mechanism indeed operates in energy-coupling membranes. However, we shall see that a study of the extensive literature concerning the effects of protonophores on proton motive systems reveals a much more complex state of affairs than may be adduced from the simple view that uncouplers uncouple simply by dissipating the bulk-phase $\Delta\tilde{\mu}_{H^+}$ by catalyzing electrogenic H^+ transfer across the bilayer portions of energy-coupling membranes. Thus we wish to know whether uncouplers uncouple by acting in the bilayer, phospholipid-containing parts of such membranes or whether some uncoupler–protein interaction(s) might be of significance to the uncoupling process. We first draw attention to a strikingly comparable problem that is being earnestly debated in the field of general anesthetics. Since a fuller discussion of this will be given elsewhere, some summary remarks will suffice.

2. A Digression: Where Do General Anesthetics Act?

As is widely understood, the structure–activity relationship among different volatile narcotics (general anesthetics) is weak almost to the point of nonexistence (e.g., Bowman et al., 1969). This and other facts have led to the view that general anesthetics do not act by virtue of specific molecular interactions, but act purely by their possession of one or more biophysical properties (e.g., Halsey, 1974; Miller and Miller, 1975; Kaufman, 1977), in particular their ability, by insertion into lipid bilayers, to expand nerve membranes above a critical volume in such a way as to inhibit neurotransmission (see e.g., Miller et al., 1973; Janoff and Miller, 1982). We refer to the critical volume hypothesis and other cognate hypotheses as *phospholipid hypotheses*. Since, in addition, it has been known for many years that narcotics can interact with soluble proteins (e.g., White, 1974), some workers, notably Franks and Lieb (1978, 1979, 1981), have adduced evidence from a variety of experimental approaches, especially one involving solvent correlations, that the site of narcotic action has both polar and apolar characteristics and may thus be assumed to be partly proteinaceous in nature. Such protein hypotheses were lent especially cogent, even conclusive, support by a very elegant, yet conceptually simple study (Fernandez et al., 1982). This study showed that while chloroform actually *stimulated* the rate of translocation of the lipophilic dipicrylamine ion across the bilayer portion of a squid axon (as expected; see Section VIII,A,3,b), it inhibited the gating currents associated with the opening and closing of the proteinaceous Na^+ channel. Lenaz (1978) also summarized extensive work from his laboratory consistent with the view that the site of general anesthetic action may lie at the interface between lipids and integral membrane proteins. Thus, there is an interesting parallel in the narcosis field with a debate of more immediate interest to us in the uncoupler field, concerning whether the site of action of such xenobiotics in biomembranes is lipidic or proteinaceous in nature. In our view, it may be concluded (see e.g., Hille, 1980; Sandorfy, 1980; Trudell, 1980; Franks and Lieb, 1981; Fernandez et al., 1982; but contrast e.g., Janoff and Miller, 1982) that the bulk of evidence and opinion has swung in favor of the view that the true site of anesthetic action is proteinaceous in nature (Franks and Lieb, 1984; Smith et al., 1984).

To return us to our discussion of uncouplers, we draw attention to a very important article by Rottenberg (1983), who found that the narcotics chloroform and halothane, at significantly supraclinical concentrations, acted to uncouple oxidative phosphorylation in rat liver mitochondria. It was demonstrated, however, that this uncoupling was *not* accompanied by any diminution in the apparent $\Delta\tilde{\mu}_{H^+}$ (judged by ion distribution methods) and was ascribed to "interference with delicate intramembrane processes that mediate direct energy transfer between the electron transport components and the ATPase" (Rottenberg, 1983). One might therefore invoke an interaction between the narcotics and putative protoneural proteins in this system. However, it was not absolutely excluded that the narcotics might induce a slip in the H^+-ATP synthases in an otherwise chemiosmotically operating system. This general topic is considered further in part C of this section.

3. Where do Protonophores Act? Bilayer and Protein Hypotheses

The great complexity of energy-coupling membranes has led many investigators to study simpler model phospholipid bilayer systems, of which the liposome (e.g., Bangham, 1968, 1972; Szoka and Paphadjopoulos, 1980) and the planar black or bilayer lipid membrane (e.g., Jain, 1972; Tien, 1974) systems remain both the most popular and the best characterized. Although, as widely recognized, studies with such systems cannot easily be extrapolated to the situation *in vivo*, many useful findings have emerged.

We take as our starting point the data discussed in the comprehensive review article by McLaughlin and Dilger (1980). Classical uncoupling (for a recent discussion, see Benz and McLaughlin, 1983) by the A^- class of uncouplers is illustrated in Fig. 7a. In general, the correlation (at a given pH

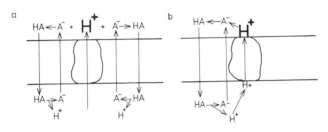

Fig. 7. Mechanisms of action of an A^- protonophore such as FCCP. (a) Classical, macroscopic picture, illustrated with reference to delocalized protonmotive activity by a single membrane-located primary proton pump. The chief feature of the uncoupler is that both charged and uncharged forms are membrane permeable. No special sites of action are required for the transmembrane passage of HA or of A^- during uncoupling, since the putative driving force is delocalized. (b) A more coarse-grained analysis, illustrated with reference to the concept (for a primary proton pump) of localized coupling. Localized charge separation occurs due to the activity of the primary proton pump. A *nearby* A^- molecule can then be driven across the membrane by the local field in a region of high permittivity at the protein–phospholipid interface. This causes the primary proton pump to lose its energization, releasing the proton which had been pumped. The protonophoric cycle is then completed as in (a). Note that the delocalized coupling model *permits* preferential movement of protonophore molecules near *any* membrane protein; localized coupling *requires* the rate-limiting steps of uncoupling to be driven only near the *energized* proteins. Within the framework of localized coupling, uncouplers may also possibly act to cause molecular slip (Westerhoff *et al.*, 1983c, 1984b); this is not illustrated in this figure. For further discussion, see text.

value) between (a) the effectiveness of uncoupling in rat liver mitochondria and (b) the ability of a compound to increase the conductance of BLM (which is generally ohmic up to supraphysiological values of transmembrane voltages), is extremely good (Skulachev, 1971; McLaughlin and Dilger, 1980). An important apparent exception in this context, however, is picric acid.

a. *Uncoupling by Picric Acid.* Hanstein and Hatefi (1974a) found that picrate could not uncouple mitochondria but was a good uncoupler of some submitochondrial particles (SMP, which are topologically inverted). They proposed that picrate was a (relatively) membrane-impermeant substance that exerted its effect by binding to the matrix side of the inner mitochondrial membrane. However, McLaughlin and colleagues (1978) showed that picrate is indeed protonophoric and permeates BLM in both the neutral and charged forms. They pointed out that since the presumed membrane potentials of mitochondria and SMPs are of opposite polarity, consideration of a typical kinetic model for an A^--type uncoupler can lead, in fact, to an analysis quite consistent with the chemiosmotic model. However, there are some severe quantitative problems with this view. In the model of McLaughlin *et al.* (1978) detectable uncoupling by picrate (10 μM) in SMP is observed (in the model) under conditions in which the bulk phase membrane potential was taken to be 175 mV and the pH gradient 1 unit. A similar model was used by Michels and Bakker (1981) in *E. coli*. However, it would seem that (a) the bulk-to-bulk phase proton permeability induced across the membrane of SMPs by this concentration of picrate is at least 1.5 orders of magnitude too low to account for the observed uncoupling in strictly chemiosmotic terms (Hanstein and Kiehl, 1981) and (b) uncoupling concentrations of picrate (10 μM) *do* decrease the membrane potential rather markedly, as judged by oxonol-VI fluorescence in SMP (Hanstein and Kiehl, 1981) or by SCN^- uptake in everted *E. coli* vesicles (Michels and Bakker, 1981). Similar remarks may be made concerning the nonprotonophorous tetraphenyl borate ion (Phelps and Hanstein, 1977), which is a significantly *more* potent uncoupler than is picrate in everted *E. coli* vesicles (Michels and Bakker, 1981). Thus examples seem to exist which clearly indicate that results in BLM are not easily extrapolated to native energy-coupling membranes, and that, by implication, some kind of uncoupler–protein interaction may be of importance in causing uncoupling.

b. *Effect of Dielectric Permittivity on Uncoupler Fluxes.* McLaughlin and Dilger (1980) drew attention to two interesting anomalies between the protonophorous properties of weak acid uncouplers in BLM and their ability to uncouple in mitochondria. The first is that although the correlation between these two types of property among different compounds is generally excellent, weak acid protonophores are two orders of magnitude more effective in mitochondria than in BLM. The second is that a weak acid for which the bilayer conductance depends quadratically on the carrier concentration (i.e., an HA_2^--type protonophore) produces an uncoupling effect that depends linearly upon concentration (Dilger and McLaughlin, 1979). It was proposed (Dilger and McLaughlin, 1979; McLaughlin and Dilger, 1980) that the apparent anomalies are due to "a region of the bilayer component of the mitochondrial membrane having a higher dielectric constant than that of an artificial bilayer (p. 379)." We consider this extremely important and suggestive idea in some detail.

It is both reasonable and usual to treat a BLM as a homogeneous condenser of static permittivity ε_2; however, the extent to which this macroscopic simplification retains its utility in the heterogeneous milieu of a biomembrane is far from clear.

As indicated by a number of authors (e.g., Läuger and Neumcke, 1973; Andersen and Fuchs, 1975; Parsegian, 1975; McLaughlin and Dilger, 1980), the Born energy W required to transfer a monovalent, spherical, non-polarizable ion of charge e and radius a from an aqueous phase of permittivity ε_1 into the center of a bilayer of permittivity ε_2 and thickness d is:

$$W = \frac{e^2}{8\pi\varepsilon_0 a}\left(\frac{1}{\varepsilon_2} - \frac{1}{\varepsilon_1}\right) - \frac{e^2}{4\pi\varepsilon_0\varepsilon_2 d}\ln\left(\frac{2\varepsilon_1}{\varepsilon_1 + \varepsilon_2}\right), \tag{4}$$

where ε_0 is the permittivity of free space. Thus the Born energy depends both on the thickness and the permittivity of the membrane. It was most elegantly shown (Dilger and McLaughlin, 1979; McLaughlin and Dilger, 1980; Benz and McLaughlin, 1983), by using the somewhat polar chlorodecane in the BLM-forming procedure, that the protonophore-induced conductivity of the BLM could be raised by approximately two orders of magnitude; changes in bilayer thickness played only a minor role. Further, under these conditions, the HA_2^- uncoupler DTFB (Cohen et al., 1977) indeed behaved as an A^--type protonophore, since, from the Born charging equation, the relative decrease in Born energy induced by increasing the BLM permittivity will be greater for the smaller A^- ion. Similar observations on the membrane permeability were made using the membrane-permeable SCN^- and ClO_4^- ions (Dilger et al., 1979; McLaughlin, 1981).

Now, it has of course been known for many years (see e.g., Pauly and Packer, 1960; Packham et al., 1978; Pethig, 1979; Harris and Kell, 1983) that the *macroscopic* static permittivity of biomembranes is greater than that of BLM of comparable thickness by a factor of approximately two. Let us then take a coarse-grained approach (Fig. 7b). Given the great heterogeneity of energy-coupling membranes, it may then be stated quite clearly that, in view of the foregoing considerations alone, protonophores will tend to act preferentially in the regions of highest permittivity, that is, adjacent to membrane proteins. This does not of itself require that *specific* interactions with proteins are part of the uncoupling process, nor does it, for instance, indicate of itself that the high-energy intermediate is confined to these preferential regions. To examine these questions, we must turn away from work with BLM.

c. *Uncoupler-Binding Proteins in Energy-Coupling Membranes.* We saw in the last section that there tended to be a preferential partitioning of charged uncoupling molecules into regions of highest local permittivity. Might some *specific* protein–uncoupler interactions be of significance to the uncoupling process? The structure–function relationships of homologous protonophores (e.g., Terada, 1981) do not lend much support to the idea that factors other than hydrophobicity and acid dissociation constant play a significant role in inducing uncoupling activity, so that steric effects, which might also play a role in putative protein–uncoupler interactions, are assigned a rather minor role (but see Büchel and Draber, 1972).

Now it is, of course, well known that uncouplers *can* interact with a great many proteins, and it is worth pointing out that delipidation of mitochondria with $CHCl_3$-methanol has a negligible effect upon the ability of mitochondrial proteins to bind a number of uncouplers (Weinbach and Garbus, 1965, 1969; Hanstein et al., 1979). In general, energy-coupling membranes can bind much more uncoupler than is necessary to give maximal uncoupling, so that the question arises as to what fraction of the total bound uncoupler is actually uncoupling.

Hanstein and Hatefi (1974b) initiated an important approach to this question by synthesizing 2-azido-4-nitrophenol (NPA), a photoaffinity analogue of DNP with an uncoupling potency two to three times greater. In the dark, NPA bound to mitochondria both specifically and nonspecifically, the number of binding sites approximating one per ATPase. The specific binding was noncooperative and independent of the degree of energization of the mitochondria, and was competitively inhibited by other uncouplers such as DNP, PCP, CCCP, and S-13. When radiolabeled NPA was photolyzed in the presence of mitochondria, more than 90% of the label was associated with protein, and predominantly with (a) the F_1-ATPase and (b) one or more hydrophobic proteins of ~ 21 kDa, located in or near the F_o-ATPase, and referred to as the uncoupler-binding protein (Hanstein, 1978; Hanstein et al., 1979; Drosdat et al., 1982). This protein is located in carefully prepared complex V, an $F_o F_1$-ATPase preparation with *high* ATP-P_i exchangease activity (Hatefi, 1975; Kiehl and Hanstein, 1981; cf. Berden and Hennecke, 1981).

The photoaffinity labeling strategy was adopted by Kurup and Sanadi (1977) using NPA and by Katre and Wilson (1977, 1978, 1980), who synthesized 2-nitro-4-azidocarbonyl cyanide phenylhydrazone (N_3CCP). These latter workers obtained broadly comparable results to those of Hanstein and colleagues (op.cit.). However, the polypeptides labeled with photolyzed N_3CCP apparently differed considerably from those labeled by NPA, raising the possibility that "... the ability of the weak acid protonophores to bind to proteins, and indeed to any hydrophobic surface, may be an epiphenomenon unrelated to their mechanism of action as uncouplers" (McLaughlin and Dilger, 1980). Competition between N_3CCP and NPA does not yet seem to have been tested, nor do protease inhibitors seem to have been included in the labeling experiments. However, it is probable that N_3CCP, being more hydrophobic, is more catholic in its binding to membrane proteins than is the relatively hydrophilic NPA (Hanstein, 1978). Since mitochondria covalently labeled with N_3CCP were completely uncoupled (Katre and Wilson, 1978), it seems improbable that the compound could have been acting as a mobile protonophore under these conditions. Nevertheless, it is probably fair to conclude that the very interesting observations made with these compounds do not alone yet allow us to assess the exact importance of the uncoupler-binding protein(s) in the mechanism of uncoupling by protonophores.

d. Resistance to Uncouplers. Resistance of mutant strains of biological cell systems to typical protonophores has so far been reported in *Bacillus subtilis* (Decker and Lang, 1977, 1978; Lang and Decker, 1978; Guffanti *et al.*, 1981), in *Escherichia coli* (Ito and Ohnishi, 1981; Ito *et al.*, 1983), and in the mitochondria of yeast (Griffiths *et al.*, 1972, 1974) and of Chinese hamster cells (Freeman *et al.*, 1980, 1983). In each of these studies, it was shown that the resistance was nontrivial, and the mutation seems to affect a protein lying in or near the F_o part of the ATPase as judged by cross-resistance to F_o inhibitors, increased ATP hydrolase activity, etc. In some cases (e.g., Griffiths *et al.*, 1972; Freeman *et al.*, 1980, 1983) mutant cells are as much as 10 times more resistant to selected protonophores than their wild-type parents, and Hitchens and Kell (unpublished observations, 1982) obtained similar data in *Rps. capsulata*. It seems, therefore, that *some* type of specific protein–uncoupler interaction, albeit transient (Hitchens and Kell, 1982c, 1983a,b), must be of functional significance in determining uncoupling by protonophorous uncouplers. This conclusion would be greatly strengthened by comparison of the titer of uncoupler-binding protein in various strains with the degree of resistance they possess to appropriate uncouplers. It may also be worth drawing attention to the ability of a variety of respiratory bacteria to grow on pentachlorophenol (Rao, 1978).

It has been proposed that uncouplers may cause redox-linked proton pumps to exhibit exacerbated slip (Pietrobon *et al.*, 1981, 1982; Walz, 1983), a proposal which is consistent with uncoupler titrations discussed above. These proposals (Pietrobon *et al.*, 1981, 1982) were based upon experiments measuring the apparent protonmotive force under various conditions. The general conception of slip has been further strengthened (Pietrobon et al., 1983) in experiments which contraindicated a role for a non-ohmic leak across the inner mitochondrial membrane. However, the latter experiments showed that since slip is not *only* a function of the apparent proton motive force then the energy coupling should not be presumed to be proceeding *via* the apparent proton motive force either.

It may be concluded that uncoupling does require some transient, specific, protein–uncoupler interactions. The substoichiometry of uncoupler–ATP synthase "binding" is simply explained by the fast *lateral* mobility of uncouplers (Hitchens and Kell, 1982c, 1983a,b; Kell and Hitchens, 1983).

B. Ionophorous Uncouplers

Most of what we have written in the previous section may be applied to the ionophorous type of uncoupler. For reasons of space, we will not give a detailed exposition, save to note that the synergism often found in uncoupling by electrogenic and electroneural ionophores reflects simply a requirement for a replenishment of the inner phase of the membrane vesicle suspension with the appropriate ion. The uncoupling step could be ascribed to a dissipation of the *localized* fields set up across the membrane in response to proton motive activity (Fig. 7b). Far more interesting from our point of view are a number of compounds, which we shall refer to as *decouplers*, which exhibit certain uncoupling properties yet which are not protonophorous.

C. Decouplers

A number of compounds have been described which, while not energy-transfer (ATPase) inhibitors, have the ability to inhibit ATP synthesis, to inhibit state 3 but not state 4 respiration, but not to release respiratory control in the sense that protonophores do. There does not seem to be an adequate role for such compounds in the chemiosmotic model, and they are thus generally ignored. We may mention biguanides (Schäfer and Rowohl-Quishoudt, 1975; Schäfer, 1981), fluorescein mercuric acetate (Southard et al., 1974), tetraphenyl borate (Phelps and Hanstein, 1977; Michels and Bakker, 1981), and a variety of lipophilic cations often used in attempts to measure the membrane potential across energy-coupling membranes (Higuti et al., 1978, 1980, 1981, and references therein; Zaritsky and MacNab, 1981). These types of compound, which should be distinguished from certain other types with comparable properties except that uncoupling *is* sensitive to ATPase inhibitors (Mai and Allison, 1983), also, in most cases, have the effect of increasing the number of H^+ translocated in the bulk phase external to the system. Authors who have worked with this type of compound have offered a variety of interpretations of their findings. Just as with Rottenberg's (1983) study with anesthetics, however, we wish to stress the probability, discussed in more detail elsewhere (Kell and Hitchens, 1982), that this type of compound is interfering with the operation of the protoneural proteins, an interpretation greatly strengthened by work with an azido derivative of ethidium bromide (Higuti *et al.*, 1981) demonstrating specific labeling of certain hydrophobic proteins in the inner mitochondrial membrane. Unfortunately, however, the protein(s) identified by Higuti *et al.* (1981), which are distinct from the so-called DCCD-binding protein, have not yet been further characterized, although it may be mentioned that the apparent molecular weight given in the paper of Higuti *et al.* (1981) does not correspond to those putatively encoded by the URFs in the bovine mitochondrial genome (Anderson *et al.*, 1982).

D. Conclusions and Summary of Uncoupling Mechanisms

The more classical protonophorous types of uncouplers seem to require transient interactions with, *inter alia*, proteins in or near the F_o portion of the H^+-ATP synthase in energy-coupling membranes. Such interactions are required in some localized coupling theories, which are thus consistent with the available data. A variety of *nonprotonophorus* compounds also "uncouple" in some way; most workers believe that they interfere with the transfer of electron transport-derived free energy to the proton motive sinks, but that they cannot do so by affecting $\Delta\tilde{\mu}_{H^+}$, and they are not energy-transfer inhibitors *sensu stricto*. They could function by inhibiting the conformational transitions of the protoneural proteins which may be required in energy coupling.

IX. CONTROL THEORETICAL VIEW OF ENERGY COUPLING

Presently there is a revival of interest (see e.g., Groen *et al.*, 1982; Porteous, 1983; Tager *et al.*, 1983; Westerhoff, 1983; Westerhoff *et al.*, 1983b, 1984a; Westerhoff and Van Dam, 1985) in the metabolic control theories of Kacser and Burns (1973) and Heinrich and Rapoport (1974). One important application is the analysis of which enzymes control mitochondrial respiration. Groen and colleagues (1982b) demonstrated that this process is not controlled by one enzyme alone, but by at least three of the participating enzymes. Moreover, the control (coefficient) (or flux control) of respiration by, for instance, the adenine nucleotide translocase varied with the work load imposed on the mitochondria.

A crucial aspect of this approach is that the flux control coefficient is a mathematically defined, but experimentally readily determined, parameter indicating the percentage of the control on a metabolic flux exercised by a certain enzyme (for review, see Westerhoff *et al.*, 1984a). One operational definition of this parameter is the percentage by which the flux is reduced upon a 1% inhibition of that enzyme. Importantly, the sum of the flux control (coefficient)(s) by all the enzymes exercised on a given flux, must, in this theory, equal one. As a consequence, flux control is shared among enzymes, and enzymes compete for flux control. This competition is the basis for the principle that when one enzyme in a system becomes rate-limiting [i.e., exercises a flux control (coefficient) close to one] the other enzymes must sacrifice their share of the flux control. Since partial inhibition of an enzyme usually increases the flux control (coefficient) by that enzyme, such inhibition is expected to reduce the flux control (coefficient) by the other enzymes in the system and thus the titers of inhibitors of those other enzymes. It is the latter principle that underlies the predictions of the results of dual inhibitor titrations of free energy transduction by the hypothesis of delocalized chemiosmosis [see Section IV,E and Fig. 5b(ii)].

Here we reach a point of special significance to the present volume. With respect to dual inhibitor titrations, the predictions of the mosaic protonic coupling hypothesis are at variance with those of delocalized chemiosmotic coupling and, as we have just seen, also with the predictions of control theory. Moreover, the experimental results are in line with the predictions of the hypothesis of mosaic protonic coupling (Section IV,E).

Control theory models metabolic systems exclusively as consisting of enzymes that are mutually independent, except through their interactions with common metabolites, which form pools. The pool property of the metabolites implies that a metabolite molecule is not confined to react with one individual enzyme molecule, but can react with all enzymes in the system with a given catalytic specification (e.g., an ATP molecule can react with all the H^+-ATPase molecules present in the system). In such a system, a 1% increase in the activities of all the enzymes will lead to unaltered concentrations of all metabolites and, consequently, to a steady state that is identical to the original one, except that all fluxes will have increased by the same 1%. Consequently, the sum of all flux control coefficients in such a system is equal to one (as already stated above).

A mosaic coupling system does not conform to this modeling of a metabolic system: each energized proton can only react with one H^+-ATPase molecule. Consequently, one may wonder whether the (summation) theorem that the sum of the flux control coefficients must equal one should hold in mosaic systems (Groen *et al.*, 1982; Kell and Hitchens, 1983). In fact we can easily show that this law is not valid in a mosaic coupling system: from Eq. (3) it follows that a 1% inhibition of the electron-transfer chains or 1% inhibition of the H^+-ATPase *both* lead to a 1% inhibition of ATP synthesis, that is, both enzymes exercise a flux control (coefficient) of one, so that the sum of the flux control (coefficient)s is close to two, rather than to the one following from control theory. [It may be noted that this (in contrast to the case treated in Westerhoff and Arents, 1984) is not due to the presence of a leak with a control coefficient of minus one.]

In principle, an assessment of whether the magnitude of the sum of the flux control coefficients exceeds one can constitute an important criterion for localized, or mosaic, versus delocalized free-energy transduction or metabolism. Promising as it seems, this approach has not yet been taken by many authors. It may, however, be noted that in mitochondrial oxidative phosphorylation there was in fact a tendency for the total flux control to exceed one (Groen et al., 1982b). A similar, though different, example is found in the work of Reddy and Pardee (1983), in which a variety of specific inhibitors of enzymes isolated from an E. coli replicase complex acted additionally to inhibit other enzymes in the replicase complex.

From the above it will be clear that if localization effects are indeed present (Welch, 1977; Clegg, 1983a,b) in metabolism, they require thoughtful application of the metabolic control theories (cf. Kell and Hitchens, 1983; Westerhoff et al., 1983b): if two enzymes are operating as a coupling unit, then their flux control (coefficients) should be averaged rather than summed when summing all flux control coefficients in the system, if one wishes to establish a total flux control of one.

The implications of localization effects for the control of fluxes in biological systems are extremely important: rather than having to share the control, so that each enzyme can only control any flux to a small extent, the enzymes in a completely mosaic system all have a complete control (i.e., exercise a flux control coefficient of one) over their "own" flux (cf. Williams, 1978b; Westerhoff et al., 1983b). If, for instance in a bacterium, the transport rate of a given substance is increased, because the concentration of that substance in the external medium increases, then the use of free energy for this transport process would, in the case of delocalized chemiosmotic coupling, lead to an inhibition of all the other free-energy requiring process, because $\Delta\tilde{\mu}_{H^+}$ would drop. In the case of mosaic protonic coupling, the other free-energy-requiring processes would not necessarily be affected.

From a biotechnological standpoint, however, the metabolic control theories make one important point: if one wishes to maximize the flux through a metabolic pathway, there may be *more than one* bottleneck (rate-limiting step), and strategies to enhance productivity, even if semiempirical, should be directed at a variety of control points. A good example is given in a recent discussion of strain improvements during the development of commercial lysine fermentations (Tosaka et al., 1983).

X. ON THE "ORGANIZATION" OF ENERGY-COUPLING MEMBRANES AND OTHER ORGANIZED MULTIENZYME SYSTEMS

There is an epoch in the growth of a science during which facts accumulate faster than theories can accommodate them.

Medawar (1982)

A. Introduction

The foregoing, rather extensive survey of current knowledge of electron-transport phosphorylation has, we regret, been somewhat iconoclastic in nature. This is due, of course, to the still widespread belief in the simplest, delocalized chemiosmotic scheme that we set up as a straw man at the beginning of our survey. Those readers who are still with us may justifiably wonder, therefore, where we can go next: obviously, we must endeavor to provide some kind of more explicit model that accounts for the data we have thus far discussed.

As foreshadowed in Section VI, we find it necessary to modify the delocalized chemiosmotic view in the sense that, while we retain the proton pumps, their mutual activity is coupled via protons (in concert with some kind of membranous phonon wave) that do not equilibrate with the proton electrochemical potential in the bulk phases to which the given faces of the coupling membrane are adjacent. We have sought to adduce evidence that there exist, in these coupling membranes, proteinaceous devices whose role is to transfer the protonic free energy along the plane of the membrane, although we are not yet in a position to speculate explicitly upon their exact biophysical mechanism. Let us remind ourselves of the salient structural features of these energy-coupling membranes.

The key feature is that, although these membranes are organized as fluid mosaics, there is now a significant amount of evidence that the disposition of proteins involved in free-energy transduction, although they are isolatable as individual complexes, is such that the lateral movement of these complexes is not easily viewed as occurring by free diffusion under conditions approximating those *in vivo*.

We have seen that even a single turnover of the primary, redox-linked proton pumps causes an initial, localized charge separation across the plane of the membrane. The energetic magnitude of this field may be calculated from simple electrostatic considerations (Malpress, 1981a,b) (Section VI) and turns out to be more than adequate to permit ATP synthesis given reasonable assumptions about the magnitude of the "local" permittivity. Such a static field will affect the conformational state of any dipoles to which it is adjacent, the change in free energy being dependent, *inter alia*, upon the dipole moment of the induced protein in a very simple fashion (Schwarz, 1978a,b). Since typical energy-transducing membrane proteins have very large dipole moments, typically in excess of 400 D (e.g., Petersen and Cone, 1975; Keszthelyi, 1980; Todorov et al., 1982; Tsuji and Neumann, 1983), it is easy to see how localized fields could transfer free energy directly between spatially adjunct protein complexes. However, life is not so simple; let us consider bacteriorhodopsin, with which three of these studies were concerned, in more detail.

It is assumed, in the chemiosmotic coupling model, that the steady-state proton electrochemical potential difference across a working proton pump itself is as great as that between the two bulk phases to which the surfaces of the proton pump are adjacent. In the case of isolated bacteriorhodopsin sheets, these two aqueous phases are one. Two very interesting studies measured the average surface potential of such sheets (Carmeli et al., 1980; Tokutomi et al., 1980) using a spin-labeled probe. As expected, the surface potential measured with this method varied, in the dark, in exact accordance with the requirements of Gouy–Chapman theory, as the ionic strength was varied. However, the *light-dependent changes* in surface potential were *ionic strength independent*, under macroscopically stationary-state conditions. This indicates that free energy is stored in the protein under illuminated conditions (Kell and Griffiths, 1981) and is thus not simply dissipated when the protein is taken from the otherwise insulating membrane in which it is normally embedded. Thus, during the photocycle, as elaborated in more detail elsewhere, this "energized" protein will have the effect of altering the standard chemical potential of any ligand with which it happens to interact (see Welch and Kell, 1985; Somogyi et al., 1984). Welch and Berry (1983 and this volume) have considered the consequences of such behavior for modeling the energy metabolism in cells.

Consider for a moment a protein isolated in a heat bath, in thermal equilibrium with its surroundings. Such a protein might be an isolated bacteriorhodopsin sheet. The addition of energy to this system, let us say, for bacteriorhodopsin, by absorption of a photon of wavelength 570 nm, evidently leads to some kind of nonequilibrium state of the protein system, which may then relax "forward" to the ground state via a catalytic cycle (see e.g., Westerhoff, 1983; Westerhoff and Dancshazy, 1984). In bacteriorhodopsin, the half-time for this relaxation is of the order of a few milliseconds, although certain treatments may prolong this for as much as 30 s. Nevertheless (Fig. 1), as discussed, for instance, by Careri et al. (1979), many relaxational modes, such as H^+ transfer and thermal exchange with bound and free solvent molecules, take place on a time scale of nanoseconds to microseconds. How, then, can an energized protein escape a much more rapid relaxation to its ground state via a back-reaction? Obviously, one possible mechanism is that of a cascade of chemical reactions with rate constants of 100/s and greater. An indication for this is the occurrence of a succession of different spectral forms of photoexcited bacteriorhodopsin (Lozier *et al.*, 1975). It should be noted, of course, that visible spectroscopy probes only the environment of the chromophoric retinal group itself, while knowledge of the conformational state(s) of the entire system is required for true thermokinetic analyses. It cannot therefore be excluded that the actual mechanism is much less classical than this.

A number of workers (see e.g., Del Giudice *et al.*, 1982; Blumenfeld, 1983; Scott, 1983; Davydov, 1983, and references therein) have pointed out that there exist a variety of methods of transferring such free energy by the formation and utilization of solitary excitations, which exhibit greatly increased lifetimes and a correspondingly increased ability to transport free energy over biologically significant distances. The formation of such solitary waves is absolutely dependent upon the nonlinear character of interatomic forces between different parts of the system of interest. In this regard, we would draw attention to a long-standing general model of Fröhlich (see Fröhlich, 1980, for a review) concerning the possibility of long-distance energy transport (coherent excitations) in biology and dependent upon appropriate nonlinear interactions between biomembrane proteins and their surrounding heat bath.

In particular, the considerations of Fröhlich (1980, and references therein) offer us an escape route from the dilemma of why an energized charged membrane protein may not fulfil the requirements of Gouy–Chapman theory. Fröhlich considered the relationship between the probable velocity of sound (phonons) in a membrane protein (~ 1000 m/s) and a typical membrane thickness of 10^{-8} m, concluding that interactions between the electrical and acoustic modes of such a charged, transmembrane protein might have a frequency of the order of 10^{11} Hz. At these frequencies, electrostatic screening is of course impossible. Thus, in contrast to what might be expected from static considerations alone, these "unexpected" findings are perfectly compatible with well-known physical laws.

Thus, as Scott (1983) and others have noted, one should expect to see important changes upon energization in the spectra of proteins (such as bacteriorhodopsin) in the range 10^{11} to 2×10^{12} Hz. It is probably also worth remarking that the conceptual subdivision between (a) the activities of the *primary* proton pump(s) and (b) the subsequent long-range transfer of free energy along the coupling membrane bear striking analogies to (i) the initial and (ii) later processes discussed in the generation of ferroelectric states in this context by Bilz *et al.* (1981). As discussed elsewhere (Kell and Hitchens, 1983), this type of general model is fully consistent with the data presently available on the systems catalyzing electron-transport phosphorylation. However, we will not pursue these issues further here. Our purpose is thus only to point out the possibility of the excitation of nonlinear transitions in vibrational modes of membrane proteins during proton motive activity. We may therefore, in contrast to our earlier, iconoclastic survey, allow ourselves a rather more optimistic conclusion: that the recognition of these possibilities may usher in a new and exciting phase of research in membrane bioenergetics.

B. Biotechnological Considerations

The foregoing discussion has concentrated on what we may refer to as "pure scientific" aspects of membrane bioenergetics. In the spirit of this volume, however, we wish finally to mention some consequences of the behavior of such membranous systems for work with immobilized cells and artificial enzyme membranes.

The most striking consequence of metabolic microcompartmentation may be stated as follows. Imagine a permeabilized, immobilized microbial cell catalyzing a multistep reaction such as glycolysis. The volume taken up by the enzymes in relation to the total volume is negligible. Thus, if each intermediate is allowed to become "free," so that its effective activity approximates the number of molecules divided by the total volume of the system, the transit time of the system will be enormous, and the number of substrate molecules existing as intermediates, rather than products, might form a substantial fraction of the whole. However, if the enzymes are arranged so that intermediate molecules remain bound to the multienzyme complexes, this number will be equal only to the number of enzymic active sites present. Evidently, the latter situation is far more satisfactory from a biotechnological standpoint, and an understanding of this type of behavior *in vivo* is of great importance to the design of synthetic and analytical systems based upon immobilized biocatalysts. Kasche (1983) has reviewed the consonance between the behavior of natural and artificial systems in the diffusion-limited regime and found it excellent, and the modulation of immobilized enzyme activities by microenvironmental effects is discussed by several other authors in this volume. The improved control structures possible in mosaic systems have been discussed in Section IX. From the standpoint of the role of cellular membranes in biotechnology, it is worth drawing attention to their role in slowing the egress of products, both of low and high molecular weight. A classic example of this occurs in the glutamic acid fermentation catalyzed by *Cornyebacterium glutamicum*; growth of this organism under biotin-limited conditions markedly enhances the membrane permeability of this organism and glutamate productivity of this fermentation (see e.g., Dulaney, 1967). The means by which this organism catalyzes oxidative phosphorylation under these membrane-leaky conditions does not, however, seem to be a point of discussion in the biotechnological literature.

Our final purpose is therefore to consider one neglected possible consequence of the coherent behavior of cellular metabolism and multienzyme aggregates.

C. Thermophilic Microorganisms

The idea explicit in the use of the term *coherent excitations* in the previous section is that the motions of particles in one part of a system are dependent upon the motions of particles in other, spatially separate parts of the system, so that there is thus a functional linkage between the two.

Now, as pointed out, for instance, by Klibanov (1983), by far the most important means of enzyme inactivation is thermal, and Lapange (1978), Mozhaev and Martinek (1982), and Sonnleitner and Fiechter (1983) have given useful reviews of the thermally based routes of enzyme denaturation/ inactivation. Thus, a great many workers have sought to exploit the greater thermostability shown by most, *though not all*, enzymes isolated from thermophilic microorganisms in biotechnological applications. From a bioenergetic standpoint, an excellent example is provided by Kagawa's studies on the H^+-ATPase synthase of the thermophilic *Bacillus* PS3 (see, e.g., Kagawa, 1979).

Nevertheless, a consideration of the literature on thermophily (see, e.g., Amelunxen and Murdock, 1977, 1978; Kushner, 1978; Shilo, 1978; Friedman, 1978; Zeikus, 1979) reveals that although a variety of modifications in the structure of isolated macromolecules have been exploited by thermophiles, none offers an easy explanation for the remarkable (Baross and Deming, 1983) thermostability exhibited *in vivo*. Qualitatively, it is obvious that the organization of enzymes into multienzyme aggregates, involving the formation of extra noncovalent bonds, can in principle greatly increase the thermostability of the individual enzymes, provided that the enzymes in the complexes act as mutual thermal buffers. Thus, it would be predicted from such considerations that the thermostability of the enzymes and other macromolecules in thermophiles rests additionally upon the organization of the *milieu intérieur* of such organisms. How might one seek to test such a notion? It has been known for many years (e.g., Pauly and Schwan, 1966; Marquis and Carstensen, 1973) that the internal electrical conductivity of a variety of cells is significantly less, usually between one-third and one-half, of that expected (and found, upon cell disruption) on the basis of the ionic content of the cells. We may refer to this finding as the conductance deficit. It seems, then, to be a logical corollary of the foregoing considerations that the conductance deficit should be greater in thermophiles than in comparable mesophiles if the functional linkage between their enzymes is greater than that expected on the basis that all are freely diffusing in the membranes or cytoplasm of such cells. Unfortunately, the relevant experiments to test this conception do not yet appear to have been performed.

D. Concluding Remarks

One could, without too much exaggeration, say of review articles what Mark Twain is supposed to have said about the weather—everybody talks about it but nobody *does* anything about it.

Garfield (1977)

The evidence for catalytic facilitation through organizing single enzymes into functional multienzyme aggregates is now very great, as discussed by other contributors to this volume. We hope that we have been able to show that the same holds true for enzymes catalyzing free energy transduction in biomembranes. While we have largely confined ourselves to the process of electron-transport phosphorylation, the interpretations we have given may be taken to apply to a variety of other processes occurring in energy-coupling membranes. The consequences of this for an integrated understanding of cellular metabolism and bioenergetics will provide a most interesting phase in the evolution of our understanding of the behavior of such systems.

ACKNOWLEDGMENTS

As will be obvious to all who read this, we have benefitted enormously from many stimulating discussions with all our scientific colleagues, who are collectively too numerous to thank explicitly. We wish, however, to acknowledge in particular enlightening discussions with Drs. Licio Azzone, Jim Clegg, Stuart Ferguson, Andrea Melandri, Catia Sorgato, Karel van Dam, Giovanni Venturoli, and Rick Welch. DBK is indebted to the Science and Engineering Research Council, UK, for financial support through the previous eight years' evolution of his thinking, and to the SERC Biotechnology Directorate for current financial support. We are extremely grateful to Jane Watts for her unfailing assistance in the preparation of this manuscript.

REFERENCES

Amelunxen, R. E., and Murdock, A. L. (1977). *Dev. Ind. Microbiol.* 18, 233–248.
Amelunxen, R. E., and Murdock, A. L. (1978). *CRC Crit. Rev. Microbiol.* 6, 343–393.
Anderson, J. M. (1975). *Biochim. Biophys. Acta* 416, 191–235.
Anderson, J. M. (1981). *FEBS Lett.* 124, 1–10.
Anderson, J. M., and Andersson, B. (1982). *Trends Biochem. Sci.* 7, 288–292.
Andersen, O. S., and Fuchs, M. (1975). *Biophys. J.* 15, 795–830.
Anderson, S., Bankier, A. T., Barrell, B. G., de Bruijn, M. H. L., Coulson, A. R., Drouin, J., Eperon, I. C., Nierlich, D. P., Roe, B. A., Sanger, F., Schreier, P. H., Smith, A. J. H., Staden, R., and Young, I. G. (1981). *Nature (London)* 290, 457–470.
Anderson, S., de Bruijn, M. H. L., Coulson, A. R., Eperon, I. C., Sanger, F., and Young, I. G. (1982). *J. Mol. Biol.* 156, 683–717.

126   DOUGLAS B. KELL AND HANS V. WESTERHOFF

Anderson, W. M., Fowler, W. T., Pennington, R. M., and Fisher, R. R. (1981). *J. Biol. Chem.* **256,** 1888–1895.

Andersson, B., and Haehnel, W. (1982). *FEBS Lett.* **146,** 13–17.

Archbold, G. L., Farrington, C. L., Gill, B. J. O., and Malpress, F. H. (1974). *Biochem. Soc. Trans.* **2,** 751–754.

Archbold, G. P. R., Farrington, C. L., Lappin, S. A., McKay, A. M., and Malpress, F. H. (1979). *Biochem. J.* **180,** 161–174.

Arntzen, C. J. (1975). *In* "Bioenergetics of Photosynthesis" (Govindjee, ed.), pp. 51–113. Academic Press, New York.

Arntzen, C. J. (1978). *Curr. Top. Bioenerg.* **8,** 111–160.

Arvidson, G., Lindblom, G., and Drakenberg, T. (1975). *FEBS Lett.* **54,** 249–252.

Attardi, G., Cantatore, P., Chomyn, A., Crews, S., Gelfand, R., Merkel, C., Montoya, J., and Ojala, D. (1982). *In* "Mitochondrial Genes" (P. Slonimski, P. Borst, and G. Attardi, eds.), pp. 51–71. Cold Spring Harbor Laboratory, Cold Spring Harbor, New York.

Axelrod, D. (1983). *J. Membr. Biol.* **75,** 1–10.

Azzi, A. (1980). *Biochim. Biophys. Acta* **594,** 231–252.

Azzone, G. F., Massari, S., and Pozzan, T. (1977). *Mol. Cell. Biochem.* **17,** 101–112.

Azzone, G. F., Pozzan, T., and Massari, S. (1978). *Biochim. Biophys. Acta* **501,** 307–316.

Azzone, G. F., Pietrobon, D., and Zoratti, M. (1984). *Curr. Top. Bioenerg.* **14,** 1–77.

Baccarini-Melandri, A., Casadio, R., and Melandri, B. A. (1977). *Eur. J. Biochem.* **78,** 389–402.

Baccarini-Melandri, A., Casadio, R., and Melandri, B. A. (1981). *Curr. Top. Bioenerg.* **12,** 197–258.

Bachmann, B. J. (1983). *Microbiol. Rev.* **47,** 180–230.

Baird, B. A., and Hammes, G. G. (1979). *Biochim. Biophys. Acta* **549,** 31–53.

Bangham, A. D. (1968). *Prog. Biophys. Mol. Biol.* **18,** 29–95.

Bangham, A. D. (1972). *Annu. Rev. Biochem.* **41,** 753–776.

Barber, J. (1980). *FEBS Lett.* **118,** 1–10.

Barber, J. (1982a). *Biosci. Rep.* **2,** 1–13.

Barber, J. (1982b). *Annu. Rev. Plant Physiol.* **33,** 261–295.

Barber, J. (1983). *Plant Cell Environ.* **6,** 311–322.

Baross, J. A., and Deming, J. W. (1983). *Nature (London)* **303,** 423–426.

Baum, H. (1978). *In* "The Molecular Biology of Membranes" (S. Fleischer, Y. Hatefi, D. H. MacLennan, and A. Tzagoloff, eds.), pp. 243–262. Plenum, New York.

Baum, H., Hall, G. S., Nalder, J., and Beechey, R. B. (1971). *In* "Energy Transduction in Respiration and Photosynthesis" (E. Quagliariello, S. Papa, and C. S. Rossi, eds.), pp. 747–755. Adriatica Editrice, Bari, Italy.

Bendich, A. J. (1982). *In* "Mitochondrial Genes" (P. Slonimski, P. Borst, and G. Attardi, eds.), p. 477. Cold Spring Harbor Laboratory, Cold Spring Harbor, New York.

Benz, R., and McLaughlin, S. (1983). *Biophys. J.* **41,** 381–398.

Benz, R., and Zimmermann, U. (1981). *Biochim. Biophys. Acta* **640,** 169–178.

Berden, J. A., and Hennecke, M. A. C. (1981). *FEBS Lett.* **126,** 211–214.

Berden, J. A., Hennecke, M. A. C., van der Bend, R. L., and van Dam, K. (1981). *In* "Vectorial Reactions in Electron and Ion Transport in Mitochondria and Bacteria" (F. Palmieri, E. Quagliariello, N. Siliprandi, and E. C. Slater, eds.), pp. 209–212. Elsevier, Amsterdam.

Bibb, M. J., van Etten, R. A., Wright, C. T., Walberg, M. W., and Clayton, D. A. (1981). *Cell* **26,** 167–180.

Bilz, H., Büttner, H., and Fröhlich, H. (1981). *Z. Naturforsch.* **36B,** 208–212.

Blaurock, A. E. (1982). *Biochim. Biophys. Acta* **650,** 167–207.

Blobel, G. (1980). *Proc. Natl. Acad. Sci. U.S.A.* **77,** 1496–1500.

Blumenfeld, L. A. (1983). "Physics of Bioenergetic Processes." Springer-Verlag, Berlin and New York.

Blumenthal, R., and Klausner, R. D. (1982). *In* "Membrane Reconstitution" (G. Poste and G. L. Nicolson, eds.), pp. 43–82. Elsevier, Amsterdam.

Boork, J. and Baltscheffsky, M. (1982). *EBEC Rep.* **2**, 365.

Booth, I. R., and Kroll, R. G. (1983). *Biochem. Soc. Trans.* **11**, 70–72.

Booth, I. R., Kroll, R. G., Findlay, M. G., Stewart, L. M. D., and Rowland, G. C. (1984). *Biochem. Soc. Trans.* **12**, 409–411.

Borst, P., and Grivell, L. A. (1981). *Nature (London)* **290**, 443–444.

Bowman, W. C., Rand, M. J., and West, G. B. (1969). "Textbook of Pharmacology." Blackwell, Oxford.

Brewer, G. J. (1976). *Biochemistry* **15**, 1387–1392.

Brock, C. J., and Tanner, M. J. A. (1982). *In* "Biological Membranes" (D. Chapman, ed.), Vol. 4, pp. 75–130. Academic Press, New York.

Brown, M. F., Ribeiro, A. A., and Williams, G. D. (1983). *Proc. Natl. Acad. Sci. U.S.A.* **80**, 4325–4329.

Brunori, M., and Wilson, M. T. (1982). *Trends Biochem. Sci.* **7**, 295–299.

Büchel, K. H., and Draber, W. (1972). *ACS Adv. Chem. Ser.* **114**, 141–154.

Buetow, D. E., and Gilbert, C. W. (1982). *In* "Cell Function and Differentiation, Part B" (G. Akoyunoglou, A. E. Evangelopoulous, J. Georgatsos, G. Palaiologos, A. Trakatellis, and C. P. Tsiganos, eds.), pp. 139–148. Liss, New York.

Capaldi, R. A. (1982a). *Trends Biochem. Sci.* **7**, 292–295.

Capaldi, R. A. (1982b). *Biochim. Biophys. Acta* **694**, 291–306.

Capaldi, R. A., Komai, H., and Hunter, D. R. (1973). *Biochem. Biophys. Res. Commun.* **55**, 655–659.

Careri, G., Fasella, P., and Gratton, E. (1979). *Annu. Rev. Biophys. Bioeng.* **8**, 69–97.

Carmeli, C., Quintanilha, A. T., and Packer, L. (1980). *Proc. Natl. Acad. Sci. U.S.A.* **77**, 4707–4711.

Casadio, R., Baccarini-Melandri, A., and Melandri, B. A. (1978). *FEBS Lett.* **87**, 323–328.

Casadio, R., Venturoli, G., and Melandri, B. A. (1982). *EBEC Rep.* **2**, 185–186.

Casey, R. P., Thelen, M., and Azzi, A. (1980). *J. Biol. Chem.* **255**, 3994–4000.

Chapman, D., Gómez-Fernández, J. C., and Goni, F. M. (1982). *Trends Biochem. Sci.* **7**, 67–70.

Cherry, R. J. (1979). *Biochim. Biophys. Acta* **559**, 289–327.

Ching, E., and Attardi, G. (1982). *Biochemistry* **21**, 3188–3195.

Clarke, D. J., Morley, C. D., Kell, D. B., and Morris, J. G. (1982). *Eur. J. Biochem.* **127**, 105–116.

Clegg, J. S. (1983a). *In* "Coherent Excitations in Biological Systems" (H. Fröhlich and F. Kremer, eds.), pp. 162–177. Springer-Verlag, Berlin and New York.

Clegg, J. S. (1983b). *Am. J. Physiol.* **246**, R133–R151.

Clementi, E., and Sarma, R. H., eds. (1983). "Structure and Dynamics: Nucleic Acids and Proteins." Adenine, Guilderland, New York.

Cleveland, M. v. B., Slatin, S. Finkelstein, A., and Levinthal, C. (1983). *Proc. Natl. Acad. Sci. U.S.A.* **80**, 3706–3710.

Cogdell, R. J., and Crofts, A. R. (1974). *Biochim. Biophys. Acta* **347**, 264–272.

Cohen, F. S., Eisenberg, M., and McLaughlin, S. (1977). *J. Membr. Biol.* **37**, 361–396.

Conover, T. E., and Azzone, G. F. (1981). *In* "Mitochondria and Microsomes" (C.-P. Lee, G. Schatz, and G. Dallner, eds.), pp. 481–518. Addison-Wesley, Reading, Massachusetts.

Cotton, N. P. J., and Jackson, J. B. (1983). *FEBS Lett.* **161**, 93–99.

Cotton, N. P. J., Clark, A. J., and Jackson, J. B. (1981). *Arch. Microbiol.* **129**, 94–99.

Cramer, W. A., Dankert, J. R., and Uratani, Y. (1983). *Biochim. Biophys. Acta* **737**, 173–193.

Criddle, R. S., Bock, R. M., Green, D. E., and Tisdale, H. (1962). *Biochimistry* **1**, 827–842.

Crilly, J. F., and Earnshaw, J. C. (1983). *Biophys. J.* **41**, 197–210.

Crofts, A. R., and Wood, P. M. (1978). *Curr. Top. Bioenerg.* **7**, 175–244.

Cullis, P. R., and de Kruijff, B. (1979). *Biochim. Biophys. Acta* **559**, 399–420.

Cullis, P. R., de Kruijff, B., Hope, M. J., Nayar, R., Rietveld, A., and Verleij, A. J. (1980). *Biochim. Biophys. Acta* **600**, 625–635.

Dandeu, J. P., Billault, A., and Barbu, E. (1969). *C. R. Hebd. Séances Acad. Sci.* **269**, 2044–2047.

Darwin, C. (1903). *In* "More Letters" (F. Darwin and A. C. Seward, eds.). Murray, London.

Davenport, J. W., and McCarty, R. E. (1980). *Biochim. Biophys. Acta* **589**, 353–357.

Davies, R. W., Scazzocchio, C., Waring, R. B., Lee, S., Grisi, E., McPhail Berks, M., and Brown, T. A. (1982). *In* "Mitochondrial Genes" (P. Slonimski, P. Borst, and G. Attardi, eds.), pp. 405–410. Cold Spring Harbor Laboratory, Cold Spring Harbor, New York.

Davis, J. H. (1983). *Biochim. Biophys. Acta* **737**, 117–171.

Davydov, A. S. (1983). *In* "Structure and Dynamics: Proteins and Nucleic Acids" (E. Clementi and R. H. Sarma, eds.), pp. 377–387. Adenine, Guilderland, New York.

Decker, S. J., and Lang, D. R. (1977). *J. Biol. Chem.* **252**, 5936–5938.

Decker, S. J., and Lang, D. R. (1978). *J. Biol. Chem.* **253**, 6738–6743.

de Kouchovsky, Y., and Haraux, F. (1981). *Biochem. Biophys. Res. Commun.* **99**, 205–212.

de Kouchovsky, Y., Haraux, F., and Sigalat, C. (1982). *FEBS Lett.* **139**, 245–249.

de Kouchovsky, Y., Haraux, F., and Sigalat, C. (1983). *Bioelectrochem. Bioenerg.* **11** (in press).

de Kruijff, B., Nayar, R., and Cullis, P. F. (1982). *Biochim. Biophys. Acta* **684**, 47–52.

Delalic, Z., Takashima, S., Adachi, K., and Asakura, T. (1983). *J. Mol. Biol.* **168**, 659–671.

Del Giudice, E., Doglia, S., and Milani, M. (1982). *Phys. Scr.* **26**, 232–238.

De Pierre, J. W., and Ernster, L. (1977). *Annu. Rev. Biochem.* **46**, 201–262.

Dierstein, R., Schumacher, A., and Drews, G. (1981). *Arch. Microbiol.* **128**, 376–383.

Dilger, J., and McLaughlin, S. (1979). *J. Membr. Biol.* **46**, 359–384.

Dilger, J. P., McLaughlin, S. G. A., McIntosh, T. A., and Simon, S. A. (1979). *Science* **206**, 1196–1198.

Dilley, R. A., Prochaska, L. J., Baker, G. M., Tandy, N. E., and Millner, P. A. (1982). *Curr. Top. Membr. Trans.* **16**, 345–369.

Dixon, L. K., Leaver, C. J., Brettell, R. I. S., and Gengenbach, B. G. (1982). *Theor. Appl. Genet.* **63**, 75–80.

Drews, G. (1982). *In* "From Cyclotrons to Cytochromes" (N. O. Kaplan and A. Robinson, eds.), pp. 355–366. Academic Press, New York.

Drosdat, H., Kiehl, R., Hoffman-Posorske, E., Kordt, S., and Hanstein, W. G. (1982). *EBEC Rep.* **2**, 19–20.

Dulaney, E. L. (1967). *In* "Microbial Technology" (H. J. Peppler, ed.), pp. 308–343. Academic Press, New York.

Dutton, P. L., and Prince, R. C. (1978). *In* "The Photosynthetic Bacteria" (R. K. Clayton and W. R. Sistrom, eds.), pp. 525–570. Plenum, New York.

Edidin, M. (1974). *Symp. Soc. Exp. Biol.* **28**, 1–14.

Edwards, C. (1981). "The Microbial Cell Cycle." Nelson, London.

Edwardson, J. R. (1970). *Bot. Rev.* **36**, 341–420.

Engleman, D. M., Henderson, R., McLachlan, A. D., and Wallace, B. A. (1980). *Proc. Natl. Acad. Sci. U.S.A.* **77**, 2023–2027.

Ernster, L., and Nordenbrand, K. (1974). *BBA Libr.* **13**, 283–288.

Etemadi, A. H. (1980). *Biochim. Biophys. Acta* **604**, 347–422, 423–475.

Evans, E. A., and Hochmuth, R. M. (1978). *Curr. Top. Membr. Trans.* **10**, 1–64.

Feingold, D. S. (1970). *J. Membr. Biol.* **3**, 372–386.

Felle, H., Stetson, D. L., Long, W. S., and Slayman, C. L. (1978). *In* "Frontiers of Biological Energetics" (P. L. Dutton, J. S. Leigh, and A. Scarpa, eds.), Vol. 2, pp. 1399–1407. Academic Press, New York.

Felle, H., Porter, J. S., Slayman, C. L., and Kaback, H. R. (1980). *Biochemistry* **19**, 3585–3590.

Ferguson, S. J., and Sorgato, M. C. (1982). *Annu. Rev. Biochem.* **51**, 185–217.

Ferguson, S. J., John, P., Lloyd, W. J., Radda, G. K., and Whatley, F. R. (1976). *FEBS Lett.* **62**, 272–275.

Ferguson, S. J., Jones, O. T. G., Kell, D. B., and Sorgato, M. C. (1979). *Biochem. J.* **180**, 75–85.

Fernandez, J. M., Bezanilla, F., and Taylor, R. E. (1982). *Nature (London)* **297**, 150–152.

Fields, K. L., and Luria, S. E. (1969a). *J. Bacteriol.* **97**, 64–77.

Fields, K. L., and Luria, S. E. (1969b). *J. Bacteriol.* **97**, 57–63.

Fillingame, R. H. (1980). *Annu. Rev. Biochem.* **49**, 1079–1113.

Finean, J. B., Coleman, R., and Michell, R. H. (1978). "Membranes and Their Cellular Functions," 2nd Ed. Blackwell, Oxford.

Fleischer, S., Fleischer, B., and Stoeckenins, W. (1967). *J. Cell Biol.* **32**, 193–208.

Forde, B. G., and Leaver, C. J. (1980). *In* "The Plant Genome" (D. R. Davies and D. A. Hopwood, eds.), pp. 131–146. John Innes Charity, Norwich.

Forde, B. G., Oliver, R. J. C., and Leaver, C. J. (1978). *Proc. Natl. Acad. Sci. U.S.A.* **75**, 3841–3845.

Fraley, R. T., Jameson, D. M., and Kaplan, S. (1978). *Biochim. Biophys. Acta* **511**, 52–69.

Franks, N. P., and Lieb, W. R. (1978). *Nature (London)* **274**, 339–342.

Franks, N. P., and Lieb, W. R. (1979). *J. Mol. Biol.* **133**, 469–500.

Franks, N. P., and Lieb, W. R. (1981). *Nature (London)* **292**, 248–251.

Franks, N. P., and Lieb, W. R. (1984). *Nature (London)* **310**, 599–601.

Freeman, K. B., Yatscoff, R. W., and Mason, J. R. (1980). *In* "The Organisation and Expression of the Mitochondrial Genome" (A. M. Kroon and C. Saccone, eds.), pp. 343–346. Elsevier, Amsterdam.

Freeman, K. B., Yatscoff, R. W., Mason, J. R., Patel, H. V., and Buckle, M. (1983). *Eur. J. Biochem.* **134**, 215–222.

Friedman, S. M., ed. (1978). "Biochemistry of Thermophily." Academic Press, New York.

Fröhlich, H. (1980). *Adv. Electron. Electron Phys.* **53**, 85–152.

Fuller, S. D., Capaldi, R. A., and Henderson, R. (1979). *J. Mol. Biol.* **134**, 305–327.

Gabellini, N., Bowyer, J. R., Hurt, E., Melandri, B. A., and Hauska, G. (1982). *Eur. J. Biochem.* **126**, 105–111.

Galante, Y. M., Wong, S. Y., and Hatefi, Y. (1979). *J. Biol. Chem.* **254**, 12372–12378.

Garcia, A. F., Drews, G., and Reidl, H. H. (1981). *J. Bacteriol.* **145**, 1121–1128.

Garfield, E. (1977). *In* "Essays of an Information Scientist" (Vol. 2, pp. 170–171). Inst. for Scientific Information, Philadelphia, Pennsylvania.

Gennis, R. B., and Jonas, A. (1977). *Annu. Rev. Biophys. Bioeng.* **6**, 195–238.

Gibson, F. (1983). *Biochem. Soc. Trans.* **11**, 229–240.

Goodchild, D. J., Duniec, J. T., and Anderson, J. M. (1983). *FEBS Lett.* **154**, 243–246.

Gorringe, D. M., and Moses, V. (1980). *Int. J. Biol. Macromol.* **2**, 161–173.

Gould, J. M. (1979). *J. Bacteriol.* **138**, 176–184.

Gould, J. M., and Cramer, W. A. (1977a). *J. Biol. Chem.* **252**, 5491–5497.

Gould, J. M., and Cramer, W. A. (1977b). *J. Biol. Chem.* **252**, 5875–5882.

Graan, T., Flores, S., and Ort, D. R. (1981). *In* "Energy Coupling in Photosynthesis" (B. Selman and S. Selman-Reimer, eds.), pp. 25–34. Elsevier, Amsterdam.

Gräber, P. (1981). *Curr. Top. Membr. Trans.* **16**, 215–245.

Gräber, P., and Witt, H. T. (1976). *Biochim. Biophys. Acta* **423**, 141–163.

Gregory, P., Earle, E. D., and Gracen, V. E. (1977). *ACS Symp. Ser.* **62**, 90–114.

Gregory, P., Earle, E. D., and Gracen, V. E. (1980). *Plant Physiol.* **66**, 477–481.

Griffiths, D. E., Avner, P. R., Lancashire, W. E., and Turner, J. R. (1972). *In* "Biochemistry and Biophysics of Mitochondrial Membranes" (G. F. Azzone, E. Carafoli, A. L. Lehninger, E. Quagliariello, and N. Siliprandi, eds.), pp. 505–521. Academic Press, New York.

Griffiths, D. E., Houghton, R. L., and Lancashire, W. E. (1974). *In* "The Biogenesis of Mitochondria" (A. M. Kroon and C. Saccone, eds.), pp. 215–223. Academic Press, New York.

Groen, A. K., van der Meer, R., Westerhoff, H. V., Wanders, R. J. A., Akerboom, T. P. M., and Tager, J. M. (1982a). *In* "Metabolic Compartmentation" (H. Sies, ed.), pp. 9–37. Academic Press, New York.

Groen, A. K., Wanders, R. J. A., Westerhoff, H. V., van der Meer, R., and Tager, J. M. (1982b). *J. Biol. Chem.* **257**, 2754–2758.

Guffanti, A. A., Susman, P., Blanco, R., and Krulwich, T. A. (1978). *J. Biol. Chem.* **253**, 708–715.

Guffanti, A. A., Blumenfeld, H., and Krulwich, T. A. (1981). *J. Biol. Chem.* **256**, 8418–8421.

Gurd, F. R. N., and Rothgeb, T. M. (1979). *Adv. Prot. Chem.* **33**, 73–165.

Gutman, M. (1980). *Biochim. Biophys. Acta* **594**, 53–84.

Gutman, M., Huppert, D., Pines, E., and Nachliel, E. (1981). *Biochim. Biophys. Acta* **642**, 15–26.

Gutman, M., Huppert, D., and Nachliel, E. (1982). *Eur. J. Biochem.* **121**, 637–642.

Gutman, M., Nachliel, E., Gershon, E., and Giniger, R. (1983). *Eur. J. Biochem.* **134**, 63–69.

Hackenbrock, C. R. (1972). *Ann. N.Y. Acad. Sci.* **195**, 492–505.

Hackenbrock, C. R. (1976). *In* "The Structure of Biological Membranes" (S. Abrahamsson and Pascher, eds.), pp. 199–234. Plenum, New York.

Hackenbrock, C. R. (1981). *Trends. Biochem. Sci.* **6**, 151–154.

Hackenbrock, C., Höchli, M., and Chau, R. M. (1976). *Biochim. Biophys. Acta* **455**, 466–484.

Haehnel, W. (1982). *Biochim. Biophys. Acta* **682**, 245–257.

Haines, T. H. (1979). *J. Theor. Biol.* **80**, 307–323.

Haines, T. H. (1982). *Biophys. J.* **37**, 147–148.

Halsey, J. J. (1974). *In* "Anesthetic Uptake and Action" (E. I. Eger, ed.), pp. 45–76. Williams & Wilkins, Baltimore, Maryland.

Hamamoto, T., Ohno, K., and Kagawa, Y. (1982). *J. Biochem.* **91**, 1759–1766.

Hangarter, R. P., and Good, N. E. (1982). *Biochim. Biophys. Acta* **681**, 397–404.

Hangarter, R. P., and Good, N. E. (1984). *Biochemistry* **23**, 122–130.

Hanstein, W. G. (1976a). *Biochim. Biophys. Acta* **456**, 129–148.

Hanstein, W. G. (1976b). *Trends Biochem. Sci.* **1**, 65–67.

Hanstein, W. G. (1978). *In* "Methods in Enzymology" (S. Fleischer and L. Packer, eds.), Vol. 56, pp. 653–683. Academic Press, New York.

Hanstein, W. G., and Hatefi, Y. (1974a). *Proc. Natl. Acad. Sci. U.S.A.* **71**, 288–292.

Hanstein, W. G., and Hatefi, Y. (1974b). *J. Biol. Chem.* **249**, 1356–1372.

Hanstein, W. G., and Kiehl, R. (1981). *Biochem. Biophys. Res. Commun.* **100**, 1118–1125.

Hanstein, W. G., Hatefi, Y., and Kiefer, H. (1979). *Biochemistry* **18**, 1019–1025.

Haraux, F., Sigalat, C., Moreau, A., and de Kouchovsky, Y. (1983). *FEBS Lett.* **155**, 248–252.

Harris, C. M., and Kell, D. B. (1983). *Bioelectrochem. Bioenerg.* **11**, 15–28.

Hatefi, Y. (1975). *J. Supramol. Struct.* **3**, 201–213.

Hatefi, Y. (1976). *In* "The Enzymes of Biological Membranes" (A. Martonosi, ed.), pp. 3–41. Plenum, New York.

Hauska, G., Samoray, D., Orlich, G., and Nelson, N. (1980). *Eur. J. Biochem.* **111**, 535–543.

Hauska, G., Hurt, E., Gabellini, N., and Lockau, W. (1983). *Biochim. Biophys. Acta* **726**, 97–133.

Heber, U., and Kirk, M. R. (1975). *Biochim. Biophys. Acta* **376**, 136–150.

Heinrich, R., and Rapoport, T. A. (1974). *Eur. J. Biochem.* **42**, 89–95, 97–102.

Heinz, E., Westerhoff, H. V., and van Dam, K. (1981). *Eur. J. Biochem.* **115**, 107–113.

Henderson, R., and Unwin, P. N. T. (1975). *Nature (London)* **257**, 28–32.

Henderson, R., Capaldi, R. A., and Leigh, J. S. (1977). *J. Mol. Biol.* **112**, 631–648.

Hengge, R., and Boos, W. (1983). *Biochim. Biophys. Acta* **737**, 443–478.

Heron, C., Ragan, C. I., and Trumpower, B. (1978). *Biochem. J.* **174**, 791–800.

Herweijer, M. A., Berden, J. A., and Kemp, A. (1984). *EBEC Rep.* **3**, 241–242.

Heytler, P. G. (1963). *Biochemistry* **2**, 357–361.

Heytler, P. G. (1979). *In* "Methods in Enzymology" (S. Fleischer and L. Packer, eds.), Vol. 55, pp. 462–472. Academic Press, New York.

Higuti, T., Yokota, M., Arakaki, N., Hattori, A., and Tani, L. (1978). *Biochim. Biophys. Acta* **503**, 211–222.

Higuti, T., Arakaki, N., Niimi, S., Nakasima, S., Saito, R., Tani, I., and Ota, F. (1980). *J. Biol. Chem.* **255**, 7631–7636.

Higuti, T., Ohe, T., Arakaki, N., and Kotera, Y. (1981). *J. Biol. Chem.* **256**, 9855–9860.

Hill, T. L. (1977). "Free Energy Transduction in Biology." Academic Press, New York.

Hille, B. (1980). *Prog. Anesthesiol.* **2**, 1–5.

Hinkle, P. C., Tu, Y. L., and Kim, J. J. (1975). *In* "Molecular Aspects of Membrane Phenomena" (H. R. Kaback, H. Neurath, G. K. Radda, R. Schwyzer, and W. R. Wiley, eds.), pp. 222–232. Springer-Verlag, Berlin and New York.

Hirata, H. J., Fukui, S., and Ishikawa, S. (1969). *J. Biochem.* **65**, 843–847.

Hitchens, G. D., and Kell, D. B. (1982a). *Biochem. J.* **206**, 351–357.

Hitchens, G. D., and Kell, D. B. (1982b). *Biochem. Soc. Trans.* **10**, 261.

Hitchens, G. D., and Kell, D. B. (1982c). *Biosci. Rep.* **2**, 743–749.

Hitchens, G. D., and Kell, D. B. (1983a). *Biochem. J.* **212**, 25–30.

Hitchens, G. D., and Kell, D. B. (1983b). *Biochim. Biophys. Acta* **723**, 308–316.

Hitchens, G. D., and Kell, D. B. (1984). *Biochim. Biophys. Acta* **766**, 222–223.

Hitchens, G. D., Kell, D. B., and Morris, J. G. (1982). *J. Gen. Microbiol.* **128**, 2207–2209.

Höchli, M., and Hackenbrock, C. R. (1976). *Proc. Natl. Acad. Sci. U.S.A.* **73**, 1636–1640.

Höchli, M., and Hackenbrock, C. R. (1977). *J. Cell. Biol.* **72**, 278–291.

Holland, I. B. (1975). *Adv. Microbial. Physiol.* **12**, 55–139.

Hong, J.-S. (1977). *J. Biol. Chem.* **252**, 8582–8588.

Hong, J.-S., Haggerty, D. L., and Lieberman, M. A. (1977). *Antimicrob. Agents Chemother.* **11**, 881–887.

Hong, Y. Q., and Junge, W. (1983). *Biochim. Biophys. Acta* **722**, 197–208.

Houslay, M. D., and Stanley, K. K. (1982). "Dynamics of Biological Membranes." Wiley, New York.

Howard, J. C. (1981). *Nature (London)* **290**, 441–442.

Huang, H. W. (1973). *J. Theor. Biol.* **40**, 11–17.

Israelachvili, J. N., Marcelja, S., and Horn, R. G. (1980). *Q. Rev. Biophys.* **13**, 121–200.

Ito, M., and Ohnishi, Y. (1981). *FEBS Lett.* **136**, 225–230.

Ito, M., Ohnishi, Y., Itoh, S., and Nishimura, M. (1983). *J. Bacteriol.* **153**, 310–315.

Jackson, J. B., Venturoli, G., Baccarini-Melandri, A., and Melandri, B. A. (1981). *Biochim. Biophys. Acta* **636**, 1–8.

Jain, M. K. (1972). "The Biomolecular Lipid Membrane." Van Nostrand-Reinhold, Princeton, New Jersey.

Jaffe, L. (1977). *Nature (London)* **265**, 600–601.

Jagendorf, A. T., and Uribe, E. G. (1966). *Proc. Natl. Acad. Sci. U.S.A.* **55**, 170–177.

Jähnig, F. (1983). *Proc. Natl. Acad. Sci. U.S.A.* **80**, 3691–3695.

Janoff, A. S., and Miller, K. W. (1982). *In* "Biological Membranes" (D. Chapman, ed.), Vol. 4, pp. 417–476. Academic Press, New York.

Jardetzky, O., and Roberts, G. C. K. (1981). "NMR in Molecular Biology," pp. 537–579. Academic Press, New York.

Ji, S. (1976). *J. Theor. Biol.* **59**, 319–330.

Johannsson, A., Keightley, C. A., Smith, G. A., Richards, C. R., Hesketh, T. R., and Metcalfe, J. C. (1981a). *J. Biol. Chem.* **256**, 1643–1650.

Johannsson, A., Smith, G. A., and Metcalfe, J. C. (1981b). *Biochim. Biophys. Acta* **641**, 416–421.

John, P., and Whatley, F. R. (1977). *Biochim. Biophys. Acta* **463**, 129–153.

Junge, W., and Jackson, J. B. (1982). *In* "Photosynthesis: Energy Conversion by Plants and Bacteria" (Govindjee, ed.), pp. 589–646. Academic Press, New York.

Kacser, H., and Burns, J. A. (1973). *Symp. Soc. Exp. Biol.* **32**, 65–104.

Kagawa, Y. (1972). *Biochim. Biophys. Acta* **245**, 297–338.

Kagawa, Y. (1979). *J. Bioenerg.* **11**, 39–78.

Kagawa, Y., and Racker, E. (1971). *J. Biol. Chem.* **246**, 5477–5487.

Kaiser, I., and Oelze, J. (1980). *Arch. Microbiol.* **126**, 187–194.

Kasche, V. (1983). *Enzyme Microb. Technol.* **5**, 2–13.

Katre, N. V., and Wilson, D. F. (1977). *Arch. Biochem. Biophys.* **184**, 578–585.

Katre, N. V., and Wilson, D. F. (1978). *Arch. Biochem. Biophys.* **191**, 647–656.

Katre, N. V., and Wilson, D. F. (1980). *Biochim. Biophys. Acta* **593**, 224–229.

Kaufman, R. D. (1977). *Anesthesiol.* **46**, 49–62.

Kaufmann, N., Reidl, H.-H., Golecki, J. R., Garcia, A. F., and Drews, G. (1982). *Arch. Microbiol.* **131**, 313–322.

Kayalar, C., and Luria, S. E. (1979). *In* "Membrane Bioenergetics" (C.-P. Lee, G. Schatz, and L. Ernster, eds.), pp. 297–305. Addison-Wesley, Reading, Massachusetts.

Kell, D. B. (1979). *Biochim. Biophys. Acta* **549**, 55–99.

Kell, D. B. (1983). *Bioelectrochem. Bioenerg.* **11**, 405–415.

Kell, D. B. (1984). *Trends Biochem. Sci.* **9**, 86–88.

Kell, D. B., and Griffiths, A. M. (1981). *Photobiochem. Photobiophys.* **2**, 105–110.

Kell, D. B., and Hitchens, G. D. (1982). *Faraday Discuss. Chem. Soc.* **74**, 377–388.

Kell, D. B., and Hitchens, G. D. (1983). *In* "Coherent Excitations in Biological Systems" (H. Fröhlich and F. Kremer, eds.), pp. 178–198. Springer-Verlag, Berlin and New York.

Kell, D. B., and Morris, J. G. (1981). *In* "Vectorial Reactions in Electron and Ion Transport in Mitochondria and Bacteria" (F. Palmieri, E. Quagliariello, N. Siliprandi, and E. C. Slater, eds.), pp. 339–347. Elsevier, Amsterdam.

Kell, D. B., Ferguson, S. J., and John, P. (1978a). *Biochem. Soc. Trans.* **6**, 1292–1295.

Kell, D. B., John, P., and Ferguson, S. J. (1978b). *Biochem. J.* **174**, 257–266.

Kell, D. B., Ferguson, S. J., and John, P. (1978c). *Biochim. Biophys. Acta* **502**, 111–126.

Kell, D. B., Burns, A., Clarke, D. J., and Morris, J. G. (1981a). *Specul. Sci. Technol.* **4**, 109–120.

Kell, D. B., Clarke, D. J., and Morris, J. G. (1981b). *FEMS Microbiol. Lett.* **11**, 1–11.

Kenyon, C. N. (1978). *In* "The Photosynthetic Bacteria" (R. K. Clayton and W. R. Sistrom, eds.), p. 281. Plenum, New York.

Kepes, A., and Autissier, F. (1972). *Biochim. Biophys. Acta* **265**, 443–469.

Keszthelyi, L. (1980). *Biochim. Biophys. Acta* **598**, 429–436.

Khan, S., and Berg, H. C. (1983). *J. Biol. Chem.* **258**, 6709–6712.

Kiehl, R., and Hanstein, W. G. (1981). *In* "Vectorial Reactions in Electron and Ion Transport in Mitochondria and Bacteria" (F. Palmieri, E. Qualiariello, N. Siliprandi, and E. C. Slater, eds.), pp. 217–222. Elsevier, Amsterdam.

Kimelberg, H. K., and Papahadjopoulos, D. (1971). *Biochim. Biophys. Acta* **233**, 805–809.

Klibanov, A. M. (1983). *Biochem. Soc. Trans.* **11**, 19–20.

Konisky, J. (1978). *In* "The Bacteria" (L. N. Ornston and J. R. Sokatch, eds.), Vol. 6, pp. 71–136. Academic Press, New York.

Konisky, J. (1982) *Annu. Rev. Microbiol.* **36**, 125–144.

Konisky, J., and Tokuda, H. (1979). *Zbl. Bakteriol. Parasit. Infekt Hyg Abt. 1 Orig. Reihe A* **244**, 1105–1120.

Konisky, J., Gilchrist, M. J. R., Nieva-Gomez, D., and Gennis, R. B. (1975). *In* "Molecular Aspects of Membrane Phenomena" (H. R. Kaback, H. Neurath, G. K. Radda, R. Schwyzer, and W. R. Wiley, eds.), pp. 193–215. Springer-Verlag, Berlin and New York.

Kono, Y., Takeuchi, S., Kawarda, A., Daly, J. M., and Knoche, H. W. (1980). *Tetrahedron Lett.* **21**, 1537–1540

Kornberg, R. D., and McConnell, H. M. (1971). *Biochemistry* **10**, 1111–1120.

Kröger, A., and Klingenberg, M. (1970). *Vitam. Horm.* **28**, 533–674.

Kröger., A., and Klingenberg, M. (1973). *Eur. J. Biochem.* **39**, 313–323.

Kurup, C. K. R., and Sanadi, D. R. (1977). *J. Bioenerg. Biomembr.* **9**, 1–15

Kusher, D, J., ed. (1978). "Microbial Life in Extreme Environments." Academic Press, New York.

Lam, E., and Malkin R. (1982). *Proc. Natl. Acad. Sci. U.S.A.* **79**, 5494–5498.

Lang, D. R., and Decker, S. J. (1978). *In* "Spores VII" (G. Chambliss and J. C. Vary, eds.), American Society for Microbiology, Washington, D. C. pp. 265–270.

Lapange, S. (1978). "Physicochemical Aspects of Protein Denaturation." Wiley, New York.

Lardy, H., Reed, P., and Lin, C. C.-H. (1975). *Fed. Proc. Fed. Am. Soc. Exp. Biol.* **34**, 1707–1710.

Läuger, P., and Neumcke, B. (1973). *In* "Membranes: A series of advances", G. Eisenman, ed.). pp. 1–59. Dekker, New York.

Leaver, C. J., and Gray, M. W. (1982). *Annu. Rev. Plant Physiol.* **33**, 373–402

Leaver, C. J., Forde, B. G., Dixon, L. K., and Fox, T. D. (1982). *In* "Mitochondrial Genes" (P. P. Slonimski, P. Borst, and G. Attardi, eds.). Cold Spring Harbor Laboratory, Cold Spring Harbor, New York.

Lee, C. P., Ernster, L., and Chance, B. (1969). *Eur. J. Biochem.* **8**, 153–163.

Lehninger, A. L. (1982). "Principles of Biochemistry." Worth, New York.

Lenaz, G. (1978). *In* "The Molecular Biology of Membranes" (S. Fleischer, Y. Hatefi, D. H. MacLennan, and A. Tzagoloff, eds.), pp. 137–162. Plenum, London.

Lieberman, M. A., and Hong, J.-S. (1974). *Proc. Natl. Acad. Sci. U.S.A.* **71**, 4395–4399.

Lieberman, M. A., and Hong, J.-S. (1976). *J. Bacteriol.* **125**, 1024–1031.

Lieberman, M. A., Simon, M., and Hong, J.-S. (1977). *J. Biol. Chem.* **252**, 4056–4067.

Ling, G. N. (1981). *Physiol. Chem. Phys.* **13**, 29–96.

Lloyd, D., Poole, R. K., and Edwards, S. W. (1982). "The Cell Division Cycle: Temporal Organization and Control of Cellular Growth and Reproduction." Academic Press, New York.

Loomis, W. F., and Lipmann, F. (1948). *J. Biol. Chem.* **173**.

Lozier, R. H., Bogomolni, R. A., and Stoeckenius, W. (1975). *Biophys. J.* **15**, 955–962.

Ludwig, B., Prochaska, L., and Capaldi, R. A. (1980). *Biochemistry* **19**, 1516–1523.

Luria, S. E. (1964). *Ann. Inst. Pasteur.* **107**, 67–73.

Luria, S. E. (1973). *In* "Bacterial Membranes and Walls" (I. Leive, ed.), pp. 293–320. Dekker, New York.

Luria, S. E. (1982). *J. Bacteriol.* **149**, 386.

McCammon, J. A., and Karplus, M. (1983). *Acc. Chem. Res.* **16**, 187–193.

McCarthy, J. E. G., and Ferguson, S. J. (1982). *Biochem. Biophys. Res. Commun.* **107**, 1406–1411.

McElhaney, R. N. (1974). *J. Mol. Biol.* **84**, 145–157.

McLaughlin, S. (1981). *In* "Chemiosmotic Proton Circuits in Biological Membranes" (V. P. Skulacher and P. Hinkle, eds.), pp. 601–609. Addison-Wesley, Reading, Massachusetts.

McLaughlin, S. G. A., and Dilger, J. P. (1980). *Physiol. Rev.* **60**, 825–863.

McLaughlin, S., Elsenberg, M., Cohen, F., and Dilger, J. (1978). *In* "Frontiers of Biological Energetics" (P. L. Dutton, J. S. Leigh, and A. Scarpa, eds.), Vol. 2, pp. 1205–1214. Academic Press, New York.

Mai, M., and Allison, W. S. (1983). *Arch. Biochem. Biophys.* **221**, 467–476.

Malenkova, I. V., Kupriov, S. P., Davydov, R. M., and Blumenfeld, L. A. (1982). *Biochim. Biophys. Acta* **682**, 179–183.

Maloff, B. L., Scordilis, S. P., and Tedeschi, H. (1978). *J. Cell Biol.* **78**, 214–226.

Maloney, P. C. (1982). *J. Membr. Biol.* **67**, 1–12.

Malpress, F. H. (1981a). *Nature (London)* **289**, 355.
Malpress, F. H. (1981b). *J. Theor. Biol.* **92**, 255–265.
Mandolino, G., De Santis, A., and Melandri, B. A. (1983). *Biochim. Biophys. Acta* **723**, 428–439.
Mar, T., Picorel, R., and Gingras, G. (1982). *Biochim. Biophys. Acta* **682**, 354–363.
Marquis, R. E., and Carstensen, E. L. (1973). *J. Bacteriol.* **116**, 1273–1279.
Marsh, D. (1983). *Trends Biochem. Sci.* **8**, 330–333.
Masters, C. J. (1981). *CRC Crit. Rev. Biochem.* **11**, 105–144.
Maynard-Smith, J. (1977). *In* "The Encyclopedia of Ignorance, Vol. 2, Life Sciences and Earth Sciences" (R. Duncan and M. Weston-Smith, eds.), pp. 235–242. Pergamon, Oxford.
Mechler, B., and Steim, J. M. (1976). *Annu. Rev. Biophys. Bioeng.* **5**, 205–238.
Medawar, P. (1982). *In* "Pluto's Republic," p. 29. Oxford Univ. Press, London and New York.
Melandri, B. A., Venturoli, G., de Santis, A., and Baccarini-Melandri, A. (1980). *Biochim. Biophys. Acta* **592**, 38–52.
Melandri, B. A., Venturoli, G., Casadio, R., Azzone, G. F., Kell, D. B., and Westerhoff, H. V. (1983). *Proc. Int. Photosynthesis Congr. 6th, Brussells.*
Michels, M., and Bakker, E. P. (1981). *Eur. J. Biochem.* **116**, 513–519.
Miller, I. R. (1981). *Top. Bioelectrochem. Bioenerg.* **4**, 162–224.
Miller, J. C., and Miller, K. W. (1975). *MTP Int. Rev. Sci. Biochem.* **12**, 33–75.
Miller, K. W., Paton, W. D. M., Smith, R. A., and Smith, E. B. (1973). *Mol. Pharmacol.* **9**, 131–143.
Mitchell, P. (1966). *Biol. Rev.* **41**, 445–502.
Mitchell, P. (1967). *Fed. Proc., Fed. Am. Soc. Exp. Biol.* **26**, 1370–1379.
Mitchell, P. (1968). "Chemiosmotic Coupling and Energy Transduction." Glynn Research, Bodmin, Cornwall.
Mitchell, P. (1976). *J. Theor. Biol.* **62**, 327–367.
Mitchell, P. (1977a). *Symp. Soc. Gen. Microbiol.* **27**, 383–423.
Mitchell, P. (1977b). *FEBS Lett.* **78**, 1–20.
Mitchell, P. (1979). *Science* **206**, 1148–1159.
Mitchell, P. (1981). *In* "Of Oxygen, Fuels and Living Matter" (G. Semenza, ed.), pp. 1–160. Wiley, New York.
Mitchell, P., and Moyle, J. (1967). *Biochem. J.* **105**, 1147–1162.
Mitchell, P., Moyle, J., and Mitchell, R. (1979). *In* "Methods in Enzymology" (S. Fleischer and L. Packer, eds.), Vol. 55, pp. 627–640.
Montecucco, C., Smith, G. A., Dabbeni-Sala, F., Johannsson, A., Galante, Y. M., and Bisson, R. (1982). *FEBS Lett.* **144**, 145–148.
Morris, J. F., and Newman, E. B. (1980). *J. Bacteriol.* **143**, 1504–1505.
Mozhaev, V. V., and Martinek, K. (1982). *Enzyme Microb. Technol.* **4**, 299–309.
Muller, M., Krebs, J. J. R., Cherry, R. J., and Kawato, S. (1982). *J. Biol. Chem.* **257**, 117–120.
Murakami, S., Torres-Pereira, J., and Packer, L. (1975). *In* "Bioenergetics of Photosynthesis" (Gorindjee, ed.), pp. 555–618. Academic Press, New York.
Muraoka, S., and Terada, H. (1972). *Biochim. Biophys. Acta* **275**, 271–275.
Murato, N., and Fork, D. C. (1975). *Plant Physiol.* **56**, 791–796.
Newman, E. B., Morris, J. F., Walker, C., and Kapook, V. (1981). *Mol. Gen. Genet.* **182**, 143–147.
Newman, E. B., Malik, N., and Walker, C. (1982). *J. Bacteriol.* **150**, 710–715.
Nicholls, D. G. (1982). "Bioenergetics." Academic Press, New York.
Nicolay, K., Lolkema, J., Hellingwerf, K. J., Kaptein, R., and Konings, W. N. (1981). *FEBS Lett.* **123**, 319–323.
Nishimura, M., and Pressman, B. C. (1969). *Biochemistry* **8**, 1360–1370.
Nomura, M., and Maeda, A. (1965). *Zbl. Bakteriol. Abt. 1 Orig.* **196**, 216–239.
Oelze, J., and Drews, G. (1981). *In* "Organisation of Prokaryotic Cell Membranes" (B. K. Ghosh, ed.), Vol. 2, pp. 131–195. CRC Press, Boca Raton, Florida.

Oren, R., Weiss, S., Garty, H., Caplan, S. R., and Gromet-Elhanan, Z. (1980). *Arch. Biochem. Biophys.* **205**, 503–509.

Ort, D. R., and Melandri, B. A. (1982). *In* "Photosynthesis; Energy Conversion by Plants and Bacteria" (Govindjee, ed.), pp. 537–587. Academic Press, New York.

Ort, D. R., Dilley, R. A., and Good, N. E. (1976). *Biochim. Biophys. Acta* **449**, 108–124.

Ovchinnikov, Y. A., Abdulaev, N. G., Feigina, M. Y., Kiselev, A. V., and Lobanov, N. A. (1979). *FEBS Lett.* **100**, 219–224.

Packham, N. K., and Barber, J. (1983). *Biochim. Biophys. Acta.* **723**, 247–255.

Packham, N. K., Berriman, J. A., and Jackson, J. B. (1978). *FEBS Lett.* **89**, 205–210.

Packham, N. K., Tiede, D. M., Mueller, P., and Dutton, P. L. (1980). *Proc. Natl. Acad. Sci. U.S.A.* **77**, 6339–6343.

Padan, E. and Rottenberg, H. (1973). *Eur. J. Biochem.* **40**, 431–437.

Padan, E., Zilberstein, D., and Schuldiner, S. (1981). *Biochim. Biophys. Acta* **650**, 151–166.

Papa, S. (1976). *Biochim. Biophys. Acta* **456**, 39–84.

Parsegian, V. A. (1975). *Ann. N.Y. Acad. Sci.* **264**, 161–174.

Parsonage, D., and Ferguson, S. J. (1982). *Biochem. Soc. Trans.* **10**, 257–258.

Pattus, F., Cavard, D., Verger, R., Lazdunski, C., Rosenbusch, J., and Schindler, H. (1983). *In* "Physical Chemistry of Transmembrane Ion Motions" (G. Spach, ed.), pp. 407–413. Elsevier, Amsterdam.

Pauly, H., and Packer, L. (1960). *J. Biophys. Biochem. Cytol.* **7**, 603–612.

Pauly, H., and Schwan, H. P. (1966). *Biophys. J.* **6**, 621–639.

Pedersen, P. L. (1982). *Ann. N.Y. Acad. Sci.* **402**, 1–20.

Pedersen, P. L., Schwerzmann, K., and Cintron, N. (1981). *Curr. Top. Bioenerg.* **11**, 149–199.

Petersen, D. C., and Cone, R. A. (1975). *Biophys. J.* **15**, 1181–1200.

Pethig, R. (1979). "Dielectric and Electronic Properties of Biological Materials." Wiley, New York.

Phelps, D. C., and Hanstein, W. G. (1977). *Biochem. Biophys. Res. Commun.* **79**, 1245–1254.

Phelps, D. C., and Hatefi, Y. (1981). *J. Biol. Chem.* **256**, 8217–8221.

Pick, U., and Racker, E. (1979). *J. Biol. Chem.* **254**, 2793–2799.

Pietrobon, D., Azzone, G. F., and Walz, D. (1981). *Eur. J. Biochem.* **117**, 389–394.

Pietrobon, D., Zoratti, M., Azzone, G. F., Stucki, J. W., and Walz, D. (1982). *Eur. J. Biochem.* **127**, 483–494.

Pietrobon, D., Zoratti, M., and Azzone, G. F. (1983). *Biochim. Biophys. Acta* **723**, 317–321.

Plate, C. A. (1976). *J. Bacteriol.* **125**, 467–474.

Plate, C. A. (1979). *In* "Microbiology—1979" (D. Schlessinger, ed.), pp. 58–61. Am. Soc. Microbiol., Washington, D.C.

Plate, C. A., and Suit, J. L. (1981). *J. Biol. Chem.* **256**, 12974–12980.

Poo, M.-m. (1981). *Annu. Rev. Biophys. Bioeng.* **10**, 245–276.

Poo, M., and Cone, R. A. (1973). *Nature (London)* **247**, 438–441.

Poo, M.-m., and Robinson, K. R. (1977). *Nature (London)* **205**, 602–604.

Poole, R. K. (1981). *In* "Diversity of Bacterial Respiratory Chains" (C. W. Jones, ed.), pp. 87–114. CRC Press, Boca Raton, Florida.

Porteous, J. W. (1983). *Trends Biochem. Sci.* **8**, 200–202.

Pressman, B. C. (1972). *In* "Biochemistry and Biophysics of Mitochondrial Membranes" (G. F. Azzone, E. Carafoli, A. L. Lehninger, E. Quagliariello, and N. Siliprandi, eds.), pp. 591–602. Academic Press, New York.

Price, B. D., and Brand, M. D. (1983). *Eur. J. Biochem.* **132**, 595–601.

Prince, R. C., O'Keefe, D. P., and Dutton, P. L. (1982). *In* "Electron Transport and Photophosphorylation" (J. Barber, ed.), pp. 197–248. Elsevier, Amsterdam.

Racker, E. (1976). "A New Look at Mechanisms in Bioenergetics." Academic Press, New York.

Racker, E., and Kandrach, A. (1971). *J. Biol. Chem.* **246**, 7069–7071.

Racker, E., and Stoeckenius, W. (1974). *J. Biol. Chem.* **249**, 662–663.

Ragan, C. I., and Racker, E. (1973). *J. Biol. Chem.* **248**, 2563–2569.

Raison, J. K. (1973). *Symp. Soc. Exp. Biol.* **27**, 485–512.

Rao, K. R. (1978). "Pentachloroplenol. Chemistry, Pharmacology and Environmental Toxicology." Plenum, New York.

Reddy, G. P. V., and Pardee, A. B. (1983). *Nature (London)* **304**, 86–88.

Reed, P. W., and Lardy, H. A. (1975). *J. Biol. Chem.* **250**.

Reynafarje, B., Brand, M. D., Alexandre, A., and Lehninger, A. L. (1979). *In* "Methods in Enzymology" (S. Fleischer and L. Packer, eds.), Vol. 55, pp. 640–656. Academic Press, New York.

Rich, P. R. (1982). *Faraday Discuss. Chem. Soc.* **74**, 349–364.

Richardson, S. H., Hultin, H. O., and Fleischer, S. (1964). *Arch. Biochem. Biophys.* **105**, 254–260.

Robertson, R. N. (1983). "The Lively Membranes." Cambridge Univ. Press, London and New York.

Roscoe, T. J., and Ellis, R. J. (1982). *In* "Methods in Chloroplast Molecular Biology" (M. Edelman, R. B. Hallick, and N.-H. Chua, eds.), pp. 1015–1028. Elsevier, Amsterdam.

Rosen, B. P., and Kashket, E. R. (1978). *In* "Bacterial Transport" (B. P. Rosen, ed.), pp. 559–620. Dekker, New York.

Rottenberg, H. (1973). *Biophys. J.* **13**, 503–511.

Rottenberg, H. (1979). *In* "Methods in Enzymology" (S. Fleischer and L. Packer eds.), Vol. 55, pp. 547–569. Academic Press, New York.

Rottenberg, H. (1983). *Proc. Natl. Acad. Sci. U.S.A.* **80**, 3313–3317.

Russell, B. (1966). "Human Knowledge, Its Scope and Limits," p. 213. Allen & Unwin, London.

Ryrie, I. J., Critchley, C., and Tillberg, J.-E. (1978). *Arch. Biochem. Biophys.* **198**, 182–194.

Sabet, S. F. (1976). *J. Bacteriol.* **126**, 601–608.

Sandorfy, C. (1980). *Prog. Anesthesiol.* **2**, 353–359.

Saraste, M. (1983). *Trends Biochem. Sci.* **8**, 139–142.

Schäfer, G. (1981). *In* "Inhibitors of Mitochondrial Function" (M. Erecinska and D. F. Wilson, eds.), pp. 165–185. Pergamon, Oxford.

Schäfer, G., and Rowohl-Quisthoudt. (1975). *FEBS Lett.* **59**, 48–51.

Schatz, G., and Saltzgaber, J. (1970). *In* "Electron Transport and Energy Conservation" (J. M. Tager, S. Papa, E. Quagliariello, and E. C. Slater, eds.), pp. 273–280. Adriatica Editrice, Bari, Italy.

Schein, S. J., Kagan, B. L., and Finkelstein, B. A. (1979). *Nature (London)* **276**, 159–163.

Schlodder, E., Gröber, P., and Witt, H. T. (1982). *In* "Electron Transport and Photophosphorylation" (J. Barber, ed.), pp. 105–175. Elsevier/North-Holland, Amsterdam.

Schneider, H., Lemasters, J. J., Höchli, M., and Hackenbrock, C. R. (1980a). *Proc. Natl. Acad. Sci. U.S.A.* **77**, 442–446.

Schneider, H., Lemasters, J. J., Höchli, M., and Hackenbrock, C. R. (1980b). *J. Biol. Chem.* **255**, 3748–3756.

Schneider, H., Lemasters, J. J., and Hackenbrock, C. R. (1982). *J. Biol. Chem.* **257**, 10789–10793.

Scholes, P., and Mitchell, P. (1970). *J. Bioenerg.* **1**, 309–323.

Schreiber, U., and Del Valle-Tascon, S. (1982). *FEBS Lett.* **150**, 32–37.

Schuldiner, S. (1977). *Encycl. Plant Physiol. New Ser.* **5**, 416–422.

Schwarz, G. (1978a). *J. Membr. Biol.* **43**, 127–148.

Schwarz, G. (1978b). *J. Membr. Biol.* **43**, 149–167.

Scott, A. C. (1983). *In* "Structure and Dynamics: Proteins and Nucleic Acids" (E. Clementi, and R. H. Sarma, eds.), pp. 389–404. Adenine Press, Guilderland, New York.

Senior, A. E. (1979). *In* "Membrane Proteins in Energy Transduction" (R. A. Capaldi, ed.), pp. 233–278. Dekker, New York.

Senior, A. E., and Wise, J. G. (1983). *J. Membr. Biol.* **73**.
Shilo, M., ed. (1978). "Strategies of Microbial Lie in Extreme Environments." Verlag Chemie Weinheim.
Singer, S. J., and Nicolson, G. L. (1972). *Science,* **175,** 720–731.
Sjöstrand, F. S. (1978). *J. Ultrastruct. Res.* **64,** 217–245.
Skulachev, V. P. (1971). *Curr. Top. Bioenerg.* **4,** 127–190.
Slooten, L. and Branders, C. (1979). *Biochim. Biophys. Acta* **547,** 79–90.
Smith, D. J., Stokes, B. O., and Boyer, P. D. (1976). *J. Biol. Chem.* **251,** 4165–4171.
Smith, E. B., Bowser-Riley, F., Daniels, S., Dunbar, I. T., Harrison, C. B., and Paton, W. D. M. (1984). *Nature (London)* **311,** 56–58.
Somogyi, B., Welch, G. R., and Damjarovich, S. (1984). *Biochim. Biophys. Acta* **768,** 81–112.
Sonnleitner, B., and Fiechter, A. (1983). *Trends Biotechnol.* **1,** 74–80.
Soper, J. W., Decker, G. L., and Pedersen, P. L. (1979). *J. Biol. Chem.* **254,** 11170–11176.
Sorgato, M. C., and Ferguson, S. J. (1979). *Biochemistry* **18,** 5737–5742.
Southard, J., Nitisewojo, P., and Green, D. E. (1974). *Fed. Proc., Fed. Am. Soc. Exp. Biol.* **33,** 2147–2153.
Sowers, A., and Hackenbrock, C. R. (1980). *Fed. Proc., Fed. Am. Soc. Exp. Biol.* **39,** 1955.
Sowers, A. E., and Hackenbrock, C. R. (1981). *Proc. Natl. Acad. Sci. U.S.A.* **78,** 6246–6250.
Stryer, L. (1981). "Biochemistry." Freeman, San Francisco, California.
Stucki, J. W. (1980). *Eur. J. Biochem.* **109,** 269–283.
Stucki, J. W. (1982). *In* "Metabolic Compartmentation" (H. Sies, ed.), pp. 39–69. Academic Press, New York.
Szoka, F., and Papahadjopoulos, D. (1980). *Annu. Rev. Biophys. Bioeng.* **9,** 467–508.
Tager, J. M., Wanders, R. J. A., Groen, A. K., Kunz, W., Bohnensack, R., Küster, U., Letko, G., Böhme, G., Duszynski, J., and Wojtczak, L. (1983). *FEBS Lett.* **151,** 1–9.
Tanford, C. (1978). *Science* **200,** 1012–1018.
Tanford, C. (1980). "The Hydrophobic Effect: Formation of Micelles and Biological Membranes," 2nd Ed. Wiley, New York.
Tedeschi, H. (1975). "Mitochondria: Structure, Biogenesis and Transducing Functions." Springer-Verlag, Berlin and New York.
Tedeschi, H. (1980). *Biol. Rev.* **55,** 171–206.
Tedeschi, H. (1981). *Biochim. Biophys. Acta* **639,** 157–196.
Terada, H. (1981). *Biochim. Biophys. Acta* **639,** 225–242.
Thayer, W. S., and Hinkle, P. C. (1975). *J. Biol. Chem.* **250,** 5336–5342.
Thorbjarnardóttir, S. H., Magnúsdóttir, R. Á, Eggertson, G., Kagan, S. A., and Andrésson, O. S. (1978). *Mol. Gen. Genet.* **161,** 89–98.
Tien, H. T. (1974). "Bilayer Lipid Membranes (BLM). Theory and Practice." Dekker, New York.
Todorov, G., Sokerov, S., and Stoylov, S. P. S. (1982). *Biophys. J.* **40,** 1–5.
Tokuda, H., and Konisky, J. (1978). *Proc. Natl. Acad. Sci. U.S.A.* **75,** 2579–2583.
Tokuda, H., and Konisky, J. (1979). *Proc. Natl. Acad. Sci. U.S.A.* **76,** 6167–6171.
Tokutomi, S., Iwasa, T., Yoshizawa, T., and Ohnishi, S. (1980). *FEBS Lett.* **114,** 145–148.
Tomochika, K.-I., and Hong, J.-S. (1978). *J. Bacteriol.* **133,** 1008–1014.
Tosaka, O., Enei, H., and Hirose, Y. (1983). *Trends Biotechnol.* **1,** 70–74.
Trudell, J. R. (1980). *Prog. Anesthesiol.* **2,** 261–270.
Trumpower, B. L., ed. (1982). "Function of Quinones in Energy Conserving Systems." Academic, New York.
Tsuji, K., and Neumann, E. (1983). *Biophys. Chem.* **17,** 153–163.
Tu, S.-I., Lam, E., Ramirez, F., and Marecek, J. F. (1981). *Eur. J. Biochem.* **113,** 391–396.
Tzagoloff, A. (1982). "Mitochondria." Plenum, New York.
Ullstrup, A. J. (1972). *Annu. Rev. Phytopathol.* **10,** 37–50.

Uratani, Y., and Cramer, W. A. (1981). *J. Biol. Chem.* **256**, 4017–4023.

van Dam, K., and Tsou, C. S. (1970). *In* "Electron Transport and Energy Conservation" (J. M. Tager, S. Papa, E. Quagliariello, and E. C. Slater, eds.), pp. 421–425. Adriatica Editrice, Bari, Italy.

van Dam, K., Woelders, H., Colen, A. M., and Westerhoff, H. V. (1983). *Biochem. Soc. Trans.* **12**, 401–402.

van der Bend, R. L., Cornelissen, J.B.W.J., Berden, J.A., and van Dam, K. (1984). *Biochim. Biophys. Acta* **767**, 87–101.

Vanderkooi, G. (1978). *In* "The Molecular Biology of Membranes" (S. Fleischer, Y. Hatefi, D. H. MacLennan, and A. Tzagoloff, eds.), pp. 29–55. Plenum, New York.

Vaz, W. L. C., Derzko, Z. I., and Jacobson, K. A. (1982). *In* "Membrane Reconstitution" (G. Poste and G. L. Nicolson, eds.), pp. 83–135. Elsevier, Amsterdam.

Venturoli, G., and Melandri, B. A. (1982). *Biochim. Biophys. Acta* **680**, 8–16.

Verhoek-Köhler, B., Hampp, R., Ziegler, H., and Zimmerman, U. (1983). *Planta* **158**, 199–204.

Vignais, P. M., Henry, M.-F., Sim, E., and Kell, D. B. (1981). *Curr. Top. Bioenerg.* **12**, 115–196.

Vinkler, C., and Korenstein, R. (1982). *Proc. Natl. Acad. Sci. U.S.A.* **79**, 3183–3187.

Vinkler, C., Avron, M., and Boyer, P. D. (1980). *J. Biol. Chem.* **255**, 2263–2266.

von Jagow, G., and Engel, W. D. (1980). *Angew. Chem. Int. Ed.* **19**, 659–675.

von Stedingk, L.-V. (1967). *Arch. Biochem. Biophys.* **120**, 537–541.

von Stedingk, L.-V., and Baltscheffsky, H. (1966). *Arch. Biochem. Biophys.* **117**, 400–404.

Vredenberg, W. J. (1976). *In* "The Intact Chloroplast" (J. Barber, ed.), pp. 53–88. Elsevier, Amsterdam.

Walz, D. (1983). *In* "Biological Structures and Coupled Flows" (A. Oplatka and M. Balaban, eds.), pp. 45–60. Academic Press, New York.

Weaver, C. A., Kagan, B. L., Finkelstein, A., and Konisky, J. (1981). *Biochim. Biophys. Acta* **645**, 137–142.

Webb, W. W. (1977). *In* "Electrical Phenomena at the Biological Membrane Level" (E. Roux, ed.), pp. 119–156. Elsevier, Amsterdam.

Webb, W. W., Barak, L. S., Tank, D. W., and Wu, E.-S. (1982). *Biochem. Soc. Symp.* **46**, 191–205.

Weinbach, E. C., and Garbus, J. (1965). *J. Biol. Chem.* **240**, 1811–1819.

Weinbach, E. C., and Garbus, J. (1969). *Nature (London)* **221**, 1016–1018.

Weiss, M. J., and Luria, S. E. (1978). *Proc. Natl. Acad. Sci. U.S.A.* **75**, 2483–2487.

Welch, G. R. (1977). *Prog. Biophys. Mol. Biol.* **32**, 103–191.

Welch, G. R., ed. (1985). "The Fluctuating Enzyme." Wiley, New York.

Welch, G. R., and Berry, M. N. (1983). *In* "Coherent Excitations in Biological Systems" (H. Fröhlich and F. Kremer, eds.), pp. 95–118. Springer-Verlag, Berlin and New York.

Welch, G. R., and Keleti, T. (1981). *J. Theor. Biol.* **93**, 701–735.

Welch, G. R., and Kell, D. B. (1985). *In* "The Fluctuating Enzyme" (G. R. Welch, ed.). Wiley, New York.

Welch, G. R., Somogyi, B., and Domjanovich, S. (1982). *Progr. Biophys. Mol. Biol.* **39**, 109–146.

Wendt, L. (1970). *J. Bacteriol.* **104**, 1236–1241.

Westerhoff, H. V. (1983). Ph.D. thesis, University of Amsterdam.

Westerhoff, H. V., and Arents, J. C. (1984). *Biosci. Rep.* **4**.

Westerhoff, H. V., and Dancshazy, Zs. (1984). *Trends Biochem. Sci.* **9**, 112–117.

Westerhoff, H. V., and van Dam, K. (1985). "Mosaic Non-equilibrium Thermodynamics and the Control of Biological Free-Energy Transduction." Elsevier, Amsterdam, in press.

Westerhoff, H. V., Simonetti, A. L. M., and van Dam, K. (1981). *Biochem. J.* **200**, 193–202.

Westerhoff, H. V., de Jonge, P. C., Colen, A., Groen, A. K., Wanders, R. J. A., van den Berg, G. B., and van Dam, K. (1982). *EBEC Rep.* **2**, 267–268.

Westerhoff, H. V., Helgerson, S. L., Theg, S. M., van Kooten, O., Wilkström, M., Skulachev, V. P., and Dancshazy, Z. S. (1983a). *Acta Biol. Acad. Sci. Hung.* **18**, 125–149.

Westerhoff, H. V., Colen, A., and van Dam, K. (1983b). *Biochem. Soc. Trans.* **11**, 81–85.
Westerhoff, H. V., Melandri, B. A., Venturoli, G., Azzone, G. F., and Kell, D. B. (1983c). *FEBS Lett.* **165**, 1–5.
Westerhoff, H. V., Groen, A. K., and Wanders, R. J. A. (1983d). *Biochem. Soc. Trans.* **11**, 90–91.
Westerhoff, H. V., Groen, A. K., Wanders, R. J. A., and Tager, J. M. (1984a). *Biosci. Rep.* **4**, 1–22.
Westerhoff, H. V., Melandri, B. A., Venturoli, G., Azzone, G. F., and Kell, D. B. (1984b). *Biochim. Biophys. Acta* **768**, 257–292.
White, D. C. (1974). *In* "Molecular Mechanisms in General Anaesthesia" (M. J. Halsey, R. A. Millar, and J. A. Sutton, eds.), pp. 209–221. Churchill Livingstone, Edinburgh.
Wiechmann, A. C. H. A., Beem, E. P., and van Dam, K. (1975). *In* "Electron Transfer Chains and Oxidative Phosphorylation" (E. Quagliariello, F. Palmieri, E. C. Slater, and N. Siliprandi, eds.), pp. 335–342. North-Holland Publ., Amsterdam.
Wikström, M., and Krab, K. (1980). *Curr. Top. Bioenerg.* **10**, 51–101.
Wilkström, M., and Penttilä, T. (1982). *FEBS Lett.* **144**, 183–189.
Wikström, M., Krab, K., and Saraste, M. (1981). *Annu. Rev. Biochem.* **50**, 623–655.
Williams, R. J. P. (1961). *J. Theor. Biol.* **1**, 1–17.
Williams, R. J. P. (1978a). *FEBS Lett.* **85**, 9–19.
Williams, R. J. P. (1978b). *Biochim. Biophys. Acta* **505**, 1–44.
Williams, R. J. P. (1982). *FEBS Lett.* **150**, 1–3.
Wilson, D. F., and Forman, N. G. (1982). *Biochemistry* **21**, 1438–1444.
Winget, G. D., Kanner, N., and Racker, E. (1977). *Biochim. Biophys. Acta* **460**, 490–499.
Wraight, C. A., Cogdell, R. J., and Chance, B. (1978). *In* "The Photosynthetic Bacteria" (R. K. Clayton and W. R. Sistrom, eds.), pp. 471–511. Plenum, New York.
Young, S. H., and Poo, M.-m. (1983). *Nature (London)* **304**, 161–163.
Yu, C.-A., and Yu, L. (1981). *Biochim. Biophys. Acta* **639**, 99–128.
Zaritsky, A., and MacNab, R. M. (1981). *J. Bacteriol.* **147**, 1054–1062.
Zeikus, J. G. (1979). *Enzyme Microb. Technol.* **1**, 243–252.
Zimmermann, U. (1982). *Biochim. Biophys. Acta* **694**, 227–277.
Zimmermann, U., and Vienken, J. (1982). *J. Membr. Biol.* **67**, 165–182.
Zimmermann, U., Scheurich, P., Pilwat, G., and Benz, R. (1981). *Angew. Chem. Int. Ed.* **20**, 325–344.
Zoratti, M., Pietrobon, D., and Azzone, G. F. (1982). *Eur. J. Biochem.* **126**, 443–451.
Zukav, G. (1980). "The Dancing Wu Li Masters," p. 328. Fontana/Collins, London.

3

Dynamic Compartmentation
in Soluble Multienzyme Systems

Peter Friedrich

Institute of Enzymology
Biological Research Center
Hungarian Academy of Sciences
Budapest, Hungary

I. INTRODUCTION

Compartmentation is one of the basic consequences and prerequisites of the structural organization of living matter. It is manifested at all levels of complexity, from the organismal down to the molecular. The very existence of cells testifies to the biological advantage of segregating certain types and amounts of biological materials and processes. Like modern laboratory furniture and instruments, living systems are built up on a *modular* basis. At the

ORGANIZED
MULTIENZYME SYSTEMS

141

supracellular level modules correspond to the different tissues or circumscribed regions within a given tissue, for example, in the brain. The formation of these modules can be followed during differentiation of clearly defined compartments, as revealed by the so-called fate maps of developmental genetics.

In this chapter, however, we shall not be concerned with such *inter*cellular compartmentation phenomena, but rather with intracellular ones. An *intra*cellular metabolic compartment has been defined by Kempner (1980) as "a subcellular region of biochemical reactions kinetically isolated from the rest of cellular processes." The compartment, which is an actual or virtual reservoir, accommodates definite *pools* of substances: macromolecules, metabolites, etc. The *pool* is then a *given amount of compound(s) subject to the above kinetic isolation.* When we speak of multiple pools of a compound, we usually mean kinetically distinguishable fractions, that is, fractions of the compound that exhibit different reactivities toward a suitable reactant. Multiple pools will manifest themselves in the heterogeneous reactivity of the given compound.

II. MACROCOMPARTMENTS VERSUS MICROCOMPARTMENTS

The interior of living eukaryotic cells is subdivided by membranes into various regions, such as the nucleus, mitochondrion, a great variety of cytosomes, and, last but not least, the medium in which all intracellular organelles are embedded, the cytoplasm. These regions correspond to more or less well-defined morphological entities, recognizable under the microscope, which have their characteristic content of macromolecules. According to DeDuve's (1964) "postulate of single location," each enzyme in the cell has its unique site of occurrence, that is, a membrane-bounded space where it is found. The characteristic distribution of enzymes is, conceivably, accompanied by a similar distribution of metabolites. Although small molecules diffuse rapidly even in the highly viscous cell interior, many metabolites, coenzymes, etc. cannot penetrate the boundary membrane. Since the various organelles and the cytoplasm are large relative to the molecular dimension, we may call them *macrocompartments.* In effect, the boundary membrane determines what is let into and out of the macrocompartment, and at what rate and stoichiometry. The distribution of, for example, ATP over several macrocompartments is equivalent to the existence of multiple ATP pools in the cell.

Multiple pools, however, may arise not only as a result of membranous septa, but through further, more subtle means of metabolite sequestration, which might be called *microcompartments,* for their dimension is on the order of the size of metabolites. Such a microcompartment is a "channel" in multienzyme clusters. These channels are formed by the association of two or

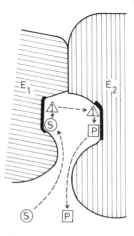

Bulk medium

Fig. 1.   Scheme of metabolite channeling in a stable two-enzyme complex. The substrate, S, of the first enzyme, E_1, can enter the microcompartment, where it is bound and transformed to I at the active site of E_1. This intermediate, after release from E_1, will be picked up by the active site of E_2, rather than diffuse out of the channel, and be transformed to P, which can leave the channel. The size of the microcompartment, relative to the size of metabolites, may be markedly smaller than depicted here, bordering the case when the facilitated diffusion of I from E_1 to E_2 becomes an intracomplex transfer effected by the concerted action of functional groups from both enzymes.

more enzymes, that is, they are spaces confined by the irregular surfaces around the active sites of enzymes (Fig. 1). The term *channeling of metabolites* has been introduced in connection with the sequestration of certain steps in amino acid biosynthesis in *Neurospora* (Vogel and Bonner, 1954; Matchett and DeMoss, 1964) on a merely functional basis. In fact, we do not yet know the precise structure, such as would be revealed by X-ray diffraction, of any such channels. Nevertheless, as illustrated in Fig. 1, it is probably true that such a microcompartment must be able to communicate with the bulk medium to allow the first substrate to enter and the last one to leave it. Sequestration of the intermediate (I in Fig. 1) is then somewhat probabilistic, insofar as the juxtaposition of active sites in the complex is such that the probability of the intermediate leaving the channel without being transformed to the final product is small, in stringent cases practically nil. This picture is valid for stable multienzyme clusters (complexes or conjugates) in which the physical juxtaposition of enzymes is permanent, or at least the lifetime of an enzyme–enzyme complex is long relative to the time required for a catalytic cycle. It may be recalled that *multienzyme cluster* is a collective name for physically associated enzymes (Welch and Gaertner, 1980): in multienzyme complexes the constituent enzymes have separate polypeptide chains, whereas in multienzyme conjugates the polypeptide chain is continuous, but the different catalytic activities are confined to more or less autonomous protein structural domains. Experimental evidence for such metabolite channeling has been provided, for example, for indole compartmentation in the tryptophan synthase complex (Yanofsky and Crawford, 1972; Matchett, 1974) and for several intermediary metabolites in the *arom* multienzyme cluster (Giles, 1978; Welch and Gaertner, 1975).

Another type of microcompartmentation is embodied by the Debye–Hückel layer (Träuble, 1976) on the surface of some biomembranes (see Hess, 1980). The solvent layer at the structural surface creates a microenvironment, that is, it behaves differently from the bulk medium. In this microenvironment metabolites may to some extent be confined; their three-dimensional diffusion is reduced to two dimensions. It is clear that such layers are not like compartments with surrounding walls, not even like the enzyme–enzyme channel, whose physically partially secluded space still exists. The Debye–Hückel layer is "open," in the sense that it has a firm boundary only on one side (i.e., the membrane) whereas the other boundary is virtual, corresponding to a moderate energy barrier. Its "micro" character refers to its thickness, for in the other two dimensions it is well over the molecular size range. Experiments with artificial matrix-bound enzyme systems indicate that for metabolite compartmentation to occur the solid support, that is, the surface layer, is crucial, rather than the physical (random) clustering of the enzymes in solution, as achieved, for example, by chemical cross-linking (Koch-Schmidt *et al.*, 1977). Microenvironmental effects in multienzyme systems are dealt with in detail by Siegbahn *et al.* in Chapter 6.

A further type of microcompartmentation is created by the existence of (possibly multiple) binding sites for metabolites within a macrocompartment. Here the physical image of a bounded space is entirely irrelevant. However, if we think of the definition of a metabolic compartment as a kinetically isolated fraction of a compound, the conceptual relevance is obvious. Although bound metabolites are, as a rule, in equilibrium with their unbound (free) fraction, the rate of equilibration may vary over a wide range depending on the on- and off-rate constants. The smaller is the off-rate constant for a metabolite–binding site complex the more pronounced will be the extent of isolation. It follows that "isolation" is not meant in an absolute sense; it is perhaps more appropriate to speak of kinetically distinguishable fractions, where distinction is a function of the resolving power of our investigative method. Clearly, powerful and sensitive analytical methods may reveal microheterogeneities in the reactivity of some compound in a living cell that are meaningless from the biological point of view. Nevertheless, it is important for any type of compartmentation phenomenon, and by the same token for many other molecular and systemic properties of biological materials, that its physiological role should be carefully examined. Later in this chapter we shall briefly consider the potential significance of microcompartmentation via binding sites in the regulation of metabolic pathways, and present experimental evidence for such microcompartmentation in erythrocyte glycolysis.

III. STATIC VERSUS DYNAMIC COMPARTMENTS

Another criterion for the classification of metabolic compartments is according to their lifespan. Static compartments are durable, that is, once formed they persist over long periods of time and are abolished by cellular processes such as proteolysis (related to protein turnover), organellar disassembly, or effector-induced dissociation. Dynamic compartments, in contrast, are transient; they form and decompose rapidly; their very existence is the result of a dynamic equilibrium. Organellar macrocompartments are practically all static, just as are stable multienzyme clusters. Layer-type and binding-site-type microcompartments do not lend themselves unequivocally to this categorization, for the following reasons. Although the area of membrane surfaces, as well as of binding sites for metabolites, is fairly constant in the living cells and therefore they would qualify as static compartments, their content, that is, the metabolite pool they harbor, may nevertheless rapidly change, which is clearly a dynamic character (see Kurganov, Chapter 5). Here we are confronted with a problem both semantic and conceptual, namely, the usage in the literature tends to treat the terms compartment and pool synonymously, although the former is the physical space in which the latter is accommodated. In the virtual microcompartments the physical space is no longer apparent. Therefore it is perhaps more appropriate to state that in the case of layer- and binding-site-type microcompartments the capacity of the compartment is fairly stable, that is, static, whereas the harbored metabolite pools may undergo rapid, dynamic changes.

As we shall describe in a later model, enzyme–enzyme channels may form not only in stable multienzyme clusters, but also transiently, in short-lived complexes of two otherwise separate enzymes. If during the lifetime of these complexes the product of the first enzyme is transferred to the second enzyme, then efficient metabolite compartmentation will ensue even though the two catalytic entities are not permanently associated. This type of hypothetical compartmentation mechanism is indeed dynamic, as it is the result of the incessant motion and collision of the enzymes molecules.

A special kind of dynamic compartmentation is generated by biochemical oscillations. Compartmentation means here the spatial heterogeneity of metabolites created by oscillations in an otherwise homogeneous biochemical system, which impose spatiotemporal order on a random mixture. As described by Nicolis and Prigogine (1977), in chemical systems far from equilibrium, the processes of diffusion and some chemical reaction may be coupled, so that diffusion tends to homogenize the reactant in the mixture, while the chemical reaction works against this equilibration. Hess and co-workers have studied glycolytic oscillations in detail (Boiteux *et al.*, 1975, 1980; Boiteux and Hess, 1978; Hess *et al.*, 1978; Hess, 1980); the pacemaker enzyme of these oscillations is phosphofructokinase. Apart from various clock functions, oscillations have been invoked to play a role in suppressing futile cycles, for example, in glycolysis by creating alternating reactant gradients (Boiteux *et al.*, 1980).

IV. EVIDENCE FOR STRUCTURAL ENZYME ORGANIZATION IN GLYCOLYSIS

There is an ever-increasing body of evidence that enzyme–enzyme interactions of functional relevance occur not only among the constituents of recognized multienzyme complexes and conjugates, but also among the so-called soluble enzymes. This operative term means that these enzymes can be extracted from ruptured cells by dilute aqueous salt solutions, and that they occur in the extract, and can be purified therefrom as individual proteins. Soluble enzymes can usually be readily crystallized and are catalytically fully active after redissolving the crystals. Our current knowledge about enzymes is mainly based upon studies with such soluble enzyme proteins.

Nevertheless, while this analytic trend of research has provided us with deep insights into the anatomy of enzyme action at the molecular and atomic levels, it has inevitably neglected the relationship to other enzymes that coexist with a given enzyme in the living cell. The highly artificial conditions of classical *in vitro* enzymology (e.g., low enzyme and high substrate concentrations and lack of other "contaminating" enzymes) did not allow one to draw conclusions as to how enzymes work *in situ*, in their natural environment. One would not expect the basic mechanism of action of an enzyme, as established *in vitro*, to change under the conditions *in vivo*. Nonetheless, the sum of minor perturbations may eventually markedly influence the output of a multienzyme pathway.

Therefore the earlier, and also current, analytic approach has to be complemented by synthetic endeavors, *viz.* to test the behavior of enzymes in each others' company, under near-physiological conditions. In a manner of speaking, what has been carefully taken apart and examined in detail has now to be put back together and examined for novel, systemic properties.

An important example of such potential phenomena is metabolite compartmentation.

There are two major macrocompartments in cells where many soluble enzymes are located: the mitochondrial matrix and the cytoplasm. The concentration of enzymes is high in both. Hackenbrock (1968) estimated the protein concentration in the mitochondrial matrix to be $\sim 56\%$ (w/w). Srere has pointed out (see Srere, 1980, 1981; Srere and Henslee, 1980) that this is close to the value calculated for the densest possible array of identical spheres, hence the packing of protein molecules in the mitochondrial matrix is quasi-crystalline. Under such conditions the enzymes are bound in such a manner that they interact with one another. The question of enzyme organization in the mitochondrial matrix is discussed in detail by P. Srere in Chapter 1.

In the cytoplasm of several cell types, a major fraction of soluble enzymes are those needed for glycolysis. For example, in yeast, glycolytic enzymes constitute about 65% of total soluble protein (Hess et al., 1969). In muscle, the space not occupied by the proteins of the contractile apparatus seems to be filled with glycolytic enzymes (Sigel and Pette, 1969). In erythrocytes, which are bags containing hemoglobin at a concentration of about 300 mg/ml glycolysis is practically the sole energy-generating pathway. In the following sections we review experimental evidence for the supramolecular organization of glycolysis in muscle and human red cells (see also Kurganov, Chapter 5).

A. Muscle Glycolysis

1. Interactions between Enzyme Pairs

During prolonged vigorous activity the considerable store of glycogen in striated muscle is broken down and lactate is formed, that is, glycogenolysis occurs. This proceeds mainly under anaerobic conditions, because oxygen supplies are insufficient in heavily working muscle. Therefore it is vital for the organism to run glycogenolysis as efficiently as possible.

Glycogenolysis starts with the phosphorylase reaction. Glycogen phosphorylase (EC 2.4.1.1) catalyzes the phosphorolytic cleavage of one glucosyl residue from glycogen to produce glucose 1-phosphate. This dimeric enzyme is the last member of a multistep enzyme cascade, a chain of reactions in which each enzyme serves as substrate for the preceding enzyme of the sequence. In muscle, phosphorylase occurs associated with glycogen; figuratively, the phosphorylase dimers sit on the polysaccharide branches as birds on a tree. In this "glycogen particle" (Meyer et al., 1970) phosphorylase coexists with the opposing enzyme glycogen synthase (EC 2.4.1.11) and its converting enzymes phosphorylase kinase (EC 2.7.1.38) and phosphorylase phosphatase (EC

3.1.3.17). Owing to the enzyme–substrate relationship, these enzymes, and several others in the cascade, necessarily interact with each other, which has been amply demonstrated. Since our interest now is focused on metabolite compartmentation, we will not describe the phosphorylase system in any more detail; the reader is instead referred to the reviews by Fletterick and Madsen (1980) and Dombrádi (1981).

From the viewpoint of enzyme–enzyme interactions, the central part of glycolysis has been analyzed most thoroughly. Several lines of evidence indicate that fructose-1,6-bisphosphate aldolase (aldolase, EC 4.1.2.13) and D-glyceraldehyde-3-phosphate dehydrogenase (GAPD, EC 1.2.1.12) cooperate in some manner, although the reports are far from unanimous about the details of the interaction. Ovádi and Keleti (1978) studied the kinetics of the coupled two-enzyme reaction by the stopped-flow technique. For the K_m of aldolase-generated glyceraldehyde 3-phosphate (GAP) they found a value 10^{-5} M, which was about one order of magnitude lower than the value obtained with GAPD alone and GAP supplied from a stock solution. Since GAP is an aldehyde, in aqueous medium it exists in the hydrated diol form [the ratio diol/oxo was found to be 29 by Trentham *et al.* (1969), and about 60, in Tris buffer, by Ovádi and Keleti (1978)] as opposed to its nascent oxo form, which is the substrate of enzymes. The coupled reaction can then be described in a simplified form as follows:

$$\text{fructose 1,6-P}_2 \xrightarrow{\text{Aldolase}} \text{GAP}_{\text{ald}} \xrightarrow{\text{GAPD}} \text{NADH}$$
$$k_1 \Big\downarrow \quad \Big\uparrow k_{-1}$$
$$\text{GAP}_{\text{diol}}$$

where the subscripts to GAP denote the aldehyde and diol forms. If k_1, the rate constant for hydration of the oxo form, is small relative to k_{cat} for the GAPD reaction, then the above K_m value refers to the oxo form of GAP. However, it was found that hydration was faster than dehydrogenation. The authors therefore suggested that the low apparent K_m value observed in the coupled reaction was due to the direct transfer (channeling) of GAP between the two enzymes, which prevented its trapping by water.

Ovádi *et al.* (1978) were able to corroborate their suggestion of an aldolase-GAPD interaction by a physicochemical method. They measured the polarization of fluorescence of fluorophores covalently attached either to aldolase or GAPD, while increasing the concentration of the unlabeled partner enzyme. In both setups fluorescence polarization was augmented, which suggests complex formation between the two enzymes. The apparent dissociation constant that could be derived was 3×10^{-7} M and the rate constant of association was surprisingly low: about 40 $M^{-1}\text{s}^{-1}$. Taking these two values, one arrives at a half-life for the complex of about 40 h. It is intriguing that such an apparently rather stable complex escaped detection by other physicochemical methods (see below).

Grazi and Trombetta (1980) adopted a somewhat circumstantial approach to test complex formation and GAP channeling between aldolase and GAPD. They measured the amount of aldolase–dihydroxyacetone phosphate intermediate and of the 3-phosphoglyceryl–GAPD intermediate in the mixture of the two enzymes. The concentration of neither of these intermediates changed during prolonged incubation and, furthermore, the concentration of the latter intermediate was unaffected if triose phosphate isomerase, which would serve as a trapping agent for GAP, was also included in the mixture. The authors have concluded that their results are compatible with slow complex formation but not with GAP channeling between the two enzymes.

Kálmán and Boross (1982) used an affinity batch system to test for enzyme–enzyme interaction. They monitored the binding of GAPD to a NAD–Sepharose 4B sorbent in the absence and presence of aldolase. They found that although GAPD bound to the sorbent with an apparent $K_d \simeq 5$–12×10^{-4} M, aldolase alone or in the presence of GAPD did not bind at all; moreover, it interfered with the binding of GAPD. The authors suggest that sorbent-bound GAPD undergoes a conformational change that renders it unsuitable for aldolase binding and vice versa; the aldolase–GAPD complex cannot bind to the affinity sorbent. On the basis of these premises the authors calculated the dissociation constant for the aldolase–GAPD complex (of $1:1$ stoichiometry) to be $\sim 10^{-6}$ M at pH 7.5 . In fact, the pH dependence of complex formation was pronounced, there being negligible association below pH 7.0 and a $K_d = 10^{-7}$ M at pH 8.5, the practical upper limit of measurements.

Patthy and Vas (1978) exploited the chemistry of the aldolase reaction to test for interaction with GAPD. Their approach was based on the observation that aldolase catalyzes a "suicide" reaction in the presence of electron acceptors [e.g., $K_3Fe(CN)_6$] when cleaving fructose 1,6-bisphosphate (Christen et al., 1976). Namely, the carbanion intermediate of aldolase–dihydroxyacetone phosphate can be oxidized to hydroxypyruvaldehyde phosphate, which is a powerful arginine-modifying reagent (Patthy, 1978). In the substrate-binding site of aldolase there are arginine residues (Patthy and Thész, 1980) and the modification of one of them, Arg-55, abolishes enzyme activity (Patthy et al., 1979). Since only the nascent oxo form of hydroxypyruvaldehyde phosphate is a potent arginine modifier, its hydrated form being practically inactive, the suicide reaction is the result of the action of nascent reagent on the parent enzyme molecule (Patthy, 1978). When GAPD was added to the aldolase suicide mixture, the former enzyme was also inactivated (Fig. 2). The plot of GAPD molecules inactivated per catalytic cycle of

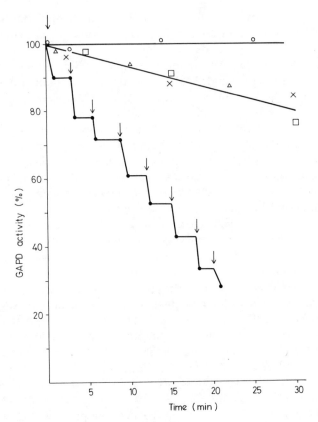

Fig. 2. Demonstration of enzyme–enzyme interaction by *syn*-catalytic inactivation. GAPD ($2 \times 10^{-5} M$) from rabbit muscle was incubated with aldolase ($10^{-4} M$), fructose 1,6-bisphosphate (5 mM), and repeated doses of hexacyanoferrate (III) at pH 7.5 and 20°C. Hexacyanoferrate doses (1 mM), added as indicated by the arrows, were reduced immediately, then GAPD activity (●) was assayed. Control experiments: ○, hexacyanoferrate (III) omitted; ×, aldolase omitted; △, fructose 1,6-bisphosphate omitted from the reaction mixture; □, GAPD added after the reduction of 8 mM hexacyanoferrate (III). (From Patthy and Vas, 1978.)

aldolase versus GAPD concentration gave a rectangular hyperbola, that is, it exhibited saturation-type behavior with an apparent dissociation constant for aldolase–GAPD (1:1 molar ratio) of about 10^{-6} M. Complex formation is strongly indicated by the fact that the solvent (H_2O)-mediated diol form of the reactant could not have been an efficient modifier, which it definitely was. In fact, whereas one out of every 68 hydroxypyruvaldehyde phosphate molecules produced inactivated one aldolase subunit, about one out of six molecules inactivated one GAPD subunit. Thus the aldolase-generated reactant was much more efficient toward GAPD than toward the parent aldolase. Later in this chapter we shall return to the possible meaning of this observation. It should be added that the syncatalytic inactivation of GAPD by aldolase did not require prolonged preincubation; it was observed practically immediately after mixing the two enzymes. Therefore the phenomenon could not be the consequence of *slow* complex formation. Similarly, Kwon and Olcott (1965) observed that the activity of tuna aldolase was enhanced by rabbit muscle GAPD, which also seemed to be an instantaneous phenomenon.

In contrast to the experiments discussed so far, in some other investigations there was no sign of physical interaction between muscle aldolase and GAPD. Földi *et al.* (1973) passed a concentrated myogen fraction containing both enzymes through a gel chromatographic column using the frontal analysis technique and monitored the elution volumes for various glycolytic enzyme activities. They found that aldolase emerged from the column at an apparent molecular weight significantly higher than its actual M_r (158,000), but the position of GAPD corresponded to its own molecular weight. Thus the association partner for aldolase was not GAPD, but some other constituent of the myogen fraction, possibly element(s) of the contractile proteins, or glycerol-3-phosphate dehydrogenase or fructose-1,6-bisphosphatase (see below).

Likewise, Masters and Winzor (1981) were unable to detect aldolase–GAPD interaction by frontal analysis gel chromatography and by velocity and equilibrium sedimentation measurements. They attributed the positive results obtained with fluorescence polarization (Ovadi *et al.*, 1978) to the "stickiness" of the fluorescent probe. In fact, Church *et al.* (1981) have shown that prolactin modified with fluorescein isothiocyanate binds to bovine serum albumin with a $K_d \simeq 10^{-7} M$, whereas no detectable association was found with native prolactin.

In conclusion, for the aldolase–GAPD coupling, the reports are contradictory both in respect to finding any interaction and, in the positive case, concerning the details of complex formation. It should be borne in mind, however, that in the detection of weak macromolecular interactions much depends on the reaction conditions, which were far from uniform in the above studies.

Aldolase has been suggested to associate with other enzymes, too. Rabbit muscle cytoplasmic glycerol-3-phosphate dehydrogenase (EC 1.1.1.8) was claimed, on the basis of the active enzyme centrifugation technique (Cohen and Mire, 1971), to combine with aldolase with an apparent $K_d \simeq 10^{-7} M$ (Batke *et al.*, 1980). Fluorescence polarization studies with glycerol-3-phosphate dehydrogenase labeled with fluoresceine isothiocyanate revealed a saturation-type increase in anisotropy with increasing aldolase concentration

Ovádi et al., 1983). Glycerol-3-phosphate dehydrogenase is a dissociable dimer, and it is less active in the monomeric state than as a dimer (Batke et al., 1980). Aldolase exhibits greater affinity toward the dimer ($K_d \simeq 0.2 \mu M$) than toward the monomer ($K_d \simeq 1 \mu M$), which could explain the activating effect of aldolase on the dehydrogenase. It is worth mentioning, not only for its historic but also for its heuristic interest, that Baranowski and Niederland (1949) found aldolase and glycerol-3-phosphate dehydrogenase cocrystallized in myogen A.

Aldolase also seems to interact with triose phosphate isomerase (EC 5.3.1.1), the enzyme that equilibrates the two triose phosphate products of the aldolase reaction. The structural connection between the two enzymes from rabbit muscle was indicated by their mutual protection against denaturation by perchloric acid (Salerno and Ovádi, 1982). These two enzymes from the flight muscle of the insect *Ceratitis capitata* were found to interact with each other by fluorescence polarization and gel permeation measurements in batch system (Gavilanes et al., 1981). The apparent K_d deduced from the fluorescence anisotropy data was 2×10^{-6} M.

Though presumably not relevant to muscle, it is worth mentioning that liver aldolase forms a complex with fructose-1,6-bisphosphatase (EC 3.1.3.11) as demonstrated by Horecker and co-workers (Pontremoli et al., 1979; MacGregor et al., 1980). The data suggested that one or two bisphosphatase tetramers and one aldolase tetramer built up the complex, which was formed rapidly. The interaction was specific: if instead of the liver enzyme the muscle enzyme was used as either partner in the mixture, no complex was found. It is of interest to note that, although complex formation was indicated with the homologous enzymes by fluorescence, Zn^{2+} binding to the bisphosphatase, gel penetration, and limited proteolytic data, no diagnostic change for complex formation was found by sedimentation and column gel chromatographic techniques.

The enzyme after GAPD in the sequence of glycolysis is 3-phosphoglycerate kinase (PGK, EC 2.7.2.3). The possible interaction between the two enzymes has been studied by several groups of workers (see also erythrocyte glycolysis, below), namely, it seemed teleologically attractive that the intermediary metabolite of the coupled reaction, 1,3-bisphosphoglycerate (1,3-P_2G), which is a rather unstable high-energy compound (see Huskins et al., 1982), be channeled between the two enzymes rather than mixed with bulk water and exposed to hydrolysis. Furthermore, PGK (M_r 45,000), the only monomeric enzyme in glycolysis, has a rather peculiar shape: it consists of two structural domains connected by a "hinge" (Blake and Rice, 1981). It has been suggested that one of these domains might serve to interact with GAPD (Scopes, 1973) or even that the two enzymes might form a unique three-dimensional structure (Ottaway and Mowbray, 1977).

Vas and Batke (1981) conducted a systematic study in an attempt to detect symptoms of interaction between PGK and GAPD but came up with a negative conclusion. Three lines of evidence indicated the absence of interactions.

1.   The inverse relationship between transient time and GAPD concentration, that is, the linearity of the $1/\tau$ versus [GAPD] plot (see below), in the coupled reaction:

$$\text{3-PG} \xrightarrow[\text{PGK}]{\text{ATP}\quad\text{ADP}} \text{1,3-P}_2\text{G} \xrightarrow[\text{GAPD}]{\text{NADH}\quad\text{NAD}} \text{GAP} + \text{P}_i$$

where 3-PG stands for 3-phosphoglycerate.

2.   The lack of any change of fluorescence anisotropy of GAPD labeled with fluorescein isothiocyanate in the presence of PGK.

3.   The lack of any shift toward higher apparent molecular weights of the elution profile of PGK on frontal analysis gel chromatography in the presence of high concentrations of GAPD.

The transient-time measurement has been recommended by Hess and Wurster (1970) as a test of metabolite compartmentation. The transient time, τ, is a parameter characterizing the speed of attaining the steady state in a reaction sequence. By definition, τ is the time required for the intermediate metabolite to reach 63% [i.e., fraction $1\text{-}(1/e)$] of its steady-state value. In a random, nonorganized, two-enzyme system $1/\tau$ is directly proportional to the concentration of the second enzyme under specified conditions (see Kuchel, Chapter 7; McClure, 1969; and also below). If the two enzymes interact so that the intermediary metabolite is channeled, τ is expected to be drastically shortened. The reason for this is that the intermediate need not accumulate to secure the appropriate velocity of the second enzyme reaction. Hess and Boiteux (1972) have used this technique to screen for enzyme–enzyme interactions among the glycolytic enzymes of yeast, with negative results.

While the above transient-time measurement with muscle PGK–GAPD by Vas and Batke (1981) also proved negative, another kinetic approach led to an apparently contradictory result. Weber and Bernhard (1982) also studied the PGK–GAPD coupled reaction in the direction of NAD production, but they started with the 1,3-P$_2$G–PGK complex rather than running the PGK reaction from 3-PG and ATP. They considered two possible mechanisms:

Mechanism I (random diffusion):

$$\text{E}_1\text{S} \underset{k_{-1}}{\overset{k_1}{\rightleftharpoons}} \text{E}_1 + \text{S} \tag{1}$$

$$\text{E}_2 + \text{S} \underset{k_{-2}}{\overset{k_2}{\rightleftharpoons}} \text{E}_2\text{S} \xrightarrow{k_3} \text{products} \tag{2}$$

Mechanism II (direct transfer):

$$E_1S + E_2 \underset{k_{-a}}{\overset{k_a}{\rightleftharpoons}} E_1SE_2 \xrightarrow{k_b} \text{products} \tag{3}$$

where E_1 and E_2 stand for PGK and GAPD, respectively, and S is 1,3-P_2G.

In mechanism I (random diffusion) S has to dissociate from E_1 and then be picked up by E_2. In mechanism II (direct transfer) the substrate for E_2 is the E_1S complex, that is, no free S is involved. Weber and Bernhard (1982) claim that the two mechanisms can be distinguished kinetically: under pseudo-first-order conditions the rate constant for the decay of E_1S, as measured by monitoring the appearance of the product NAD, will depend on the total concentrations of E_1 and E_2 in different manners. In the random mechanism it will be determined, in essence, by the ratio $E_2/(E_1 + E_2)$, whereas in the direct transfer mechanism it will be directly proportional to E_2. (A factor of uncertainty arises, though, as the relative values of products of rate constants with which the enzyme concentrations have to be weighted are unknown.) Experimental evidence supported the latter alternative, on which basis the authors concluded that 1,3-P_2G is channeled between the two enzymes.

The source of the discrepancy between the conclusions of Vas and Batke (1981) and Weber and Bernhard (1982) is not clear. It is to be noted, however, that the experimental systems of the two groups were not identical. Apart from such trivial differences as the origin of enzyme (pig muscle versus halibut) and the PGK/GAPD molar ratio ($\ll 1$ versus $\gg 1$), one important difference seems to be the high concentration (20 mM) of 3-PG present in the mixture of Vas and Batke (1981). For Weber and Bernhard (1982) observed that the rate of the overall reaction was increased 20-fold by the addition of 3-PG (half-maximal, i.e., 10-fold, activation was at 0.6 mM 3-PG). These authors proposed that 3-PG played a facilitating role in the transfer of 1,3-P_2G by inducing a structural change in PGK. Indeed, profound structural alterations have been observed in PGK on 3-PG binding by X-ray diffraction (Banks et al., 1979) and small-angle X-ray scattering (Pickover et al., 1979) (see also Blake and Rice, 1981).

Nevertheless, Weber and Bernhard (1982) suggest no explanation as to how 3-PG brings about this activation. They argue that millimolar 3-PG concentrations cannot displace any substantial amount of bound 1,3-P_2G from PGK by direct competition, owing to the great difference in the binding constants in favor of the bisphosphate derivative. Where then is the point of attack of 3-PG? If we disregard the unattractive hypothesis of a distinct (effector) 3-PG site on PGK, we are left with two alternatives: (a) 3-PG exerts its activating effect on GAPD in the complex or (b) 3-PG binds to the substrate-binding site in PGK after the dissociation and transfer to GAPD of 1,3-P_2G, and its rate-enhancing effect is due to the dissociation of the two-enzyme complex. It must be admitted that while the first alternative is unlikely,

(3-PG markedly inhibits rather than activates GAPD) (see Batke and Keleti, 1968), there is no evidence whatsoever for the second. Nevertheless, at the end of this chapter we shall consider some models that may account for the observed phenomenon.

Without the intention to make unwarranted conclusions from contradictory data, it may be pertinent here to examine the diagnostic value of transient time measurements for enzyme–enzyme interactions. The linearity of the plot of $1/\tau$ versus $[E_2]$ means that (a) there is *no* complex formation between the two enzymes

$$E_1 + E_2 \overset{K_d}{\rightleftharpoons} E_1E_2 \tag{4}$$

with a K_d value allowing considerable changes in the degree of association within the enzyme concentration ranges tested, and (b) the complex, if formed, would effect the channeling of intermediary metabolite independent of the catalytic cycle. In other words, such an association equilibrium would exist irrespective of the presence of substrates and the active sites of the two enzymes would become juxtaposed in the complex. Since the derivation of transient time (McClure, 1969; Hess and Wurster, 1970; Kuchel, Chapter 7) implies that (a) the consecutive reactions are (practically) irreversible, (b) the rate of the first enzymatic reaction, v_1, is much smaller than the V_{max} of the second enzymatic step, and (c) the steady-state concentration of the intermediate, I_{ss}, is much smaller than K_{m2}, the Michaelis constant of E_2 for I, we may write the reaction sequence in the following simplified way [Bartha and Keleti (1979) have cautioned that if $[E_1] \to 0$, the $1/\tau$ versus $[E_2]$ plot becomes hyperbolic]:

$$E_1 + S \underset{k_{-1}}{\overset{k_1}{\rightleftharpoons}} E_1S \overset{k_2}{\longrightarrow} E_1I \overset{k_3}{\longrightarrow} E_1 + I \tag{5}$$

$$E_2 + I \underset{k_{-4}}{\overset{k_4}{\rightleftharpoons}} E_2I \overset{k_5}{\longrightarrow} E_2 + P \tag{6}$$

If the enzyme–enzyme complex of Eq. (4) is formed, the sequence is modified as follows:

$$E_1E_2 + S \underset{k_{-1}}{\overset{k_1}{\rightleftharpoons}} SE_1E_2 \overset{k_2}{\longrightarrow} IE_1E_2 \overset{k_t}{\longrightarrow} E_1E_2I \overset{k_5}{\longrightarrow} E_1E_2 + P \tag{7}$$

where it is assumed that the microscopic state constants of the individual enzyme reactions are not affected by the association. Metabolite channeling then would mean that k_t, the rate constant of transfer of I within the complex, is much greater than the same parameter in the nonclustered case, where it is a complex function of k_3, k_4, k_{-4} and $[E_2]$. Only if the above-described complex formation with its functional corollaries occurs can one expect an *upward* curvature in the $1/\tau$ versus $[E_2]$ plot, indicative of enzyme–enzyme interaction.

Let us consider then the case, hypothesized earlier by Friedrich (1974; see also below) and now claimed to have been detected by Weber and Bernhard (1982), in which E_1I can react with E_2 and the intermediate I is directly transferred. Then the second reaction (Eq. 6) will modify to

$$E_1I + E_2 \xrightleftharpoons[k_{-6}]{k_6} IE_1E_2 \xrightarrow{k_i'} E_1E_2I \xrightarrow{k_7} E_1 + E_2 + P \qquad (8)$$

where k_i' may or may not be identical with k_t above. If $k_3 \ll k_6[E_2]$, the dissociation of E_1I will be negligible. One can apply then the formalism of transient-time derivation, provided that $v_1 \ll V_{max2}$, with the difference that instead of the condition $[I]_{ss} \ll K_{m_2}$, the inequality $[E_1I]_{ss} \ll K'_{m_2}$ should hold, where K'_{m_2} refers to the reaction of Eq. (8). It follows from the foregoing that formally such a one-encounter type of metabolite transfer cannot be distinguished from the random mechanism involving free I by transient time measurements.

It remains to be scrutinized, of course, whether in the PGK–GAPD coupled reaction the E_1I species can substitute for I as steady-state intermediate. The available data are insufficient to prove or disprove this possibility. However, if the dissociation rate constant of the PGK–1,3-P_2G complex is as low as stated by Weber and Bernhard (1982), $\sim 1 \text{ s}^{-1}$, then it is not unlikely that at the relatively high GAPD concentrations of the transient time measurements the condition $k_3 \ll k_6[\text{GAPD}]$ is met.

2. Interactions with the Contractile Apparatus

The juxtaposition, or even association, of soluble enzymes may markedly be promoted by a matrix that serves as a solid support for enzyme clustering (see also Kurganov, Chapter 5). In muscle, the proteins of the contractile apparatus are likely candidates for such a role. The first observations on rabbit muscle proteins by Pette and co-workers (Arnold and Pette, 1968, 1970; Arnold et al., 1971) showed that F-actin bound several glycolytic enzymes, most strongly aldolase and GAPD, and less strongly phosphofructokinase, PGK, pyruvate kinase, and lactate dehydrogenase. Association increased both the V_{max} and K_m of aldolase. Histochemical analysis localized the glycolytic enzymes in the I-band of muscle (Arnold et al., 1969; Sigel and Pette, 1969). Subsequently, Masters and his co-workers conducted extensive studies on the rabbit muscle system. By analyzing the sedimentation profiles of muscle extracts they observed apparent molecular weight increases of aldolase, lactate dehydrogenase, and pyruvate kinase (Clarke and Masters, 1973). The molecular weight shift for aldolase was also found by Földi et al. (1973) using the frontal analysis gel chromatographic method. The association partner in muscle extract of these glycolytic enzymes was later identified as the thin filament, F-actin–tropomyosin–troponin (Clarke and Masters, 1974). At high protein concentrations a considerable portion of the enzymes remained bound at 0.15 M salt concentration (Clarke and Masters, 1975; Kuter et al., 1981), which in earlier studies was not observed, that is, binding only occurred at low ionic strength, and had raised the doubt that the enzyme–muscle protein interaction was a hypotonic artifact. The kinetic properties of some of the enzymes were affected by binding and by Ca^{2+} ions (for reviews see Clarke and Masters, 1976; Masters, 1978).

Electron microscopic studies first suggested the troponin complex as the carrier of binding sites (Morton et al., 1977), later tropomyosin, and, again, actin were also implicated (Stewart et al., 1980; Walsh et al., 1980). Although the two-dimensional lattices formed from thin filaments cross-linked by aldolase (or phosphofructokinase or GAPD or pyruvate kinase) molecules are probably in vitro artifacts, they demonstrate the harboring capacity of the contractile apparatus for glycolytic enzymes. F-Actin and the thin filament altered the kinetic properties of phosphofructokinase (Liou and Anderson, 1980).

In spite of the observed kinetic changes in enzymes induced by binding to muscle proteins, the biological role of these macromolecular interactions remains to be clarified. Hypotheses as daring as a direct involvement in the motile function of cells have been proposed (Stewart et al., 1980). Apart from specific effects, it is reasonable to assume that the protein scaffolding provided by the contractile apparatus creates a microenvironment in which metabolite compartmentational phenomena may occur. Furthermore, the enzyme–enzyme interactions discussed in the previous section may be amplified in the densely packed muscle cell.

B. Erythrocyte Glycolysis

The other cell type in which the supramolecular organization of glycolysis has been extensively studied is the human erythrocyte. The mature red cell is crammed with hemoglobin, the O_2 and CO_2 transport protein. It has a rudimentary metabolic system consisting mainly of glycolysis and the pentose phosphate pathway. ATP, needed primarily for running the ion pumps and for shape maintenance, is obtained almost entirely from glycolysis. Another product of glycolysis, unique to the red cell, is 2,3-bisphosphoglycerate (2,3-P_2G) generated in the 2,3-P_2G bypass, discovered by Rapoport and Luebering (1950). 2,3-P_2G is an allosteric effector of hemoglobin, lowering its oxygen affinity. The relative simplicity of the red cell, as well as its medical relevance, has rendered it a popular subject for metabolic studies. Rapoport (1968) reasoned that it would probably be the erythrocyte, of all cells, whose metabolism would first be completely understood. While this prediction may prove correct, there seem to be complicating factors in connection with enzyme organization and metabolite compartmentation that have not yet been taken into account in descriptions of erythrocyte glycolysis (e.g., Heinrich *et al.*, 1977). Thus, this system may prove to be a harder nut to crack than anticipated.

1. Relation of Glycolytic Enzymes to the Erythrocyte Membrane

Early experiments on the incorporation of extracellular $^{32}P_i$ into glycolytic intermediates of red cells indicated that ATP was labeled more rapidly than intracellular P_i (Gourley, 1952; Prankerd and Altman, 1954; Gerlach *et al.*, 1958). Since P_i enters glycolysis at the GAPD reaction (see Fig. 3), the implication of this finding was that GAPD was located at the cell membrane near the anion channel so that it had ready access to penetrating P_i. Although

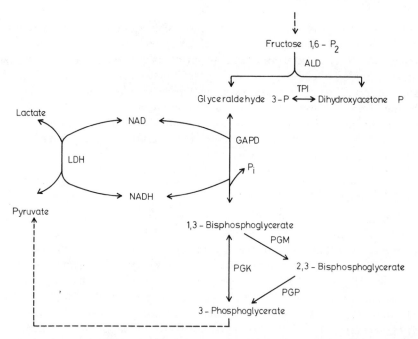

Fig. 3. Part of the glycolytic pathway in erythrocytes illustrating NAD–NADH recycling and the 2,3-bisphosphoglycerate by-pass. ALD, aldolase; TPI, triose phosphate isomerase; GAPD, glyceraldehyde-3-phosphate dehydrogenase; PGK, 3-phosphoglycerate kinase; PGM, phosphoglycerate mutase; PGP, phosphoglycerate phosphatase; LDH, lactate dehydrogenase. If exogenous glycerol-3-phosphate dehydrogenase is introduced into this system, it may compete with LDH in recycling NADH by reducing dihydroxyacetone phosphate to glycerol 3-phosphate (see Section B,2).

these data were questioned later (Rose and Warms, 1970; Schrier, 1970; Till *et al.*, 1973), more recent experiments seem to corroborate them (Niehaus and Hammerstedt, 1976).

The idea that GAPD and PGK are harbored by the cell membrane was also derived from work with "ghosts" (cell membranes), which could synthesize ATP from the appropriate glycolytic intermediates (Green *et al.*, 1965; Mitchell *et al.*, 1965; Schrier 1966, 1967; Arese *et al.*, 1974). The then novel SDS polyacrylamide gel electrophoretic technique, when applied to analyze the protein composition of red cell membrane, revealed that one of the electrophoretic bands (No. 6) in the pattern corresponded to the polypeptide of the GAPD subunit (Tanner and Gray, 1971; Carraway and Shin, 1972; Kant and Steck, 1973). Subsequent analysis disclosed many aspects of the GAPD–erythrocyte membrane interaction. The enzyme can be readily eluted by isotonic saline from the membrane, where it binds only to the cytoplasmic side (Kant and Steck, 1973). Binding is reversible and has been characterized by dissociation constants of 10^{-7} to 10^{-8} M (Kant and Steck, 1973; McDaniel *et al.*, 1974). Binding is not specific for the homologous enzyme, since rabbit and pig muscle GAPDs bind equally well (McDaniel *et al.*, 1974; Solti and Friedrich, 1976). The binding site is, at least in part, on the cytoplasmic portion of the band 3 protein (Yu and Steck, 1975a,b), which is the predominant polypeptide chain of the erythrocyte membrane. Band 3 protein dimers span the membrane and are involved in anion transport (Ho and Guidotti, 1975; Ross and McConnell, 1977; Wolosin *et al.*, 1977; Rothstein, 1979).

Binding to the red cell membrane alters the enzymatic properties of GAPD. The reports are again contradictory. Wooster and Wrigglesworth (1976) found an increase in both V_{max} and K_m and Eby and Kirtley (1979) observed an increase in specific activity, whereas Solti and Friedrich (1976) detected partial, and Tsai *et al.* (1982) complete, inhibition upon binding. The source of these discrepancies is not clear.

GAPD was eluted from the membrane by isotonic solutions and therefore, in analogy to the case of the contractile system, some authors regarded these binding phenomena as hypotonic artifacts (Maretzki *et al.*, 1974; Fujii and Sato, 1975). For this reason it was important to devise approaches by which the localization of GAPD in the intact cell could be determined. Fossel and Solomon (1977, 1978, 1979) used ^{31}P NMR to monitor extracellular effects on intracellular metabolites and suggested the existence of a submembrane GAPD–PGK–phosphoglycerate mutase complex. Although the NMR technique may prove a valuable, noninvasive tool for following metabolic processes in intact cells (e.g., Busby *et al.*, 1978), the interpretation of Fossel and Solomon's experiments has been challenged (Momsen *et al.*, 1979). Szabolcsi and co-workers (Cseke *et al.*, 1978; Szabolcsi and Cseke, 1981) made use of the molecular sieving property of the cell membrane when the erythrocyte is swollen in mildly hypotonic media. When measuring the release of proteins of different size through the membrane pores, they found that some glycolytic enzymes, including GAPD, were preferentially released. This was taken to be indicative of a submembrane localization of GAPD, as opposed to homogeneous distribution in the cell contents. In a somewhat similar approach, Kliman and Steck (1980) studied the saponin-induced release of GAPD from erythrocytes. By rapid filtration they could monitor efflux from 0.5 s onwards. The data suggest that GAPD release strongly depends on the ionic strength of the external medium. The time curves of GAPD release (or, at low ionic strength, of GAPD uptake) all converged to the same point at time 0, corresponding to 35% free GAPD. This means that in the intact cell about two-thirds of GAPD molecules are bound to the membrane.

Solti *et al.* (1981) made an attempt to localize GAPD by electron microscopic autoradiography in thin sections òf fixed erythrocytes. The total GAPD content of the cells was labeled with [³H]iodoacetate under optimized conditions: the major labeled species (~50%) was GAPD and the remaining label was about equally distributed over glutathione, hemoglobin, and carbonic anhydrase. Since the amount of radioactivity incorporated per cell

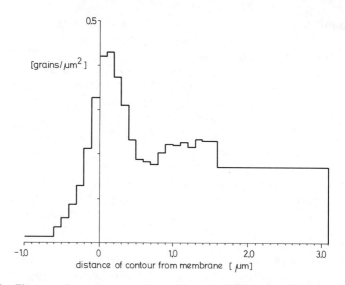

Fig. 4. Electron microscopic autoradiographic analysis of the localization of [³H]carboxymethylated GAPD in thin sections of fixed human erythrocytes. The histogram shows grain density as a function of distance from the membrane. Negative abscissa represents distances of contours outside the membrane. The peak at the membrane proved significant by a Kolmogorov–Smirnov nonparametric statistical test on the relative cumulative frequency distribution. (From Solti *et al.*, 1981.)

was small, owing to the low level of GAPD in red cells and to the practical upper limit of the specific activity of the reagent, there were too few grains over each cell to allow any definite conclusion to be drawn by simple inspection. However, evaluation of a great number of electron microscopic autoradiographs by a computer program calculating the density of grains as a function of distance from the membrane led to the distribution histogram in Fig. 4. Apparently, grains accumulate near the membrane, which indicates that the majority of GAPD is not randomly scattered in red cells but adheres to the membrane.

Among the other glycolytic enzymes, aldolase has long been known to possess affinity to the membrane (Green *et al.*, 1965; Shin and Carraway, 1973; Tillmann *et al.*, 1975). Solti and Friedrich (1976) provided the first evidence that the binding site for aldolase overlaps the binding site for GAPD (Fig. 5): aldolase previously allowed to bind to a washed membrane preparation was eluted from the membrane by an equimolar amount of GAPD. The activity loss of aldolase that occurred upon binding was reversed by GAPD. Steck and his co-workers (Strapazon and Steck, 1976, 1977; Murthy *et al.*, 1981) identified the binding site for aldolase as the cytoplasmic part of band 3 protein. Experiments by Wilson *et al.* (1982) indicate that, in spite of common elements, the binding sites for aldolase and GAPD are partially different.

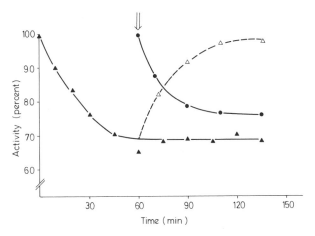

Fig. 5. Reversal of aldolase binding to the erythrocyte membrane by GAPD. Binding is indicated by the decrease of enzyme activity. Rabbit muscle aldolase (0.5 mg/ml) was incubated with white ghosts (4.2 mg/ml) at 0°C in 7 mM sodium phosphate buffer, pH 7.4, and samples were taken for aldolase assay as indicated (▲). At 60 min, as shown by the arrow, rabbit muscle GAPD (to 0.5 mg/ml concentration) was added to the suspension and the changes in aldolase (△) and GAPD (●) activities were monitored. The total amounts of enzyme activity were taken as 100%. (From Solti and Friedrich, 1976.)

Yeltman and Harris (1980) attempted to localize aldolase in the red cell by cross-linking with glutardialdehyde. They found over 90% of the red cell's aldolase content covalently fixed to the membrane. These authors demonstrated a tight complex ($K_d \simeq 10^{-9}$ M) of red cell spectrin–actin and aldolase, which was decomposed by cytochalasin B. GAPD displaced aldolase from spectrin–actin. Thus, in addition to band 3 protein, the cytoskeletal elements of erythrocyte may also participate in the binding of these glycolytic enzymes.

The enzyme preceding aldolase in the glycolytic sequence, phosphofructokinase (EC 2.7.1.11), has much in common with aldolase and GAPD in binding to the red cell membrane. Karadsheh and Uyeda (1977) demonstrated that binding occurs exclusively to the inner side of the membrane; it is sensitive to ionic strength, and the kinetic properties of the bound enzyme differ from those of the free enzyme. Higashi et al. (1979) showed that the binding site was related to the band 3 protein; both GAPD and aldolase compete with phosphofructokinase, but aldolase is more efficient than GAPD on a stoichiometric basis.

3-Phosphoglycerate kinase (PGK) has also been reported to associate with the red cell membrane (Green et al., 1965; Schrier, 1966; Schrier et al., 1975; Tillmann et al., 1975). Parker and Hoffman (1967) advanced the idea that PGK generates ATP that would directly supply the (Na^+, K^+)-ATPase. Indeed, ATP pools confined to the membrane phase have been detected (Proverbio and Hoffman, 1977; Latzkovits et al., 1972). In contrast, erythrocytes deficient in PGK could operate ion pumps normally (Segel et al., 1975). Beutler et al. (1978) and De and Kirtley (1977) found no coupling between (Na^+, K^+)-ATPase and PGK, although the latter authors confirmed earlier observations of a PGK–membrane association. However, the number of binding sites for PGK per cell was very small, ~ 500, in sharp contrast to the other enzymes described above, for which the corresponding value was 10^5–10^6.

In spite of the many unsettled questions, it seems that the enzyme sequence phosphofructokinase–aldolase–GAPD–PGK may be harbored by the cytoplasmic side of the red cell membrane. A number of hypothetical topologies have been suggested (e.g., Schrier, 1966; Tillmann et al., 1975; Proverbio and Hoffman, 1977; Salhany and Gaines, 1981), one of the most recent ones by Friedrich (1984). In the latter the above sequence of enzymes is anchored by band 3 protein dimers and spectrin–actin under the membrane, allowing microenvironmental compartmentation to facilitate vectorial metabolite flow. Clearly, the existence of such a functional unit is very difficult to prove by experiments with purified membranes and erythrocyte enzymes, especially since the high hemoglobin concentration in the red cell might be necessary for these interactions to become significant. That is, if hemoglobin–hemoglobin interactions are stronger than interactions between hemoglobin and glycolytic enzymes, then hemoglobin will gradually extrude these enzymes, necessarily to the cell periphery. Such an enzymatic shell might be advantageous for running glycolysis, but it is inevitably lost after the rupture of cells, when only some of the stronger interactions may remain detectable.

2. Test of Metabolite Compartmentation in Erythrocyte Glycolysis: the Enzyme Probe Method

One way of detecting metabolite compartmentation is to use some trapping agent for the intermediate in question. In the absence of compartmentation (channeling) the intermediate will mix in the bulk medium and be trapped by the appropriate agent, whereas in the opposite case it will not. One of the best trapping agents is an enzyme, since it is highly specific. Indeed, enzymes have been used for this purpose in several instances (e.g., Bernofsky and Pankow, 1971; Traut, 1980).

Friedrich *et al.* (1977) devised an approach by which metabolite compartmentation might be detected in case of an associating–dissociating enzyme complex in crude cell extracts. The salient feature of this approach is to add an alien enzyme that can transform the metabolite in question to the system and measure the efficiency of this probe enzyme in diverting metabolite flow as a function of overall enzyme concentration. The simple scheme is as follows:

$$A \xrightarrow{\ E_1\ } (B) \xrightarrow{\ E_2\ } C$$
$$\downarrow E_p$$
$$D$$

where A, B, and C are metabolites, E_1 and E_2 are two consecutive enzymes, and E_p is a probe enzyme that is able to transform intermediate B into D, a product different from the normally formed C. Friedrich *et al.* (1977) measured lactate production from glucose in packed red cells disrupted by sonication. In this concentrated sonicate the cell content is hardly diluted, so weak macromolecular associations may be expected to prevail. As probe enzyme they used glycerol-3-phosphate dehydrogenase, which is absent from mature red cells. Its substrates, dihydroxyacetone phosphate and NADH, are, however, present during steady-state glycolysis (see Fig. 3). Consequently, if both of the metabolites are available in the bulk medium the probe enzyme will produce glycerol 3-phosphate at the expense of lactate, whereas if either or both of these compounds are compartmented, the probe enzyme will have no effect on lactate production. As seen in Fig. 6, in the concentrated sonicate glycerol-3-phosphate dehydrogenase did not influence the rate of lactate production, but if the whole system was diluted threefold, lactate production in the presence of probe enzyme was reduced to about one-half of the value

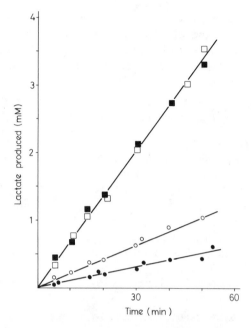

Fig. 6. Effect of exogenous probe enzyme, rabbit muscle glycerol-3-phosphate dehydrogenase, on lactate production of concentrated and diluted human erythrocyte sonicate. Lactate production was measured in concentrated (nominal hematocrit 90%) sonicate without (□) and with (■) 0.5 mg/ml glycerol-3-phosphate dehydrogenase, and in diluted (hematocrit 30%) sonicate without (○) and with (●) 0.17 mg/ml probe enzyme. (From Friedrich *et al.*, 1977.)

obtained in its absence. These data suggest that in the concentrated sonicate of human red cells dihydroxyacetone phosphate and/or NADH are sequestered in a microcompartment that is sensitive to dilution.

3. Heterogeneous Reactivity of the NAD Pool in Erythrocytes

As outlined above, distribution over free and, possibly, multiple bound fractions of a compound is another means of microcompartmentation. Solti and Friedrich (1979) studied the NAD pool in concentrated human erythrocyte sonicate by testing the pool's reactivity against a probe enzyme. Red cells happen to have NAD glycohydrolase (EC 3.2.2.5), an enzyme that cleaves NAD to nicotinamide and ADP–ribose, anchored on the outer surface of their plasma membrane. In intact cells NAD cannot be attacked by the enzyme, being separated by the membrane. However, when the cells are disrupted by sonication the enzyme gains access to the cell's NAD content. The authors followed the time course of decomposition of NAD + NADH in the concentrated sonicate. (Under the conditions used NADH was practically negligible relative to NAD.) Figure 7 shows that the complex decay curve could be resolved graphically into three first-order reactions, which corresponded to three NAD fractions (pools) of different reactivity. Pool I (35%, $t_{1/2} \simeq 1$ min) had the same reactivity toward NAD glycohydrolase as had free NAD, hence it very probably is unbound NAD in the sonicate. Pool II (23%, $t_{1/2} = 7$ min) was identified as NAD bound to GAPD. This conclusion was made by monitoring the release of firmly bound NAD from GAPD, which was found to run parallel with the depletion of pool II. The dissociation of GAPD–NAD complexes could be measured in the thick slurry by determining the relative reactivity of the active site SH group of GAPD (Harris and Perham, 1965) toward iodoacetate, which is strongly increased by bound NAD (Racker and Krimsky, 1958; Cseke and Boross, 1970). Therefore the "stripping" of GAPD mainfested itself in a decrease of reactivity toward iodoacetate, the latter being estimated in terms of the iodoacetate concentration that effected 50% inactivation of GAPD in the sonicate (Fig. 8). Interestingly, lactate production stopped by the time pool II was depleted. Pool III (42%, $t_{1/2} \simeq 4$ decomposed very slowly and was unable to maintain glycolysis. Its identity is not known; it rapidly disappears in the presence of detergents but it does not represent NAD entrapped in lipid vesicles. As Solti and Friedrich (1979) pointed out, in the concentrated red cell sonicate only pool II is needed for normal glycolysis, as the depletion of pool I did not affect the rate of lactate production.

The NAD-sequestering role of GAPD might prevail in other tissues as well. The protective effect of muscle GAPD on NAD against glycohydrolase attack has earlier been observed (Astrachan et al., 1957; Bernofsky and Pankow, 1971). It should be borne in mind, however, that from model studies such as those described above one cannot directly infer the conditions in the living cell. Nonetheless, it is more than probable that similar phenomena are also operative in vivo.

One can further question to what extent such pool heterogeneities may affect metabolic fluxes. In general, for an open system with continuous and practically infinite inflow and outflow, metabolite sequestration via binding to a macromolecule can only have control power through changing the amount of the binding macromolecule. This is not likely to happen over a short time scale. On the other hand, for conserved compound pairs, like NAD–NADH or CoA–acetyl-CoA, sequestration by binding may have significant influence on flux provided that the bound species predominates (e.g., NAD at high NAD–NADH ratio). In the opposite case, if the bound species is the minor party (e.g., NADH in the above example), no effect of binding on flux can be expected, since the free pool will reequilibrate and the small amount removed by sequestration will not affect detectably the concentration of the free species.

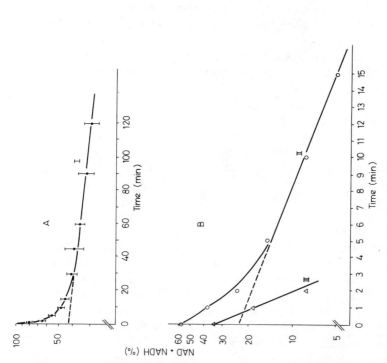

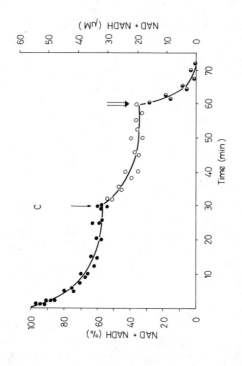

Fig. 7. Heterogeneous reactivity of the NAD pool toward NAD glycohydrolase in human erythrocyte sonicate. Packed red cells were mildly sonicated to achieve complete hemolysis (0 time). Then aliquots were taken and the NAD + NADH content was determined. (A) Semilogarithmic plot of coenzyme decay at 37°C (100% corresponds to about 50 μM). The overall decay curve (I, ●) could be resolved graphically into three first-order reactions (cf. panels A and B), which is indicative of three distinct NAD pools. Curve II (○) was derived by subtracting the extrapolated slow phase from curve I; curve III (△) was obtained in a similar way from curve II. (C) Resolution of the three NAD pools as follows: ●, incubation of sonicate at 0°C (depletion of pool I); ○, incubation at 37°C (depletion of pool II), ↓, transfer to 37°C water bath; ●, at 37°C after the addition of 0.01% Triton X-100; ⇓, addition of detergent. (From Solti and Friedrich, 1979.)

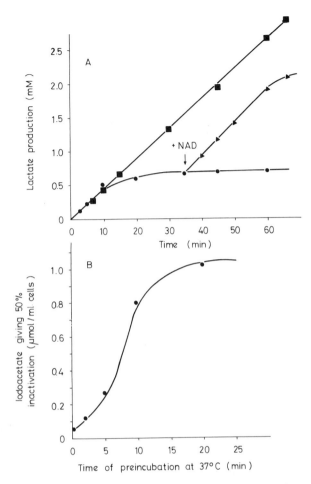

Fig. 8. Parallel decline of lactate production (A) and development of iodoacetate resistance of GAPD (B) in erythrocyte sonicate (nominal hematocrit 90%). (A) Lactate production at 37°C from 5 mM glucose by intact red cells (■), sonicate (●), and in sonicate after the addition of 100 μM exogenous NAD at 30 min (▲). (B) Increase in the amount of iodoacetate required for 50% inactivation of GAPD reflects the decrease of reactivity of the active site SH group due to the release of NAD from GAPD. (From Solti and Friedrich, 1979.)

V. MODELS OF DYNAMIC COMPARTMENTATION IN SOLUBLE ENZYME SYSTEMS

The idea that metabolite compartmentation may occur in soluble multien-zyme systems via direct metabolite transfer in transient enzyme–enzyme complexes has been raised by Friedrich (1974), who presented also some mechanistic models for such phenomena. We shall discuss them briefly.

A. Direct Metabolite Transfer in the Complementary Cage

If two consecutive enzymes of a pathway collide, they may form short-lived complexes. The residence time of one enzyme on the other is small, so that almost elastic collisions occur. The following postulates are made:

1. Around the active sites of two functionally adjacent enzymes, E_1 and E_2, there are foreign recognition sites, which are contact areas complementary to each other on the two enzymes. Therefore when the complex is formed the active sites become juxtaposed. Since most active sites are located in a cleft on the enzyme surface, a "complementary cage" is formed. The binding forces involved are small, so the enzyme–enzyme complex can rapidly form and decompose.

2. The frequency of productive, that is, complex-forming, collisions between the two enzymes is higher than the frequency of product release from E_1, which equals the metabolic flux rate.

The combination of the above two postulates may lead to the channeling of intermediary metabolite. Since in the complementary cage the intermediate is locked up, it will find the active site of E_2 with high probability. For this to happen, however, the formation of complementary cages should be properly timed, that is, a cage must form whenever the product of E_1 is about to dissociate from the parent enzyme. This is hard to achieve unless E_2 is in a large excess over E_1.

Some rough estimations can be made, for example, for the glycolytic system in yeast. Friedrich (1974) calculated from the data of Hess and Boiteux (1972) that taking an average enzyme with $k_{cat} = 100 \, s^{-1}$ and assuming that it works *in vivo* at 1–10% of its V_{max}, one gets the figure of 10^5–10^6 collisions between two functionally adjacent enzymes per catalytic cycle. The number of collisions for a single product to form in the $A + B \xrightarrow{k} AB$ reaction with $k = 10^5 \, M^{-1} \, s^{-1}$ is estimated to be 2×10^5. Since a second-order rate constant of this magnitude may readily be encountered in interprotein reactions, the required frequency of formation of complementary cages does not seem to be unrealistic.

B. Facilitation by Complex Formation with Partner Enzyme

The efficiency of compartmentation is greatly enhanced if the productive collision (complex formation) not only confines the intermediate in the complementary cage, but also facilitates the overall reaction. The following postulates are made for the two-enzyme system:

1. E_1 can assume three different conformations: E_1^o, the state of the unliganded enzyme, E_1^s, and E_1^i, the states characteristic of the E_1S and E_1I complexes, respectively.

2. The foreign recognition sites on E_1 for E_2 are present only in the E_1^i state.

3. Complex formation with E_2 and/or with the appropriate metabolite induces a conformational change in E_1.

The reaction of the first enzyme can then be described, without interference of E_2, as follows:

$$E_1^o + S \underset{k_{-1}}{\overset{k_1}{\rightleftharpoons}} E_1^s S \underset{k_{-2}}{\overset{k_2}{\rightleftharpoons}} E_1^i I \underset{k_{-3}}{\overset{k_3}{\rightleftharpoons}} E_1^o + I \tag{9}$$

Complex formation with the second enzyme may facilitate the overall reaction in several ways.

1. Catalytic Facilitation by Partner Enzyme

Collision with E_2 promotes the $E_1^s S \rightarrow E_1^i I$ transition in an induced-fit manner:

$$E_1^s S + E_2 \underset{k'_{-2}}{\overset{k'_2}{\rightleftharpoons}} E_1^i I E_2 \dashrightarrow E_1^o + E_2 + P \tag{10}$$

Catalytic facilitation means that $k'_2[E_2] > k_2$. In addition, the circumstance that E_2 promotes the formation of I will markedly increase the efficiency of metabolite channeling.

2. Transfer Facilitation by Partner Enzyme

Collision with E_2 promotes the dissociation of I from $E_1^i I$:

$$E_1^i I + E_2 \underset{k_{-a}}{\overset{k_a}{\rightleftharpoons}} IE_1^{i'} E_2 \underset{k'_{-3}}{\overset{k'_3}{\rightleftharpoons}} E_1^{i'} E_2 I \underset{k_{-4}}{\overset{k_4}{\rightleftharpoons}} E_1^o + E_2 I \tag{11}$$
$$\underset{k_{-5}}{\overset{k_5}{\rightleftharpoons}} E_2 + P$$

where $E_1^{i'}$ is a conformation of E_1 that releases I more rapidly than does E_1^i. In addition to the triggered release of I ($k'_3 > k_3$), I is also channeled to E_2.

3. Dissociation of Enzyme–Enzyme Complex Induced by Substrate

A special case of the preceding mechanism can be when the dissociation rate constant k_4 is kinetically significant, that is, it (partially) determines the overall reaction rate. Then the binding of S to E_1 may accelerate the reaction:

$$E_1^i I + E_2 \underset{k'_{-a}}{\overset{k'_a}{\rightleftharpoons}} IE_1^i E_2 \underset{k''_{-3}}{\overset{k''_3}{\rightleftharpoons}} E_1^i E_2 I \tag{12}$$

$$S + E_1^i E_2 I \underset{k_{-b}}{\overset{k_b}{\rightleftharpoons}} SE_1^s E_2 I \underset{k'_{-4}}{\overset{k'_4}{\rightleftharpoons}} E_1^s S + E_2 I \underset{k_{-5}}{\overset{k_5}{\rightleftharpoons}} E_2 + P \tag{13}$$

In this mechanism E_2 does not trigger the transfer of I by inducing a conformational change in E_1 (therefore $k_a \neq k'_a$ and $k'_3 \neq k''_3$), it rather forms a fairly stable complementary cage with E_1 that is then disrupted by S. At high [S] it is likely that $k_b[S] \gg k'_4$ and then the inequality $k'_4 > k_4$ expresses the difference compared with the preceding case [Eq. (11)].

It should be added that the above three cases are neither the only possible mechanisms nor are they mutually exclusive. In fact, they may combine, giving rise to a variety of mechanisms for facilitation through transient enzyme–enzyme complex formation.

C. Alternating Complementarity in an Enzyme Sequence

In a three-enzyme system (E_1, E_2, and E_3) both the substrate and the product of E_2 can be sequestered if E_2 has foreign recognition sites for both E_1 and E_3. Although a rigid structure may also have this property, it is more attractive to invoke a conformational change that occurs during catalysis. The following postulates are made:

1. E_2 can assume three different conformations: E_2^0, the state of the unliganded enzyme, and E_2^s and E_2^p, the states characteristic of the E_2S and E_2P complexes, respectively. (Here S and P are the substrate and product, respectively, of E_2.)

2. The foreign recognition sites on E_2 for E_1 and E_3 are formed when the enzyme assumes the E_2^s and E_2^p conformations, respectively.

3. The conformational transitions are promoted by the appropriate metabolite and/or enzyme partner.

The concept of alternating complementarity is illustrated in Fig. 9. The model does not state that conformations E_2^s and E_2^p are unavailable to the

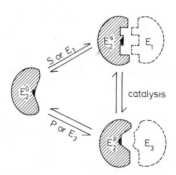

Fig. 9. Conformational transitions and alternating complementarity of enzyme E_2 in a three-enzyme sequence. S and P are the substrate and product of E_2, respectively, whereas E_1 and E_3 are the preceding and subsequent enzymes to E_2 in the sequence. The active site on E_2 is indicated by a dark spot. (From Friedrich, 1974.)

ligand-free enzyme, that is, that there cannot be a preequilibrium among the states in the absence of metabolites and enzyme partners. It is only required that, as detailed in the preceding section, the productive collision with the appropriate ligand (metabolite or enzyme) should promote transition into the respective state.

If we examine the postulates of Sections V,B and C, we can find abundant evidence that conformational changes take place during catalysis. Indeed, early models of enzyme action (Lumry, 1959; Hammes, 1964) were based on the idea that the enzyme protein undergoes a structural change in the catalytic cycle. Citri (1973) has compiled a staggering amount of data on the conformational adaptability of enzymes in the presence of their substrates and cofactors. More recent structural investigations, mainly by X-ray diffraction, corroborate the modifying role of ligands on enzyme structure and also provide us with the details of the stereochemical changes.

For the other postulates in the above models, however, there is hardly any unequivocal evidence. We have examined in detail the case of muscle PGK and GAPD, for which Weber and Bernhard (1982) suggested that the intermediate, 1,3-P_2G, was transferred directly from one enzyme to the other. The very pronounced rate-enhancing effect of 3-PG is puzzling in the authors' interpretation, as pointed out earlier in this chapter. We would rather envisage that for this system case 3 in Section V,B above might be valid, that is, that 3-PG binding to PGK promotes the breakup of the transient PGK–GAPD complex. Obviously, more experiments are needed to clarify the picture. We hope that the considerations presented here might be helpful in this effort.

Another experimental observation, also discussed above (see Fig. 2), that might be indicative of a direct metabolite transfer between soluble enzymes triggered by enzyme collisions was made by Patthy and Vas (1978). The syncatalytic enzyme inactivation by aldolase-generated nascent hydroxypyruvaldehyde phosphate is about 10 times as efficient for GAPD as it is for aldolase, the parent molecule. Although the authors did not invoke such a mechanism, it may well be that collision of aldolase with GAPD triggers the formation and/or release of the reactive intermediate and the preference of GAPD to aldolase in respect of modification is at least partly due to this phenomenon.

D. Nonrandom Distribution of Soluble Enzymes in Macrocompartments

Let us consider a three-enzyme sequence with the following properties:

1. E_1 and E_3 bind weakly but selectively to different structural elements in the cell, the A wall and Z wall, respectively.

2. E_2 tends to self-associate, that is, it carries self-recognition sites on its surface.

If this enzyme system is "sandwiched" in a fairly narrow space, that is, the A and Z walls are close to each other, then the distribution of enzymes in that space will be nonrandom; E_1 and E_3 will accumulate at the A and Z walls, respectively, and E_2 will concentrate in the middle. Thus an enzyme gradient develops where "useless" collisions, such as between E_1 and E_3, will be less frequent than if the enzymes were randomly distributed. Any of the above collision models will work more efficiently in such an arrangement. Nonrandom distribution of this type cannot prevail over a long distance, but this is probably not needed. The high surface/volume ratio in cellular macrocompartments (on the average 60 μm^{-1}; see Sitte, 1980) ensures that the postulated walls are close to each other. Srere (1981) has pointed out that in the mitochondrial matrix, especially in the "closed phase," the spaces between the foldings of cristae do not exceed the width of a few enzyme molecules (see Srere, Chapter 1).

If the above recognition sites are formed as a consequence of ligation with substrates and products, one may envisage enzyme gradients whose development is governed by the metabolic flux. In such a case the metabolic state would be reflected in the organizational pattern of enzymes. Wilson's (1978, 1980) concept of "ambiquitous enzymes," which may exist in both structure-bound and free states, is akin to this idea (see Kurganov, Chapter 5).

VI. SUMMARY AND PERSPECTIVES

In this chapter we have canvassed metabolite compartmentation in living cells, with special reference to the various microcompartmentational modes, such as enzyme–enzyme channels, microenvironments at structural surfaces, and pool heterogeneities due to multiple binding sites. We have drawn a distinction between static and dynamic compartments, referring to the lifetime of the compartment. Detailed discussion was given of the experimental evidence for enzyme interactions in glycolysis of two selected cell types, striated muscle and erythrocytes. Weighing the pros and cons, one cannot come up with a conclusive pattern for either of these systems. Nevertheless, the data provide us with a great deal of food for thought.

In respect to the question of supramolecular organization of soluble enzymes the biochemical community has long been divided into "believers" and "nonbelievers," the latter outnumbering the former. However, scientific questions cannot be settled by accepting or rejecting beliefs. Although the present state of the art probably does not convert all the faithless, it may not be far-fetched to say that the available evidence for interactions can no longer be simply dismissed. Some other chapters in this volume, as well as more comprehensive recent treatises (e.g., Frieden and Nichol, 1981; Friedrich, 1984) lend further support to this notion. It is true, however, that both conceptual and methodical advances are required in this difficult field. Some attempts to this end have been described in this chapter. The various models for direct metabolite transfer in a dynamically organized, but otherwise "soluble," multienzyme system may prove to have heuristic value. It stands without doubt, at least for the author, that full understanding of life processes and their regulation cannot be achieved without taking all functionally significant macromolecular interactions into account, of which metabolite compartmentation in multienzyme systems is but one example.

REFERENCES

Arese, P., Bosia, A., Pescarmona, G. P., and Till, U. (1974). *FEBS Lett.* **49**, 33–36.
Arnold, H., and Pette, D. (1968). *Eur. J. Biochem.* **6**, 163–171.
Arnold, H., and Pette, D. (1970). *Eur. J. Biochem.* **15**, 360–366.
Arnold, H., Nolte, J., and Pette, D. (1969). *J. Histochem.* **17**, 314–320.
Arnold, H., Henning, R., and Pette, D. (1971). *Eur. J. Biochem.* **22**, 121–126.
Astrachan, L., Colowick, S. P., and Kaplan, N. O. (1957). *Biochim. Biophys. Acta* **24**, 141–154.
Banks, R. D., Blake, C. C. F., Evans, P. R., Haser, R., Rice, D. W., Hardy, G. W., Merrett, M., and Phillips, A. W. (1979). *Nature (London)* **279**, 773–777.
Baranowski, T., and Niederland, T. R. (1949). *J. Biol. Chem.* **180**, 543–551.
Bartha, F., and Keleti, T. (1979). *Oxid. Commun.* **1**, 75–84.
Batke, J., and Keleti, T. (1968). *Acta Biochim. Biophys. Acad. Sci. Hung.* **3**, 385–395.
Batke, J., Asbóth, G., Lakatos, S., Schmitt, B., and Cohen, R. (1980). *Eur. J. Biochem.* **107**, 389–394.
Bernofsky, C., and Pankow, M. (1971). *Biochim. Biophys. Acta* **242**, 437–440.
Beutler, E., Guinto, E., Kuhl, W., and Matsumoto, F. (1978). *Proc. Natl. Acad. Sci. U.S.A.* **75**, 2825–2828.
Blake, C. C. F., and Rice, D. W. (1981). *Philos. Trans. R. Soc. London Ser. B* **293**, 93–104.
Boiteux, A., and Hess, B. (1978). *In* "Frontiers in Biological Energetics" (P. L. Dutton, J. Leigh, and A. Scarpa, eds.), pp. 789–798. Academic Press, New York.
Boiteux, A., Goldbeter, A., and Hess, B. (1975). *Proc. Natl. Acad. Sci. U.S.A.* **72**, 3829–3833.
Boiteux, A., Hess, B., and Sel'kov, E. E. (1980). *Curr. Top. Cell. Regul.* **17**, 171–203.
Busby, S. J. W., Gadian, D. G., Radda, G. K., Richards, R. E., and Seeley, P. J. (1978). *Biochem. J.* **170**, 103–114.
Carraway, K. L., and Shin, B. C. (1972). *J. Biol. Chem.* **247**, 2102–2108.
Christen, P., Cogoli-Greuter, M., Healy, M. J., and Lubini, D. (1976). *Eur. J. Biochem.* **63**, 223–231.
Church, W. R., Rawitch, A. B., and Ebner, K. E. (1981). *Arch. Biochem. Biophys.* **206**, 285–290.
Citri, N. (1973). *Adv. Enzymol.* **37**, 397–648.
Clarke, F. M., and Masters, C. J. (1973). *Biochim. Biophys. Acta* **327**, 223–226.
Clarke, F. M., and Masters, C. J. (1974). *Biochim. Biophys. Acta* **358**, 193–207.
Clarke, F. M., and Masters, C. J. (1975). *Biochim. Biophys. Acta* **381**, 37–46.
Clarke, F. M., and Masters, C. J. (1976). *Int. J. Biochem.* **7**, 359–365.
Cohen, R., and Mire, M. (1971). *Eur. J. Biochem.* **23**, 267–275.
Cseke, E., and Boross, L. (1970). *Acta Biochim. Biophys. Acad. Sci. Hung.* **5**, 385–397.

Cseke, E., Váradi, A., Szabolcsi, G., and Biszku, E. (1978). *FEBS Lett.* **96**, 15–18.
De, B. K., and Kirtley, M. E. (1977). *J. Biol. Chem.* **252**, 6715–6720.
De Duve, C. (1964). *J. Theor. Biol.* **6**, 33–59.
Dombrádi, V. (1981). *Int. J. Biochem.* **13**, 125–139.
Eby, D., and Kirtley, M. E. (1979). *Arch. Biochem. Biophys.* **198**, 608–613.
Fletterick, R. J., and Madsen, N. B. (1980). *Annu. Rev. Biochem.* **49**, 31–61.
Földi, J., Szabolcsi, G., and Friedrich, P. (1973). *Acta Biochim. Biophys. Acad. Sci. Hung.* **8**, 263–265.
Fossel, E. T., and Solomon, A. K. (1977). *Biochim. Biophys. Acta* **464**, 82–92.
Fossel, E. T., and Solomon, A. K. (1978). *Biochim. Biophys. Acta* **510**, 99–111.
Fossel, E. T., and Solomon, A. K. (1979). *Biochim. Biophys. Acta* **553**, 142–153.
Frieden, C., and Nichol, L. W., eds. (1981). "Protein-Protein Interactions." Wiley, New York.
Friedrich, P. (1974). *Acta Biochim. Biophys. Acad. Sci. Hung.* **9**, 159–173.
Friedrich, P. (1984). "Supramolecular Enzyme Organization: Quaternary Structure and Beyond." Akadémiai Kiadó, Budapest; Pergamon, Oxford.
Friedrich, P., Apró-Kovács, V. A., and Solti, M. (1977). *FEBS Lett.* **84**, 183–186.
Fujii, T., and Sato, M. (1975). *J. Clin. Chem.* **3**, 453–459.
Gavilanes, F., Salerno, C., and Fasella, P. (1981). *Biochim. Biophys. Acta* **660**, 154–156.
Gerlach, E., Fleckenstein, A., Cross, E., and Lübben, K. (1958). *Pflügers Arch.* **266**, 528–555.
Giles, N. H. (1978). *Am. Nat.* **112**, 641–657.
Gourley, D. R. H. (1952). *Arch. Biochem. Biophys.* **40**, 1–12.
Grazi, E., and Trombetta, G. (1980). *Eur. J. Biochem.* **107**, 369–373.
Green, D. E., Murer, E., Hultin, H. O., Richardson, S. H., Salmon, B., Brierley, G. P., and Baum, H. (1965). *Arch. Biochem. Biophys.* **112**, 635–647.
Hackenbrock, C. R. (1968). *Proc. Natl. Acad. Sci. U.S.A.* **61**, 598–605.
Hammes, G. G. (1964). *Nature (London)* **204**, 342–343.
Harris, J. I., and Perham, R. N. (1965). *J. Mol. Biol.* **13**, 876–884.
Heinrich, R., Rapoport, S. M., and Rapoport, T. A. (1977). *Prog. Biophys. Mol. Biol.* **32**, 1–82.
Hess, B. (1980). *In* "Cell Compartmentation and Metabolic Channeling" (L. Nover, F. Lynen, and K. Mothes, eds.), pp. 75–92. Fischer, Jena, Elsevier, Amsterdam.
Hess, B., and Boiteux, A. (1972). *In* "Protein-Protein Interaction" (R. Jaenicke and E. Helmreich, eds.), pp. 271–297. Springer-Verlag, Berlin and New York.
Hess, B., and Wurster, B. (1970). *FEBS Lett.* **9**, 73–77.
Hess, B., Boiteux, A., and Krüger, J. (1969). *Advances Enzyme Regul.* **7**, 149–167.
Hess, B., Goldbeter, A., and Lefever, R. (1978). *Adv. Chem. Phys.* **38**, 363–413.
Higashi, T., Richards, C. S., and Uyeda, K. (1979). *J. Biol. Chem.* **254**, 9542–9550.
Ho, M. K., and Guidotti, G. (1975). *J. Biol. Chem.* **250**, 675–683.
Huskins, K. R., Bernhard, S. A., and Dahlquist, F. W. (1982). *Biochemistry* **21**, 4180–4188.
Kálmán, M., and Boross, L. (1982). *Biochim. Biophys. Acta* **704**, 272–277.
Kant, J. A., and Steck, T. L. (1973). *J. Biol. Chem.* **248**, 8457–8464.
Karadsheh, N. S., and Uyeda, K. (1977). *J. Biol. Chem.* **252**, 7418–7420.
Kempner, E. S. (1980). *In* "Cell Compartmentation and Metabolic Channeling" (L. Nover, F. Lynen, and K. Mothes, eds.), pp. 211–224. Fischer, Jena; Elsevier, Amsterdam.
Kliman, H. J., and Steck, T. L. (1980). *J. Biol. Chem.* **255**, 6314–6321.
Koch-Schmidt, A. C., Mattiasson, B., and Mosbach, K. (1977). *Eur. J. Biochem.* **81**, 71–78.
Kuter, M. R., Masters, C. J., Walsh, T. P., and Winzor, D. J. (1981). *Arch. Biochem. Biophys.* **212**, 306–310.
Kwon, T.-W., and Olcott, H. S. (1965). *Biochem. Biophys. Res. Commun.* **19**, 300–305.
Latzkovits, L., Szentistványi, I., and Fajszi, Cs. (1972). *Acta Biochim. Biophys. Acad. Sci. Hung.* **7**, 55–66.

Liou, R.-S., and Anderson, S. (1980). *Biochemistry* **19**, 2684–2688.
Lumry, R. (1959). *In* "The Enzymes" (P. D. Boyer, H. Lardy, and K. Myrbäck, eds.), Vol. 1, pp. 157–231. Academic Press, New York.
McClure, W. R. (1969). *Biochemistry* **8**, 2782–2786.
McDaniel, C. F., Kirtley, M. E., and Tanner, M. J. A. (1974). *J. Biol. Chem.* **249**, 6478–6485.
MacGregor, J. S., Singh, V. N., Davoust, S., Melloni, E., Pontremoli, S., and Horecker, B. L. (1980). *Proc. Natl. Acad. Sci. U.S.A.* **77**, 3889–3892.
Maretzki, D., Groth, J., Tsamaloukas, A. G., Gründel, M., Krüger, S., and Rapoport, S. M. (1974). *FEBS Lett.* **39**, 83–87.
Masters, C. J. (1978). *Trends Biochem. Sci.* **3**, 206–208.
Masters, C. J., and Winzor, D. J. (1981). *Arch. Biochem. Biophys.* **209**, 185–190.
Matchett, W. H. (1974). *J. Biol. Chem.* **249**, 4041–4049.
Matchett, W. H., and DeMoss, J. A. (1964). *Biochim. Biophys. Acta* **86**, 91–99.
Meyer, F., Heilmeyer, L. M. G., Haschke, R. H., and Fischer, E. H. (1970). *J. Biol. Chem.* **245**, 6642–6648.
Mitchell, C. D., Mitchell, W. B., and Hanahan, D. J. (1965). *Biochim. Biophys. Acta* **104**, 348–358.
Momsen, G., Rose, Z. B., and Gupta, R. K. (1979). *Biochem. Biophys. Res. Commun.* **91**, 651–657.
Morton, D. J., Clarke, F. M., and Masters, C. J. (1977). *J. Cell Biol.* **74**, 1016–1023.
Murthy, S. N. P., Liu, T., Kaul, R. K., Köhler, H., and Steck, T. L. (1981). *J. Biol. Chem.* **256**, 11203–11208.
Nicolis, G., and Prigogine, I. (1977). "Self-Organization in Nonequilibrium Systems." Wiley, York.
Niehaus, W. G., and Hammerstedt, R. H. (1976). *Biochim. Biophys. Acta* **443**, 515–524.
Ottaway, J. H., and Mowbray, J. (1977). *Curr. Top. Cell. Regul.* **12**, 107–208.
Ovádi, J., and Keleti, T. (1978). *Eur. J. Biochem.* **85**, 157–161.
Ovádi, J., Salerno, C., Keleti, T., and Fasella, P. (1978). *Eur. J. Biochem.* **90**, 499–503.
Ovádi, J., Mohamed Osman, I. R., and Batke, J. (1983). *Eur. J. Biochem.* **133**, 433–437.
Parker, J. C., and Hoffman, J. F. (1967). *J. Gen. Physiol.* **50**, 893–916.
Patthy, L. (1978). *Eur. J. Biochem.* **88**, 191–196.
Patthy, L., and Thész, J. (1980). *Eur. J. Biochem.* **105**, 387–393.
Patthy, L., and Vas, M. (1978). *Nature, (London)* **276**, 94–95.
Patthy, L., Váradi, A., Thész, J., and Kovács, K. (1979). *Eur. J. Biochem.* **99**, 309–313.
Pickover, C. A., McKay, D. B., Engelman, D. M., and Steitz, T. A. (1979). *J. Biol. Chem.* **254**, 11323–11329.
Pontremoli, S., Melloni, E., Salamino, F., Sparatore, B., Michetti, M., Singh, V. N., and Horecker, B. L. (1979). *Arch. Biochem. Biophys.* **197**, 356–363.
Prankerd, T. A. J., and Altman, K. I. (1954). *Biochem. J.* **58**, 662–633.
Proverbio, F., and Hoffman, J. F. (1977). *J. Gen. Physiol.* **69**, 605–632.
Racker, E., and Krimsky, I. (1958). *Fed. Proc., Fed. Am. Soc. Exp. Biol.* **17**, 1135–1141.
Rapoport, S. M., (1968). *In* "Essays in Biochemistry" (P. N. Campbell and G. D. Greville, eds.), Vol. 4, pp. 69–103. Academic Press, New York.
Rapoport, S. M., and Luebering, J. (1950). *J. Biol. Chem.* **183**, 507–516.
Rose, I. A., and Warms, J. V. B. (1970). *J. Biol. Chem.* **245**, 4009–4015.
Ross, A. H., and McConnell, H. M. (1977). *Biochem. Biophys. Res. Commun.* **74**, 1318–1325.
Rothstein, M. (1979). *Mech. Ageing Dev.* **9**, 197–202.
Salerno, C., and Ovádi, J. (1982). *FEBS Lett.* **138**, 270–272.
Salhany, J. M., and Gaines, K. C. (1981). *Trends Biochem. Sci.* **6**, 13–15.
Schrier, S. L. (1966). *Am. J. Physiol.* **210**, 139–143.
Schrier, S. L. (1967). *Biochim. Biophys. Acta* **135**, 591–598.
Schrier, S. L. (1970). *J. Lab. Clin. Med.* **75**, 422–434.

Schrier, S. L., Ben-Bassat, I., Junga, I., Seeger, M., and Grumet, F. C. (1975). *J. Lab. Clin. Med.* **85,** 797–810.

Scopes, R. K. (1973). *In* "The Enzymes" (P. D. Boyer, ed.), Vol. 8, pp. 335–351. Academic Press, New York.

Segel, G. B., Feig, S. A., Glader, B. E., Müller, A., Dutcher, P., and Nathan, D. G. (1975). *Blood* **46,** 271–278.

Shin, B. C., and Carraway, K. L. (1973). *J. Biol. Chem.* **248,** 1436–1444.

Sigel, P., and Pette, D. (1969). *J. Histochem. Cytochem.* **17,** 225–237.

Sitte, P. (1980). *In* "Cell Compartmentation and Metabolic Channeling" (L. Nover, F. Lynen, and K. Mothes, eds.), pp. 17–32. Fischer, Jena; Elsevier, Amsterdam.

Solti, M., and Friedrich, P. (1976). *Mol.Cell. Biochem.* **10,** 145–152.

Solti, M., and Friedrich, P. (1979). *Eur. J. Biochem.* **95,** 551–559.

Solti, M., Bartha, F., Halász, N., Sirokmán, F., and Friedrich, P. (1981). *J. Biol. Chem.* **256,** 9260–9265.

Srere, P. A. (1980). *Trends Biochem. Sci.* **5,** 120–121.

Srere, P. A. (1981). *Trends Biochem. Sci.* **6,** 4–7.

Srere, P. A., and Henslee, J. G. (1980). *In* "Cell Compartmentation and Metabolic Channeling" (L. Nover, F. Lynen, and K. Mothes, eds.), pp. 159–168. Fischer, Jena; Elsevier, Amsterdam.

Stewart, M., Morton, D. J., and Clarke, F. M. (1980). *Biochem. J.* **186,** 99–104.

Strapazon, E., and Steck, T. L. (1976). *Biochemistry* **15,** 1421–1424.

Strapazon, E., and Steck, T. L. (1977). *Biochemistry* **16,** 2966–2971.

Szabolcsi, G., and Cseke, E. (1981). *Acta Biol. Med. Ger.* **40,** 471–477.

Tanner, M. J. A., and Gray, W. R. (1971). *Biochem. J.* **125,** 1109–1117.

Till, U., Köhler, W., Ruschke, I., Köhler, A., and Lösche, W. (1973). *Eur. J. Biochem.* **35,** 167–178.

Tillmann, W., Cordua, A., and Schröter, W. (1975). *Biochim. Biophys. Acta* **382,** 157–171.

Träuble, H. (1976). *In* "Structure of Biological Membranes" (S. Abrahamsson and I. Pascher, eds.), pp. 509–550. Plenum, New York.

Traut, T. W. (1980). *Arch. Biochem. Biophys.* **200,** 590–594.

Trentham, D. R., McMurray, C. H., and Pogson, C. I. (1969). *Biochem. J.* **114,** 19–24.

Tsai, I.-H., Murthy, S. N. P., and Steck, T. L. (1982). *J. Biol. Chem.* **257,** 1438–1442.

Vas, M., and Batke, J. (1981). *Biochim. Biophys. Acta* **660,** 193–198.

Vogel, H. J., and Bonner, D. M. (1954). *Proc. Natl. Acad. Sci. U.S.A.* **40,** 688–694.

Walsh, T. P., Winzor, D. J., Clarke, F. M., Masters, C. J., and Morton, D. J. (1980). *Biochem. J.* **186,** 89–98.

Weber, J. P., and Bernhard, S. A. (1982). *Biochemistry* **21,** 4189–4194.

Welch, G. R., and Gaertner, F. H. (1975). *Proc. Natl. Acad. Sci. U.S.A.* **72,** 4218–4222.

Welch, G. R., and Gaertner, F. H. (1980). *Trends Biochem. Sci.* **5,** VII.

Wilson, J. E. (1978). *Trends Biochem. Sci.* **3,** 124–125.

Wilson, J. E. (1980). *Curr. Top. Cell. Regul.* **16,** 1–44.

Wilson, J. E., Reid, S., and Masters, C. J. (1982). *Arch. Biochem. Biophys.* **215,** 610–620.

Wolosin, J. M., Ginsburg, H., and Cabantchik, Z. I. (1977). *J. Biol. Chem.* **252,** 2419–2427.

Wooster, M. S., and Wrigglesworth, J. M. (1976). *Biochem. J.* **159,** 627–631.

Yanofsky, C., and Crawford, I. P. (1972). *In* "The Enzymes" (P. D. Boyer, ed.), Vol. 7, pp. 1–31. Academic Press, New York.

Yeltman, D. R., and Harris, B. G. (1980). *Arch. Biochem. Biophys.* **199,** 186–196.

Yu, J., and Steck, T. L. (1975a). *J. Biol. Chem.* **250,** 9170–9175.

Yu, J., and Steck, T. L. (1975b). *J. Biol. Chem.* **250,** 9176–9184.

4

Organized Polymeric Enzyme Systems: Catalytic Properties

Jacques Ricard

*Centre de Biochimie et de Biologie Moléculaire
du C.N.R.S
Marseille, France*

I. INTRODUCTION

The living cell may be considered as an exceedingly complex enzyme bioreactor. Enzymes in the cell are not free in solution but loosely or tightly associated with other enzymes or with membranes. Direct interactions with other proteins, diffusional resistances of reactants, and electrostatic partition effects may dramatically alter the behavior of an enzyme with respect to that it would have if it were free in solution.

Polymeric enzymes made up with identical subunits probably represent the simplest organized system where identical polypeptide chains may modify the functioning of an enzyme's active site. There is already a vast scientific literature devoted to the structure–function relationships of polymeric enzymes. This work has been aptly analyzed in a number of review articles (Kirschner, 1968; Koshland, 1969, 1970; Koshland and Neet, 1968; Levitzki and Koshland, 1976; Neet, 1980; Neet and Ainslie, 1980; Ricard, 1980; Sanwal, 1970; Sanwal et al., 1971; Stadtman, 1970; Stadtman and Ginsburg, 1974; Whitehead, 1970; Wyman, 1972) and even in two books (Kurganov, 1982; Levitzki, 1978).

The aim of this chapter however is somewhat different. It is to understand, on a physical basis, how the clustering of folded polypeptide chains into a complex molecular edifice may completely change the properties of the active sites borne by these polypeptide chains. In other words, this goal is to relate the behavior of a protein subunit in a polymeric enzyme to the behavior this subunit should have if it were isolated.

Most of the literature pertaining to the cooperativity of polymeric enzymes is devoted to the ligand-binding properties of these molecular edifices. This is a very important matter indeed. Still, the most important property of an enzyme is not to bind a substrate but to catalyze the conversion of this substrate into a product. If binding of a substrate to the active site is obviously required for its chemical transformation, ligand binding and catalysis are quite distinct processes. One of the aims of this chapter is therefore to understand how subunit interactions, quaternary constraints, and conformation changes may control not only the binding of the substrate but also its conversion to a product.

Evolution occupies a key position in biology in the sense that any important problem may be considered in an evolutionary perspective. In the logic of neo-Darwinian evolution, one may speculate that the emergence of flexible polymeric enzyme systems probably represents a "functional advantage" for living cells. One may therefore wonder which kind of "functional advantage" is offered by the clustering of biologically active polypeptide chains. This evolutionary perspective will be briefly discussed as well.

The above general considerations define the framework of this chapter and justify that many important problems relevant to polymeric enzyme systems have not been considered here. The emphasis has been put on general physical concepts and on their mathematical formulation. The specific properties of individual enzyme systems have only occasionally been covered in this chapter and the interested reader is referred to the general references given above.

II. PROTEIN FLEXIBILITY AS THE *PRIMA RATIO* OF ALLOSTERIC ENZYME BEHAVIOR

Catalysis and regulation of biochemical reactions have long been considered independent attributes of enzyme molecules. In a certain sense these two functions seem somewhat antagonistic, for it is often considered that enzyme catalysis, with its remarkable efficiency and specificity, requires a strict complementarity between the active site and the substrate, according to the classical key–lock hypothesis. This often postulated complementarity implies the enzyme to be a rather rigid, templatelike structure.

On the other hand, it has been understood from the very early days of the concept of enzyme regulation (Monod *et al.*, 1963, 1965) that fine tuning of enzyme activity and information transfer from one position on the enzyme molecule to another are explainable only by assuming that the protein is capable of changing its conformation. It has certainly been the merit of Koshland (1958, 1959) to show that enzyme catalysis of necessity requires the enzyme to be a flexible entity as well. This concept of flexibility is therefore at the center of modern enzymology since it may allow understanding of the physical bases of the amazing catalytic power and regulation of enzymes. Any modern discussion on enzyme regulation must therefore integrate some basic features of enzyme catalysis.

A. Enzyme Flexibility and Cooperativity between Identical and Nonidentical Sites

The term allostery (from the Greek αλλοσ, different, and στερεόσ, spatial) was coined in the 1960s by Monod *et al.* (1963, 1965) to express the idea that regulatory enzymes must be made up of several subunits bearing active and regulatory sites. Each subunit possesses an active and/or a regulatory site. The occupancy of one of these sites affects the capability of the others to bind a ligand. This change of affinity of a site upon the occupancy of another site can only be understandable if the protein changes its conformation. The nonhyperbolic binding isotherms and reaction rate curves, as well as feedback activation and inhibition, are understood easily on this basis. In the early days of allostery theory however, the idea that a regulatory enzyme could bear several identical or nonidentical sites was obtained from rather indirect arguments. This idea was deduced from qualitative analysis of the complex kinetic patterns of several enzyme systems, and from interpretations already available from hemoglobin studies extrapolated to polymeric enzymes. Moreover, this early conclusion has been confirmed in a large number of cases by direct binding data and even by crystallographic studies which show directly the existence of multiple binding sites and conformational changes (Evans and Hudson, 1979; Biesecker *et al.*, 1977; Buehner *et al.*, 1974; Levine *et al.*, 1978; Johnson *et al.*, 1978; Weber *et al.*, 1978; Steitz *et al.*, 1977).

A dimeric allosteric enzyme may be viewed as shown in Fig. 1. The two active sites are located on the two subunits which are made up of one or several polypeptide chains. In many cases these subunits are identical and are termed protomers. These subunits are bound together by several amino acid residues which may be identical or different. If these residues are identical on the two subunits, they constitute a set called an isologous binding domain. If they are different, the binding domain is termed heterologous. Quite often allosteric enzymes possess at least one symmetry axis. Either of these two active sites is termed allosteric with respect to the other. This means that the binding of a substrate to one of these sites modifies the affinity of the other site for the same substrate. This change of affinity is obviously mediated through a conformation change of the protein. The binding of the two substrate molecules is therefore a cooperative process. If substrate binding to the first site enhances

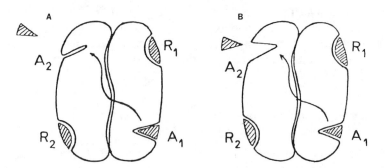

Fig. 1. A dimeric regulatory enzyme (from Ricard, 1980). A_1 and A_2 are the active sites, R_1 and R_2 the regulatory sites. (A) Negative cooperativity between the two active sites (the binding of the substrate at the A_1 site inhibits the binding at the A_2 site). (B) Positive cooperativity between the two active sites (the binding of the substrate at the A_1 site promotes the binding at the A_2 site). The arrow visualizes information transfer from one site to the other.

the affinity of the substrate for the second site, this cooperativity is defined as positive. If alternatively the occupancy of the first site diminishes the affinity of the substrate for the second site, the cooperativity is negative. Positive cooperativity corresponds to an all-or-none response of the enzyme (but not necessarily to an all-or-none conformational transition) to a slight change of substrate concentration. With respect to the hyperbolic binding isotherm, this type of cooperativity corresponds to an increased sensitivity to a signal (the change of substrate concentration). Conversely, negative cooperativity is equivalent to a decreased sensitivity to a change of substrate concentration.

A dimeric enzyme may only exhibit positive cooperativity, negative cooperativity, or no cooperativity. The situation may become more complex if the enzyme is made up of more than two subunits. Then the cooperativity may well be mixed, i.e., positive and negative. Two sites, for instance, may display positive cooperativity whereas two others are negatively cooperative. Mixed cooperativity results in the binding isotherm exhibiting "bumps" or inter-mediary plateaus (Teipel and Koshland, 1969; Hill *et al.*, 1977; Bardsley, 1976; Bardsley and Childs, 1976; Bardsley and Crabbe, 1976). A change of the affinity of the substrate for the active site obviously generates a change of the rate of product appearance. Contrary to a common belief however, there is no proportionality between the degree of saturation of the enzyme, under equilibrium conditions, and the steady-state enzyme reaction rate. This point will be discussed at length later on.

In addition to the active sites in interaction, allosteric enzymes may bear regulatory sites which bind modifiers of the enzyme reaction, for instance, feedback activators or inhibitors. Any regulatory site may be considered as allosteric with respect to an active site or to another regulatory site. Active and regulatory sites may be located on different polypeptide chains or on the same polypeptide chain. Aspartate transcarbamylase of *Escherichia coli* belongs to the first group. This enzyme may be dissociated into two groups of subunits with p-chloromercuribenzoate (Gerhart and Schachman, 1965, 1968). The first set contains subunits C, which bind the substrate, and the second set contains subunits R, which bind allosteric effectors ATP and CTP. Subunits C are the catalytic subunits, whereas subunits R are regulatory. The probable quaternary structure of this enzyme is $C_6 R_6$, and its molecular weight is 310,000. After dissociation of the regulatory subunits, the set of six catalytic subunits binds six substrate molecules noncooperatively, has full catalytic activity, and is totally insensitive to allosteric modifiers ATP and CTP (Wiley and Lipscomb, 1968; Weber, 1968; Kerbiriou and Hervé, 1972, 1973; Kerbiriou *et al.*, 1977; Thiry and Hervé, 1978). A similar situation occurs with a protein kinase where active and regulatory sites are located on different polypeptide chains. However, in the case of ribonucleoside diphosphate reductase of *E. coli* (von Dobeln and Reichard, 1976), both the active and the regulatory sites sit on the same subunit. The same occurs for glutamate dehydrogenase and apparently for many other enzymes (Engel, 1973).

Extreme negative cooperativity may generate the so called half-of-the-sites or flip-flop reactivity (Lazdunski et al., 1971; Lazdunski, 1972; Levitzki and Koshland, 1976; Levitzki, 1973, 1974). Alkaline phosphatase of E. coli has been claimed to possess this behavior. The enzyme behaves as a dimer and binds an organic phosphate noncovalently. This binding process induces a conformation change that turns off this subunit for a while (extreme negative cooperativity). The covalent interaction of phosphate with a serine residue of this subunit turns the other subunit on and allows the noncovalent binding of another molecule of organic phosphate to the subunit which has been turned on. Half-of-the-sites reactivity has been claimed to occur with dehydrogenases (Bernhard et al., 1970; Dunn and Bernhard, 1971; McFarland and Bernhard, 1972; Dunn, 1974; Luisi and Favilla, 1972; Luisi and Bignetti, 1974; McFarland and Chu, 1975; Dunn and Hutchinson, 1973), but this claim has been questioned (Pettersson, 1976; Kvassman and Pettersson, 1976; Hadorn et al., 1975). Half-of-the-sites reactivity may also be the consequence of a preexisting asymmetry of the enzyme molecule (Matthews and Bernhard, 1973; Seydoux and Bernhard, 1974; Seydoux et al., 1973; Viratelle and Seydoux, 1975). The concept of allosteric interaction between sites may be extended to more complex situations such as interactions within multifunctional enzymes (Wetlaufer and Ristow, 1973; Kirschner and Bisswanger, 1976; Welch, 1977; De Moss, 1962; Matchett, 1974) or membranes (Changeux et al., 1967).

B. Enzyme Flexibility and Catalysis

The idea that it is protein flexibility that explains that an allosteric enzyme is much more than the sole collection of independent subunits represents the very basis of enzyme regulation studies. Moreover the same idea is now required to explain the mechanisms of enzyme catalysis.

There are a vast number of experimental results showing that the mechanisms of enzyme catalysis are basically the same as those occurring in organic chemistry, namely acid–base and covalent catalysis. Therefore, trying to explain the amazing efficiency of enzyme catalysis by the static structure of the active site as revealed by enzyme crystallography has been somewhat disappointing. In most cases this type of study has simply confirmed the existence of catalytic mechanisms which earlier chemical studies had already suggested. The observation that the active site is small with respect to the total enzyme molecule suggests that efficient catalysis and control require the entirety of the protein molecule and not just the residues that participate in the chemical act of catalysis. As pointed out by Atkinson (1970), to try to understand the amazing efficiency of enzyme catalysis by the structure of the active site alone would be equivalent to trying to understand the genius of Michelangelo by a detailed anatomical study of his hand!

In fact, two types of effects, which do not exist to the same extent with organic reactions, largely contribute to the efficiency of enzyme catalysis: an entropy effect and a destabilization effect (Koshland, 1962; Storm and Koshland, 1970, 1972a,b; Koshland *et al.*, 1972; Jencks, 1969, 1975). The entropy effect, which will not be discussed here, corresponds to the orientation and approximation of the reactants in the enzyme's active site. The destabilization effect comprises two types of phenomena: a geometric destabilization, or distortion, and an electrostatic destabilization, or desolvation, of the substrate. Both require a conformational change of the enzyme.

Very early, Haldane (1930) remarked that in a one-substrate, one-product enzyme reaction, the free enzyme could not be complementary to both the substrate and the product, since these two ligands have a different structure. As soon as an enzymatic reaction is viewed as bidirectional, elementary logic shows the key–lock hypothesis to be inadequate. Haldane suggested that the active site in the free enzyme should be complementary to the transition state of the reaction, and that it must have a loose structure, midway between that of the substrate and that of the product.

The requirement for complementarity between the transition state and the active site of the free enzyme may be deduced from simple thermodynamic principles (Lienhard *et al.*, 1972). Let there be the simple one-substrate, one-product reaction

$$E \underset{k_{-1}}{\overset{k_1[S]}{\rightleftharpoons}} ES \underset{k_{-e}}{\overset{k_e}{\rightleftharpoons}} EP \underset{k_{-2}[P]}{\overset{k_2}{\rightleftharpoons}} E$$

Comparison of the rate constant k_e of the enzyme-catalyzed reaction to that, k_{ne}, of the corresponding uncatalyzed reaction

$$S \underset{k_{-ne}}{\overset{k_{ne}}{\rightleftharpoons}} P$$

is easily done by using the thermodynamic box of Fig. 2. Since free energies are state functions one must have

$$\Delta G_{ne}^{\ddagger} - \Delta G_e^{\ddagger} = \Delta G_x - \Delta G_s.$$

Since $\Delta G_{ne}^{\ddagger} \gg \Delta G_e^{\ddagger}$ it follows that $\Delta G_x \gg \Delta G_s$. Therefore, the lower is the affinity of the substrate for the enzyme the larger is the forward catalytic constant. The difference between the binding energies of the transition state and of the substrate to the enzyme is understandable only if this enzyme is more complementary to the transition state than to the substrate. The possible upper limit of the catalytic rate constant will be reached when the free enzyme

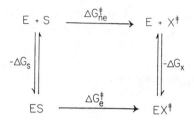

Fig. 2.   A thermodynamic box that relates the binding free energies of the substrate and of the transition state. (From Ricard, 1980.)

is perfectly complementary to the transition state $X^{\ddagger}$. Clearly, rate accelerations brought about by the enzyme are the consequence of its flexibility and of substrate destabilization.

Although it is impossible to measure directly the binding energy of a transition state to the enzyme, the validity of the above idea has been verified experimentally. Wolfenden (1969), Lienhard (1971), and Secemski (1972) have shown that transition-state analogues bind very tightly to the active site. This binding is so tight that they can be used as "super-inhibitors" of enzyme reactions. This is exactly what is expected from the above theoretical considerations. Probably the most direct experimental evidence for the validity of the above ideas stems from the study of the lysozyme reaction. Lysozyme is a rather small protein which catalyzes the hydrolysis of polymers of N-acetylglucosamine. From detailed crystallographic studies (Phillips, 1967; Ford et al., 1974), the lysozyme molecule appears as an ellipsoid bearing in its active site six subsites A, B, C, D, E, and F. The enzymatic reaction is basically an acid–base catalysis occurring between subsites D and E. The increments of the binding energy that can be measured upon binding of sugar residues to subsites A, B, and C are -1.8, -3.7, and -5.7 kcal/mol, respectively. Binding at subsite D, however, requires an unfavorable increment of binding energy of $+2.9$ kcal/mol (Chipman et al., 1967). These values suggest that the productive binding of a sugar residue to subsite D necessitates distortion of the substrate and of the enzyme as well. However, the binding energy of a lactone tetrasaccharide to the ABCD subsites is increased by a factor of 6×10^3 compared to that of a tetramer of N-acetylglucosamine (Secemski et al., 1972). This is a direct proof that the conformation of the active site is more complementary to the transition state than to the substrate.

Destabilization of the substrate may also be effected by suppressing its interaction with the solvent molecules. Again this desolvation requires a conformational change of the enzyme molecule. Models derived from bioorganic chemistry clarify how substrate desolvation coupled to enzyme conformational change may dramatically increase the reaction rate. The

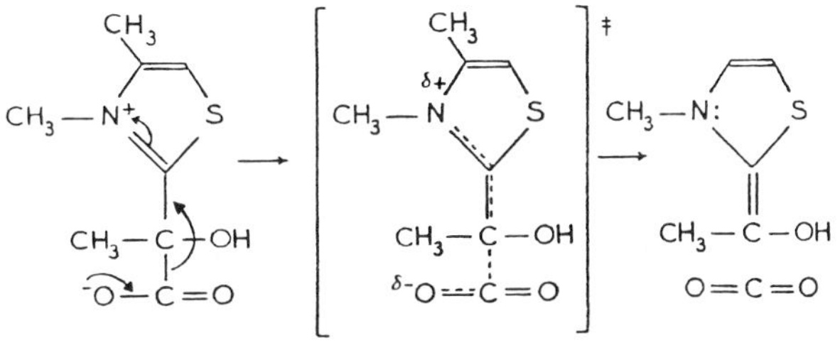

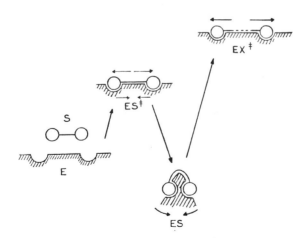

Scheme 1

adduct of pyruvate and thiamine, or thiamine pyrophosphate, spontaneously decarboxylates (see Scheme 1). In water, however, this decarboxylation is a slow process, for carboxylate ion is strongly solvated. In organic solvents this decarboxylation reaction is accelerated by a factor of 10^5. This is exactly what is occurring with thiamine pyrophosphate enzymes. Pyruvate is covalently bound to the ylide structure of thiamine pyrophosphate located on the enzyme's surface. This binding process induces a conformational change of the enzyme, which results in burying the adduct in a hydrophobic cleft. This in turn produces desolvation of the carboxylate ion and accelerates the reaction by a factor of 10^5. The view that the enzyme is strained in the enzyme–substrate complex, whereas it is not in the enzyme–transition state complex, provides a simple explanation of the mechanism of catalysis (Lumry, 1959; Jencks, 1969). The driving force required for overcoming the energy barrier comes, at least in part, from the enzyme that reaches an unstrained conformation at the top of the barrier (Fig. 3).

Fig. 3. Strain of the enzyme as a source of catalytic power. The enzyme is complementary to the transition state $X^‡$ and is strained in the $ES^‡$ and ES complexes. (From Ricard, 1980.)

The idea that efficient catalysis is obtained when the free enzyme is complementary to the transition state, $X^{\ddagger}$, must obviously be extended to polymeric enzymes as well. This implies that a given subunit bound to a transition state must have, in a polymeric enzyme, an unstrained conformation, that is, the conformation the subunit would have if subunit interactions and quaternary constraints were not occurring. The important idea that strain and quaternary constraints, within the allosteric enzyme molecule, must be released in a transition state has fundamental implications that will be developed at length later.

C. Preequilibrium, Induced-Fit, and Cooperativity

There are two very different mechanisms to explain enzyme flexibility and its effect upon ligand binding. The first possibility is to assume that the free enzyme preexists in different states in equilibrium, having different affinities for the ligand. Ligand binding to one of these conformational states shifts this equilibrium toward one of these states. It does not create the conformational change but selects one of the preexisting conformations. The second extreme possibility is to consider that the free enzyme exists in only one conformation and that the collision of the reaction partners induces a fast conformational change which allows ligand binding. These two concepts form the basis of the models of Monod et al. (1965) and Koshland et al. (1966), and have, for some time, been considered mutually exclusive. In fact, they are not if it is considered that the number and the nature of the conformational states are not infinite, which is supported by evidence. Both the preequilibrium and the induced-fit concepts are limiting cases of a more general scheme, which is shown in Fig. 4 for a dimeric enzyme. The situation shown in Fig. 4 is still quite simple for it is assumed that any subunit may exist in only two conformations, namely, A (the circle) and B (the square).

The model of Fig. 4 clearly shows that there are two limiting cases of subunit coupling. The first one is exemplified in the diagonal of the model (Fig. 4). It postulates that protomers are loosely coupled and that a subunit may change its conformation without propagating this conformation change

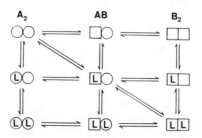

Fig. 4. Induced-fit and preequilibrium for a dimeric enzyme (see text).

to the other subunits. There exists no quaternary constraint in the dimeric protein. This model, which is associated with loose coupling between subunits, is called the "simple" sequential model by Koshland et al. (1966). When loose coupling prevails, symmetry of the enzyme molecule is not conserved during the conformational transition, for some molecules exist in the AB state (Fig. 4).

Another limiting case of subunit coupling is observed when subunits are coupled so tightly that any local conformational change propagates to the rest of the molecule. The conformational transition is then of the all-or-none type. Symmetry is conserved during such a "concerted" conformational change. The model of Monod et al. (1965) typically corresponds to a concerted symmetry model. The existence of a concerted conformational transition implies that strong quaternary constraints exist and that any subunit is in a strained state. We have already seen that in a transition state, at the top of the energy barrier, a subunit cannot exist in a strongly strained state. This means that quaternary constraints, within tightly coupled subunits of an allosteric enzyme, must be released from time to time, at least when a transition state is reached along the reaction coordinate. If it were not so, no chemical reaction could ever be controlled by this polymeric enzyme. The tridimensional structure of an enzyme, whether its subunits are tightly coupled or not, must always be viewed as a dynamic entity (Karplus and Weaver, 1976, 1979; Rossky et al., 1979; McCammon et al., 1977, 1979; McCammon and Karplus, 1977; Gelin and Karplus, 1975; Szabo and Karplus, 1975; Rossky and Karplus, 1979; Austin et al., 1975; Alberding et al., 1978a–c; Beece et al., 1979, 1980; Frauenfelder et al., 1979; Hartmann et al., 1982).

Finally, it may occur that the coupling between subunits is not strong enough to produce a concerted conformational transition upon substrate binding. In this case, the model of subunit interaction may be considered as partially concerted and a given subunit may exist, within the polymeric structure, in more than just the two states A and B.

III. PHENOMENOLOGY OF MULTIPLE LIGAND BINDING TO ENZYMES

The basic allosteric properties of multiple-subunit enzymes stem from the analysis of their multiple ligand-binding properties.

A. The Generalized Adair Equation as an Expression of Cooperativity

Let us consider the equilibrium between an n-sited enzyme and a ligand L:

$$E_0 \; \overset{K_1}{\rightleftharpoons} \; E_1 \; \overset{K_2}{\rightleftharpoons} \; E_2 \cdots E_{n-1} \; \overset{K_n}{\rightleftharpoons} \; E_n \tag{1}$$

where E_0 is the free enzyme, E_1 the enzyme that has bound one molecule of ligand, and so on. The fractional saturation, $\bar{v}$, of the enzyme by the ligand L is defined as

$$\bar{v} = n\bar{Y} = \frac{[E_1] + 2[E_2] + \cdots + n[E_n]}{[E_0] + [E_1] + [E_2] + \cdots + [E_n]} \tag{2}$$

where $\bar{Y}$ is the scaled fractional saturation. Setting

$$\psi_1 = K_1, \qquad \psi_2 = K_1 K_2, \qquad \cdots, \qquad \psi_n = K_1 K_2 \cdots K_n, \tag{3}$$

Eq. (2) assumes the form

$$\bar{v} = n\bar{Y} = \frac{\sum_{i=1}^{n} i\psi_i [L]^i}{1 + \sum_{i=1}^{n} \psi_i [L]^i}, \tag{4}$$

which is the generalized expression of the Adair equation (1925). A plot of $\bar{v}$ (or $\bar{Y}$) as a function of [L] may give a complex curve, where complexity is the expression of cooperativity between sites. It is therefore of interest to understand the phenomenological nature of this cooperativity.

If it is assumed that the binding sites are all equivalent, any macroscopic binding constant K_i allows definition of a corresponding microscopic constant K_i'. Thus,

$$K_i = \left[\binom{n}{i} \Big/ \binom{n}{i-1} \right] K_i' = \frac{n-i+1}{i} K_i'. \tag{5}$$

Any microscopic binding constant refers to a specified binding site. Equation (5) above becomes meaningless if the sites are nonequivalent. In the following it will always be assumed that the sites are equivalent.

Recalling that $d(\ln x) = dx/x$, one has

$$d\left\{ \ln\left(1 + \sum_{i=1}^{n} \psi_i [L]^i\right) \right\} = \frac{\left[\sum_{i=1}^{n} i\psi_i [L]^{i-1}\right] d[L]}{1 + \sum_{i=1}^{n} \psi_i [L]^i}, \tag{6}$$

and

$$d\ln[L] = d[L]/[L]. \tag{7}$$

Therefore,

$$\bar{v} = n\bar{Y} = \frac{d[\ln(1 + \sum_{i=1}^{n} \psi_i [L]^i)]}{d\ln[L]}. \tag{8}$$

Moreover, for n equivalent sites

$$\psi_i = K_1 \cdots K_i = \prod_{j=1}^{i} \frac{n-j+1}{j} K_j'. \tag{9}$$

Since

$$\prod_{j=1}^{i} \frac{n-j+1}{j} = \frac{n!}{i!(n-i)!} = \binom{n}{i}, \tag{10}$$

Eq. (9) assumes the form

$$\psi_i = \binom{n}{i} K'_1 K'_2 \cdots K'_i, \tag{11}$$

and

$$1 + \sum_{i=1}^{n} \psi_i[L]^i = \prod_{i=1}^{n}(1 + K'_i[L]). \tag{12}$$

Therefore the Adair equation becomes

$$\bar{v} = n\bar{Y} = \sum_{i=1}^{n} \frac{d\{\ln(1 + K'_i[L])\}}{d\ln[L]}. \tag{13}$$

If the sites are not only equivalent but also independent there exists only one microscopic binding constant, K', and Eq. (12) becomes

$$1 + \sum_{i=1}^{n} \psi_i[L]^i = (1 + K'[L])^n. \tag{14}$$

Inserting this expression into Eq. (8) yields

$$\bar{v} = n\bar{Y} = \frac{nK'[L]}{1 + K'[L]}, \tag{15}$$

and the binding equation becomes hyperbolic. Clearly, if the binding sites on the enzyme are all equivalent and independent there may exist no cooperativity in ligand binding. Cooperativity stems from nonequivalence, nonindependence, or both nonequivalence and nonindependence of the binding sites.

B. The Hill Plot and Its Significance

Let us assume now that n molecules of ligand are bound to the enzyme in virtually one step. The Adair binding equation becomes

$$\bar{Y} = \frac{\bar{v}}{n} = \frac{\psi_n[L]^n}{1 + \psi_n[L]^n}, \tag{16}$$

which can be rearranged to

$$\ln\{\bar{Y}/(1 - \bar{Y})\} = \ln \psi_n + n\ln[L]. \tag{17}$$

This is the classical Hill formulation. When plotting $\ln\{\bar{Y}/(1 - \bar{Y})\}$ as a function of $\ln[L]$, one should obtain a straight line of slope n. Obviously the binding of a ligand does not occur in one step in practice. Therefore, when plotting $\ln\{\bar{Y}/(1 - \bar{Y})\}$ as a function of $\ln[L]$, one obtains a curve and by analogy with Eq. (17), one may define a so-called Hill coefficient n as

$$n = \frac{d\ln\{\bar{Y}/(1 - \bar{Y})\}}{d\ln[L]} = \frac{[L]}{\bar{Y}(1 - \bar{Y})}\frac{d\bar{Y}}{d[L]}. \tag{18}$$

The value of the Hill coefficient will be maximum at half-saturation of the enzyme by the ligand, that is, when $\bar{Y} = 0.5$. This value of n is defined as the maximum Hill coefficient n_H:

$$n_H = 4[L]_{0.5}\frac{d\bar{Y}}{d[L]_{0.5}}. \tag{19}$$

The value of $[L]_{0.5}$ for a dimeric enzyme may be obtained algebraically by solving the equation

$$\bar{Y}_{0.5} = \left(\frac{1}{2}\right)\frac{\psi_1[L]_{0.5} + 2\psi_2[L]_{0.5}^2}{1 + \psi_1[L]_{0.5} + \psi_2[L]_{0.5}^2} = \frac{1}{2}, \tag{20}$$

and one finds

$$[L]_{0.5} = 1/\sqrt{\psi_2}. \tag{21}$$

Differentiating the binding isotherm $\bar{Y}$ and inserting Eq. (21) into Eq. (19) yield

$$n_H = \frac{4}{2 + \psi_1/\sqrt{\psi_2}}. \tag{22}$$

This expression shows that the maximum Hill coefficient cannot be greater than 2. If there is no cooperativity, the two ψ parameters and the macroscopic binding constants may be expressed as

$$\psi_1 = K_1 = 2K', \qquad \psi_2 = K_1K_2 = K'^2.$$

Inserting these values into Eq. (22) above shows that $n_H = 1$.

A typical Hill plot pertaining to the Adair equation is shown in Fig. 5. At very low and very high ligand concentration, the slope of the plot is close to unity. These regions of the plot correspond to the binding of the first and the last ligand, respectively.

As stressed by Wyman (1964) and Whitehead (1970, 1974), Hill plots have an important thermodynamic significance. Let us consider the free energy of dissociation, K_D, of a ligand L from a site:

$$\Delta G = -RT\ln K_D. \tag{23}$$

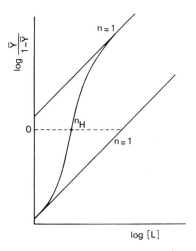

Fig. 5. A typical Hill plot. The slope of the plot at half-saturation of the enzyme by the ligand corresponds to the maximum Hill plot n_{H}.

Since

$$K_{\mathrm{D}} = [\mathrm{L}](1 - \bar{Y})/\bar{Y}, \tag{24}$$

Eq. (23) assumes the form

$$\Delta G = RT \ln\{\bar{Y}/(1 - \bar{Y})\} - RT \ln[\mathrm{L}]. \tag{25}$$

Wyman (1964) defined a free energy of interaction as the free energy required to change the degree of saturation of the protein. This free energy of interaction may be expressed as the gradient of free energy that can be determined when either the saturation $\bar{Y}$ or the ligand concentration $[\mathrm{L}]$ is changed. One may therefore define two interaction free energy gradients as $d\Delta G/d\bar{Y}$ and $d\Delta G/d[\mathrm{L}]$. Differentiating Eq. (25) with respect to $\ln\{\bar{Y}/(1 - \bar{Y})\}$ yields

$$\frac{d\Delta G}{d\ln\{\bar{Y}/(1 - \bar{Y})\}} = RT - RT\frac{d\ln[\mathrm{L}]}{d\ln\{\bar{Y}/(1 - \bar{Y})\}} = RT\left(1 - \frac{1}{n}\right), \tag{26}$$

which can be rearranged to

$$\frac{d\Delta G}{d\bar{Y}} = \frac{RT}{\bar{Y}(1 - \bar{Y})}\left(1 - \frac{1}{n}\right). \tag{27}$$

Similarly, differentiating Eq. (25) with respect to $\ln[\mathrm{L}]$ yields

$$\frac{d\Delta G}{d\ln[\mathrm{L}]} = RT\frac{d\ln\{\bar{Y}/(1 - \bar{Y})\}}{d\ln[\mathrm{L}]} - RT, \tag{28}$$

which can be rearranged to

$$\frac{d\Delta G}{d[L]} = \frac{RT}{[L]}(n - 1).$$

(29)

Equations (27) and (29) represent simple relations between the Hill coefficient and the free energy gradients of interaction. Obviously, when there is no cooperativity, $n = 1$ and these two gradients are null.

C. Linked Functions and the Binding of Two Different Ligands

Let there be two ligands L and M, for instance, a substrate and a product, which can bind to an enzyme. If there exist n binding sites for L and m binding sites for M, one may define the two binding functions $\bar{Y}_L$ and $\bar{Y}_M$ as

$$\bar{Y}_L = \frac{\sum_{i=1}^{n}[EL_i] + \sum_{i=1}^{n}\sum_{j=1}^{m}[EL_iM_j]}{n\{[E_0] + \sum_{i=1}^{n}[EL_i] + \sum_{j=1}^{m}[EM_j] + \sum_{i=1}^{n}\sum_{j=1}^{m}[EL_iM_j]\}},$$

$$\bar{Y}_M = \frac{\sum_{j=1}^{m}[EM_j] + \sum_{i=1}^{n}\sum_{j=1}^{m}[EL_iM_j]}{n\{[E_0] + \sum_{i=1}^{n}[EL_i] + \sum_{j=1}^{m}[EM_j] + \sum_{i=1}^{n}\sum_{j=1}^{m}[EL_iM_j]\}}.$$

(30)

Moreover, if $[E]_T$ is the total enzyme concentration and $[E_0]$ the unliganded enzyme concentration, one may express their ratio ρ as

$$\rho = \frac{[E]_T}{[E_0]} = 1 + \sum_{i=1}^{n} \psi_{i0}[L]^i + \sum_{j=1}^{m} \psi_{0j}[M]^j + \sum_{i=1}^{n}\sum_{j=1}^{m} \psi_{ij}[L]^i[M]^j,$$

(31)

where $\ln\rho$ is the binding polynomial (Wyman 1964). The two binding isotherms may be expressed as

$$n\bar{Y}_L = \left(\frac{\partial\ln\rho}{\partial\ln[L]}\right)_{[M]},$$

(32)

and

$$m\bar{Y}_M = \left(\frac{\partial\ln\rho}{\partial\ln[M]}\right)_{[L]}.$$

(33)

One must then have

$$d\ln\rho = \left(\frac{\partial\ln\rho}{\partial\ln[L]}\right)_{[M]} d\ln[L] + \left(\frac{\partial\ln\rho}{\partial\ln[M]}\right)_{[L]} d\ln[M].$$

(34)

Taking into account Equations (32) and (33), Eq. (34) above assumes the simpler form

$$d\ln\rho = n\bar{Y}_L\, d\ln[L] + m\bar{Y}_M\, d\ln[M].$$

(35)

Cross-differentiation of this equation yields

$$n\left(\frac{\partial \bar{Y}_L}{\partial \ln[M]}\right)_{[L]} = m\left(\frac{\partial \bar{Y}_M}{\partial \ln[L]}\right)_{[M]}. \tag{36}$$

This is the basic linkage equation. It shows that the variation of $\bar{Y}_L$ upon changing [M] at constant [L] is simply related to the variation $\bar{Y}_M$ would exhibit when changing [L] by the same amount at constant [M].

IV. THE CONCERTED AND THE INDUCED-FIT MODELS OF LIGAND BINDING

So far cooperativity has been discussed in a purely formal and phenomenological fashion. The main goal of modern theories of allostery is to relate, in terms of molecular parameters, the dynamic changes of quaternary structure to the expression of the ligand binding function. Both the concerted model of Monod *et al.* (1965) and the sequential induced-fit model of Koshland *et al.* (1966) offer an interpretation of binding isotherms and cooperativity in either molecular or structural terms.

A. The Concerted Model of Monod–Wyman–Changeux

The concerted (or symmetry) model was set up by Monod *et al.* (1965) and its analytical treatment has been refined by Buc (1967), Buc and Buc (1967), Blangy *et al.* (1968), Rubin and Changeux (1966), Whitehead (1970), and Wyman (1972). These treatments rest on the theory of multiple equilibria (Wyman, 1948, 1964, 1967, 1968), briefly discussed in Section III,B. The symmetry, or concerted, model rests on several postulates which can be expressed as follows:

1. The protomers are associated in such a way that they occupy equivalent positions. This implies that the molecule possesses at least one symmetry axis.
2. The free enzyme occurs under two different conformations having different affinities for the substrate or the effectors. As a consequence, the binding of a ligand, substrate, or effector shifts the preequilibrium between these two enzyme forms.
3. During the shift of this preequilibrium, the symmetry of the enzyme molecule is conserved.
4. The binding of the substrate to one of the conformation states does not induce any further conformational change, and the sites are equivalent and independent.

To illustrate these ideas let us consider a tetrameric enzyme. Any protomer may take two conformations represented by a circle (conformation A) and a square (conformation B). The free enzyme exists in two states in which the subunits are all in either the A or the B conformation (in Monod's formulation these conformations were designated R and T). The symmetry postulate does not allow the existence of hybrid states in which some subunits are in the A conformation and others in the B conformation. This in turn implies that the subunits are tightly coupled and that the conformational transition is concerted, or of the all-or-none type. The basic features of the model are shown in Fig. 6.

Since it is postulated that there exist only two conformational states of the protein and that the conformational changes are concerted, any of the binding steps may be described with one microscopic binding constant, either K_A^* (for the A_4 state) or K_B^* (for the B_4 state). Once in the A_4 or B_4 state the sites are equivalent and independent. Therefore, the mathematical treatment of this situation directly stems from the multiple ligand binding theory briefly discussed in Section III,B. The binding isotherm may be written as

$$\bar{Y} = \frac{1}{4}\frac{[A_4S_1] + 2[A_4S_2] + \cdots + 4[A_4S_4] + [B_4S_1] + 2[B_4S_2] + \cdots + 4[B_4S_4]}{[A_4] + [A_4S_1] + [A_4S_2] + \cdots + [A_4S_4] + [B_4] + [B_4S_1] + [B_4S_2] + \cdots + [B_4S_4]}.$$

(37)

If the allosteric transconformation constant of the free enzyme is defined as

$$L = [B_4]/[A_4],$$

(38)

the expression of the binding isotherm assumes the form

$$\bar{Y} = \frac{K_A^*[S](1 + K_A^*[S])^3 + LK_B^*[S](1 + K_B^*[S])^3}{(1 + K_A^*[S])^4 + L(1 + K_B^*[S])^4}.$$

(39)

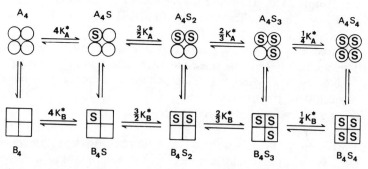

Fig. 6.   The concerted Monod–Wyman–Changeux model for a tetrameric enzyme.

This equation has several important implications. First it shows that if $L = 0$, that is, if the protein preexists in one state, the cooperativity vanishes and the binding isotherm becomes hyperbolic:

$$\bar{Y} = \frac{K_A^*[S]}{1 + K_A^*[S]}. \tag{40}$$

Cooperativity is thus clearly the consequence of preequilibrium between free enzyme forms. Second, if the substrate has the same intrinsic affinity, K^*, for the A_4 and B_4 states, the binding equation reduces to that of a hyperbola, namely

$$\bar{Y} = \frac{K^*[S](1 + K^*[S])^3(1 + L)}{(1 + K^*[S])^4(1 + L)} = \frac{K^*[S]}{1 + K^*[S]}. \tag{41}$$

Third, if the binding is exclusive, e.g., $K_B^* = 0$, the binding equation reduces to

$$\bar{Y} = \frac{K_A^*[S](1 + K_A^*[S])^3}{L + (1 + K_A^*[S])^4} \tag{42}$$

and the binding process remains cooperative.

This reasoning is indeed general and may be applied to the case of an n-sited enzyme:

$$\bar{Y} = \frac{K_A^*[S](1 + K_A^*[S])^{n-1} + LK_B^*[S](1 + K_B^*[S])^{n-1}}{(1 + K_A^*[S])^n + L(1 + K_B^*[S])^n}. \tag{43}$$

The expression for the maximum Hill coefficient has been derived (Rubin and Changeux, 1966) for this general equation [Eq. (43)] but it is quite complex. In order to illustrate the usefulness of Eq. (22), the maximum Hill coefficient is derived here for the case of a dimeric enzyme. The expression of the binding equation for a dimeric enzyme using the concerted model of Monod et al. (1965) may be cast in the form of an Adair equation. The relevant ψ coefficients are now expressed as

$$\psi_1 = \frac{2(K_A^* + LK_B^*)}{1 + L}, \qquad \psi_2 = \frac{K_A^{*2} + LK_B^{*2}}{1 + L}. \tag{44}$$

Inserting these expressions into Eq. (22) yields

$$n_H = \frac{2}{[(K_A^* + LK_B^*)/\sqrt{(1 + L)(K_A^{*2} + LK_B^{*2})}] + 1}. \tag{45}$$

This equation shows that n_H is necessarily greater than or equal to 1. When $L > 0$, cooperativity is positive and n_H is less than 2. When $L = 0$, $n_H = 1$ and there is no cooperativity. In any case, cooperativity cannot be negative. These conclusions are valid in general.

An interesting prediction of the model of Monod *et al.* (1965) is that activators and inhibitors of substrate binding act solely by displacing the equilibrium between the A_4 and B_4 states. If the A conformation has a higher affinity for the substrate than the B conformation, the activators will bind preferentially to the A conformation and the inhibitors to the B conformation. Since neither the inhibitor nor the activator induces a conformational change of the enzyme, binding Eqs. (39) and (43) are still applicable, but the transconformation constant L has to be replaced, for a tetramer, by an apparent constant L', defined as

$$L' = L\frac{(1 + K_i^*[X_i])^4}{(1 + K_a^*[X_a])^4},$$ (46)

where X_i and X_a are the inhibitory and the activatory ligands, respectively, and K_i^* and K_a^* the relevant intrinsic binding constants. Clearly, from Eq. (46), when the concentration of the activator X_a is increased, the value of L' decreases; therefore, both the sigmoidicity and the cooperativity decrease. Alternatively, when the concentration of inhibitor X_i is increased, L' is increased and both sigmoidicity and cooperativity increase. This conclusion is general. The advantage of the concerted model of Monod *et al.* (1965) is its obvious simplicity.

B. The Sequential Induced-Fit Models of Koshland–Nemethy–Filmer

It may well occur that within a polymeric enzyme molecule coupling between subunits is not tight enough to produce a concerted conformational transition when one of the subunits changes its conformation. In order to express how the strength of the quaternary constraints controls ligand binding, Koshland *et al.* (1966) proposed structural molecular models based on concepts apparently different from those used in the Monod model.

The Koshland models rest on the following simple and sensible postulates:

1. The unliganded enzyme exists in one conformation only.
2. Ligand binding to a protomer induces a conformational change of that protomer.
3. If subunits are loosely coupled, this conformational change is not propagated to the other subunits. The corresponding structural models are then termed simple.
4. Alternatively, if the subunits are tightly coupled, the conformational transition becomes concerted for all the subunits. The corresponding structural models are called fully concerted.
5. The coupling strength between subunits may be such that a conformational change is only in part propagated to the rest of the enzyme

molecule. The corresponding models are then termed partially concerted.

6. At least as a simplifying assumption, it is postulated that only one subunit conformation binds the substrate.

In most of what follows in this section it will be assumed that subunit interactions are very loose or very tight in such a way that simple or fully concerted models are applicable. The formulation of these structural models rests on the definition of molecular parameters that are required to relate the binding properties of a protein to its quaternary structure.

The Koshland theory has been refined and applied to several enzyme systems (Cornish-Bowden and Koshland, 1970a,b, 1975; Conway and Koshland, 1968; Cook and Koshland, 1969, 1970; Levitzki and Koshland, 1969, 1972a,b; Neet and Koshland, 1968).

1. Definition and Thermodynamic Significance of the Molecular Parameters

Let us assume that any subunit may exist in two conformations, the circle (A) and the square (B). Moreover, the substrate is assumed to be bound to subunit conformation B, but not to the conformation A. Let there be ΔG_s, the free energy of substrate binding to conformation B, and ΔG_t, the free energy of transconformation of state A to state B (Fig. 7). Now, let ΔG_{AA} be the free energy of association of state A with state A, with ΔG_{AB} and ΔG_{BB} being the free energies of the complex processes shown in Fig. 7. It follows from the definitions exemplified in Fig. 7 that the free energies of association of state A with state B or of state B with state B are $\Delta G_{AB} + \Delta G_{AA}$ and $\Delta G_{BB} + \Delta G_{AA}$, respectively. These ideal energy contributions allow the definition of ideal equilibrium constants, namely,

$$K_s = \exp(-\Delta G_s/RT), \qquad K_t = \exp(-\Delta G_t/RT),$$

$$K_{AA} = \exp(-\Delta G_{AA}/RT), \qquad K_{AB} = \exp(-\Delta G_{AB}/RT), \qquad (47)$$

$$K_{BB} = \exp(-\Delta G_{BB}/RT).$$

Any phenomenological binding constant may now be expressed in structural terms with the molecular parameters [Eqs. (47)] above. This obviously requires some prior knowledge of the various types of subunit interactions existing in the molecule. For example, in a tetramer it may occur that each subunit is associated, through identical contacts, with only two other subunits. This constitutes the so-called "square" model of subunit interactions. If subunits are loosely coupled (simple "square" model) the structural expression of the first substrate binding constant to the free enzyme assumes the form

$$K_1 = K_s K_t K_{AB}^2. \qquad (48)$$

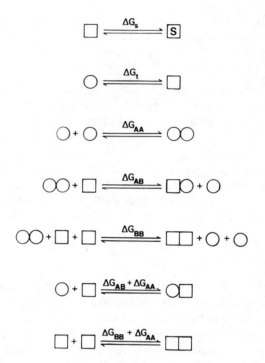

Fig. 7. Thermodynamic significance of the molecular parameters of Koshland–Nemethy–Filmer.

This expression is the direct consequence of the ideal thermodynamic loop shown in Fig. 8. Clearly, if $K_{AB}^2 < 1$, subunit association decreases the binding of substrate to the tetramer with respect to what would be expected with an ideally isolated protomer. Alternatively, if $K_{AB}^2 > 1$, subunit interactions facilitate (with respect to the ideal protomer) substrate binding. With this formalism the expression for K_{AA} disappears from the expression of phenomenological binding constants, which is equivalent to expressing K_{AB} and K_{BB} with respect to K_{AA}.

If subunits are tightly coupled in such a way that the conformational transition of all subunits is concerted (Fig. 8B), the expression for the first binding constant becomes

$$K_1 = K_s K_t^4 K_{BB}^4. \qquad (49)$$

This type of formalism may be applied to any type of binding step. Indeed the formulation of the binding constants depends upon molecular parameters related to the quaternary structure of the protein.

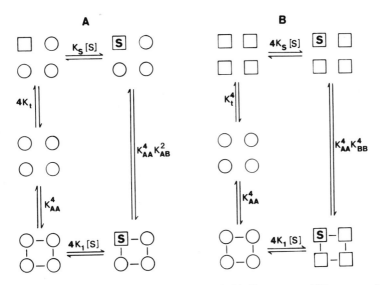

Fig. 8. Molecular parameters and the expression of a binding constant. (A) Loose coupling of subunits; (B) tight coupling of subunits.

2. Substrate Binding to Loosely and Tightly Coupled Subunit Enzymes

The first case to be considered here is the process of ligand binding to a tetrameric enzyme using the simple "tetrahedral" model. *Tetrahedral* here means a tetrameric molecule composed of subunits making three identical contacts with their neighbors (Fig. 9A). The expression for the binding isotherm assumes the form

$$\bar{v} = \frac{4K_{AB}^3 K_s K_t[S] + 12K_{AB}^4 K_{BB} K_s^2 K_t^2[S]^2 + 12K_{AB}^3 K_{BB}^3 K_s^3 K_t^3[S]^3 + 4K_{BB}^6 K_s^4 K_t^4[S]^4}{1 + 4K_{AB}^3 K_s K_t[S] + 6K_{AB}^4 K_{BB} K_s^2 K_t^2[S]^2 + 4K_{AB}^3 K_{BB}^3 K_s^3 K_t^3[S]^3 + K_{BB}^6 K_s^4 K_t^4[S]^4}.$$

$$(50)$$

If the tetrameric enzyme molecule exhibits a "square" type of subunit interactions, there exist two different types of $A_2B_2S_2$ complexes (Fig. 9B) and the binding isotherm assumes a different equation, namely,

$$\bar{v} = \frac{4K_{AB}^2 K_s K_t[S] + 4(K_{AB}^4 + 2K_{AB}^2 K_{BB})K_s^2 K_t^2[S]^2 + 12K_{AB}^2 K_{BB}^2 K_s^3 K_t^3[S]^3 + 4K_{BB}^4 K_s^4 K_t^4[S]^4}{1 + 4K_{AB}^2 K_s K_t[S] + 2(K_{AB}^4 + 2K_{AB}^2 K_{BB})K_s^2 K_t^2[S]^2 + 4K_{AB}^2 K_{BB}^2 K_s^3 K_t^3[S]^3 + K_{BB}^4 K_s^4 K_t^4[S]^4}.$$

$$(51)$$

If now the subunits are very tightly coupled in such a way that the first substrate binding step induces a concerted transition in all the subunits

A

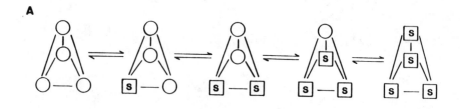

B

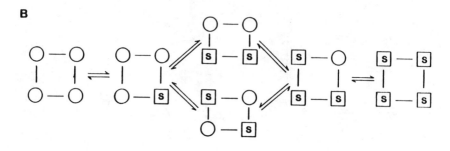

C

$$\begin{matrix} O-O \\ | \quad | \\ O-O \end{matrix} \rightleftharpoons \begin{matrix} \square-\square \\ | \quad | \\ \square-\boxed{s} \end{matrix} \rightleftharpoons \begin{matrix} \square-\square \\ | \quad | \\ \boxed{s}-\boxed{s} \end{matrix} \rightleftharpoons \begin{matrix} \boxed{s}-\square \\ | \quad | \\ \boxed{s}-\boxed{s} \end{matrix} \rightleftharpoons \begin{matrix} \boxed{s}-\boxed{s} \\ | \quad | \\ \boxed{s}-\boxed{s} \end{matrix}$$

Fig. 9. Subunit interactions in a tetrameric enzyme (see text). (A) Tetrahedral simple sequential; (B) square simple sequential; (C) square fully concerted.

(Fig. 9C), the binding equation may be cast into the following form:

$$\bar{v} = \frac{4K_s[S](1 + K_s[S])^3}{\lambda + (1 + K_s[S])^4},$$
(52)

where

$$\lambda = (1 - K_t^4 K_{BB}^4)/K_t^4 K_{BB}^4.$$
(53)

If

$$K_t^4 K_{BB}^4 < 1,$$
(54)

then $\lambda > 0$ and Eq. (52) is reminiscent of the basic equation of the concerted model of Monod *et al.* (1965), with λ being formally equivalent to an allosteric transition constant. Inequality (54) above necessarily holds. As seen in Fig. 10, this inequality simply means that the conformational state B_4 must be destabilized with respect to the state A_4, otherwise the free enzyme would have existed in the B_4 state. Assuming (as is done in the context of induced-fit

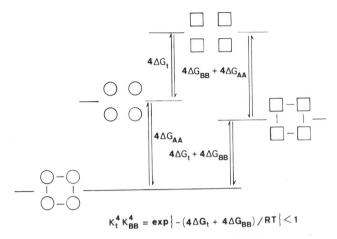

$$K_t^4 K_{BB}^4 = \exp\{-(4\Delta G_t + 4\Delta G_{BB})/RT\} < 1$$

Fig. 10.   Thermodynamic evidence that $K_t^4 K_{AB}^4 < 1$.

theory) that the free enzyme exists in the A_4 state is equivalent to assuming that the free energy of this state is lower than that of the B_4 state. The existence of this energy difference requires inequality (54) to hold.

3. Cooperativity in Induced-Fit Structural Models

The simple "tetrahedral" and "square" sequential models may generate positive, negative, and mixed cooperativity since the macroscopic constants may take any positive value. Owing to the possible existence of mixed cooperativity, the determination of the maximum Hill coefficient is of little interest. In the case of simple sequential models, this parameter is meaningful only for dimeric enzymes. For a dimeric enzyme exhibiting loose coupling between subunits the ψ parameters of the Adair equation are

$$\psi_1 = 2K_s K_t K_{AB}, \qquad \psi_2 = K_s^2 K_t^2 K_{BB}. \tag{55}$$

Inserting these expressions into Eq. (22) the maximum Hill coefficient assumes the form

$$n_H = \frac{2}{K_{AB}/\sqrt{K_{BB}} + 1}. \tag{56}$$

If $K_{BB} > K_{AB}^2$ cooperativity is positive and if $K_{BB} < K_{AB}^2$ cooperativity is negative.

If tight coupling between subunits occurs (fully concerted model), the ψ coefficients assume the form

$$\psi_1 = 2K_t^2 K_{BB} K_s, \qquad \psi_2 = K_t^2 K_{BB} K_s^2, \tag{57}$$

and λ is defined by

$$\lambda = (1 - K_t^2 K_{BB})/K_t^2 K_{BB}. \tag{58}$$

The maximum Hill coefficient is then

$$n_H = \frac{2}{\sqrt{1/(1 + \lambda)} + 1} = \frac{2}{K_t\sqrt{K_{BB}} + 1}. \tag{59}$$

Clearly, when the model is fully concerted cooperativity can be only positive.

4. Constraints between Parameters of Adair Equations as a Possible Tool for Distinguishing between Structural Models

All structural models considered thus far may be expressed by an Adair equation of the type

$$\bar{v} = \frac{\psi_1[S] + 2\psi_2[S]^2 + 3\psi_3[S]^3 + 4\psi_4[S]^4}{1 + \psi_1[S] + \psi_2[S]^2 + \psi_3[S]^3 + \psi_4[S]^4}. \tag{60}$$

This equation can be solved if the four ψ coefficients have been estimated from experimental data. The mathematical evaluation of these ψ coefficients depends on the hypotheses which have been formulated about enzyme quaternary structure. The structural formulation of simple "tetrahedral" and "square" models requires three independent molecular parameters, namely, K_{AB}, K_{BB}, and the product $K_s K_t$ (K_{AA} has been set equal to one). The fully concerted model requires two independent parameters, namely, K_s and the product $K_t^4 K_{BB}^4$, as well as the concerted exclusive model of Monod et al. (1965) whose parameters are L and K^*. The comparative formulation of the ψ coefficients for the four models is given in Table I. Since for any of these models the number of structural (or molecular) parameters is smaller than the number of phenomenological parameters ψ, there must exist constraints between the coefficients, which cannot all be independent. The existence of a given type of constraint may be used as a diagnostic tool for distinguishing among structural models.

Comparison of the expressions given in Table I shows that for simple "tetrahedral" and "square" models, one must have

$$\psi_1^2 \psi_4 = \psi_3^2. \tag{61}$$

This relation does not hold for concerted models. In the case of the two concerted models one must have

$$2\psi_2/3\psi_1 = 3\psi_3/2\psi_2 = 4\psi_4/\psi_3. \tag{62}$$

It must be stressed again that there is no fundamental difference between the

TABLE I

COMPARISON OF ψ COEFFICIENTS FOR FOUR STRUCTURAL MODELS

	Tetrahedral	Square	Fully concerted (induced-fit)	Fully concerted (preequilibrium)
ψ_1	$4K_{AB}^3 K_s K_t$	$4K_{AB}^2 K_s K_t$	$4K_t^4 K_{BB}^4 K_s$	$4\dfrac{K^*}{1+L}$
ψ_2	$6K_{AB}^4 K_{BB} K_s^2 K_t^2$	$2K_{AB}^4 + 4K_{BB}K_{AB}^2 K_s^2 K_t^2$	$6K_t^4 K_{BB}^4 K_s^2$	$6\dfrac{K^{*2}}{1+L}$
ψ_3	$4K_{AB}^3 K_{BB}^3 K_s^2 K_t^2$	$4K_{AB}^2 K_{BB}^2 K_s^3 K_t^3$	$4K_t^4 K_{BB}^4 K_s^3$	$4\dfrac{K^{*3}}{1+L}$
ψ_4	$K_{BB}^6 K_s^4 K_t^4$	$K_{BB}^4 K_s^4 K_t^4$	$K_t^4 K_{BB}^4 K_s^4$	$\dfrac{K^{*4}}{1+L}$

concerted models based on the concepts of enzyme preequilibrium and of induced fit. Both give rise to similar binding equations. The basic idea which must be kept in mind is that of coupling strength between subunits. For loose coupling between subunits, cooperativity may be positive, negative, or mixed. As the coupling strength between subunits is decreased, cooperativity tends to become positive. For a fully concerted model cooperativity can only be positive.

V. PHENOMENOLOGICAL KINETICS OF ALLOSTERIC ENZYMES

Thus far, the effects of conformation changes and subunit interactions on substrate affinity have been studied under equilibrium conditions. Still, the point of importance is to understand how changes of quaternary structure control the rate of product appearance. A common practice has been to use the binding equations derived from the models of Monod *et al.* (1965) or Koshland *et al.* (1966) to fit rate data. This practice is incorrect for several reasons, however. The steady-state rate of product appearance cannot be taken as a measure of the saturation of a multisited enzyme by its substrate. The aim of this section is to discuss briefly some features of the phenomenological kinetics of allosteric enzymes.

A simple kinetic version of the Monod model is shown in Fig. 11A. Its correct steady-state treatment is intractable in practice. It is when the system is in a pseudo-equilibrium state or when only the free enzyme may undergo the conformational transition that the rate equation may be derived (Dalziel, 1968; Whitehead, 1976, 1979). The use of the Monod model in its kinetic

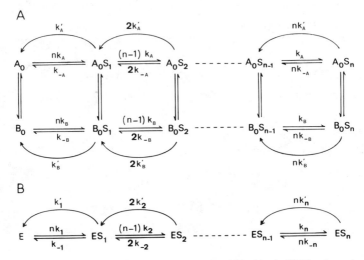

Fig. 11. Kinetic version of the models of Monod and Koshland. (A) Kinetic version of the Monod–Wyman–Changeux model; (B) kinetic version of the Koshland–Nemethy–Filmer model.

version (Fig. 11A) raises a difficult problem. Since it is assumed that the enzyme exists in two states throughout the reaction, there must exist only two microscopic catalytic rate constants. The catalytic rate constants originating from the same type of conformation should all be the same (Fig. 11). However, this assumption is hardly compatible with the present knowledge of the mechanisms of enzyme catalysis. It has already been mentioned that when a transition state $X^{\ddagger}$ is reached, at the top of the energy barrier, strain and quaternary constraints of the enzyme must be released. It then becomes difficult to imagine that the enzyme conformation will be the same, in any of these transition states, whatever the degree of saturation of the enzyme by the substrate, when strain and quaternary constraints are released. In aspartate transcarbamylase, the existence of "local" movements of the enzyme, in addition to the concerted allosteric transition, seems to give experimental support to this view (Kerbiriou *et al.*, 1977; Thiry and Hervé, 1978). It then becomes somewhat illogical to postulate that the microscopic catalytic constants originating from the same type of conformation state will be the same, whatever the degree of saturation of the enzyme by the substrate.

The kinetic version of an Adair model is shown in Fig. 11B and the corresponding equation is

$$\frac{v}{[E]_0} = \frac{\sum_{i=1}^{n} i k'_i \bar{\psi}_i [S]^i}{1 + \sum_{i=1}^{n} \bar{\psi}_i [S]^i},$$ (63)

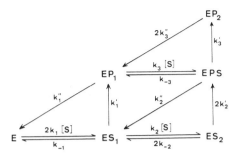

Fig. 12. Kinetic version of the Adair model for a one-substrate, one-product dimeric enzyme.

where $[E]_0$ is the total enzyme concentration. The apparent Adair constants $\bar{\psi}_i$ are now the products of the apparent macroscopic binding constants, namely,

$$\bar{\psi}_i = \bar{K}_1 \bar{K}_2 \cdots \bar{K}_i \qquad (i = 1, \ldots, n). \tag{64}$$

Any of these apparent macroscopic constants, $\bar{K}_i$, is defined as

$$\bar{K}_i = k_i/(k_{-i} + k_i'), \tag{65}$$

where k_i' are the macroscopic catalytic constants.

If this kinetic version of the Adair equation were acceptable, the degree of the substrate-binding isotherm and of the corresponding rate equation would be the same, and this point raises another difficulty. In fact, in Eq. (63) it has been implicitly postulated that no hybrid enzyme–substrate–product complex exists. There is no reason to believe that this assumption is always valid.

As a matter of fact, the simplest realistic model for a one-substrate, one-product dimeric enzymatic reaction is shown in Fig. 12. The corresponding steady-state rate equation is 3 : 3 in substrate concentration, which means that the steady-state rate is the ratio of the polynomials in $[S]^3$, but the substrate-binding equation is 2:2.[1] When the number of protomers of the enzyme increases, the corresponding rate equation becomes exceedingly complex. For instance, for a tetrameric one-substrate enzyme, the corresponding rate equation is already 10:10!

In view of the complexity of the rate equations one may be surprised by the rather smooth shape of the velocity plots obtained for allosteric enzymes. Although there is no doubt that some enzymes exhibit "wavy" or "bumpy" rate curves (Bardsley, 1977a,b; Hill et al., 1977; Teipel and Koshland, 1969) this is rather rare and this effect is usually not of large amplitude. Indeed, complex equations may generate smooth curves (Solano-Munoz et al., 1981; Bardsley et al., 1980). Still one may wonder whether some constraints between

[1] In a more general way, an n : n rate equation is defined by the ratio of two polynomials in $[S]^n$.

rate constants may not decrease the degree of many rate equations. There may exist three types of constraints: constraints imposed by thermodynamics, fortuitous constraints, and constraints generated by a certain type of quaternary structure and symmetry of the enzyme molecule (Ricard and Noat, 1982). The constraints imposed by thermodynamics correspond to microscopic reversibility which states that the rate constants involved in a closed loop of a complex reaction mechanism cannot be independent. This obvious point will not be discussed any further here.

Another type of constraint, which at first sight appears fortuitous, may greatly simplify the derivation of rate equations (Whitehead, 1976). This constraint is the so-called generalized microscopic reversibility and implies that the ratio of rate constants for product desorption and catalysis may remain constant along a reaction sequence. As will be seen later, there may exist some physical reasons for the existence of such constraints.

Finally, subunit interactions and symmetry of the enzyme molecule may create additional constraints between rate constants. For instance, in the model of Fig. 12, if

$$k_2' k_1'' k_2(k_{-3} + k_3') = k_1' k_2'' k_3(k_{-2} + k_2'), \tag{66}$$

the corresponding rate equation reduces to a degree 2:2. It will be seen later (Section VI) that simple assumptions about subunit coupling precisely generate Eq. (66) above. The most important idea that must be kept in mind is that the kinetics of allosteric enzymes is extremely complex and cannot be reduced to the analysis of binding isotherms.

VI. THERMODYNAMICS OF SUBUNIT INTERACTIONS AND THE PRINCIPLES OF STRUCTURAL KINETICS

The classical induced-fit and symmetry models relate some simple features of polymeric enzyme structure to the mathematical structure of ligand-binding equations. Similarly one may hope to relate enzyme quaternary structure to the mathematical expression of the velocity of an enzyme reaction step. The term structural kinetics has been given to this attempt to express how subunit interactions and quaternary constraints control the rate of an enzymatic process (Ricard et al., 1974a; Nari et al., 1974; Ricard, 1978).

Indeed, one may consider in all logic that it is function that is the driving force of neo-Darwinian Evolution and that a given type of quaternary structure has been selected because it exhibits some type of functional advantage. The most obvious of these functional advantages is an improvement of the catalytic efficiency of the enzyme molecule. It is therefore an important matter to examine, in a very simple case, what the physical nature of this improvement could be.

A. Functional Efficiency of an Enzymatic Reaction

Let there be the simple one-substrate, one-product enzyme reaction

$$E \underset{k_{-1}}{\overset{k_1[S]}{\rightleftharpoons}} ES \underset{k_{-c}}{\overset{k_c}{\rightleftharpoons}} EP \underset{k_{-2[P]}}{\overset{k_2}{\rightleftharpoons}} E \tag{67}$$

and the corresponding energy profile shown in Fig. 13. An important aspect of this energy profile is the existence of extra costs of energy, ε, in going from the initial state to any of the transition states along the reaction coordinate (Ricard, 1978; Albery and Knowles, 1976). These extra costs of energy correspond to the free energy difference between the initial state and any of the transition states. One must have

$$\varepsilon_{s\bar{s}} = \Delta G_1^{\ddagger},$$

$$\varepsilon_{s\bar{x}} = \Delta G_1^{\ddagger} + \Delta G_c^{\ddagger} - \Delta G_{-1}^{\ddagger}, \tag{68}$$

$$\varepsilon_{s\bar{p}} = \Delta G_1^{\ddagger} + \Delta G_c^{\ddagger} + \Delta G_2^{\ddagger} - \Delta G_{-1}^{\ddagger} - \Delta G_{-c}^{\ddagger}.$$

In these expressions $\varepsilon_{s\bar{s}}$, $\varepsilon_{s\bar{x}}$, and $\varepsilon_{s\bar{p}}$ are the extra costs between the initial state and the $ES^{\ddagger}$, $EX^{\ddagger}$, and $EP^{\ddagger}$ transition states, respectively. Each of these extra costs may be split into three energy contributions: a positive contribution equal to the energy required to transform the substrate into the relevant transition state ($S^{\ddagger}$, $X^{\ddagger}$, or $P^{\ddagger}$), another positive contribution equal to the energy required to bring the enzyme from its initial state to a state

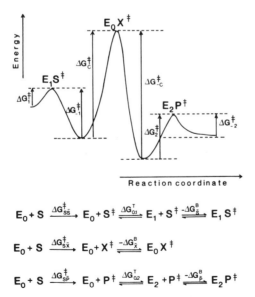

Fig. 13.   Free energy profile for a one-substrate, one-product enzyme reaction (see text).

complementary to these transition states (ΔG_{01}^T and ΔG_{02}^T in Fig. 13), and a negative contribution equal to the free energy of binding of transition states to the complementary enzyme forms ($-\Delta G_{\bar{s}}^B$, $-\Delta G_{\bar{x}}^B$, $-\Delta G_{\bar{p}}^B$). If it is assumed that the free enzyme is complementary to the central transition state $X^\ddagger$, one must have

$$\varepsilon_{s\bar{s}} = \Delta G_{s\bar{s}}^\ddagger + \Delta G_{01}^T - \Delta G_{\bar{s}}^B,$$

$$\varepsilon_{s\bar{x}} = \Delta G_{s\bar{x}}^\ddagger - \Delta G_{\bar{x}}^B, \tag{69}$$

$$\varepsilon_{s\bar{p}} = \Delta G_{s\bar{p}}^\ddagger + \Delta G_{02}^T - \Delta G_{\bar{p}}^B.$$

The significance of these relationships is illustrated in Fig. 13.

In order to maximize the enzymatic reaction rate, selective pressure has to minimize the extra costs of energy ε. Selective pressure has obviously no effect on the $\Delta G_{s\bar{s}}^\ddagger$, $\Delta G_{s\bar{x}}^\ddagger$, and $\Delta G_{s\bar{p}}^\ddagger$ contributions and is very likely not to change dramatically the binding energies of the transition states. Therefore, the improvement of the catalytic efficiency may be achieved, in the course of evolution, when the contributions ΔG_{01}^T and ΔG_{02}^T are negligible. Expressions (69) then reduce to

$$\varepsilon_{s\bar{s}} = \Delta G_{s\bar{s}}^\ddagger - \Delta G_{\bar{s}}^B, \qquad \varepsilon_{s\bar{x}} = \Delta G_{s\bar{x}}^\ddagger - \Delta G_{\bar{x}}^B, \qquad \varepsilon_{s\bar{p}} = \Delta G_{s\bar{p}}^\ddagger - \Delta G_{\bar{p}}^B. \tag{70}$$

This implies that, in any of the transition states, the enzyme has reached a conformation whose energy is equal to that of the free enzyme. In other words uniqueness of enzyme conformation, in the various transition states along the reaction coordinate, is the *prima ratio* of catalytic efficiency improvement. This uniqueness is indeed a limiting case that can only be approached in practice.

In addition to uniqueness of enzyme conformation in the transition states, evolution must favor either reversibility or irreversibility of the catalytic step, for we know examples of enzymes belonging to these two types. One may indeed wonder whether there is a relation between this reversibility and the energies of enzyme conformations stabilized by substrate and product.

Enzyme catalysis is exerted in the forward and reverse directions thanks to driving forces which allow the energy barrier of the central transition state $X^\ddagger$ to be passed. These thermodynamic forces $X_{s\bar{x}}$ and $X_{p\bar{x}}$ may be expressed as

$$X_{s\bar{x}} = \Delta G_T + \Delta G_{\bar{x}}^B, \cdot \qquad X_{p\bar{x}} = \Delta G_T' + \Delta G_{\bar{x}}^B, \tag{71}$$

where $X_{s\bar{x}}$ is the force exerted in the forward direction and $X_{p\bar{x}}$ the force exerted in the backward direction. The significance of the components of these forces is illustrated in Fig. 14A; $-\Delta G_T$ and $-\Delta G_T'$ are the free energies released when the enzyme conformations stabilized by the substrate and by the product relapse to the conformational state of the free enzyme; $-\Delta G_{\bar{x}}^B$ is the free energy of binding of the transition state $X^\ddagger$ to the complementary free enzyme form.

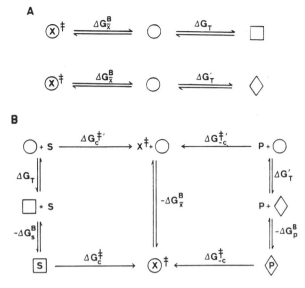

Fig. 14. Driving forces of catalysis. (A) Significance of the driving forces; (B) driving forces as defined by a "thermodynamic box" (see text).

As shown in the thermodynamic boxes of Fig. 14B, thermodynamics imposes that the relation between the free energies of activation, $\Delta G_c^{\ddagger}$ and $\Delta G_{-c}^{\ddagger}$, for the catalyzed reaction and those, $\Delta G_c^{\ddagger\prime}$ and $\Delta G_{-c}^{\ddagger\prime}$, for the corresponding uncatalyzed reaction be

$$\Delta G_c^{\ddagger} = \Delta G_c^{\ddagger\prime} + \Delta G_s^{B} - X_{s\bar{x}}, \qquad \Delta G_{-c}^{\ddagger} = \Delta G_{-c}^{\ddagger\prime} + \Delta G_p^{B} - X_{p\bar{x}}, \qquad (72)$$

where $-\Delta G_s^{B}$ and $-\Delta G_p^{B}$ are the binding energies of substrate and product to the enzyme. Since the values of $\Delta G_c^{\ddagger\prime}$ and $\Delta G_{-c}^{\ddagger\prime}$ do not depend on selective pressure, minimization of $\Delta G_c^{\ddagger}$ and $\Delta G_{-c}^{\ddagger}$ will be attained by minimizing the binding energies of substrate and product as well as by maximizing the thermodynamic forces $X_{s\bar{x}}$ and $X_{p\bar{x}}$.

If the binding energies $-\Delta G_s^{B}$ and $-\Delta G_p^{B}$ of substrate and product to the complementary enzyme conformations have similar values in the course of evolution, three situations may occur.

1. $X_{s\bar{x}} > X_{p\bar{x}}$, which means that the enzyme exerts a force mostly in the forward direction. This has two interesting implications: first, $\Delta G_T > \Delta G_T'$, i.e., (see Fig. 15A) the free energy level of the square conformation (the one stabilized by the substrate) is higher than that of the rhombus conformation (the one stabilized by the product); second, $\Delta G_c^{\ddagger}$ is decreased with respect to $\Delta G_{-c}^{\ddagger}$ and the energy level of the enzyme–product complex tends to be

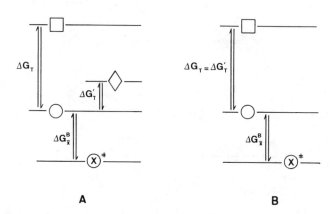

Fig. 15. Respective values of the driving forces $X_{s\bar{x}}$ and $X_{p\bar{x}}$. (A) $X_{s\bar{x}} > X_{p\bar{x}}$ (see text); (B) $X_{s\bar{x}} = X_{p\bar{x}}$ (see text).

lowered with respect to that of the enzyme–substrate complex (Eq. 72). Catalysis is then facilitated in the forward direction and product desorption may become the limiting step of the overall process.

2. $X_{s\bar{x}} < X_{p\bar{x}}$; this is the symmetrical situation and catalysis is favored in the backward direction.

3. $X_{s\bar{x}} = X_{p\bar{x}}$, which means the existence of a "catalytic balance." The forces exerted by the enzyme are the same in both the forward and the backward directions. The existence of a "catalytic balance" implies that $\Delta G_T = \Delta G'_T$ (Fig. 15B). Therefore, the substrate and the product stabilize the same enzyme conformation (Fig. 15B) or energetically indistinguishable enzyme conformations.

These rather general ideas, which have been formulated for monomeric enzymes, may indeed be applied to subunits of polymeric enzyme molecules. This matter will be considered later.

B. Energy Diagrams and the Principles of Structural Kinetics

The basic idea of structural kinetics is that interactions may have two effects on an enzymatic reaction step. Owing to protomer arrangement and association, they can modify the rate of a conformational transition associated with this reaction step. This change does not require distortion of the active site and may be termed protomer arrangement energy contribution (Ricard and Noat, 1982, 1984). However, subunit interactions may also generate some form of protomer distortion, and this energy contribution is termed the quaternary constraint energy contribution. Both the protomer arrangement and quaternary constraint contributions usually stabilize the polymeric enzyme (with respect to the ideal isolated subunit) in its ground state.

Another basic idea of structural kinetics is that strain of the active site and quaternary constraints tend to be released in the transition states. This important idea cannot be avoided if it is assumed that in a transition state a subunit has the conformation it should have if it were isolated from the other subunits. For a reaction involving the conversion of a substrate S into a product P, an ideally isolated subunit may exist under three different conformations: the unliganded conformation (the circle A), the substrate liganded conformation (the square B) and the product liganded conformation (the rhombus C). There may therefore exist, in the ground and transition states, interactions between subunits of the AA, AB, BB, AC, and CC types. If the principle of maximum catalytic efficiency applies, the subunit bearing any of the transition states should have the A conformation, namely, that of the unliganded and unstrained subunit.

Let there be a chemical process (binding or release of a ligand, catalysis, or protein isomerization) occurring on a polymeric enzyme; the free energy of activation of that process, $\Delta G^{\ddagger}$, may be split up into three basic energy components:

1. The intrinsic energy contribution, $\Delta G^{\ddagger *}$, which corresponds to what the activation free energy of the chemical process would be if subunit interactions were not occurring. This is the activation free energy of the process controlled by an ideally isolated protomer.

2. The protomer arrangement contribution, $\sum(^{\alpha}\Delta G^{int})$, which expresses quantitatively how association of subunits, apart from any distortion of the active sites, may modify the rate of the conformational transition associated with the chemical process. This energy contribution corresponds to the free energy associated with the polymerization of the ideally isolated subunits on assuming that no conformation change of the subunits would occur during this process. If there exists l subunits in conformation A, m in conformation B, and n in conformation C in the same ground or transition state, simple combinatorial analysis allows calculation of the corresponding expression of $\sum(^{\alpha}\Delta G^{int})$. If all the interactions are allowed, one finds for either the ground or the transition states:

$$\sum(^{\alpha}\Delta G^{int}) = \binom{l}{2}(^{\alpha}\Delta G_{AA}) + \binom{m}{2}(^{\alpha}\Delta G_{BB}) + \binom{n}{2}(^{\alpha}\Delta G_{CC})$$

$$+ lm(^{\alpha}\Delta G_{AB}) + ln(^{\alpha}\Delta G_{AC}) + mn(^{\alpha}\Delta G_{BC}). \tag{73}$$

3. The quaternary constraint contribution, $\sum(^{\sigma}\Delta G^{int})$, expresses quantitatively how site distortion (strain) through quaternary constraints may modify the rate of the considered chemical process. It may be expressed by an equation analogous to Eq. (73). It is worth noting that this energy component corresponds to the difference between the actual free energy of subunit

association and what this energy would be if no strain of the subunits were occurring upon their association.

Therefore, one may write

$$\Delta G^{\ddagger} = \Delta G^{\ddagger*} + \sum({}^{\alpha}\Delta G^{int}) + \sum({}^{\sigma}\Delta G^{int}). \tag{74}$$

Since quaternary constraints are assumed to be released in the transition states, the sum $\Sigma({}^{\sigma}\Delta G^{int})$ must be taken over the ground states, whereas the sum $\Sigma({}^{\alpha}\Delta G^{int})$ must be extended over both the ground states and the transition states. This will be illustrated later.

The intrinsic free energy of activation, $\Delta G^{\ddagger*}$, allows one to define an intrinsic rate constant, $k*$ for the considered chemical process, namely,

$$k* = \frac{k_B T}{h}\exp\{-\Delta G^{\ddagger*}/RT\}, \tag{75}$$

where k_B, h, T, and R are the Boltzman constant, the Planck constant, the absolute temperature, and the gas constant, respectively. Any component of the protomer arrangement contribution (${}^{\alpha}\Delta G^{int}$), allows definition of a corresponding interaction coefficient α, namely,

$$\alpha = \exp\{-({}^{\alpha}\Delta G^{int})/RT\}. \tag{76}$$

Similarly, any component of the quaternary constraint contribution allows one to define the corresponding strain coefficient σ, namely,

$$\sigma = \exp\{-({}^{\sigma}\Delta G^{int})/RT\}. \tag{77}$$

Indeed, both α and σ are dimensionless coefficients formally analogous to equilibrium constants.

In order to illustrate the use of the principles of structural kinetics, let us express how subunit arrangements and quaternary constraints in a tetramer with "square" geometry may control the value of a forward catalytic constant k. A possible energy diagram pertaining to this situation is depicted in Fig. 16a. One subunit has bound the substrate S. For the ideal unstrained tetramer in its ground state, three subunits exist in the A (circle) conformation whereas one has the B (square) conformation. Therefore, there exist two AA and two AB interactions. If any of these interactions tend to stabilize the polymer with respect to the ideally isolated protomer, the protomer arrangement contribution has to be $2({}^{\alpha}\Delta G_{AA}) + 2({}^{\alpha}\Delta G_{AB})$. This contribution is then negative, otherwise the enzyme would tend to dissociate. Owing to quaternary constraints between subunits in the real strained tetramer, none of these subunits has the A or the B conformation. The quaternary constraint energy contribution is $2({}^{\sigma}\Delta G_{AA}) + 2({}^{\sigma}\Delta G_{AB})$ in Fig. 16a. If the distorted conformations of the unliganded and liganded subunits are A' and B', this

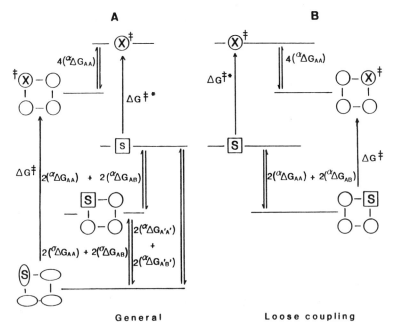

Fig. 16. Structural formulation of a catalytic rate constant (see text) for a tetrameric enzyme. (A) Strain is exerted between subunits; (B) strain is not exerted between subunits (loose coupling).

quaternary constraint contribution is equal to (see Fig. 16)

$$-\{(^{\sigma}\Delta G_{AA}) + (^{\sigma}\Delta G_{AB})\} = \{(^{\alpha}\Delta G_{AA}) + (^{\alpha}\Delta G_{AB})\} - \{(^{\alpha}\Delta G_{A'A'}) + (^{\alpha}\Delta G_{A'B'})\}. \tag{78}$$

Protomer arrangement and quaternary constraints usually stabilize, in the ground state, the tetramer with respect to the ideally isolated protomer. In the transition state, the four subunits are in the A conformation. Since the enzyme does not dissociate during catalysis the protomer arrangement energy contribution is $4(^{\alpha}\Delta G_{AA})$. No $(^{\sigma}\Delta G^{int})$ term can exist for the transition state since it is assumed that quaternary constraints are released in this state.

The free energy diagram of Fig. 16a forms a thermodynamic box which allows the free energy of activation of the catalytic process, $\Delta G^{\ddagger}$, to be expressed as a function of the corresponding intrinsic free energy process, $\Delta G^{\ddagger *}$. One must have

$$\Delta G^{\ddagger} = \Delta G^{\ddagger *} + 2(^{\sigma}\Delta G_{AA}) + 2(^{\sigma}\Delta G_{AB}) + 2(^{\alpha}\Delta G_{AA}) + 2(^{\alpha}\Delta G_{AB}) - 4(^{\alpha}\Delta G_{AA}), \tag{79}$$

which reduces to

$$\Delta G^{\ddagger} = \Delta G^{\ddagger *} + 2(^{\sigma}\Delta G_{AA}) + 2(^{\sigma}\Delta G_{AB}) + 2(^{\alpha}\Delta G_{AB}) - 2(^{\alpha}\Delta G_{AA}). \tag{80}$$

The structural expression of the catalytic constant then becomes

$$k = k^*(\alpha_{AB}^2/\alpha_{AA}^2)\sigma_{AA}^2\sigma_{AB}^2. \tag{81}$$

This expression can be considered as "structural" because it associates a rate constant with thermodynamic parameters related to structure and structural changes of the polymeric enzyme.

This kind of formalism may be applied to any reaction process controlled by a polymeric enzyme molecule. However, in the general case, it does not result in any simplification of the steady-state kinetic equations, which, in most cases, remain intractable for experimental purposes. This formalism and the resulting rate equations, however, become beautifully simple in two limiting cases that are often encountered in nature: the case of loosely coupled subunits and the case of tightly coupled subunits.

The equations of structural kinetics obviously rely on the hypotheses made about interactions existing between subunits in the transition states. So far, there has been no experimental possibility for studying directly such interactions in these short-lived intermediates. The hypothesis, which has been made above, is that quaternary constraints are abolished in the transition states but that the subunits remain in interaction. If it is not so, the sum $\Sigma(^\alpha\Delta G^{int})$ must be taken over the ground states only and the α coefficients pertaining to the transition states must all be equal to unity. Expression (81) then becomes

$$k = k^*\alpha_{AB}^2\alpha_{AA}^2\sigma_{AB}^2\sigma_{AA}^2. \tag{82}$$

This assumption, which is probably an oversimplification, previously has been made implicitly (Ricard et al., 1974b). Unless specified, this assumption will not be made in this chapter.

C. Loose Coupling between Subunits and Structural Kinetics

As previously mentioned, loose coupling between subunits in a polymeric enzyme molecule means that any subunit may undergo a conformational change that does not propagate to the rest of the enzyme molecule. The so-called "simple sequential" model of Koshland precisely describes a loose coupling of subunits in the polymer. When it is so,

$$\sum(^\sigma\Delta G^{int}) = 0 \tag{83}$$

and all strain coefficients become equal to unity. In the case of loose coupling between subunits the energy diagram of Fig. 16a would simplify as shown in Fig. 16b and the corresponding expression of the rate constant would reduce to

$$k = k^*\alpha_{AB}^2/\alpha_{AA}^2. \tag{84}$$

If loose coupling is associated with the unicity of conformation of the subunits that have bound a transition state and with a "catalytic balance," there exist constraints between rate constants. In other words the rate constants are not all independent and this may result in a degeneration, and therefore in a simplification, of the rate equation.

Let us consider, as an example, the kinetic reaction scheme of Fig. 12. If the two subunits are loosely coupled, if there exists a "catalytic balance," and if there is uniqueness of conformation in the transition states, at any stage of the enzyme reaction subunits exist in conformation A or B. The energy diagrams associated with rate constants k_1' and k_1'' are shown in Fig. 17. Clearly, the principles of structural kinetics lead to the following equations:

$$k_1' = k^* \alpha_{AB}/\alpha_{AA}, \qquad k_1'' = k^*_{-p} \alpha_{AB}/\alpha_{AA}. \tag{85}$$

Similar reasoning may be applied to the two corresponding pairs of rate constants, namely, k_2' and k_2'', and k_3' and k_3'':

$$k_2' = k^* \alpha_{BB}/\alpha_{AB}, \qquad k_2'' = k^*_{-p} \alpha_{BB}/\alpha_{AB},$$
$$k_3' = k^* \alpha_{BB}/\alpha_{AB}, \qquad k_3'' = k^*_{-p} \alpha_{BB}/\alpha_{AB}. \tag{86}$$

Therefore,

$$k_1''/k_1' = k_2''/k_2' = k_3''/k_3' = k^*_{-p}/k^*, \tag{87}$$

A   **B**

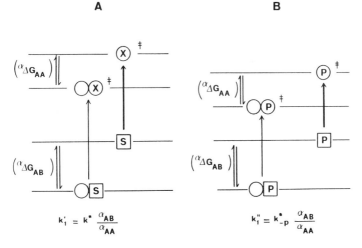

$$k_1' = k^* \frac{\alpha_{AB}}{\alpha_{AA}} \qquad\qquad k_1'' = k^*_{-p} \frac{\alpha_{AB}}{\alpha_{AA}}$$

Fig. 17. Energy diagrams for a dimeric enzyme exhibiting a "catalytic balance" (see text). The scheme shows the constancy of the ratio between a catalytic rate constant and the corresponding constant of product desorption. (A) Energy diagram for catalysis; (B) energy diagram for product release.

where k^* and k^*_{-p} are indeed the intrinsic rate constants of catalysis and product desorption, respectively. Occurrence of condition (87) generates the constraint condition (66) which simplifies the 3:3 rate equation into a 2:2 rate equation. These equations will be presented and discussed in Section VII,A. In addition to constraint condition (87), the principles of structural kinetics allow one to write the following equalities

$$k'_2 = k'_3, \qquad k''_2 = k''_3, \qquad k_1 = k_2 = k_3 = k^*, \qquad k_{-2} = k_{-3}. \qquad (88)$$

The important conclusion of the above reasoning is that loose coupling between subunits associated with high catalytic efficiency and with "catalytic balance" results in a degenerate and simplified rate equation.

D. Tight Coupling between Subunits and Structural Kinetics

It may well occur that the protomers are so tightly coupled that any local conformational change propagates to the entire molecule. All the subunits may acquire the same conformation. Subunit interactions are then of the all-or-none type and symmetry is conserved during the conformational transition.

An example of an energy diagram pertaining to a catalytic process controlled by a tetramer with "square" geometry is shown in Fig. 18A.

It is possible that the intersubunit quaternary constraints are released when the polymeric enzyme is in either the unliganded or the fully liganded state (Fig. 18B). If any subunit exists in the A or the B conformation, there exists a simple relation between the quaternary constraint contribution, the strain of the individual subunits, and the protomer arrangement contribution. Let us consider as an example a tetrameric enzyme state of the type B_4S_2 (Fig. 19). If $(^\sigma \Delta G_B^*)$ is the intrinsic free energy required to strain the ideally isolated subunit, then, from the energy diagram of Fig. 19,

$$(^\sigma \Delta G_{BB}) + (^\sigma \Delta G_{AA}) + 2(^\sigma \Delta G_{AB}) + (^\sigma \Delta G_{AA}) + 2(^\sigma \Delta G_{AB}) = 3(^\sigma \Delta G_{BB}) - 2(^\sigma \Delta G_B^*), \qquad (89)$$

which is equivalent to

$$\sigma_{BB}\sigma_{AA}\sigma_{AB}^2 = (\alpha_{BB}^3/\alpha_{AA}\alpha_{AB}^2)(1/\sigma_B^{*2}). \qquad (90)$$

The assumption made above is certainly a very interesting one for it allows simplification of the mathematical expression for the fully concerted models. This important point will be discussed later. The kinetic version of the model of Monod et al. (1965), shown in Fig. 11, is indeed an example of tight coupling between subunits. Constraints between rate constants, as previously described, may also occur for tight coupling of subunits.

A   B

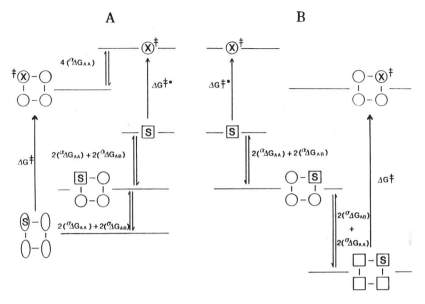

Fig. 18. Structural formulation of a catalytic rate constant for a tetrameric enzyme exhibiting tight coupling of subunits. (A) Quaternary constraints are not released in the liganded state; (B) Quaternary constraints are released in the liganded state.

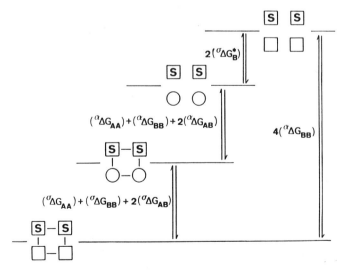

Fig. 19. Thermodynamic relationship between the individual strain of subunits and the global strain of the polymeric enzyme molecule (see text).

VII. STRUCTURAL FORMULATION OF STEADY-STATE ENZYME REACTION RATES AND EQUILIBRIUM LIGAND BINDING ISOTHERMS

The structural formalism allows structural rate equations as well as substrate binding isotherms to be written very easily. Its main virtue is therefore to show how subunit interactions may affect, in different ways, substrate binding and the overall rate of product appearance.

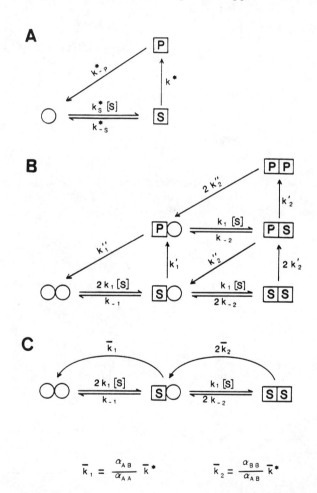

Fig. 20. Structural kinetic model for a dimeric enzyme exhibiting "catalytic balance" and loose coupling of subunits. (A) Kinetic model for an ideally isolated subunit (intrinsic process); (B) kinetic model for the dimeric enzyme; (C) kinetic model for the dimeric enzyme after graph reduction (see text).

A. Structural Steady-State Rate Equations for Loose and Tight Coupling of Subunits

Let us consider again a dimeric enzyme exhibiting loose coupling between subunits, uniqueness of enzyme conformation in the transition states, and "catalytic balance." If this enzyme catalyzes the conversion of one substrate into one product, the relevant reaction scheme may be depicted as shown in Fig. 20B. Owing to constraint conditions occurring between rate constants, only 7 different rate constants have to be defined instead of the 12 which appear in the reaction scheme. Figure 20A shows the so-called intrinsic process, that is, the kinetic scheme of an ideally isolated subunit. The kinetic parameters associated with the steady-state kinetic behavior of this ideally isolated subunit are indeed the apparent intrinsic affinity constant $\bar{K}^*$ (or its reciprocal, the apparent Michaelis constant) and the apparent catalytic constant $\bar{k}^*$. One has

$$\bar{K}^* = \frac{k_s^*(k^* + k_{-p}^*)}{k_{-p}^*(k^* + k_{-s}^*)}, \qquad \bar{k}^* = \frac{k^* k_{-p}^*}{k^* + k_{-p}^*}, \tag{91}$$

where k_s^*, k_{-s}^*, k^*, and k_{-p}^* are the four intrinsic rate constants pertaining to substrate binding, substrate release, catalysis, and product release. One of the aims of structural kinetics is to determine precisely whether it is possible to derive the steady-state equation pertaining to a polymeric enzyme in terms of intrinsic kinetic parameters $\bar{K}^*$ and $\bar{k}^*$ and structural thermodynamic parameters α. The seven rate constants may be expressed in structural terms as

$$k_1 = k_s^*, \qquad\qquad k_{-1} = k_{-s}^* \alpha_{AB}/\alpha_{AA},$$

$$k_1' = k^* \alpha_{AB}/\alpha_{AA}, \qquad k_{-2} = k_{-s}^* \alpha_{BB}/\alpha_{AB}, \tag{92}$$

$$k_2' = k^* \alpha_{BB}/\alpha_{AB}, \qquad k_1'' = k_{-p}^* \alpha_{AB}/\alpha_{AA},$$

$$k_2'' = k_{-p}^* \alpha_{BB}/\alpha_{AB}.$$

Steady-state treatment of the kinetic scheme of Fig. 20B allows derivation of the steady-state rate equation in structural terms, namely,

$$\frac{v}{[E]_0} = \frac{2\bar{k}^* \bar{K}^*[S] + 2\bar{k}^*(\alpha_{AA}/\alpha_{AB})\bar{K}^{*2}[S]^2}{1 + 2(\alpha_{AA}/\alpha_{AB})\bar{K}^*[S] + (\alpha_{AA}/\alpha_{BB})\bar{K}^{*2}[S]^2}, \tag{93}$$

where v and $[E]_0$ are the initial steady-state rate and total enzyme concentration, respectively. If the assumption is made that no subunit interaction occurs in the transition states, the steady-state expression Eq. (93) becomes

$$\frac{v}{[E]_0} = \frac{2\bar{k}^* \alpha_{AA} \bar{K}^*[S] + 2\bar{k}^* \alpha_{AA} \bar{K}^{*2}[S]^2}{1 + 2(\alpha_{AA}/\alpha_{AB})\bar{K}^*[S] + (\alpha_{AA}/\alpha_{BB})\bar{K}^{*2}[S]^2}. \tag{94}$$

Equation (93) is the mathematical description of the model in Fig. 20C. The important conclusion of the above developments is that the model of Fig. 20B is equivalent to that of Fig. 20C. Owing to the loose interactions between subunits, uniqueness of subunit conformation in the transition states, and "catalytic balance," the existence of enzyme–substrate–product and enzyme–(product)$_2$ complexes may be totally ignored. Therefore, the principles of structural kinetics, as well as the concepts of loose coupling, maximum catalytic efficiency, and "catalytic balance," represent a physical basis for the common practice of ignoring the enzyme–product complexes in the phenomenological expressions of steady-state rate equations of allosteric enzymes. Clearly, when subunit interaction coefficients all equal unity, Eq. (53) reduces to

$$\frac{v}{[E]_0} = \frac{2\bar{k}^*\bar{K}^*[S]}{1 + \bar{K}^*[S]}, \tag{95}$$

and the polymeric enzyme does not display any cooperativity.

This kind of reasoning may easily be extended to a more complex polymeric enzyme structure. For a tetrameric enzyme with "tetrahedral" geometry and loose coupling between subunits, the phenomenological steady-state equation is of the form

$$\frac{v}{[E]_0} = \frac{\psi_1[S] + 2\psi_2[S]^2 + 3\psi_3[S]^3 + 4\psi_4[S]^4}{1 + \psi_1'[S] + \psi_2'[S]^2 + \psi_3'[S]^3 + \psi_4'[S]^4}, \tag{96}$$

where ψ and ψ' are groupings of rate constants. The structural expression of this equation assumes the form

$$\frac{v}{[E]_0} = \frac{4\bar{k}^*\bar{K}^*[S] + 12\frac{\alpha_{AA}^3}{\alpha_{AB}^3}\bar{k}^*\bar{K}^{*2}[S]^2 + 12\frac{\alpha_{AA}^5}{\alpha_{AB}^4\alpha_{BB}}\bar{k}^*\bar{K}^{*3}[S]^3 + 4\frac{\alpha_{AA}^6}{\alpha_{AB}^3\alpha_{BB}^3}\bar{k}^*\bar{K}^{*4}[S]^4}{1 + 4\frac{\alpha_{AA}^3}{\alpha_{AB}^3}\bar{K}^*[S] + 6\frac{\alpha_{AA}^5}{\alpha_{AB}^4\alpha_{BB}}\bar{K}^{*2}[S]^2 + 4\frac{\alpha_{AA}^6}{\alpha_{AB}^3\alpha_{BB}^3}\bar{K}^{*3}[S]^3 + \frac{\alpha_{AA}^6}{\alpha_{BB}^6}\bar{K}^{*4}[S]^4}, \tag{97}$$

and reduces to a classical Michaelis–Menten equation when $\alpha_{AA} = \alpha_{AB} = \alpha_{BB} = 1$.

Equation (96) contains eight kinetic parameters, ψ and ψ', and structural Eq. (97) allows these to be expressed in terms of only five parameters and coefficients, namely, the three interaction coefficients α and the apparent intrinsic affinity and catalytic constants. $\bar{k}^*$ and $\bar{K}^*$. Simple inspection of Eq. (97) shows that

$$\psi_2/\psi_1' = \tfrac{3}{8}\psi_1, \qquad \psi_3/\psi_2' = \tfrac{1}{6}\psi_1, \qquad \psi_4/\psi_3' = \tfrac{1}{16}\psi_1. \tag{98}$$

The occurrence of these constraints allows one to test the validity of the kinetic version of the tetrahedral model of Koshland *et al.* (1966). As previously noted, a similar reasoning has been used to test the validity of the various models of subunit interactions through binding equations.

If subunits are tightly coupled and if the free enzyme preexists under two conformational states that cannot both bind the substrate (exclusive binding) the relevant equation for a tetrameric enzyme (see Fig. 11),

$$\frac{v}{[E]_0} = \frac{4k'\bar{K}^*[S](1 + \bar{K}[S])^3}{L + (1 + \bar{K}[S])^4}, \tag{99}$$

may be rewritten in the form of Eq. (96), but now the constraints between kinetic coefficients ψ and ψ' are

$$\psi_2/\psi_1' = \tfrac{3}{8}(L + 1)\psi_1, \quad \psi_3/\psi_2' = \tfrac{1}{6}(L + 1)\psi_1, \quad \psi_4/\psi_3' = \tfrac{1}{16}(L + 1)\psi_1, \tag{100}$$

which are different from expressions (98). In principle, it is therefore possible to distinguish a symmetry model with exclusive binding from an induced-fit tetrahedral model. For the symmetry preequilibrium model the structural formalism brings no additional information over the classical formulation (Eq. 99) since sequential binding of substrate and catalysis do not result in any change of enzyme conformation.

If tight coupling of subunits occurs together with induced fit and catalytic balance a corresponding possible model is shown in Fig. 21A. The structural formulation for the rate constants is

$$
\begin{aligned}
k_1 &= k_s^*\sigma_{AA}, & k_{-1} &= k_{-s}^*(\alpha_{AB}/\alpha_{AA})\sigma_{AB}, \\
k_2 &= k_s^*\sigma_{AB}, & k_{-2} &= k_{-s}^*(\alpha_{BB}/\alpha_{AB})\sigma_{BB}, \\
k_1' &= k^*(\alpha_{AB}/\alpha_{AA})\sigma_{AB}, & k_2' &= k^*(\alpha_{BB}/\alpha_{AB})\sigma_{BB}, \\
k_1'' &= k_{-p}^*(\alpha_{AB}/\alpha_{AA})\sigma_{AB}, & k_2'' &= k_{-p}^*(\alpha_{BB}/\alpha_{AB})\sigma_{BB},
\end{aligned}
\tag{101}
$$

and the structural rate equation assumes the form

$$\frac{v}{[E]_0} = \frac{2\bar{k}^*\sigma_{AA}\bar{K}^*[S] + 2\bar{k}^*\sigma_{AA}(\alpha_{AA}/\alpha_{AB})\bar{K}^{*2}[S]^2}{1 + 2(\alpha_{AA}/\alpha_{AB})(\sigma_{AA}/\sigma_{AB})\bar{K}^*[S] + (\alpha_{AA}/\alpha_{BB})(\sigma_{AA}/\sigma_{BB})\bar{K}^{*2}[S]^2}. \tag{102}$$

It must be stressed that this equation is still $2:2$ in substrate concentration. Moreover, when $\sigma_{AA} = \sigma_{AB} = \sigma_{BB} = 1$, this equation reduces to Eq. (93).

If quaternary constraints are released in the unliganded and fully liganded enzyme and if the subunits can exist only in the A or the B conformation, the fully concerted model thus obtained is shown in Fig. 21B. Then one must have

$$\sigma_{AA} = \sigma_{BB} = 1. \tag{103}$$

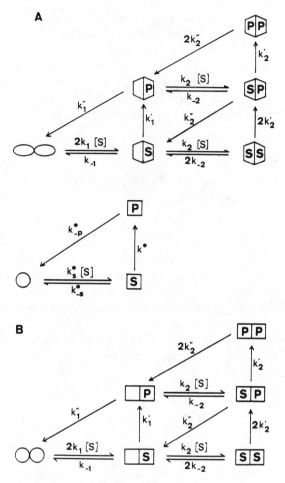

Fig. 21. Structural kinetic model for a dimeric enzyme exhibiting "catalytic balance" and tight coupling of subunits. (A) Quaternary constraints are not relieved in the nonliganded and fully liganded states; (B) quaternary constraints are relieved in the nonliganded and fully liganded states. The intrinsic process is shown between A and B.

Moreover, one must have (see Fig. 19)

$$\sigma_{AB} = (\alpha_{BB}/\alpha_{AB})(1/\sigma_B^*), \tag{104}$$

where σ_B^* is an intrinsic strain coefficient identical to the parameter K_t of Koshland. Equation (102) reduces to

$$\frac{v}{[E]_0} = \frac{2\bar{k}^*\bar{K}^*[S] + 2\bar{k}^*(\alpha_{AA}/\alpha_{AB})\bar{K}^{*2}[S]^2}{1 + 2(\alpha_{AA}/\alpha_{BB})\sigma_B^*\bar{K}^*[S] + (\alpha_{AA}/\alpha_{BB})\bar{K}^{*2}[S]^2}. \tag{105}$$

B. Substrate Binding Cooperativity and Curvature of Kinetic Reciprocal Plots

It may be of interest now to compare the main features of the Koshland formalism and the structural formalism, since both allow the binding equations to be expressed in structural terms.

The Koshland formalism expresses ligand binding to a flexible polymeric enzyme in terms of binding to an ideal, rigid, templatelike subunit. The ligand binding properties of a rigid subunit are the same whether this subunit is ideally isolated or not. One may therefore conclude that the Koshland formalism relates the binding properties of a *real, flexible* polymeric enzyme to the corresponding properties of an *ideal, rigid* polymeric enzyme.

The so-called structural formalism expresses ligand binding to a *flexible* polymeric enzyme in terms of the corresponding binding process occurring on an *ideally isolated flexible* subunit.

It is thus the "reference system" which is different in the two formalisms. In the Koshland formalism, the "reference system" is the ideal *rigid* polymer (or monomer). In the structural formalism, the "reference system" is the ideal *flexible* subunit.

The difference in the two formalisms may be easily understood when deriving the structural expression of the binding constant for a dimer having tightly coupled subunits. Figure 22A makes it obvious that the parameter K_s of Koshland is the binding constant of a rigid isolated subunit, or the microscopic constant of a rigid polymer. Figure 22B shows that the binding constant K_1 may be expressed, in the two formalisms, as

$$K_1 = K^*(\alpha_{AA}/\alpha_{AB})\sigma_{AB} = K^*(\alpha_{AA}/\alpha_{BB})\sigma_B^* = K_s K_t^2 K_{BB}. \tag{106}$$

Indeed, the molecular parameters α and K are related. As is obvious from their definition, the α parameters are equivalent to dissociation constants; α_{AA} is the dissociation constant of two subunits in conformation A, α_{AB} the dissociation constant of two subunits having conformations A and B, and α_{BB} the dissociation constant of two subunits both in conformation B. It follows that

$$\alpha_{AA} = 1/K_{AA}. \tag{107}$$

The thermodynamic boxes of Fig. 23 show that

$$K_{AB} = \alpha_{AA}/\alpha_{AB}, \qquad K_{BB} = \alpha_{AA}/\alpha_{BB}. \tag{108}$$

Since the "reference system" of the structural formalism is the ideal flexible subunit, there is no difficulty in using this formalism for the analysis of the effect of subunit interactions on enzyme catalysis. Any formalism, such as the one of Koshland, which uses a rigid enzyme as a "reference system" is

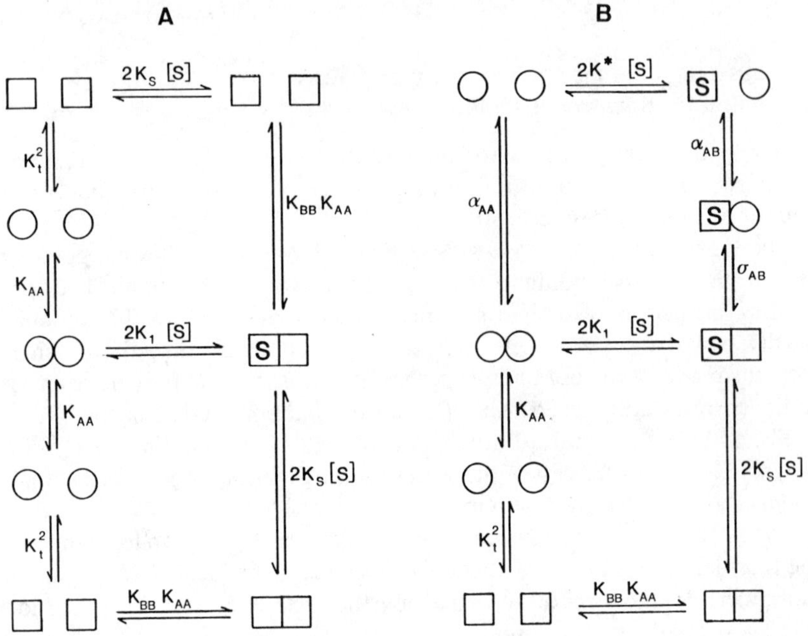

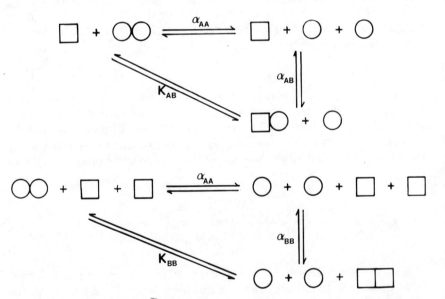

Fig. 22. Comparative expressions of a binding constant with the (A) Koshland and (B) structural formulations (see text).

Fig. 23. Relationship between the molecular parameters Ks of Koshland and the α coefficients (see text).

obviously not well suited for the analysis of enzyme catalysis which requires the subunits to be flexible.

For a dimeric enzyme following the "simple" model (loose coupling) of subunit interactions the substrate binding isotherm may be written as

$$\bar{v} = 2\bar{Y} = \frac{2(\alpha_{AA}/\alpha_{AB})K^*[S] + 2(\alpha_{AA}/\alpha_{BB})K^{*2}[S]^2}{1 + 2(\alpha_{AA}/\alpha_{AB})K^*[S] + (\alpha_{AA}/\alpha_{BB})K^{*2}[S]^2}. \tag{109}$$

Obviously, this expression is different from Eq. (92) and cannot be reduced to it. The maximum Hill coefficient at half-saturation of the protein by the substrate assumes the form [see Eq. (22)]

$$n_H = \frac{2}{\sqrt{\alpha_{AA}\alpha_{BB}/\alpha_{AB}^2} + 1}. \tag{110}$$

Clearly, if

$$\alpha_{AB}^2 > \alpha_{AA}\alpha_{BB}, \tag{111}$$

then $n_H > 1$ and the cooperativity is positive. Alternately, if

$$\alpha_{AB}^2 < \alpha_{AA}\alpha_{BB}, \tag{112}$$

then $n_H < 1$ and the cooperativity is negative. Equations (111) and (112) are indeed equivalent to those already discussed with the Koshland formalism.

It is obviously of major importance to determine how the reciprocal kinetic plots associated with the steady-state rate equation are affected by positive or negative cooperativity. The reciprocal rate equation for a "simple" dimeric enzyme [see Eq. (93)] is

$$\frac{[E]_0}{v} = \frac{(1/[S]^2) + 2(\alpha_{AA}/\alpha_{AB})\bar{K}^*(1/[S]) + (\alpha_{AA}/\alpha_{BB})\bar{K}^{*2}}{2\bar{k}^*\bar{K}^*(1/[S]) + 2\bar{k}^*\bar{K}^{*2}(\alpha_{AA}/\alpha_{AB})}. \tag{113}$$

Application of the Descartes rule of signs (see for instance, Botts, 1958; Durand, 1960) to the first derivative (with respect to $1/[S]$) of Eq. (113) shows that this equation can have at most a maximum, a minimum, and no inflection point. This minimum is obtained if

$$\alpha_{AB}^2 > 2\alpha_{AA}\alpha_{BB}. \tag{114}$$

The occurrence of this minimum implies that the enyzmatic reaction is inhibited by an excess of the substrate (Fig. 24). The concavity of the reciprocal plots is defined by the sign of the second derivative (with respect to $1/[S]$) of Eq. (113). If the sign is positive, the plots are concave up, if it is negative the plots are concave down. After some algebraic manipulations one can show that the plots are concave up if

$$\alpha_{AB}^2 > \alpha_{AA}\alpha_{BB}, \tag{115}$$

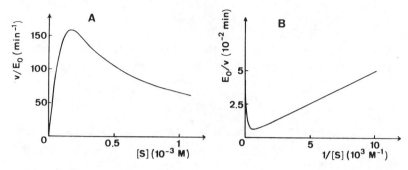

Fig. 24. Substrate inhibition generated by subunit interactions in a dimeric enzyme. The numerical values of the parameters are $\alpha_{AA} = 2$, $\alpha_{AB} = 50$, $\alpha_{BB} = 5$, $\bar{K}^* = 10^4 \, M^{-1}$, $\bar{k}^* = 100$ min^{-1}. (A) Direct plot; (B) reciprocal plot.

and they are concave down if

$$\alpha_{AB}^2 < \alpha_{AA}\alpha_{BB}. \tag{116}$$

These inequalities are precisely those which generate positive and negative cooperativity of substrate binding [Eqs. (111) and (112)]. Therefore, if substrate binding to the enzyme is positively cooperative the reciprocal plots of the corresponding kinetic equation will be concave up or exhibit a minimum. If binding of the substrate displays a negative cooperativity, the reciprocal steady-state kinetic plots will be concave down and cannot have any maximum or minimum.

It is therefore not illogical to use the term "kinetic cooperativity" to express the view that, owing to subunit interactions, the reciprocal kinetic plots may be concave up or down. When the reciprocal plots are concave up the "kinetic cooperativity" is defined as positive. Alternately, "kinetic cooperativity" is negative when the reciprocal plots are concave down.

If subunits are tightly coupled the relevant binding isotherm assumes the form

$$\bar{v} = 2\bar{Y} = \frac{2(\alpha_{AA}/\alpha_{AB})(\sigma_{AA}/\sigma_{AB})K^*[S] + 2(\alpha_{AA}/\alpha_{BB})(\sigma_{AA}/\sigma_{BB})K^{*2}[S]^2}{1 + 2(\alpha_{AA}/\alpha_{AB})(\alpha_{AA}/\sigma_{AB})K^*[S] + (\alpha_{AA}/\alpha_{BB})(\alpha_{AA}/\sigma_{BB})K^{*2}[S]^2}, \tag{117}$$

and the maximum Hill coefficient becomes

$$n_H = \frac{2}{\sqrt{\alpha_{AA}\sigma_{AA}\alpha_{BB}\sigma_{BB}/\alpha_{AB}^2\sigma_{AB}^2} + 1}. \tag{118}$$

Cooperativity is positive ($n_H > 1$) if

$$\alpha_{AB}^2\sigma_{AB}^2 > \alpha_{AA}\sigma_{AA}\alpha_{BB}\sigma_{BB}. \tag{119}$$

If

$$\alpha_{AB}^2 \sigma_{AB}^2 < \alpha_{AA} \sigma_{AA} \alpha_{BB} \sigma_{BB}, \tag{120}$$

then cooperativity is negative ($n_H < 1$). The reciprocal structural rate equation pertaining to this model of subunit interaction is

$$\frac{[E]_0}{v} = \frac{\dfrac{1}{[S]^2} + 2 \dfrac{\alpha_{AA}}{\alpha_{AB}} \dfrac{\sigma_{AA}}{\sigma_{AB}} \bar{K}^* \dfrac{1}{[S]} + \dfrac{\alpha_{AA}}{\alpha_{BB}} \dfrac{\sigma_{AA}}{\sigma_{BB}} \bar{K}^{*2}}{2\bar{k}^* \sigma_{AA} \bar{K}^* \dfrac{1}{[S]} + 2\bar{k}^* \sigma_{AA} \dfrac{\alpha_{AA}}{\alpha_{AB}} \bar{K}^{*2}}. \tag{121}$$

The reciprocal plots, which cannot exhibit any inflection point, have a minimum if

$$\sigma_{AB} \alpha_{AB}^2 > 2\alpha_{AA} \alpha_{BB} \sigma_{BB}. \tag{122}$$

The reciprocal plots will be concave down or up depending on the sign of the expression

$$(\alpha_{AA}^2/\alpha_{AB}^2)(1 - 2\sigma_{AA}/\sigma_{AB}) + (\sigma_{AA}/\sigma_{BB})(\alpha_{AA}/\alpha_{BB}) \gtrless 0. \tag{123}$$

If quaternary constraints are released in the unliganded and fully liganded enzyme, $\sigma_{AA} = \sigma_{BB} = 1$ and $\sigma_{AB} = \sigma_{BB}/\alpha_{AB}\sigma_B^*$; therefore, the binding Eq. (117) reduces to

$$\bar{v} = 2\bar{Y} = \frac{2(\alpha_{AA}/\alpha_{BB})\sigma_B^* K^*[S] + 2(\alpha_{AA}/\alpha_{BB})K^{*2}[S]^2}{1 + 2(\alpha_{AA}/\alpha_{BB})\sigma_B^* K^*[S] + (\alpha_{AA}/\alpha_{BB})K^{*2}[S]^2}, \tag{124}$$

and the reciprocal rate equation to

$$\frac{[E]_0}{v} = \frac{(1/[S]^2) + 2(\alpha_{AA}/\alpha_{BB})\sigma_B^* \bar{K}^*(1/[S]) + (\alpha_{AA}/\alpha_{BB})\bar{K}^{*2}}{2\bar{k}^* \bar{K}^*(1/[S]) + 2\bar{k}^*(\alpha_{AA}/\alpha_{AB})\bar{K}^{*2}}. \tag{125}$$

It is worth noting that no α_{AB} term appears in the expression of the binding equation, whereas it does in the rate equation. This simply reflects the release of enzyme conformational constraints when the substrate is transformed into the transition state $X^\ddagger$.

The maximum Hill coefficient assumes the form

$$n_H = \frac{2}{\sqrt{\alpha_{AA}/\alpha_{BB}} \, \sigma_B^* + 1}. \tag{126}$$

With a different formalism, this equation is equivalent to Eq. (59). The condition for stability of conformation A_2 with respect to B_2 is indeed

$$\alpha_{AA} \sigma_B^{*2}/\alpha_{BB} < 1. \tag{127}$$

This is exemplified in Fig. 10. In this case cooperativity can only be positive.

Under the same conditions (release of quaternary constraints in unliganded and fully liganded states), the curvature of the reciprocal kinetic plots is defined by the expression

$$\alpha_{AA}/\alpha_{AB}^2 + 1/\alpha_{BB} - 2(\alpha_{AA}/\alpha_{AB}\alpha_{BB})\sigma_B^* \gtrless 0. \tag{128}$$

Although this is not obvious at once, when Eq. (128) is negative, Eq. (127) must necessarily hold. Since inequality (127) is compulsory, one may conclude that tight coupling of subunits is associated with positive substrate binding cooperativity and with apparent positive "kinetic cooperativity."

C. Physical Significance of Generalized Microscopic Reversibility for Allosteric Enzymes

It has already been outlined that for several enzyme forms from which both substrate desorption and catalysis can occur, the ratio of these two rate constants is invariant. Whitehead (1976) has coined the term generalized microscopic reversibility to express the view that under nonequilibrium conditions, some constraints may exist between rate constants that would simplify the algebraic expression of the steady-state reaction rate. Indeed these constraints may be purely fortuitous, but one may wonder whether the existence of these constraints could not be understandable on the basis of simple physical arguments.

Let us consider, as an example, the kinetic version of the square "simple" model of Koshland *et al.* (1966) (Fig. 25). If no assumption is made as to the respective values of rate constant pairs k_{-i} and $\bar{k}_i$, the rate equation should be 5:5 in substrate concentration. However, if it is assumed that (a condition of generalized microscopic reversibility)

$$\frac{\bar{k}_1}{k_{-1}} = \frac{\bar{k}_2}{k_{-2}} = \frac{\bar{k}_3}{k_{-3}} = \frac{\bar{k}_4}{k_{-4}} = \frac{\bar{k}_5}{k_{-5}} = \frac{\bar{k}_6}{k_{-6}}, \tag{129}$$

the steady-state rate equation become 4:4 in substrate concentration. Since subunits are assumed to be loosely coupled, the structural formulation of the rate constants above leads to

$$\frac{\bar{k}_1}{k_{-1}} = \frac{\bar{k}_2}{k_{-2}} = \cdots = \frac{\bar{k}_6}{k_{-6}} = \frac{k^*}{k_{-s}^*} = \frac{k^*k_{-p}^*}{k_{-s}^*(k^* + k_{-p}^*)}, \tag{130}$$

where the ks with asterisks are intrinsic rate constants. This is so because subunit interactions affect substrate and product desorption, as well as catalysis, to the same extent. Therefore, it appears that generalized microscopic reversibility is the natural and obligatory consequence of loose coupling between subunits.

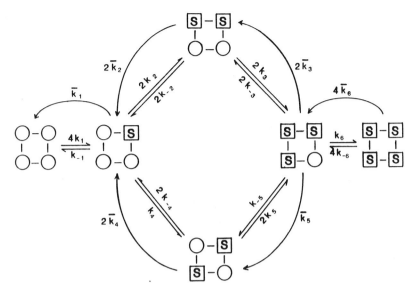

Fig. 25. Kinetic version of the square simple model of Koshland and significance of generalized microscopic reversibility.

VIII. THE ANTAGONISM BETWEEN SUBSTRATE BINDING AND ENZYME REACTION RATE

It is now obvious that the main interest of the structural formalism is to be able to study, with the same method, the differential effect of subunit interactions on substrate binding and on the rate of product appearance. Moreover, it is a common belief that "regulation" of ligand binding, that is the sigmoidal shape of the binding isotherm, is exerted at the expense of the affinity of the ligand. This belief rests on the nature of the Monod *et al.* (1965) equation in which it is evident that the greater is the value of the allosteric constant L, the lower is the affinity of the ligand for the protein. Clearly, the all-or-none response of the protein to a change of ligand concentration is paid for by a decrease of ligand affinity. Comparison of oxygen affinity for hemoglobin and myoglobin (Rossi-Fanelli and Antonini, 1958; Roughton, 1964) confirms this theoretical prediction. However, one may wonder whether some type of subunit interaction might not increase, and not always decrease, the affinity of the substrate, or the enzyme reaction rate, with respect to what would be expected if subunit interactions were not occurring.

Let us consider the dimensionless binding equation ($\bar{Y}$) and the dimensionless structural rate equation ($\bar{V}$) for a dimeric enzyme with loose coupling of subunits and catalytic balance:

$$\bar{Y} = \frac{(\alpha_{AA}/\alpha_{AB})c + (\alpha_{AA}/\alpha_{BB})c^2}{1 + 2(\alpha_{AA}/\alpha_{AB})c + (\alpha_{AA}/\alpha_{BB})c^2},$$ (131)

and

$$\bar{V} = \frac{c + (\alpha_{AA}/\alpha_{AB})c^2}{1 + 2(\alpha_{AA}/\alpha_{AB})c + (\alpha_{AA}/\alpha_{BB})c^2},$$ (132)

with

$$c = K^*[S] \quad [\text{Eq. (131)}] \quad \text{or} \quad \bar{K}^*[S] \quad [\text{Eq. (132)}],$$
$$\bar{Y} = \bar{v}/2, \qquad \bar{V} = v/2\bar{k}^*[E]_0.$$ (133)

In the absence of subunit interactions, Eqs. (131) and (132) reduce to

$$\bar{Y}^* = \bar{V}^* = c/(1 + c).$$ (134)

This equation may be considered as the intrinsic substrate binding isotherm or the intrinsic rate equation, that is, the binding isotherm or the rate equation pertaining to an ideally isolated subunit.

The comparative analysis of substrate binding to either the cooperative dimer or to the ideally isolated subunit may be effected by simple inspection of the difference between the two binding isotherms, namely,

$$\Delta_B = \bar{Y} - \bar{Y}^* = \frac{(\alpha_{AA}/\alpha_{AB} - 1)c + \alpha_{AA}(1/\alpha_{BB} - 1/\alpha_{AB})c^2}{\{1 + 2(\alpha_{AA}/\alpha_{AB})c + (\alpha_{AA}/\alpha_{BB})c^2\}(1 + c)}.$$ (135)

Clearly, this difference is associated with the modulation of substrate binding exerted by subunit interactions. When Δ_B is negative, subunit interactions decrease the affinity of the substrate for the dimer, with respect to the ideally isolated subunit. Alternatively, when Δ_B is positive, subunit interactions enhance, with respect to the ideal subunit, the binding of the substrate.

A comparative analysis of the reaction rate controlled by the cooperative dimer, or by the ideally isolated subunit, may be effected in the same manner, that is, by inspection of the sign of the difference Δ_R, given by

$$\Delta_R = \bar{V} - \bar{V}^* = \frac{(1 - \alpha_{AA}/\alpha_{AB})c^2 + \alpha_{AA}(1/\alpha_{AB} - 1/\alpha_{BB})c^3}{\{1 + 2(\alpha_{AA}/\alpha_{AB})c + (\alpha_{AA}/\alpha_{BB})c^2\}(1 + c)}.$$ (136)

This difference represents the rate modulation, with respect to the ideal subunit, exerted by subunit interactions in the dimer. When this difference is negative, subunit interactions inhibit the rate of product appearance and when it is positive, subunit interactions increase this reaction rate.

That the coefficients of the numerator terms in Eqs. (135) and (136) are similar, but opposite in sign, implies that a subunit interaction that enhances substrate affinity for the enzyme decreases the steady-state reaction rate. Alternatively, if this type of subunit interaction decreases the affinity of the substrate for the enzyme, it increases the reaction rate. It appears, therefore, that subunit interactions in polymeric enzymes create antagonistic effects on substrate binding and enzyme reaction rate.

Four cases of subunit interactions may be encountered with dimeric enzymes.

1. $\alpha_{AA} < \alpha_{AB} < \alpha_{BB}$. The cooperativity may be positive or negative. Whatever the substrate concentration, subunit interactions increase the affinity of the substrate for the enzyme but decrease the reaction rate.

2. $\alpha_{AA} < \alpha_{AB} > \alpha_{BB}$. The cooperativity is necessarily positive. At low substrate concentrations, subunit interactions decrease the affinity of the substrate and increase the rate, whereas at higher concentrations subunit interactions increase the substrate affinity and decrease the reaction rate. This is illustrated in Fig. 26.

3. $\alpha_{AA} > \alpha_{AB} < \alpha_{BB}$. The cooperativity is necessarily negative. Subunit interactions increase substrate affinity for low substrate concentrations and decrease substrate affinity for high substrate concentrations. The effect on reaction rate is symmetrical (Fig. 27).

4. $\alpha_{AA} > \alpha_{AB} > \alpha_{BB}$. The cooperativity may be positive or negative. Whatever the substrate concentration, subunit interactions generate a decrease of substrate affinity and an increase of reaction rate.

The important idea which has been established in this section is that subunit aggregation may represent a functional advantage by increasing the reaction

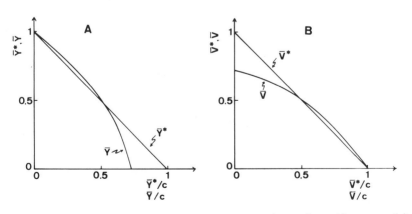

Fig. 26.   Antagonism of substrate binding and enzyme reaction rate for positive cooperativity. The figure shows Scatchard (A) and Eadie (B) plots. The numerical values of the parameters are $\alpha_{AA} = \alpha_{BB} = 3.5$, $\alpha_{AB} = 5$, $K^* = \bar{K}^* = 10^3$.

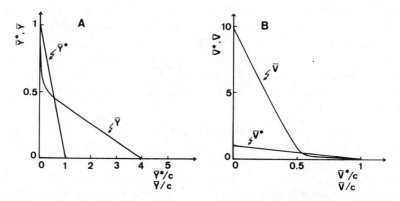

Fig. 27. Antagonism of substrate binding and enzyme reaction rate for negative cooperativity. The figure shows Scatchard (A) and Eadie (B) plots. The numerical values of the parameters are $\alpha_{AA} = 20$, $\alpha_{AB} = 5$, $\alpha_{BB} = 50$, $K^* = \bar{K}^* = 10^3$.

rate of an enzyme process. This important matter will be discussed in an evolutionary perspective in Section IX.

IX. THE EVOLUTION OF ENZYME FLEXIBILITY AND COOPERATIVITY

The concept of evolution occupies a central position in biology. Any biological problem must, at one stage or another, be viewed in an evolutionary perspective. The aim of this section is precisely to consider the concepts of flexibility and cooperativity in such an evolutionary perspective.

Enzyme evolution is often identified with the evolution of enzyme primary structure. Yet it is function rather than structure that is the driving force of evolution. Therefore, one may discuss on a physical basis how an enzyme can optimize its catalytic and regulatory function. As mentioned earlier there is no reason to consider these two functions as independent since both require protein flexibility.

Koshland (1976) proposed that the primordial catalysts were rigid nonspecific structures. Catalysis of chemical reactions was effected, thanks to an entropy decrease, by a better orientation and positioning of the reactants. These entropy effects, which still represent today an essential aspect of enzyme catalysis, required the existence of large catalyst molecules. The large molecule's functional advantage over a small one is that it allows optimization of catalytic activity in an incremental way without introducing strain, distortion, and additional degrees of rotational freedom. The demand for catalytic optimization over the course of evolution would solve the apparent paradox existing in the difference in size of the active site with respect to the rest of the enzyme molecule.

Flexibility would have appeared next as a simple response to three basic requirements: increase of enzymatic activity, specificity, and fine tuning of this activity. It has already been pointed out in this chapter that in order to be an efficient catalyst, an enzyme has to have a flexible structure. Release of strain in this structure certainly represents the main driving force that allows the passage of the energy barrier associated with a transition state. There is little doubt that flexibility has emerged in the course of evolution as a simple response to the demand for increased efficiency. The additional advantage brought about by the flexibility of the catalyst is the specificity of the reaction. The following example makes it easy to understand how the demand for specificity implies flexibility. Hexokinase phosphorylates glucose very efficiently. In the absence of glucose, ATP hydrolysis by this enzyme is negligible. Still, the nucleophilicity of alcoholic hydroxyl groups in hexoses is quite comparable to that of water. There is therefore no simple chemical explanation of why the transfer of a phosphoryl group takes place to glucose but not to water. The only simple explanation for this experimental result is to assume that glucose (but not water) induces a conformational change in hexokinase that allows the correct alignment of the catalytic groups. This induced conformational change has been observed experimentally (De la Fuente et al., 1970; Roustan et al., 1974).

An apparent "kinetic cooperativity," that is, reciprocal kinetic plots concave up or down, may well occur with one-sited monomeric enzymes (Kosow and Rose, 1971; Shill and Neet, 1971, 1975; Meunier et al., 1974; Ricard et al., 1977; Buc et al., 1977; Storer and Cornish-Bowden, 1976, 1977; Rübsamen et al., 1974). This effect, which is beyond the scope of this chapter, is explainable by "slow" conformational changes of the enzyme (hysteresis) occurring far from pseudoequilibrium conditions (Ainslie et al., 1972; Ricard et al., 1974b; Neet and Ainslie, 1980). However, one of the main features of this apparent "cooperativity" due to "slow" conformational transitions is that it is antagonistic to efficiency. In other words, a monomeric enzyme with "kinetic cooperativity" cannot be very efficient and, conversely, an efficient monomeric enzyme cannot display apparent "kinetic cooperativity" (Ricard, 1978). It therefore appears that this "cooperativity" is "paid for" at the expenses of catalytic efficiency. This result gives a sensible reason for the appearance, in the course of evolution, of cooperative polymeric enzymes.

Surprisingly, grouping polypeptide chains to form a polymeric enzyme may result in an increased activity of the individual subunits, together with a modulation of their response. Probably the main advantage, in neo-Darwinian terms, of cooperative polymeric enzymes is that "kinetic cooperativity" is no longer antagonistic to efficiency. For several reasons, which were developed earlier, subunit interactions may enhance the catalytic activity of an individual subunit. It is extremely tempting to consider this the main reason why cooperative polymeric enzymes have appeared in living organisms during evolution.

An apparent paradox, which has already been discussed, is that most of the allosteric enzymes display a rather smooth curve when plotting the steady-state rate as a function of substrate concentration. On the other hand, steady-state rate equations for these enzymes must be exceedingly complex. In other words, the potential kinetic complexities offered by allosteric enzymes are only rarely encountered in nature. These enzymes are therefore said to display a functional simplicity.

Since the improvement of catalytic efficiency is the driving force of evolution, one wonders whether the appearance of functional efficiency may not be the obligatory consequence of improved catalytic efficiency. Maximal catalytic efficiency of a polymeric enzyme rests on at least two requirements. The first requirement is the uniqueness of conformation of a subunit in the state of a complex with either of the transition states. If that conformation is that of the unliganded subunit after the quaternary constraints have been released, the driving force exerted by the enzyme to overcome the energy barrier will be maximized.

Another requirement for a high functional efficiency is the existence of a "catalytic balance." This implies that the subunit must exert the same rate enhancement for conversion of substrate to product as it does for conversion of product to substrate. If "catalytic balance" is reached, substrate and product stabilize the same conformation, or indistinguishable conformations. When these two conditions occur, rate equations become simpler and the enzyme has attained a state of functional simplicity (Ricard and Noat, 1982).

On the other hand, if these conditions are not met, rate equations are usually very complex. Kinetic complexity such as "bumps" predicted by higher degree rate equations may be considered as manifestations of an ancestral character of enzymes.

Another important reason for these evolutionary considerations is to attempt to understand the origin of the regulatory sites. As discussed earlier, some allosteric enzymes have separate subunits that bear regulatory sites, whereas others bear active and regulatory sites on the same subunits. This is the case of glutamate dehydrogenase. This enzyme is hexameric and contains only one type of polypeptide chain (Smith et al., 1970). In addition to the active site, each chain bears a regulatory site that binds ADP and GTP (Goldin and Frieden, 1971; Frieden, 1959a,b; Engel and Dalziel, 1969). Moreover, NADPH, one of the substrates of the enzyme, can be bound to both the active and the regulatory sites (Goldin and Frieden, 1971; Jallon and Iwatsubo, 1971; Di Prisco, 1971; Koberstein and Sund, 1971; Melzi d'Eril and Dalziel, 1972).

This obviously implies that both sites exhibit structural similarities. Engel (1973) has proposed an attractive hypothesis based on these results that may explain the origin of regulatory sites. As already stated, in the course of evolution, catalysis preceded control, and the regulatory sites probably originated from the active sites. This implies a duplication of the gene coding for the enzyme. Sequencing the enzyme has effectively shown a partial gene duplication. The Engel hypothesis is that ancestral glutamate dehydrogenase had smaller subunits than the actual enzyme and was lacking the regulatory sites. Later, gene duplication resulted in an enzyme having larger subunits and containing part of the catalytic site repeated. Whereas one of these sites has been maintained catalytically active by selection pressure, the other has evolved independently and has become the regulatory site.

X. CONCLUSIONS

In the course of this chapter some rather novel and unexpected ideas have been presented. These will be summarized now.

Contrary to common belief, no basic difference exists between the preequilibrium and the induced-fit concerted models. The physical concept which allows one to distinguish between structural models of ligand binding to proteins is the coupling strength between subunits. The formalism of Koshland et al. (1966) may be viewed as expressing ligand binding properties of a flexibile polymeric enzyme in terms of ligand binding on the corresponding ideal rigid enzyme.

Changing some basic features of Koshland's principles allows one to set up the so-called structural formalism; this may allow the effects of subunit interactions, strain, and quaternary constraints on the fine tuning of substrate binding and enzyme catalysis to be expressed in thermodynamic terms. This structural formalism allows kinetic parameters of an allosteric enzyme to be related to intrinsic kinetic parameters of an ideally isolated subunit.

When subunits are loosely coupled, both substrate binding cooperativity and apparent "kinetic cooperativity" may be positive or negative. As the subunits become tightly coupled, both substrate binding cooperativity and "kinetic cooperativity" become positive. The effects of quaternary structural changes of allosteric enzymes are different depending on how they are exerted on substrate binding or on the rate of product formation. There exists an antagonism between substrate affinity for the enzyme and the velocity of product appearance. The weaker is the substrate affinity for the enzyme, the greater is the rate of product appearance. Moreover, by affecting the rate of conformational transitions, subunit interactions may well enhance the rate of product appearance with respect to the corresponding rate that would be expected in the absence of any cooperativity.

Improvement of enzyme function is the driving force of evolution. Primordial enzymes were probably templatelike and catalysis was effected thanks to entropy effects. A major step in enzyme evolution was probably the emergence of flexibility, which resulted in increased catalytic efficiency, narrow specificity, and a fine tuning of the catalytic activity. Owing to flexibility, a regulatory behavior may exist with monomeric one-sited enzymes, but this apparent "kinetic cooperativity" is antagonistic to catalytic efficiency. The appearance of organized enzyme systems has resulted in an increased catalytic activity which is not antagonistic to cooperativity. Most of the allosteric enzymes today exhibit a functional simplicity, that is, a smooth reaction rate curve. This functional simplicity is believed to have emerged in the course of evolution as a trend toward an improved functional efficiency. Therefore, enzymes that have not reached this state usually exhibit rather complex kinetic behavior. Wavy curves, "bumps," and turning points may be considered as indicative of an ancestral enzyme.

The origin of the regulatory sites borne by the same subunit that contains the active site is viewed as the consequence of the duplication of a vestigial gene. All of the complex molecular devices which have evolved in living cells allow a fine tuning of enzymatic reactions, which are then integrated into a highly controlled and homeostatic network.

ACKNOWLEDGMENT

I would like to thank Dr. Georges Noat for helpful discussions and for carefully reading the manuscript.

REFERENCES

Adair, G. S. (1925). *J. Biol. Chem.* **63**, 529–545.
Ainslie, R. E., Shill, J. P., and Neet, K. E. (1972). *J. Biol. Chem.* **247**, 7088–7096.
Alberding, N., Chan, S., Eisenstein, L., Frauenfelder, H., Good, D., Gunsalus, I. C., Nordlung, T. H., Perutz, M. F., and Reynolds, A. H. (1978a). *Biochemistry* **17**, 43–51.
Alberding, N., Frauenfelder, H., and Hanggi, P. (1978b). *Proc. Natl. Acad. Sci. U.S.A.* **75**, 26–29.
Alberding, N., Austin, R. H., Chan, S., Eisenstein, L., Frauenfelder, H., Good, D., Kaufman, K., and Nordlung, T. H. (1978c). *Biophys. J.* **24**, 319–334.
Albery, W. J., and Knowles, J. R. (1976). *Biochemistry* **15**, 5631–5640.
Atkinson, D. E. (1970). *In* "The Enzymes" (P. D. Boyer, ed.), 3rd Ed. Vol. 1, pp. 461–489. Academic Press, New York.
Austin, R. H., Beeson, K. W., Eisenstein, L., Frauenfelder, H., and Gunsalus, I. C. (1975). *Biochemistry* **14**, 5355–5373.
Bardsley, W. G. (1976). *Biochem. J.* **153**, 101–117.
Bardsley, W. G. (1977a). *J. Theor. Biol.* **65**, 281–316.
Bardsley, W. G. (1977b). *J. Theor. Biol.* **67**, 121–139.

Bardsley, W. G., and Childs, R. E. (1976). *Biochem. J.* **149**, 313–328.

Bardsley, W. G. and Crabbe, M. J. C. (1976). *Eur. J. Biochem.* **68**, 611–619.

Bardsley, W. G., Leff, P., Kavanagh, J. P., and Waight, R. D. (1980). *Biochem. J.* **187**, 739–765.

Beece, D., Eisenstein, L., Frauenfelder, H., Good, D., Harden, M. C., Reinisch, L., Reynolds, A. H., Sorenson, L. B., and Yue, K. T. (1979). *Biochemistry* **18**, 3221–3223.

Beece, D., Eisenstein, L., Frauenfelder, H., Good, D., Reinisch, L., Reynolds, A. H., Sorensen, L. B., Yue, K. T., and Harden, M. C. (1980). *Biochemistry* **19**, 5147–5157.

Bernhard, S. A., Dunn, M. F., Luisi, P. L., and Schack, P. (1970). *Biochemistry* **9**, 185–192.

Biesecker, G., Harris, J. I., Thierry, J. C., Walker, J. E., and Wonacott, A. J. (1977). *Nature (London)* **266**, 328–333.

Blangy, D., Buc, H., and Monod, J. (1968). *J. Mol. Biol.* **31**, 13–35.

Botts, J. (1958). *Trans. Faraday Soc.* **54**, 593–604.

Buc, H. (1967). *Biochem. Biophys. Res. Commun.* **28**, 59–64.

Buc, H., and Buc, M. H. (1967). *In* "Regulation of Enzyme Activity and Allosteric Interactions" (E. Kvamme and A. Phil, eds.), pp. 109–130. Academic Press, New York.

Buc, J., Meunier, J. C., and Ricard, J. (1977). *Eur. J. Biochem.* **80**, 593–601.

Buehner, M., Ford, G. C., Moras, D., Olsen, K. W., and Rossmann, M. G. (1974). *J. Mol. Biol.* **90**, 25–49.

Changeux, J. P., Thiery, J. C., Tung, Y., and Kittel, C. (1967). *Proc. Natl. Acad. Sci. U.S.A.* **57**, 335–341.

Chipman, D. M., Grisario, V., and Sharon, N. (1967). *J. Biol. Chem.* **242**, 4388–4394.

Conway, A., and Koshland, D. E. (1968). *Biochemistry* **7**, 4011–4023.

Cook, R. A., and Koshland, D. E. (1969). *Proc. Natl. Acad. Sci. U.S.A.* **64**, 247–254.

Cook, R. A., and Koshland, D. E. (1970). *Biochemistry* **9**, 3337–3342.

Cornish-Bowden, A., and Koshland, D. E. (1970a). *J. Biol. Chem.* **245**, 6241–6250.

Cornish-Bowden, A., and Koshland, D. E. (1970b). *Biochemistry.* **9**, 3325–3336.

Cornish-Bowden, A., and Koshland, D. E. (1975). *J. Mol. Biol.* **95**, 201–212.

Dalziel, K. (1968). *FEBS Lett.* **1**, 346–348.

De La Fuente, G., Lagunas, R., and Sols, A. (1970). *Eur. J. Biochem.* **16**, 226–233.

De Moss, J. A. (1962). *Biochim. Biophys. Acta* **62**, 279–293.

Di Prisco, G. (1971). *Biochemistry* **10**, 585–589.

Dunn, M. F. (1974). *Biochemistry* **13**, 1146–1151.

Dunn, M. F., and Bernhard, S. A. (1971). *Biochemistry* **10**, 4569–4575.

Dunn, M. F., and Hutchinson, J. S. (1973). *Biochemistry* **12**, 4882–4892.

Durand, E. (1960). "Solutions Numériques des Équations Algebriques," Vol. 1. Masson, Paris.

Engel, P. C. (1973). *Nature (London)* **241**, 118–120.

Engel, P. C., and Dalziel, K. (1969). *Biochem. J.* **11**, 621–631.

Evans, P. R., and Hudson, P. J. (1979). *Nature (London)* **279**, 500–504.

Ford, L. O., Johnson, L. N., Machin, P. A., Phillips, D. C., and Tsian, R. (1974). *J. Mol. Biol.* **88**, 349–371.

Frauenfelder, H., Petsko, G. A., and Tsernoglou, D. (1979). *Nature (London)* **280**, 558–563.

Frieden, C. (1959a). *J. Biol. Chem.* **234**, 809–814.

Frieden, C. (1959b). *J. Biol. Chem.* **234**, 815–820.

Gelin, B. R., and Karplus, A. (1975). *Proc. Natl. Acad. Sci. U.S.A.* **72**, 2002–2006.

Gerhart, J. C., and Schachman, H. K. (1965). *Biochemistry* **4**, 1054–1062.

Gerhart, J. C., and Schachman, H. K. (1968). *Biochemistry* **7**, 538–552.

Goldin, B. R., and Frieden, C. (1971). *Curr. Top. Cell. Regul.* **4**, 77–117.

Hadorn, M., John, V. A., Meier, F. K., and Dutler, H. (1975). *Eur. J. Biochem.* **54**, 65–73.

Haldane, J. B. S. (1930). "Enzymes." Longmans, London.

Hartmann, H., Parak, F., Steigemann, W., Petsko, G. A., Pouzi, D. R., and Frauenfelder, H. (1982). *Proc. Natl. Acad. Sci. U.S.A.* **79,** 4967–4971.

Hill, C. H., Waight, R. D., and Bardsley, W. G. (1977). *Mol. Cell. Biochem.* **15,** 173–178.

Jallon, J. M., and Iwatsubo, M. (1971). *Biochem. Biophys. Res. Commun.* **45,** 964–971.

Jencks, W. P. (1969). "Catalysis in Chemistry and Enzymology." McGraw-Hill, New York.

Jencks, W. P. (1975). *Adv. Enzymol.* **43,** 219–410.

Johnson, L. N., Weber, I. T., Wild, D. L., Wilson, K. S., and Yeates, D. G. R. (1978). *J. Mol. Biol.* **118,** 579–591.

Karplus, M., and Weaver, D. L. (1976). *Nature (London)* **260,** 404–406.

Karplus, M., and Weaver, D. L. (1979). *Biopolymers* **18,** 1421–1437.

Kerbiriou, D., and Hervé, G. (1972). *J. Mol. Biol.* **64,** 379–392.

Kerbiriou, D., and Hervé, G. (1973). *J. Mol. Biol.* **78,** 687–702.

Kerbiriou, D., Hervé, G., and Griffin, J. H. (1977). *J. Biol. Chem.* **252,** 2881–2890.

Kirschner, K. (1968). *In* "Regulation of Enzyme Activity and Allosteric Interactions" (E. Kvamme and A. Phil, eds.), pp. 39–58. Academic Press, New York.

Kirschner, K., and Bisswanger, H. (1976). *Annu. Rev. Biochem.* **45,** 143–166.

Koberstein, R., and Sund, H. (1971). *FEBS Lett.* **19,** 149–151.

Koshland, D. E. (1958). *Proc. Natl. Acad. Sci. U.S.A.* **44,** 98–104.

Koshland, D. E. (1959). *J. Cell. Comp. Physiol.* **54,** 245–258.

Koshland, D. E. (1962). *J. Theor. Biol.* **2,** 75–86.

Koshland, D. E. (1969). *Curr. Top. Cell. Regul.* **1,** 1–27.

Koshland, D. E. (1970). *In* "The Enzymes" (P. D. Boyer, ed.), 3rd Ed., Vol. 1, pp. 341–396. Academic Press, New York.

Koshland, D. E., (1976). *Fed. Proc. Fed. Am. Soc. Exp. Biol.* **35,** 2104–2111.

Koshland, D. E., and Neet, K. E. (1968). *Annu. Rev. Biochem.* **37,** 359–410.

Koshland, D. E., Nemethy, G., and Filmer, D. (1966). *Biochemistry* **5,** 365–385.

Koshland, D. E., Carraway, K. W., Dafforn, G. A., Gass, J. D., and Storm, D. R. (1972). *Cold Spring Harbor Symp. Quant. Biol.* **36,** 13–19.

Kosow, D. P., and Rose, I. A. (1971). *J. Biol. Chem.* **246,** 2618–2625.

Kvassman, J., and Pettersson, G. (1976). *Eur. J. Biochem.* **69,** 279–287.

Kurganov, B. I. (1982). "Allosteric Enzymes. Kinetic Behavior." Wiley, New York.

Lazdunski, M. (1972). *Curr. Top. Cell. Regul.* **6,** 267–310.

Lazdunski, M., Petitclerc, C., Chappelet D., and Lazdunski, C. (1971). *Eur J. Biochem.* **20,** 124–134.

Levine, M., Muirhead, H., Stammers, D. K., and Stuart, D. I. (1978). *Nature (London)* **271,** 626–630.

Levitzki, A. (1973). *Biochem. Biophys. Res. Commun.* **54,** 889–893.

Levitzki, A. (1974). *J. Mol. Biol.* **90,** 451–458.

Levitzki, A. (1978). "Quantitative Aspects of Allosteric Mechanisms." Springer-Verlag, Berlin and New York.

Levitzki, A., and Koshland, D. E. (1969). *Proc. Natl. Acad. Sci. U.S.A.* **62,** 1121–1128.

Levitzki, A., and Koshland, D. E. (1972a). *Biochemistry* **11,** 241–246.

Levistki, A., and Koshland, D. E. (1972b). *Biochemistry* **11,** 247–253.

Levitzki, A., and Koshland, D. E. (1976). *Curr. Top. Cell. Regul.* **10,** 2–40.

Lienhard, G. E., Secemski, I. I., Koehler, K. A., and Lindquist, R. N. (1972). *Cold Spring Harbor Symp. Quant. Biol.* **36,** 45–51.

Luisi, P. L., and Bignetti, E. (1974). *J. Mol. Biol.* **88,** 653–670.

Luisi, P. L., and Favilla, R. (1972). *Biochemistry* **11,** 2303–2310.

Lumry, R. (1959). *In* "The Enzymes" (P. D. Boyer, H. Lardy, and K. Myrbäck, eds.), Vol. 1, pp. 157–231. Academic Press, New York.

McCammon, J. A., and Karplus, M. (1977). *Nature (London)* **268**, 765–766.

McCammon, J. A., Gelin, B. R., and Karplus, M. (1977). *Nature (London)* **267**, 585–590.

McCammon, J. A., Wolynes, P. G., and Karplus, M. (1979). *Biochemistry* **18**, 927–942.

McFarland, J. T., and Bernhard, S. A. (1972). *Biochemistry* **11**, 1486–1493.

McFarland, J. T., and Chu, Y. H. (1975). *Biochemistry* **14**, 1140–1146.

Matchett, W. H. (1974). *J. Biol. Chem.* **249**, 4041–4049.

Matthews, B. W., and Bernhard, S. A. (1973). *Annu. Rev. Biophys. Bioeng.* **2**, 257–317.

Melzi d' Eril, G., and Dalziel, K. (1972). *Biochem. J.* **130**, 3P.

Meunier, J. C., Buc, J., Navarro, A., and Ricard, J. (1974). *Eur. J. Biochem.* **49**, 209–223.

Monod, J., Changeux, J. P., and Jacob, F. (1963). *J. Mol. Biol.* **6**, 306–329.

Monod, J., Wyman, J., and Changeux, J. P. (1965). *J. Mol. Biol.* **12**, 88–118.

Nari, J., Mouttet, C., Fouchier, F., and Ricard, J. (1974). *Eur. J. Biochem.* **41**, 479–497.

Neet, K. (1980). *In* "Methods in Enzymology" (D. L. Purich, ed.), Vol. 64 (Part B), pp. 139–192. Academic Press, New York.

Neet, K., and Ainslie, G. R. (1980). *In* "Methods in Enzymology" (D. L. Purich, ed.), Vol. 64 (Part B), 192–226. Academic Press, New York.

Neet, K., and Koshland, D. E. (1968). *J. Biol. Chem.* **243**, 6392–6401.

Pettersson, G. (1976). *Eur. J. Biochem.* **69**, 273–278.

Phillips, D. C. (1967). *Proc. Natl. Acad. Sci. U.S.A.* **57**, 484–495.

Ricard, J. (1978). *Biochem. J.* **175**, 779–791.

Ricard, J. (1980). *In* "The Biochemistry of Plants" (D. D. Davies, ed.), Vol. 2, pp. 31–80. Academic Press, New York.

Ricard, J., and Noat, G. (1982). *J. Theor. Biol.* **96**, 347–365.

Ricard, J., and Noat, G. (1984). *J. Theor. Biol.* **111**, 737–753.

Ricard, J., Meunier, J. C., and Buc, J. (1974a). *Eur. J. Biochem.* **49**, 195–208.

Ricard, J., Mouttet, C. and Nari, J. (1974b). *Eur. J. Biochem.* **41**, 479–497.

Ricard, J., Buc, J., and Meunier, J. C. (1977). *Eur. J. Biochem.* **80**, 581–592.

Rossi-Fanelli, A., and Antonini, E. (1958). *Arch. Biochem. Biophys.* **77**, 478–492.

Rossky, P. J., and Karplus, M. (1979). *J. Am. Chem. Soc.* **101**, 1901–1937.

Rossky, P. J., Karplus, M., and Rahman, A. (1979). *Biopolymers* **18**, 825–854.

Roughton, F. J. W. (1964). "Oxygen in the Animal Organism." Pergamon, Oxford.

Roustan, C., Brevet, A., Pradel, L. A., and Van Thoai, N. (1974). *Eur. J. Biochem.* **44**, 353–358.

Rubin, M. M., and Changeux, J. P. (1966). *J. Mol. Biol.* **21**, 265–274.

Rübsamen, H., Khandker, R., and Witzel, H. (1974). *Hoppe-Seyler's Z. Physiol. Chem.* **355**, 687–699.

Sanwal, B. D. (1970). *Bacteriol. Rev.* **34**, 20–39.

Sanwal, B. D., Kapoo, M., and Duckworth, H. W. (1971). *Curr. Top. Cell. Regul.* **3**, 1–115.

Secemski, I. I., and Lienhard, G. E. (1971). *J. Am. Chem. Soc.* **93**, 3549–3550.

Secemski, I. I., Lehrer, S. S., and Lienhard, G. E. (1972). *J. Biol. Chem.* **247**, 4740–4748.

Seydoux, F., and Bernhard, S. A. (1974). *Biophys. Chem.* **1**, 161–174.

Seydoux, F., Bernhard, S. A., Pfenninger, O., Payne, M., and Malhotra, O. P. (1979). *Biochemistry* **12**, 4290–4300.

Shill, J. P., and Neet, K. E. (1971). *Biochem. J.* **123**, 283–285.

Shill, J. P., and Neet, K. E. (1975). *J. Biol. Chem.* **250**, 2259–2268.

Smith, E. L., Landon, M., Piszkiewics D., Brattin, W. J., Langley, T. J., and Melamed, M. D. (1970). *Proc. Natl. Acad. Sci. U.S.A.* **67**, 724–730.

Solano-Munoz, F., Mc Ginglay, P. B., Woolfson, R., and Bardsley, W. G. (1981). *Biochem. J.* **193**, 339–352.

Stadtman E. R. (1970), *In* "The Enzymes" (P. D. Boyer, ed.), 3rd Ed., Vol. 1, pp. 397–459. Academic Press, New York.

Stadtman, E. R., and Ginsburg, A. (1974). *In* "The Enzymes" (P. D. Boyer, ed.), 3rd Ed., Vol. 10, pp. 755–807. Academic Press, New York.

Steitz, T. A., Anderson, W. F., Fletterick, R. J., and Anderson, C. M. (1977). *J. Biol. Chem.* **252,** 4494–4500.

Storer, A. C., and Cornish-Bowden, A. (1976). *Biochem. J.* **159,** 7–14.

Storer, A. C., and Cornish-Bowden, A. (1977). *Biochem. J.* **167,** 61–69.

Storm, D. R., and Koshland, D. E. (1970). *Proc. Natl. Acad. Sci. U.S.A.* **66,** 445–452.

Storm, D. R., and Koshland, D. E. (1972a). *J. Am. Chem. Soc.* **94,** 5805–5814.

Storm, D. R., and Koshland, D. E. (1972b). *J. Am. Chem. Soc.* **94,** 5815–5825.

Szabo, A., and Karplus, M. (1975). *Biochemistry* **14,** 931–940.

Teipel, J., and Koshland, D. E. (1969). *Biochemistry* **8,** 4656–5663.

Thiry, L., and Hervé, G. (1978). *J. Mol. Biol.* **125,** 515–534.

Viratelle, O. M., and Seydoux, F. (1975). *J. Mol. Biol.* **92,** 193–205.

Von Dobeln, U., and Reichard, P. (1976). *J. Biol. Chem.* **251,** 3616–3622.

Weber, I. T., Johnson, L. N., Wilson, K. S., Yeates, D. G. R., Wild, D. L., and Jenkins, J. A. (1978). *Nature (London)* **274,** 433–437.

Weber, K. (1968). *Nature (London)* **218,** 1116–1119.

Welch, G. R. (1977). *Prog. Biophys.* **32,** 103–191.

Wetlaufer, D. B., and Ristow, S. (1973). *Annu. Rev. Biochem.* **42,** 135–158.

Whitehead, E. P. (1970). *Prog. Biophys.* **21,** 321–397.

Whitehead, E. P. (1974). *Acta Biol. Med. Germ.* **31,** 227–258.

Whitehead, E. P. (1976). *Biochem. J.* **159,** 449–456.

Whitehead, E. P. (1979). *J. Theor. Biol.* **80,** 355–381.

Wiley, D. C., and Lipscomb, W. N. (1968). *Nature (London)* **218,** 1119–1121.

Wolfenden, R. (1969). *Nature (London)* **223,** 704–705.

Wyman, J. (1948). *Adv. Protein Chem.* **4,** 407–531.

Wyman, J. (1964). *Adv. Protein Chem.* **19,** 223–286.

Wyman, J. (1967). *J. Am. Chem. Soc.* **89,** 2202–2218.

Wyman, J. (1968). *Q. Rev. Biophys.* **1,** 35–80.

Wyman, J. (1972), *Curr. Top. Cell. Regul.* **6,** 209–226.

5

Control of Enzyme Activity in Reversibly Adsorptive Enzyme Systems

Boris I. Kurganov

The All-Union Vitamin Research Institute
Moscow, U.S.S.R.

I. INTRODUCTION

One of the most significant factors affecting metabolic regulation in the intact cell is the interaction between enzymes and intracellular structures. The development of hypotheses about the role of reversible complexing of enzymes with intracellular membranes and structural proteins in the regulation of the rates of metabolic processes (e.g., in skeletal muscle) is in progress. The adsorption of enzymes to subcellular structures provides not only control of enzyme activity, but also the linkage of metabolic processes that proceed in

ORGANIZED
MULTIENZYME SYSTEMS

241

the different compartments of the cell (for example, in cytosol and mito-chondrion) (see also Friedrich and Srere, this volume).

In this chapter the general properties of biological adsorptive enzyme systems and the physiological significance of reversible adsorption of the enzymes to subcellular structures are discussed.

II. EXPERIMENTAL DATA SUPPORTING THE CONCEPT THAT ENZYMES MAY BE ADSORBED TO SUBCELLULAR STRUCTURES UNDER PHYSIOLOGICAL CONDITIONS

A great number of enzymes that are not classified as membrane-integrated proteins (such as the system of oxidative phosphorylation) are found, nevertheless, to be associated with particulate material after cell fragmentation under hypoosmotic conditions (Green et al., 1965; Schrier, 1967; Clarke and Masters, 1975a; Masters, 1981). These include glycolytic enzymes such as hexokinase (EC 2.7.1.1) (Rose and Warms, 1967), fructose-1,6-bisphosphate aldolase (EC 4.1.2.13) (Strapazon and Steck, 1977), phosphoglycerate kinase (EC 2.7.2.3) (Tillman et al., 1975), and glyceraldehyde-3-phosphate dehydrogenase (EC 1.2.1.12) (Tanner and Gray, 1971; Carraway and Shin, 1972; Kant and Steck, 1973). For example, 60–80% of glyceraldehyde-3-phosphate dehydrogenase remains associated with erythrocyte membrane after lysis of erythrocytes under hypoosmotic conditions (Keokitichai and Wrigglesworth, 1980; Shin and Carraway, 1973). The question arises whether the association of the enzymes with subcellular structures is the result of osmotic stress [i.e., an artifactual phenomenon as proposed by Hanahan (1973)], or whether this association occurs in vivo. To answer this question we need methods that make it possible to fix the enzymes in the intact cell and to analyze their distribution after cell fragmentation.

Keokitichai and Wrigglesworth (1980) and Chen et al. (1981) used the bifunctional reagent glutaraldehyde to cross-link intracellular proteins to the inner surface of erythrocyte membrane. Intact erythrocytes were exposed to low concentrations of glutaraldehyde for 5 min at 4°C. The erythrocyte ghosts obtained by lysis under hypoosmotic conditions were incubated in 300 mM NaCl solution in order to elute glyceraldehyde-3-phosphate dehydrogenase, which could become bound to the membrane during hemolysis of erythrocytes. The determination of enzyme activity of the membrane fraction showed that about half of total glyceraldehyde-3-phosphate dehydrogenase activity was associated with erythrocyte membrane. It would follow from these results that glyceraldehyde-3-phosphate dehydrogenase is located close enough to the plasma membrane to form cross-links with membrane proteins. The analogous conclusion is drawn for fructose-1,6-bisphosphate aldolase.

Solti *et al.* (1982) used another approach for elucidation of the localization of glyceraldehyde-3-phosphate dehydrogenase in intact erythrocytes. Human erythrocytes were treated with highly tritiated [^{3}H]iodoacetate under conditions such that half of the label became attached to glyceraldehyde-3-phosphate dehydrogenase. This common alkylating reagent is, to some extent, an affinity label for glyceraldehyde-3-phosphate dehydrogenase, by virtue of the fact that its carboxylate group favorably orients it in the active center of the enzyme to react with the SH group of Cys-149. Iodoacetate reacts with the Cys-149 SH group in the holoenzyme about three orders of magnitude faster than with cysteine or glutathione. After fixation, the cells were subjected to electron microscopic autoradiography. Statistical analysis of data showed the adherence of significant numbers of grains to the cell membrane. The results support the view that glyceraldehyde-3-phosphate dehydrogenase is localized near the membrane in the intact erythrocyte. (For a more complete discussion of the erythrocyte system, see Dr. Friedrich's article in this volume.)

The experimental data discussed here indicate the possibility of adsorption of enzymes to subcellular structures under physiological conditions. Therefore there are good reasons to investigate the general behavior of adsorptive enzyme systems and the role of adsorption in the regulation of enzyme activity and the rates of metabolic processes.

III. TYPES OF REVERSIBLY ADSORPTIVE ENZYME SYSTEMS

In this chapter we shall consider adsorptive enzyme systems in which structural proteins of skeletal muscle, inner surfaces of cytoplasmic membranes, and the surface of the membranes of cytoplasmic organelles play the role of matrix support. (Some such examples are considered also by Dr. Friedrich in this volume.)

Histochemical studies have established that the majority of the glycolytic enzymes (hexokinase being a prominent exception) are localized near, perhaps within, the I-band of the muscle fibers (Dolken *et al.*, 1975; Sigel and Pette, 1969). Biochemical investigations have shown that a number of glycolytic enzymes interact strongly with F-actin, the major structural protein of the I-band (Arnold and Pette, 1968, 1970; Arnold *et al.*, 1971). Strong binding is observed for fructose-1,6-bisphosphate aldolase, glyceraldehyde-3-phosphate dehydrogenase, and pyruvate kinase. F-actin has less affinity for lactate dehydrogenase (EC 1.1.1.27) and phosphoglycerate kinase. Subsequent studies by Clarke and Masters (1975a) have demonstrated the possibility of binding of glycolytic enzymes to the filaments containing F-actin under conditions similar to those in muscle fibers. Adsorption is greatly influenced by including in the filaments the regulatory proteins of the I-band, tropomyosin and troponin (Clarke and Morton, 1976; Clarke and Masters, 1976; Stewart *et al.*, 1980).

Enzymes that interact reversibly with intracellular membranes and desorb rather easily by an increase of ionic strength or addition of specific ligands are termed peripheral membrane-bound enzymes. Recent works have indicated that the binding of peripheral enzymes is due to their attachment to anchor proteins, which are embedded in the membrane; the participation of integral membrane proteins provides the high specificity of adsorption (see a review by Singer, 1974). Glyceraldehyde-3-phosphate dehydrogenase and fructose-1,6-bisphosphate aldolase are bound selectively by a glycoprotein of the erythrocyte membrane, the so-called Band 3 protein (the name reflects its position among the proteins of the erythrocyte membrane separated by SDS polyacrylamide gel electrophoresis) (Yu and Steck, 1975; Tsai *et al.*, 1982). Band 3 is the predominant polypeptide and the purported mediator of anion transport in the human erythrocyte membrane. This integral glycoprotein appears to exist as a noncovalent dimer of two ~93,000-Da chains that span the membrane asymmetrically. Based on data obtained by proteolysis at the two sides, it appears that a segment of about 17,000 Da spans the membane, whereas segments of about 35,000 Da containing the C terminus and of about 40,000 Da containing the N terminus are exposed on the outside and inside faces of the membrane, respectively (see Steck, 1978; Rothstein *et al.*, 1978). The 17,000-Da segment contains on its outside region the binding site for 4,4'-diisothiocyano-2,2'-stilbene disulfonate, a specific inhibitor of anion transport. Proteolytic cleavage of the 35,000 and 40,000 Da segments does not result in any reduction of anion fluxes. Thus the 17,000-Da segment of Band 3 is essential for anion transport. S-Cyanylation of a cysteine residue cleaves a 23,000-Da piece from the cytoplasmic amino-terminal region of Band 3. That very segment contains the binding site for glyceraldehyde-3-phosphate dehydrogenase (Tsai *et al.*, 1982). The stoichiometry of interaction of glyceraldehyde-3-phosphate dehydrogenase with Band 3 protein corresponds to binding of one molecule of the enzyme to one Band 3 polypeptide of molecular mass 93,000 (Yu and Steck, 1975).

The specificity of the interaction of hexokinase with the outer membrane of mitochondria is provided by the presence of a hexokinase-binding protein in the mitochondrial membrane. This fact has been established by Felgner *et al.* (1979). Fiek *et al.* (1982) reported that the hexokinase-binding protein is identical to the mitochondrial porin. This channel-forming protein is responsible for the permeability of the mitochondrial outer membrane for saccharides and ADP.

According to Lusis *et al.* (1976) glucuronidase (EC 3.2.1.31) is anchored to microsomal membrane by specific, noncovalent complexing with egasyn, an integral membrane protein.

IV. CHARACTERISTICS OF REVERSIBLY ADSORPTIVE ENZYME SYSTEMS

To understand how the adsorptive mechanism of regulation of enzyme activity becomes operative, we must discuss the main peculiarities of biological adsorptive enzyme systems.

A. Reversibility of Adsorption–Desorption Processes

The most important property of biological adsorptive enzyme systems is the reversibility of the adsorption–desorption processes. In classical work carried out by Rose and Warms (1967) it has been shown that hexokinase solubilized by glucose 6-phosphate can be bound again to previously eluted mitochondria. The reversibility of adsorption of glyceraldehyde-3-phosphate dehydrogenase to erythrocyte membranes has been demonstrated by Kant and Steck (1973), McDaniel *et al.* (1974), and other investigators.

The classical works by Arnold and Pette (1968, 1970) support the reversible character of adsorption of fructose-1,6-bisphosphate aldolase and glyceraldehyde-3-phosphate dehydrogenase to F-actin.

I shall give examples of reversible binding of the enzymes to structural proteins of skeletal muscle and biological membranes in the next sections. I should like to stress that the reversibility of adsorption in biological adsorptive enzyme systems is experimentally proved, and this must be considered when explaining experimental results and developing theoretical models.

B. Character of Equilibrium Binding Isotherms

We shall consider first the reversible adsorption of peripheral enzymes to biological membranes. Such adsorption is based on the specific interactions of peripheral enzymes with anchor proteins that are embedded in the membrane and traverse the lipid bilayer. The integral membrane proteins tend to have an oligomeric (usually dimeric) structure, in which the oligomer is arranged asymmetrically across the membrane with the axis of symmetry perpendicular to the membrane plane (see the review by Klingerberg, 1981). The existence of anchor proteins in oligomeric form predetermines the character of the adsorption of the peripheral enzyme, so that the latter must have an oligomeric structure in the adsorbed state. If the peripheral enzyme is an oligomer, one can visualize a situation in which the enzyme molecule and the anchor membrane protein form a symmetrical complex with stoichiometry of 1:1. In this case the adsorption–desorption process may be represented by Eq. (1),

$$E + \Omega \quad E\Omega \qquad (1)$$

in which E is the peripheral enzyme, Ω is the enzyme-binding site on the membrane and $E\Omega$ is the complex. Let K be the constant of the equilibrium between free and bound enzyme forms: $K = [E\Omega]/[E][\Omega]$. For characterization of the degree of binding we introduce the magnitude r, which represents the ratio of the concentration of bound enzyme to the concentration of the adsorbent: $r = [E]_{bound}/[adsorbent]$. It is evident that the ratio of the amount of bound enzyme to the amount of the adsorbent may be used when calculating the magnitude r. For Eq. (1), the magnitude r is a hyperbolic function of the concentration of free enzyme $[E]$:

$$r = r_{max} K[E]/(1 + K[E])$$ (2)

where r_{max} is the adsorption capacity of the membrane with respect to the enzyme. Scatchard's equation may be used for determination of the parameters r_{max} and K:

$$r/[E] = Kr_{max} - Kr$$ (3)

One may now discuss the situation in which the peripheral enzyme is a monomer. The attachment of one molecule of peripheral enzyme to an oligomeric anchor protein results in the formation of a complex with an asymmetrical structure. One may expect the building of this complex to continue by attachment of additional molecules of peripheral enzyme until the formation of a symmetrical structure. The total number of enzyme molecules in the final complex depends on the type of symmetry of the anchor membrane protein. If the dimeric anchor protein $(\Omega\Omega)$ embedded in the membrane binds successively two molecules of peripheral enzyme, the binding process is given by the following scheme:

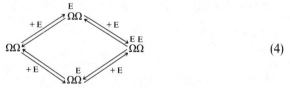

(4)

Let K_1 and K_2 be the microscopic association constants for binding of the first and second molecule of the enzyme, respectively:

$$K_1 = \frac{[E\Omega\Omega]}{[E][\Omega\Omega]} \quad \text{and} \quad K_2 = \frac{[E\Omega\Omega E]}{[E][E\Omega\Omega]}$$ (5)

We can get the following expression for r:

$$r = r_{max} \frac{K_1[E] + K_1 K_2 [E]^2}{1 + 2K_1[E] + K_1 K_2 [E]^2}$$ (6)

The dependence of r on $[E]$ is determined by the relationship between K_1 and K_2. Let α be the ratio K_2/K_1. Analysis of Eq. (6) shows that when $\alpha = 1$

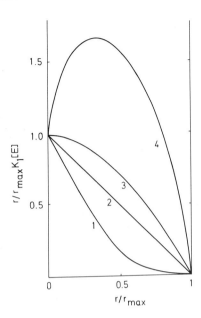

Fig. 1. The theoretical binding isotherms corresponding to Eq. 4. The dependence of $r/r_{max}K_1[E]$ on r/r_{max} (Scatchard plots) is calculated from Eq. (6) at the following values of $\alpha = K_2/K_1$: curve 1, $\alpha = 0.1$; curve 2, $\alpha = 1$; curve 3, $\alpha = 2$; and curve 4, $\alpha = 10$.

(no interaction between adsorbed enzyme molecules) the dependence of r on E is hyperbolic. For $\alpha > 1$, the attachment of the first molecule of the enzyme facilitates the binding of the adjacent enzyme molecule to the enzyme-binding site $\Omega\Omega$ (positive cooperative interactions between adsorbed enzyme molecules). The dependences of $r/[E]$ on r are convex (Fig. 1). When $\alpha > 2$ these dependences pass through a maximum, whereas the dependences of r on [E] are S-shaped. For $\alpha < 1$, the attachment of the first molecule of the enzyme hampers the binding of subsequent enzyme molecules to the $\Omega\Omega$ site (negative cooperative interactions between adsorbed enzyme molecules). In this case the dependences of $r/[E]$ on r are concave, whereas the dependences of r on [E] are characterized by delayed attainment of the level corresponding to r_{max}.

If experimental data permit the estimation of the limiting value of r (r_{max}) by extrapolation to infinitely high enzyme concentrations, the linear anamorphosis in Eq. (7) may be used for determination of parameters K_1 and K_2.

$$\frac{(r/r_{max})}{[E](1 - 2r/r_{max})} = K_1 + K_1K_2\frac{[E](1 - r/r_{max})}{(1 - 2r/r_{max})} \qquad (7)$$

It should be noted that in the general case E in Eq. (4) may represent not only monomeric enzymes but also oligomeric molecules. Therefore interactions between adsorbed enzyme molecules possessing subunit structure are possible. Such a situation has been observed for the adsorption of glyceraldehyde-3-phosphate dehydrogenase to Band 3 protein, which exists in the erythrocyte membrane in dimeric form. Each enzyme-binding site in the membrane is able to accept two molecules of the enzyme, since one Band 3 polypeptide binds one molecule of glyceraldehyde-3-phosphate dehydrogenase (Yu and Steck, 1975). One would expect that the interactions of the molecules of glyceraldehyde-3-phosphate dehydrogenase in the adsorbed state would cause deviations from the hyperbolic dependences of the amount of bound enzyme on equilibrium concentration of free enzyme. Actually, Yu and Steck (1975) observed that the graph of the function of saturation of erythrocyte ghosts by glyceraldehyde-3-phosphate dehydrogenase has an S-shaped form (Fig. 2). The scheme (Eq. 4) proposed for the binding of the enzyme to a dimeric enzyme binding site $\Omega\Omega$, which accepts two molecules of the enzyme, may be used for quantitative description of the experimental data obtained by Yu and Steck. In Fig. 2 the points are experimental values of the amount of bound enzyme. The solid line is drawn in correspondence with Eq. (6) with $K_1 = 4.7 \times 10^6 \ M^{-1}$, $K_2 = 34 \times 10^6 \ M^{-1}$, and $r_{max} = 1.90 \times 10^{-18}$ mol glyceraldehyde-3-phosphate dehydrogenase per ghost. The ratio $K_2/K_1 = \alpha$ is equal to 7.2. This value of α indicates the existence of positive cooperative interactions between adsorbed enzyme molecules and explains the S-shaped form of the dependence of r on $[E]$. The adsorption capacity of erythrocyte ghosts expressed as a number of enzyme-binding sites (Ω) per

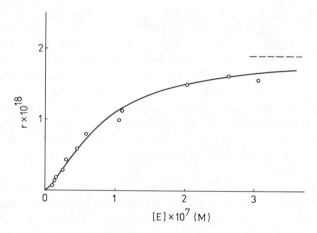

Fig. 2. Glyceraldehyde-3-phosphate dehydrogenase binding to human erythrocyte membranes; the dependence of the amount of bound enzyme (r is moles of glyceraldehyde-3-phosphate dehydrogenase bound per ghost) on equilibrium concentration of free enzyme. The points correspond to experimental data (Yu and Steck, 1975). The solid line is calculated from Eq. (6) with $K_1 = 4.7 \times 10^6 \ M^{-1}$, $K_2 = 34 \times 10^6 \ M^{-1}$, and $r_{max} = 1.90 \times 10^{-18}$ mol enzyme per ghost. Dotted line corresponds to the value of r_{max}.

ghost is equal to 1.16×10^6 and coincides with the number of Band 3 polypeptides in one ghost, $1.0–1.2 \times 10^6$, according to Fairbanks et al. (1971) and Steck (1974).

The cooperative character of the binding isotherms is observed also for adsorption of glyceraldehyde-3-phosphate dehydrogenase to reticulocyte membranes (Bohnensack and Letko, 1977), for adsorption of glucose-6-phosphate dehydrogenase (EC 1.1.1.49) to erythrocyte membranes (Benatti et al., 1978), and for adsorption of fructose-1,6-bisphosphate aldolase to microsomal membrane fraction from rat brain (Clarke and Masters, 1975b).

The binding of fructose-1,6-bisphosphate aldolase to the inner surface of erythrocyte membranes occurs, as in the case of glyceraldehyde-3-phosphate dehydrogenase, with the participation of Band 3 protein. However, experimental data obtained by Strapazon and Steck (1977) indicate the applicability of a simple scheme of adsorption of the enzyme to equivalent and noninteracting enzyme-binding sites, since the Scatchard plot is linear (Fig. 3). The adsorption capacity of erythrocyte membranes is equal to 1.2×10^6 tetrameric molecules of aldolase per ghost. The obtained value of adsorption capacity means that the dimeric enzyme binding site formed by Band 3 polypeptides attaches two molecules of aldolase. One can suppose that there are no interactions between adsorbed aldolase molecules.

The isotherm of binding of catalase (EC 1.11.1.6) to the inner surface of erythrocyte membrane also follows a hyperbolic curve (Aviram and Shaklai, 1981).

Consider now the adsorption of the enzymes to structural proteins of skeletal muscle. Arnold and Pette (1968, 1970) and Arnold et al. (1971) have shown that fructose-1,6-bisphosphate aldolase, glyceraldehyde-3-phosphate

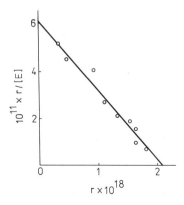

Fig. 3. Scatchard analysis of fructose-1,6-bisphosphate aldolase binding to human erythrocyte membranes (5 mM sodium phosphate, pH 7.0; 0°C). [E] is the molar concentration of free enzyme. Dimension of r is moles of the enzyme per ghost. (Reprinted with permission from Strapazon and Steck, 1977, copyright, 1977, American Chemical Society.)

dehydrogenase, pyruvate kinase, and some other enzymes of glycolysis are bound strongly to F-actin. F-Actin is known to consist of two filaments formed by molecules of G-actin, each with a molecular mass of 46,000. These two filaments are united in a superhelix. It may be assumed that the binding of any glycolytic enzyme in the groove of the superhelix would be energetically most favorable, since such a mode of enzyme attachment provides the maximum number of contacts between the enzyme molecule and G-actin monomers. Each of the enzyme molecules adsorbed in such a manner may have two neighbors at most. In order to describe quantitatively the isotherms of adsorption of the enzymes to F-actin, one can use the approaches elaborated by Schwarz (1970) and McGhee and von Hippel (1974) for the analysis of the binding of the ligand to linear polymers. These authors have taken into account the possibility of interactions between adsorbed molecules of the ligand.

Arnold and Pette (1968) have observed that low concentrations of fructose-1,6-bisphosphate aldolase are strongly bound by F-actin. Further binding of the enzyme occurs with lower affinity. The authors have suggested that such a binding isotherm is related to the presence of two nonidentical enzyme binding sites in F-actin. It should be noted that other explanations of the isotherm of binding of aldolase to F-actin are possible. Negative cooperative interactions between adsorbed enzyme molecules will result in a decrease of the affinity of F-actin as the latter becomes saturated with aldolase. In addition, binding isotherms that show a delay in achieving the limiting value of r at saturating enzyme concentrations would be observed for the binding of noninteracting ligands, if the bound ligand molecule covered (i.e., made inaccessible to another ligand) n consecutive residues of linear polymer (McGhee and von Hippel, 1974).

Chukhrai et al. (1973, 1976) were the first to analyze the character of the isotherms of protein adsorption while taking into account the association of protein molecules in the adsorbed state. These authors used experimental data on the adsorption of glucose-6-phosphate dehydrogenase to silica gel and the adsorption of hemoglobin to silica gel modified by cholesterol in order to illustrate the theoretical considerations.

C. Sensitivity of Adsorption to the Medium Composition

In order to explain the influence of the composition of the medium on the adsorption of the enzymes to the subcellular structures, we must realize that the binding of the enzymes to structural proteins of skeletal muscle and biological membranes is based on protein–protein interactions. To interpret the effects of neutral salts on the adsorption–desorption processes, one must consider the fact that salts weaken the interactions of charged groups of enzyme molecule and protein adsorbents by changing ionic strength and cause the desorption of the enzyme from the support. However, in addition to such nonspecific effects, there are striking specific influences of neutral salts on the conformation of protein molecules and protein–protein association. In accordance with their ability to stabilize or destabilize the native structure of proteins, anions and cations may be arranged in a series, first suggested by Hofmeister (Hofmeister's series or lyotropic series). This series is given below for collagen, with ions listed in order of increasing destabilizing effect (von Hippel and Schleich, 1969):

$$SO_4^{2-} < CH_3COO^- < Cl^- < Br^- < NO_3^- < ClO_4^- < I^- < SCN^- <$$
$$(CH_3)_4N^+ < NH_4^+ < Rb^+, K^+, Na^+, Cs^+ < Li^+ < Mg^{2+} < Ca^{2+} < Ba^{2+}.$$

The left-hand members of these series stabilize the native form of proteins. On the contrary, the right-hand members of the series cause the destabilization of native structure and the breakdown of protein–protein associates.

The analysis of experimental data, on the desorption of enzymes from the membranes and structural proteins of skeletal muscle under the action of neutral salts, shows that the character of the desorption curves may be explained satisfactorily if one takes into account the position of the salt in the lyotropic series. For example, as one can expect, the solubilizing efficiency of anions with respect to mitochondria-bound hexokinase increases in the order $Cl^- < NO_3^- < I^- < SCN^-$. According to Dagher and Hultin (1975) the ability of cations to elute glyceraldehyde-3-phosphate dehydrogenase from the particulate fraction of chicken breast muscle decreases in the order $Ca^{2+} > Mg^{2+} > K^+ > Na^+$.

When analyzing the influence of pH on the adsorption of enzymes to subcellular structures, it should be borne in mind that the pH profile of the degree of enzyme adsorption is based on the sensitivity of protein–protein interactions to the change in hydrogen ion concentration. The corresponding experimental data show that the breakdown of protein associates occurs at highly acid and highly alkaline pH values. The main reason for the dissociation of protein oligomers under such conditions is the mutual repulsion of the (probably charged) polypeptide chains.

The analysis of experimental data characterizing the influence of pH on the firmness of binding of the enzymes to subcellular structures shows that the transition from the slightly acid region of pH to the slightly alkaline side is accompanied, as a rule, by a decrease in the proportion of bound enzyme. Such a pattern has been observed for glyceraldehyde-3-phosphate dehydrogenase (Wilson *et al.*, 1982), fructose-1,6-bisphosphate aldolase (Strapazon and Steck, 1977; Wilson *et al.*, 1982), catalase (Aviram and Shaklai, 1981), and phosphofructokinase (EC 2.7.1.40) (Karadsheh and Uyeda, 1977) adsorbed by erythrocyte membrane and for glyceraldehyde-3-phosphate dehydrogenase

(Arnold *et al.*, 1971), fructose-1,6-bisphosphate aldolase (Arnold and Pette, 1968; Arnold *et al.*, 1971), pyruvate kinase (Arnold *et al.*, 1971), and adenylosuccinate synthetase (EC 6.3.4.4) (Ogawa *et al.*, 1978) adsorbed by F-actin. The single discrepancy is the case of the adsorption of hexokinase to mitochondria, in which the increase in pH from slightly acid values to slightly alkaline ones results in an increase in the proportion of the bound enzyme (Rose and Warms, 1967; Borrebaek, 1970; Alexakhina *et al.*, 1973). It is curious that the pH profile of the degree of enzyme desorption acquires the shape typical of other adsorptive enzyme systems if hexokinase is eluted by a specific ligand, glucose 6-phosphate (Rose and Warms, 1967).

D. Kinetics of Adsorption–Desorption Processes

To study the time factor in adsorption–desorption processes one uses two different experimental procedures. The first one is the application of high concentrations of eluting agent, which causes the *total* desorption of the enzyme from the support. The desorption process follows an exponential law. This was concluded on the basis of the analysis of experimental data on the kinetics of the desorption of hexokinase from mitochondrial membranes by 1 mM glucose 6-phosphate (Wilson, 1973; $k_1 = 5.3 \times 10^{-3}$ s^{-1}; 25°C) and the desorption of phosphofructokinase from the particulate fraction of chicken breast muscle by 5 mM ATP (Chism and Hultin, 1977; $k_1 = 4.6 \times 10^{-2}$ s^{-1}; pH 6.7; 25°C). It should be noted that such an experimental procedure gives information only about the stages of the dissociation of the enzyme from the support; it does not permit estimation of the reaction rate constants for the stages of the attachment of the enzyme to the support. The complete characterization of adsorption and desorption stages may be achieved by investigation of the kinetics of enzyme binding to the depleted support. Such an experimental procedure has been used, for example, by Wilson *et al.* (1982), who studied the kinetics of the adsorption of glyceraldehyde-3-phosphate dehydrogenase and fructose-1,6-bisphosphate aldolase to erythrocyte membranes at 0°C. The binding process is characterized by two peculiarities. First, the equilibration process is rather slow (the amount of the bound enzyme does not reach the limiting value even 4 h after the moment of mixing the components). Second, binding of both enzymes occurs in (at least) two kinetically distinguishable phases, with a rapidly equilibrating and a slowly equilibrating process being noted.

E. Combined Adsorption of Enzymes

Consider the situation when one studies the simultaneous binding of several enzymes to the support. If two enzymes compete for binding on the same site, the binding of one of the enzymes will be decreased in the presence of the other. Such a pattern is observed for the binding of glyceraldehyde-3-phosphate dehydrogenase and fructose-1,6-bisphosphate aldolase to erythrocyte membranes (Solti and Friedrich, 1976; Wilson *et al.*, 1982) and also for binding of hexokinase and glycerol kinase to mitochondria (Rose and Warms, 1967; Fiek *et al.*, 1982). The adsorption of catalase to erythrocyte membranes is suppressed by fructose-1,6-bisphosphate aldolase and glyceraldehyde-3-phosphate dehydrogenase (Aviram and Shaklai, 1981).

Situations are known in which the binding of one enzyme becomes possible after loading of the support by the other. For example, Bronstein and Knull (1981) have showed that bisphosphoglyceromutase (EC 5.4.2.1) is adsorbed to the F-actin–tropomyosin complex only if the support contains bound lactate dehydrogenase. The binding of bisphosphoglyceromutase is explained, perhaps, by its interaction with lactate dehydrogenase. This example demonstrates that adsorbed enzymes may form enzyme–enzyme complexes whose formation does not demand the interaction of each component with the support. (See also Friedrich, Chapter 3.)

V. CHANGES IN CATALYTIC AND REGULATORY PROPERTIES OF ENZYMES BY THEIR ADSORPTION

Adsorption of enzymes to subcellular structures is accompanied, as a rule, by changes in catalytic and regulatory characteristics of the enzymes. I shall consider the main causes of change in activity of the enzyme engendered by its adsorption.

1. The adsorption of the enzyme may result in steric screening of active sites with respect to substrate. It is evident that in this case an adsorbed enzyme loses the ability to catalyze the conversion of the substrate.

2. It should be borne in mind that the enzyme in the solution and the adsorbed enzyme have different microenvironments (Goldstein *et al.*, 1964; Goldstein, 1972, 1976; Maurel and Douzou, 1978; Wojtczak and Nałęcs, 1979; Ricard *et al.*, 1981). Owing to the presence of charges on the surface, the support acquires the properties of a polyelectrolyte. The summed charge on the surface of biological membranes is, as a rule, negative. The concentration of hydrogen ions near the polyelectrolyte differs from that in solution. Therefore, the adsorption of the enzyme to the polyelectrolyte surface must result in a displacement of the pH optimum for enzymatic activity. For example, the pH profile of the enzyme adsorbed to a polyanion (i.e., the dependence of the velocity of enzyme reaction on the value of pH in the solution) is displaced in the direction of higher values of pH in comparison with the pH profile for free enzyme. The interactions of ionic forms of the

substrate with the charged support may affect the kinetic characteristics revealed by adsorbed enzyme (Engasser and Horvath, 1975; Ricard et al., 1981). For example, like charges on substrate and support would lead to a decreased substrate concentration in the microenvironment of the adsorbed enzyme and an increase in the apparent K_m.

3.   Changes in catalytic properties of the enzyme by its adsorption may be caused by the conformational transitions of the enzyme molecule induced by binding to the support (the support may preferentially bind one of the conformational states of the enzyme).

4.   The adsorption of the enzyme to subcellular structures may be accompanied by a change in the oligomeric state of the enzyme. One can expect that bringing together the molecules of adsorbed enzyme will favor their association. It is well known that a change in oligomeric state of enzymes in solution affects, as a rule, enzymatic activity (see Kurganov, 1982, pp. 165–176). The same pattern must be probable for adsorbed enzymes. Therefore a list of factors determining the catalytic efficiency of adsorbed enzymes must include the association–dissociation processes of adsorbed enzyme molecules. It should be noted that the problem of association of the enzymes in an adsorption layer as a factor determining the activity of adsorbed enzyme was stated first by Poltorak in 1967. Poltorak and his colleagues (Poltorak, 1967; Poltorak and Chukhrai, 1970; Poltorak et al., 1977; Poltorak and Pryakhin, 1977; Chukhrai, 1979) analyzed in detail the influence of adsorption of enzyme molecules in an adsorption layer on dependence of specific enzyme activity relative to the degree of the occupation of the support surface.

Consider some examples that illustrate the change in catalytic properties of the enzymes by their adsorption. Tsai et al. (1982) showed that the binding of glyceraldehyde-3-phosphate dehydrogenase to erythrocyte membranes is accompanied by a decrease in catalytic activity. The degree of inhibition corresponds strictly to the proportion of bound enzyme (Fig. 4). Therefore the authors concluded that the bound enzyme does not participate in catalytic activity. In order to study the nature of inhibition of glyceraldehyde-3-phosphate dehydrogenase by its adsorption in detail, the 23,000-Da N-terminal fragment of Band 3 protein was cleaved following S-cyanylation and purified in a water-soluble form. A kinetic analysis of the inhibitory effect of the 23,000-Da fragment on the catalytic activity of glyceraldehyde-3-phosphate dehydrogenase has shown that the fragment appears to be competitive with NAD and arsenate, but noncompetitive with glyceraldehyde-3-phosphate. Analogous results have been obtained with nonfragmented Band 3 protein released into solution by treatment of erythrocyte membranes with Triton X-100.

According to Strapazon and Steck (1977), upon binding to either whole erythrocyte membranes, solubilized Band 3, or proteolytic fragments from the cytoplasmic surface part of Band 3, fructose-1,6-bisphosphate aldolase underwent a profound loss of catalytic activity.

Bustamante and Pedersen (1980) have compared the kinetic properties of mitochondria-bound and solubilized forms of hexokinase. These enzyme forms are characterized by similar affinity with respect to one of the substrates, glucose, whereas the bound enzyme reveals much higher affinity with respect to a second substrate, MgATP (K_m = 0.25 mM), in comparison with the free enzyme (K_m = 1.2 mM). Analogous conclusions were drawn earlier by Wilson (1968) on the basis of the analysis of published data. In addition, Karpatkin (1967) and Tuttle and Wilson (1970) have reported that the bound form of hexokinase is characterized by higher sensitivity to product inhibition by glucose-6-phosphate than is the solubilized enzyme.

Consider the change in catalytic characteristics of fructose-1,6-bisphosphate aldolase by its adsorption to structural proteins of skeletal muscle. According to Arnold and Pette (1970) binding of aldolase to F-actin doubles V_{max} and increases K_m for fructose 1,6-bisphosphate by almost one order of magnitude. The same type of changes in V_{max} and K_m has been observed by Walsh et al. (1977). These authors used both pure F-actin and the

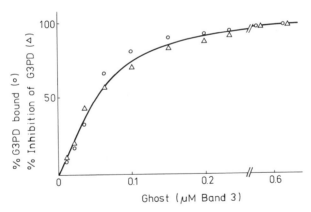

Ghost (μM Band 3)

Fig. 4. Catalytic activity of membrane-bound glyceraldehyde-3-phosphate dehydrogenase (G3PD). Ghosts were freed of endogenous enzyme and suspended at 0 to 2 × 10^8 ghosts/ml in a solution of 0.08% bovine serum albumin, 1 × 10^{-4} M NAD$^+$, 1 × 10^{-4} M sodium arsenate, 1 × 10^{-2} M imidazole acetate, 1 × 10^{-2} M sodium acetate, 10^{-4} M Na$_4$EDTA, 1 × 10^{-3} M dithiothreitol, and 6 × 10^{-8} M human erythrocyte enzyme (final concentrations). The mixture was equilibrated for 60 min on ice and then 10 min at room temperature. (○) Enzyme activity was assayed periodically at room temperature following the addition of 1 × 10^{-4} M glyceraldehyde-3-phosphate. (△) Aliquots of the same complete reaction mixtures were centrifuged periodically at 18,000 rpm for 5 min at 25°C and the supernatant fractions assayed for total enzyme at saturating levels of substrates. (Reprinted with permission from Tsai et al., 1982.)

filaments obtained from association of F-actin, tropomyosin, and troponin as adsorbents for aldolase. The study of the adsorption of aldolase to the complexes of F-actin, tropomyosin, and troponin is of special interest, since the thin filaments of isotropic bands of the muscle fiber, where (according to histochemical data) the glycolytic enzymes are localized, contain not only F-actin, but also regulatory proteins, troponin (which reveals high affinity to Ca^{2+} ions), and tropomyosin. The F-actin–tropomyosin–troponin complex provides a substantially more accurate analog *in vitro* of the thin filament of skeletal muscle than does F-actin. Walsh *et al.* (1977) showed that the kinetic characteristics of aldolase adsorbed to F-actin–tropomyosin–troponin filaments become sensitive to the presence of Ca^{2+}.

The K_m value for fructose 1,6-bisphosphate for aldolase bound to microsomal membrane fraction from rat liver was decreased from 3×10^{-4} to 6×10^{-6} M when the enzyme was dissociated from the membranes with salt (Weiss *et al.*, 1981). The value of V_{max} was unchanged by this manipulation.

Determination of kinetic parameters for adsorbed enzyme is normally accomplished as follows. One usually assumes Michaelis–Menten kinetics for the free enzyme. The corresponding linear anamorphoses (for example, the anamorphosis of Lineweaver–Burk) may be used for estimation of kinetic parameters. One constructs such linear anamorphoses for the enzyme reaction rates measured at different fixed values of the concentration of the adsorbent (Ω) and determines the apparent kinetic parameters at each value of $[\Omega]$. The procedure of extrapolation of the apparent kinetic parameters to infinitely high adsorbent concentration is used for calculation of V_{max} and K_m for adsorbed enzyme. In general, determining kinetic parameters in this manner for bound enzyme is *incorrect*, because the dependence of the enzyme reaction rate on substrate concentration does not follow Michaelis–Menten kinetics if there are both free and bound enzyme forms in the system.[1]

The *correct* determination of kinetic parameters for adsorbed enzyme may be carried out as follows. One determines the enzyme reaction rate at varying adsorbent concentrations and a fixed value of substrate concentration. The limiting value of the enzyme reaction rate observed at sufficiently high adsorbent concentrations corresponds to the value of the enzyme reaction rate for adsorbed enzyme. If we have the set of the limiting values of the enzyme reaction rate obtained at various fixed concentrations of the substrate, the kinetic parameters for bound enzyme may be calculated. If Michaelis–Menten

---

[1] The theoretical analysis shows that Michaelis–Menten kinetics can apply in the presence of adsorbent if the bound enzyme form is catalytically inactive and the equilibrium between bound and free enzyme forms is attained very slowly in comparison with the rate of enzyme reaction. However, it is evident that in this case one does not attempt to determine the kinetic parameters for adsorbed enzyme.

kinetics are fulfilled, the corresponding linear anamorphoses may be used for this computation.

Consider as an example the influence of F-actin on catalytic properties of lactate dehydrogenase. Sugrobova *et al.* (1983) showed that the addition of F-actin results in a decrease in catalytic activity of isoenzyme M_4 of lactate dehydrogenase. The authors utilized the poor substrate α-ketoglutarate in order to obtain the measurable values of the enzyme reaction rate at the high enzyme concentrations used in the experiments. The enzyme reaction rate levels off at sufficiently high concentrations of F-actin (Fig. 5A). The limiting values of the enzyme reaction rate have been used for the calculation of the kinetic parameters for adsorbed lactate dehydrogenase. Michaelis–Menten kinetics apply to both free and bound enzyme forms. The binding of lactate dehydrogenase to F-actin is deleterious to its catalytic characteristics: the value of V_{max} decreases 1.6-fold (from 2.4 to 1.5 μmol of NADH/liter min) and the value of K_m increases 2.5-fold (from 2 to 5 μM) with respect to the corresponding parameters for free enzyme (Fig. 5B).

As is known, many enzymes possessing a quaternary structure perform regulatory functions in the intact cell. The regulatory properties of oligomeric enzymes are provided by the sensitivity of the catalytic activity of the enzyme to definite metabolites–regulators and by the existence of interactions between spatially distinct active and regulatory sites in enzyme oligomer (see Ricard, Chapter 4). The latter are manifested by a distinctive pattern in the

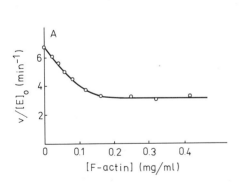
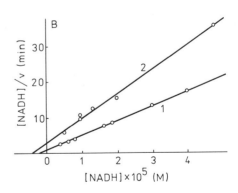

Fig. 5. The influence of F-actin on enzyme activity of pig lactate dehydrogenase (isoenzyme M_4) (Sugrobova *et al.*, 1983). (A) The dependence of specific enzyme activity of lactate dehydrogenase on F-actin concentration (10 mM Tris acetate, pH 6.0, 20°C, 0.12 mM NADH, 0.86 mM α-ketoglutarate, and 0.60 μM lactate dehydrogenase). (B) The dependence of the initial rate of enzymatic reaction on NADH concentration for free lactate dehydrogenase (line 1) and for the enzyme bound by F-actin (line 2) in the coordinates ([NADH]/v; [NADH]) (0.86 mM α-ketoglutarate and 0.25 μM lactate dehydrogenase).

kinetic behavior of oligomeric enzymes (see Kurganov, 1982). It is important to emphasize that the regulatory characteristics of enzymes may be changed by adsorption to a support. For example, Karadsheh and Uyeda (1977) have shown that the binding of phosphofructokinase (this enzyme plays an important role in the regulation of glycolysis) to erythrocyte membranes is accompanied by the disappearance of the S-shaped character of the dependence of the enzyme reaction rate on the substrate (fructose 6-phosphate) concentration and a decrease in the sensitivity of the enzyme to inhibition by 2,3-bisphosphoglycerate and high concentrations of ATP (Fig. 6). According to Liou and Anderson (1980) the adsorption of phosphofructokinase to F-

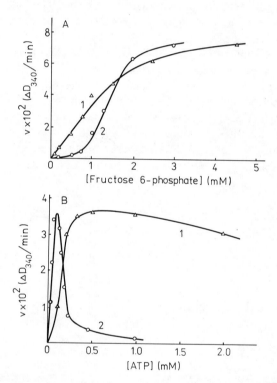

Fig. 6. The changes in activity of phosphofructokinase following its adsorption to erythrocyte membrane. (A) The dependence of the rate of enzymatic reaction (v) catalyzed by membrane-bound human erythrocyte phosphofructokinase (curve 1) and free phosphofructokinase (curve 2) on fructose 6-phosphate concentration (50 mM Tris-HCl buffer, pH 7.5, 1 mM ATP, 3 mM MgCl$_2$, 25°C). (B) The dependence of the rate of enzymatic reaction (v) catalyzed by membrane-bound phosphofructokinase (curve 1) and free enzyme (curve 2) on ATP concentration (0.2 mM fructose 6-phosphate). (Reprinted with permission from Karadsheh and Uyeda, 1977, copyright, 1977, American Society of Biological Chemists, Inc.)

actin and reconstituted thin filaments diminishes the sensitivity of the enzyme to inhibition by citrate and excess ATP.

Southard and Hultin (1972) have observed rapid decrease of the activity of mitochondria-bound hexokinase over the range of adenylate energy charge[2] from 1.0 to 0.8. Over this range, there was relatively little effect on the soluble enzyme.

AMP aminohydrolase (EC 3.5.4.6) adsorbed to subfragment 2 of myosin revealed less sensitivity to inhibition by GTP than did free enzyme (Ashby and Frieden, 1978). Shiraki et al. (1979) observed the appearance of the S-shaped character of the dependence of the enzyme reaction rate catalyzed by AMP aminohydrolase on AMP concentration (in the presence of 50 mM KCl) accompanying the adsorption of the enzyme to myosin.

The change in regulatory characteristics of the enzyme by its adsorption to the support may be caused by the same factors as the change in catalytic properties. For example, the steric screening of regulatory sites (metabolite–regulator binding sites) will result in the decrease (or disappearance) of the sensitivity to the action of metabolites–regulators. If regulatory behavior of the enzyme is determined by the existence of a mobile equilibrium between two conformational states of the enzyme having different affinities with respect to substrates and metabolites (as has been postulated in the model proposed by Monod et al., 1965), the preferential binding of one of the conformational states will change the regulatory characteristics of the enzyme (Kurganov and Loboda, 1979). In this case the support acts as "macroeffector."

VI. INFLUENCE OF SPECIFIC LIGANDS ON ADSORPTION OF ENZYMES

Numerous experimental data show that the adsorption of enzymes to subcellular structures may be regulated by cellular metabolites. Hexokinase is desorbed from the mitochondria under the influence of glucose 6-phosphate, which is the product of the enzymatic reaction catalyzed by this enzyme (Rose and Warms, 1967; Wilson, 1968, 1973; Felgner and Wilson, 1977; Fiek et al., 1982). ATP (inhibitor of hexokinase) elicits an analogous action (Rose and Warms, 1967; Wilson, 1968; Hochman and Sacktor, 1973; Bustamante and Pedersen, 1980). A substrate of the hexokinase reaction, MgATP, is not able to solubilize bound hexokinase.

---

[2] The concept of adenylate energy charge was introduced by Atkinson (see Atkinson, 1977). The following ratio of concentrations of adenine nucleotides corresponds to this term: ([ATP] + [ADP]/2)/([ATP] + [ADP] + [AMP]).

De Meis and Rubin (1970) observed the desorbing effect of ATP with respect to AMP aminohydrolase bound by microsomal membranes from rabbit skeletal muscles (in tests of the enzyme activity, ATP acts as an activator).

Glyceraldehyde-3-phosphate dehydrogenase is solubilized from erythrocyte membranes by NADH (Kant and Steck, 1973; McDaniel and Kirtley, 1975; Kliman and Steck, 1980; Keokitichai & Wrigglesworth, 1980) or inorganic phosphate (Wilson et al., 1982). If glyceraldehyde-3-phosphate dehydrogenase is adsorbed by F-actin, its substrate, glyceraldehyde 3-phosphate, may act as a desorbing agent (Arnold and Pette, 1970). The same action of glyceraldehyde 3-phosphate was observed by Dagher and Hultin (1975) for glyceraldehyde-3-phosphate dehydrogenase associated with the particulate fraction of chicken skeletal muscle.

Arnold and Pette (1970) investigated the influence of fructose 1,6-bis-phosphate on the strength of binding of fructose-1,6-bisphosphate aldolase to F-actin and showed that the substrate hinders adsorption of the enzyme. An analogous result was obtained by Clarke and Masters (1975a) in experiments in which the F-actin–tropomyosin–troponin complex was used as an adsorbent for fructose-1,6-bisphosphate aldolase. Desorbing action of fructose 1,6-bisphosphate with respect to bound fructose-1,6-bisphosphate aldolase has been observed also for the enzyme associated with erythrocyte membranes (Strapazon and Steck, 1977), the microsomal membrane fraction from rat liver (Weiss et al., 1981), and the particulate fraction of rat brain homogenate (Clarke and Masters, 1972).

For all situations discussed above, metabolites have a desorbing action on the enzyme distribution between free and bound forms. Cases are known in which a metabolite shifts this distribution toward bound enzyme. For example, ADP induces the binding of phosphofructokinase to mitochondrial membrane (Craven and Basford, 1974).

Consider some examples that illustrate the influence of specific ligands on the adsorption of the enzymes. Figure 7 shows the desorption of fructose-1,6-bisphosphate aldolase from F-actin by ATP (Arnold and Pette, 1970). It is curious that the dependence of the proportion of bound enzyme on ATP concentration has an S-shaped form. Desorption curves having such a form have been observed also for the solubilization of fructose-1,6-biphosphate aldolase from erythrocyte membranes by fructose 1,6-bisphosphate, 2,3-bisphosphoglycerate, and ATP (Strapazon and Steck, 1977), from microsomal membranes by fructose 1,6-bisphosphate and glucose 1,6-bisphosphate (Weiss et al., 1981), and for solubilization of glyceraldehyde-3-phosphate dehydrogenase from erythrocyte membranes by NADH (in the presence of NAD) (McDaniel and Kirtley, 1975).

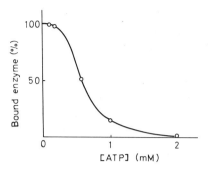

Fig. 7. Effect of ATP on the binding of fructose-1,6-bisphosphate aldolase to F-actin (pH 6.9, 30°C). (Reprinted with permission from Arnold and Pette, 1970.)

Figure 8 shows the desorption of fructose-1,6-bisphosphate aldolase from F-actin under the influence of its substrate, fructose 1,6-bisphosphate, at various fixed concentrations of the enzyme (Arnold and Pette, 1970). The desorption curves have no inflection points. The experimental data obtained by Arnold and Pette are of great interest in that they demonstrate clearly that desorption of the enzyme by specific ligands depends on the concentration of the protein in the system, an increase in protein concentration diminishing the desorbing effect of the specific ligand. This result may be predicted easily from the analysis of different schemes of enzyme adsorption (for example, the simplest scheme of the type $E + \Omega \rightleftarrows E\Omega$), but experimental data that demonstrate the character of the influence of protein concentration on the

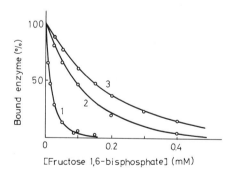

Fig. 8. Effect of fructose 1,6-bisphosphate on the binding of fructose-1,6-bisphosphate aldolase to F-actin (pH 6.9; 30°C). F-actin (1.25 mg/ml) was incubated with different concentrations of the enzyme (mg/ml): curve 1, 0.01; curve 2, 0.16; and curve 3, 0.20. (Reprinted with permission from Arnold and Pette, 1970.)

enzyme desorption by specific ligands are not numerous. In addition to the work by Arnold and Pette (1970) we would point out only the works by Melnick and Hultin (1968) and Ratner *et al.* (1974), who have shown that the increase in the concentration of the particulate fraction of homogenized chicken breast muscle is accompanied by a decrease of the solubilizing action of NADH on bound lactate dehydrogenase, and the work by Clarke and Masters (1975a), who have observed that high concentrations of the glycolytic enzymes favor the higher degree of their adsorption to F-actin and F-actin–tropomyosin–troponin complex.

What primary information can one obtain from the analysis of the dependence of the proportion of bound enzyme on specific ligand concentration registered at different fixed concentrations of the enzyme and the adsorbent? We assume that the binding of the enzyme to the adsorbent is accompanied by the steric screening of at least one of the ligand-binding sites in the enzyme molecule. In this case sufficiently high concentrations of specific ligand must cause the total desorption of the enzyme from the adsorbent. It should be noted that just the same situation takes place in all experiments in which the desorbing action of specific ligands has been studied. Consider the case in which the adsorption of the enzyme is accompanied by steric screening of one ligand-binding site in the enzyme molecule (the simplest situation is the adsorption of monomeric enzyme). The dependence of the proportion of adsorbed enzyme on specific ligand concentration $[L]$ will have no inflection points. The increase in enzyme concentration $[E]_0$ or enzyme-binding sites $[\Omega]_0$[3] must result in enhancement of the value $[L]_{0.5}$ (the ligand concentration at which the relative proportion of adsorbed enzyme is half decreased). Our anlaysis shows that at sufficiently high enzyme concentrations $[E]_0$ (or $[\Omega]_0$) the value of $[L]_{0.5}$ becomes proportional to $[E]_0$ (or $[\Omega]_0$). If the adsorption of the enzyme is accompanied by steric screening of n ligand-binding sites in the enzyme molecule ($n \geq 2$), the dependence of the proportion of bound enzyme on specific ligand concentration may acquire the S-shaped form. The value of $[L]_{0.5}$ becomes proportional to $[E]_0^{1/n}$ (or $[\Omega]_0^{1/n}$) at sufficiently high enzyme concentrations $[E]_0$ (or $[\Omega]_0$). Thus the analysis of the dependence of $[L]_{0.5}$ on $[E]_0$ (or $[\Omega]_0$) allows one to estimate the number of ligand-binding sites that are sterically screened by the adsorption of the enzyme molecule.

As one can see from the material discussed above, the degree of adsorption of enzymes to subcellular structures depends on pH, the presence of neutral salts and specific ligands, and the concentrations of the enzyme and the enzyme-binding sites. Furthermore, the binding of enzymes may be influenced by the presence of other enzymes. It is not easy to discover all the factors that determine the strength of adsorption of a given enzyme to subcellular

---

[3] One can change $[E]_0$ and $[\Omega]$ simultaneously at the constant ratio $[E]_0/[\Omega]$.

structure. Therefore it is difficult to draw a reliable conclusion about the degree of adsorption of the enzymes under physiological conditions from experiments carried out *in vitro*. Clarke and Masters (1975a), for example, have noted the fact that glycolytic enzymes do not interact with F-actin or the F-actin–tropomyosin–troponin complex under physiological ionic strength conditions if the concentrations of the enzymes are rather low. This result is in accordance with the observations of other authors. However, the increase in protein concentration permits high degrees of enzyme adsorption even under physiological ionic-strength conditions.

VII. PHYSIOLOGICAL SIGNIFICANCE OF REVERSIBLE ADSORPTION OF ENZYMES

Consider the role of the processes of reversible adsorption of the enzymes to subcellular structures in the regulation of cellular metabolism. First, I should like to emphasize once again the following important peculiarities of adsorptive enzyme systems with natural adsorbent: (a) the reversibility of the processes of enzyme adsorption–desorption; (b) the change in catalytic and regulatory properties of the enzyme triggered by its transition to an adsorbed state; and (c) the sensitivity of the equilibrium between free and bound enzyme forms to the presence of specific ligands.

When the process of binding of the enzyme to subcellular structure reveals the above characteristics we can say that an adsorptive mechanism of regulation of enzyme activity is operative. This mechanism of regulation is based on the metabolite controlled displacement of the equilibrium between free and adsorbed enzyme forms having different catalytic and regulatory properties.

The idea that the reversible binding of the enzymes to subcellular structures has regulatory significance was first suggested by Oparin (1934) (see also Kursanov, 1940). Analogous notions were developed later by Siekevitz (1959), Margreth *et al.* (1963), Wilson (1968, 1978), Masters *et al.* (1969), Hultin *et al.* (1972), Alexakhina *et al.* (1973), Hultin (1975), Clarke and Masters (1976), Masters (1977, 1978), Kurganov *et al.* (1978), Kurganov and Loboda (1979), Poglazov *et al.* (1982), and Kurganov (1982, pp. 316–320). The experimental material presented up to the present time allows us to consider the reversible adsorption of enzymes to subcellular structures as one of the important mechanisms providing the regulation of enzyme activity in the living cell.

The activity of enzymes may be regulated in different ways. Some mechanisms of enzyme regulation provide unusual kinetic characteristics and expand significantly the regulatory potentialities of the enzyme. These mechanisms include allosteric mechanisms, wherein the change in catalytic efficiency of the active site results from the conformational change in the enzyme molecule induced by metabolites–regulators [see Kurganov (1982, Ch. 3) and also Ricard, Chapter 4], and the dissociative mechanism, wherein the change in catalytic properties of the enzyme results from the change in the oligomeric state of the enzyme (see Kurganov, 1982, Ch. 4)]. The striking kinetic anomaly of allosteric and dissociative enzyme systems is the S-shaped character of the dependence of the enzyme reaction rate on substrate or metabolite–regulator concentration. The calculation of Hill's coefficient (n_H) for such dependence gives values that exceed unity. Such a kinetic dependence provides higher sensitivity of enzyme activity to variation of substrate or metabolite–regulator concentration in comparison with the hyperbolic pattern (Kurganov, 1982, p. 68). The opposite kinetic anomaly manifested by the enzyme carrying regulatory functions is the case of reduced sensitivity of enzyme reaction rate to the variation of substrate or metabolite–regulator concentration (the case in which Hill's coefficient is less than unity). For the curve of the dependence of the enzyme reaction rate on substrate concentration, this kinetic anomaly is manifested in a time-lag in reaching the limiting value of the velocity of enzyme reaction at saturating concentrations of the substrate.

The theoretical analysis carried out by Kurganov and Loboda (1979) has shown that the adsorptive enzyme systems are characterized by the same kinetic anomalies as allosteric and dissociative enzyme systems. Consider the adsorptive enzyme system that is represented by a *monomeric* enzyme interacting reversibly with the support. We assume that the binding of the enzyme is accompanied by steric screening of the substrate-binding site and that the equilibrium between free and bound enzyme forms is attained very quickly in comparison with the rate of enzymatic process. In adsorptive enzyme systems under discussion the reduced sensitivity of the enzyme reaction rate to the variation of substrate concentration may be revealed (the dependence of v versus $[S]_0$ shows a delay in reaching the limiting value of the enzyme reaction rate at saturating substrate concentrations; $n_H < 1$). However, in adsorptive enzyme systems in which the adsorption of oligomeric molecules of the enzyme to the support is accompanied by steric screening of several substrate-binding sites, the *enhanced* sensitivity of the enzyme reaction rate to the variation of substrate concentration may be revealed (the S-shaped dependence of v versus $[S]_0$; $n_H > 1$).

As has been noted, in some cases the adsorption–desorption equilibrium is attained rather slowly, the values of the half-time reaching minutes and even tens of minutes. In adsorptive enzyme systems with slow attainment of the equilibrium between free and bound enzyme forms, one can expect the appearance of deviations from the linear accumulation of the product (Pr) of enzyme reaction in time. (The linear plot of [Pr] versus time is typical for the steady-state ongoing enzymatic process.) These deviations may be manifested as lag periods or bursts on the kinetic curves, and are connected with the transitions between free and bound enzyme forms toward more or less active enzyme forms, respectively. When activating or inhibiting actions of metabolites (particularly, substrates) in adsorptive enzyme systems are time-dependent processes we can talk about hysteretic properties of these systems, using the terminology proposed by Frieden (1970, 1979). Adsorptive enzyme systems whose activity changes relatively slowly when the concentration of the metabolite–regulator is changed are able to function as buffers, preventing rapid changes in the concentrations of other metabolites in a particular metabolic pathway and ensuring the constancy of the rates of interacting metabolic pathways that utilize a common initial metabolite. Such behavior of adsorptive enzyme systems is analogous to that of allosteric and dissociative enzyme systems in which slow changes in conformational state of the enzyme take place (Kurganov, 1982, Ch. 5; and Ricard, Chapter 4).

To sum up, the reversible binding of enzymes to subcellular structures provides additional possibilities of regulation of enzyme activity in the intact cell. However, important aspects of physiological significance of the reversible adsorption of the enzymes may be overlooked if we limit ourselves to the consideration of the adsorption of individual enzymes. The simultaneous adsorption of glycolytic enzymes to thin filaments of the I-band of muscle fiber must result in their being brought together. The local enhancement of the concentration of glycolytic enzymes in the adsorption layer favors the formation of enzyme–enzyme complexes. If complexes of the enzymes that catalyze the consecutive steps of a metabolic pathway are formed, an increase in the rate of overall process may occur because of diminished time of diffusion of intermediate products. We assume that the main physiological significance of adsorption of glycolytic enzymes to structural proteins of skeletal muscles is that this provides the conditions for bringing together these enzymes in complexes, rather than because it induces a simple change in catalytic and regulatory characteristics of individual enzymes. If one holds this opinion, attention must be paid to the functioning of the complete system of coupled reactions catalyzed by adsorbed glycolytic enzymes.

It is worth noting that Green et al. (1965) drew conclusions about the existence of multienzyme complexes of glycolytic enzymes attached to membranes on the basis of investigation of the enzymatic activities of membrane fractions of hemolyzed beef erythrocytes and broken cells of Saccharomyces cerevisiae. A similar conclusion was drawn by Tillmann et al. (1975).

Consider the physiological significance of reversible adsorption of the enzymes to biological membranes. At a first glance the situation is clear. The transition of the enzyme in membrane-bound form, which is controlled by metabolites, provides additional possibilities for regulation of enzyme activity in the intact cell, since the bound form is characterized by different catalytic properties in comparison with the free one. However, such understanding of the function of peripheral membrane-bound enzymes is simplified. As stated above, the specific character of interactions of peripheral enzymes with membranes is determined by the attachment of the enzymes to anchor proteins embedded in the membrane. In the case of glyceraldehyde-3-phosphate dehydrogenase and hexokinase the corresponding anchor proteins perform transport functions in the membrane. It is natural to suppose a functional linkage between an anchor protein that participates in transport processes and the enzyme attached to this protein (see Friedrich, Chapter 3). Such an idea was first put in distinct form by Singer (1974). According to Singer, oligomeric enzyme adsorbed to anchor protein creates a combined channel with the pore of the latter (Fig. 9). Such a combined transport system acquires important new properties. First, the transport system becomes specific for certain metabolites because of the participation of the peripheral membrane-bound enzyme. Secondly, the transport system under discussion may transfer metabolites against the gradient of the concentration, owing to the coupling of transport processes with an enzymatic reaction (active transport).

Several authors have suggested that the anion transport system of erythrocyte membranes is functionally coupled with the enzymatic process catalyzed by glyceraldehyde-3-phosphate dehydrogenase. For example, Chen

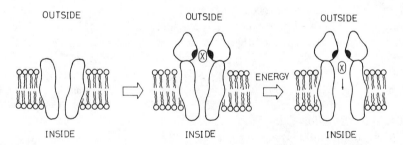

Fig. 9. A schematic mechanism for the translocation event in active transport, in which a periplasmic binding protein is obligatorily involved. It is proposed that there are present in the membrane subunit aggregates of a specific integral protein that span the membrane, forming a water-filled pore. This pore (left) is initially closed to the diffusion of molecules other than water. A binding protein with an active binding site for ligand X attaches specifically to the exposed surface of the integral protein (center), the pore remaining closed. Some energy-yielding step results in a quaternary rearrangement of the subunits of this structure (right), opening the pore and releasing X to the other side of the membrane. (Reproduced, with permission from Singer, 1974, © 1974 by Annual Reviews Inc.)

et al. (1981) believed that inorganic phosphate (one of the substrates of the enzyme reaction catalyzed by glyceraldehyde-3-phosphate dehydrogenase) reaches the active site of the enzyme through the anion transport system.

Investigations carried out by Gots *et al.* (1972), Gots and Bessman (1974), Gellerich and Augustin (1977), Inui and Ishibashi (1979), Bustamante and Pedersen (1980), and Augustin and Gellerich (1981) showed that mitochondria-bound hexokinase is able to use efficiently mitochondrially produced ATP. Such behavior of bound hexokinase provides the functional linkage between metabolic processes in cytosol and mitochondrion. Perhaps adsorbed hexokinase and mitochondrial porin create the combined channel used for specific transport of ATP.

The peripheral enzymes that participate in the transport of metabolites must have the following structural features. The peripheral enzyme must possess an oligomeric structure with an internal cavity where one molecule of the metabolite can be bound [as in the case of hemoglobin, which binds one molecule of 2,3-bisphosphoglycerate per tetrameric protein molecule (Arnone, 1972)], or its adsorbed form must associate to form a single metabolite-binding site in the region of the contact of the protein molecules (as in the case of yeast hexokinase, which forms associations with the creation of an ATP binding site located between adjacent enzyme monomers (Steitz *et al.*, 1977)].

Summing up the discussion of *the physiological significance* of the adsorption of the enzymes to subcellular structures, we see that, first, the adsorption of the enzymes provides the possibility of the regulation of enzyme activity by means of the displacement of the equilibrium between free and bound enzyme forms by metabolites; second, the binding of the enzymes to subcellular structures favors the formation of multienzyme complexes, in which preferential utilization of the products of consecutive stages is possible within the confines of the macromolecular assemblage (this results in the enhancement of the overall rate of metabolism); and third, specific complexing of the enzymes with membrane channel-forming proteins provides a functional linkage of metabolic processes that proceed in different compartments of the intact cell.

VIII. CONCLUSION

From what has been said, we feel it is possible to conclude that specific associations of enzymes with subcellular structures are based on protein–protein interactions. This circumstance allows us to apply a unique point of view to the explanation of the effects of changes in the composition of the medium (pH, neutral salts) on the strength of the binding of enzymes to subcellular structures.

When investigating the physiological significance of reversible adsorption of enzymes to subcellular structures, special attention must be paid to the formation of multienzyme complexes of adsorbed enzymes and to functional linkages between the enzymatic processes catalyzed by the peripheral enzyme and the transport of metabolites through the membrane.

REFERENCES

Alexakhina, N. V., Sitnina, N. Yu., and Scherbatykh. L. N. (1973). *Biokhimija* **38,** 915–921.
Arnold, H., and Pette, D. (1968). *Eur. J. Biochem.* **6,** 163–171.
Arnold, H., and Pette, D. (1970). *Eur. J. Biochem.* **15,** 360–366.
Arnold, H., Henning, R., and Pette, D. (1971). *Eur. J. Biochem.* **22,** 121–126.
Arnone, A. (1972). *Nature* (London) **237,** 146–149.
Ashby, B., and Frieden, C. (1978). *J. Biol. Chem.* **253,** 8728–8735.
Atkinson, D. E. (1977). "Cellular Energy Metabolism and its Regulation" pp. 88–93. Academic Press, New York.
Augustin, W., and Gellerich, F. N. (1981). *Acta Biol. Med. Germ.* **40,** 603–609.
Aviram, I., and Shaklai, N. (1981). *Arch. Biochem. Biophys.* **212,** 329–337.
Benatti, U., Morelli, A., Frascio, M., Melloni, E., Salamino, F., Sparatore, B., Pontremoli, S., and De Flora, A. (1978). *Biochem. Biophys. Res. Commun.* **85,** 1318–1324.
Bohnensack, R., and Letko, G. (1977). *Erg. Exp. Med.* **24,** 185–187.
Borrebaek, B. (1970). *Biochem. Med.* **3,** 485–497.
Bronstein, W. W., and Knull, H. R. (1981). *Can. J. Biochem.* **59,** 494–499.
Bustamante, E., and Pedersen, P. L. (1980). *Biochemistry* **19,** 4972–4977.
Carraway, K. L., and Shin, B. C. (1972). *J. Biol. Chem.* **247,** 2102–2108.
Chen, S. -M., Keokitichia, S., and Wrigglesworth, J. M. (1981). *Biochem. Soc. Trans.* **9,** 139–140.
Chism, G. W., and Hultin, H. O. (1977). *J. Food Biochem.* **1,** 75–90.
Chukhrai, E. S. (1979). *Vestnik Mosk. Univ. Ser. 2, Khim.* **19,** 235–248.
Chukhrai, E. S., and Poltorack, O. M. (1973). *Vestnik Mosk. Univ. Ser. 2, Khim.* **14,** 271–277.
Chukhrai, E. S., Veselova, M. N. Poltorak, O. M., and Pryakhin, A. N. (1976). *Vestnik Mosk. Univ., Ser. 2, Khim.* **17,** 246–284.
Clarke, F. M., and Masters, C. J. (1972). *Arch. Biochem. Biophys.* **153,** 258–265.
Clarke, F. M., and Masters, C. J. (1975a). *Biochim. Biophys. Acta* **381,** 37–46.
Clarke, F. M., and Masters, C. J. (1975b). *Int. J. Biochem.* **6,** 133–145.
Clarke, F. M., and Masters, C. J. (1976). *Int. J. Biochem.* **7,** 359–365.
Clarke, F. M., and Morton, D. J. (1976). *Biochem. J.* **159,** 797–798.
Craven, P. A., and Basford, R. E. (1974). *Biochim. Biophys. Acta* **354,** 49–56.
Dagher, S. M., and Hultin, H. O. (1975). *Eur. J. Biochem.* **55,** 185–192.
De Meis, L., and Rubin, B. M. (1970). *An. Acad. Bras. Ciênc.* **42,** 103–111.
Dolken, G., Leisner, E., and Pette, D. (1975). *Histochemistry* **43,** 113–121.
Engasser, J.-M., and Horvath, C. (1975). *Biochem. J.* **145,** 431–435.
Fairbanks, G., Steck, T. L., and Wallach, D. F. H. (1971). *Biochemistry* **10,** 2606–2617.
Felgner, P. L., and Wilson, J. E. (1977). *Arch. Biochem. Biophys.* **182,** 282–294.
Felgner, P. L., Messer, J. L., and Wilson, J. E. (1979). *J. Biol. Chem.* **254,** 4946–4949.
Fiek, C., Benz, R., Roos, N., and Brdiczka, D. (1982). *Biochim. Biophys. Acta* **688,** 429–440.
Frieden, C. (1970). *J. Biol. Chem.* **245,** 5788–5799.
Frieden, C. (1979). *Annu. Rev. Biochem.* **48,** 471–489.

Gellerich, F. N., and Augustin, H. W. (1977). *Acta Biol. Med. Germ.* **36**, 571–577.

Goldstein, L. (1972). *Biochemistry* **11**, 4072–4084.

Goldstein, L. (1976). *In* "Methods in Enzymology" (K. Mosbach, ed.), Vol. 44, pp. 379–443. Academic Press, New York.

Goldstein, L., Levin, Y., and Katchalsky, E. (1964). *Biochemistry* **3**, 1913–1916.

Gots, R. E., and Bessman, S. P. (1974). *Arch. Biochem. Biophys.* **163**, 7–14.

Gots, R. E., Gorin, F. A., and Bessman, S. P. (1972). *Biochem. Biophys. Res. Commun.* **49**, 1249–1255.

Green, D. E., Murer, E., Hultin, H. O., Richardson, S. H., Salmon, B., Brierly, C. P., and Baum, H. (1965). *Arch. Biochem. Biophys.* **112**, 635–647.

Hanahan, D. (1973). *Biochim. Biophys. Acta* **300**, 319–340.

Hochman, M. S., and Sacktor, B. (1973). *Biochem. Biophys. Res. Commun.* **54**, 1546–1553.

Hultin, H. O. (1975). *In* "Isozymes II. Physiological Function" (C. L. Markert, ed.), pp. 69–85. Academic Press, New York.

Hultin, H. O., Ehmann, J. D., and Melnick, R. L. (1972). *J. Food Sci.* **37**, 269–273.

Inui, M., and Ishibashi, S. (1979). *J. Biochem. (Tokyo)* **85**, 1151–1156.

Kant, J. A., and Steck, T. L. (1973). *J. Biol. Chem.* **248**, 8457–8464.

Karadsheh, N. S., and Uyeda, K. (1977). *J. Biol. Chem.* **252**, 7418–7420.

Karpatkin, S. (1967). *J. Biol. Chem.* **242**, 3525–3530.

Keokitichai, S., and Wrigglesworth, J. M. (1980). *Biochem. J.* **187**, 837–841.

Kliman, H. J., and Steck, T. L. (1980). *J. Biol. Chem.* **255**, 6314–6321.

Klingerberg, M. (1981). *Nature (London)* **290**, 449–454.

Kurganov, B. I. (1982). "Allosteric Enzymes. Kinetic Behavior" Wiley, New York.

Kurganov, B. I., and Loboda, N. I. (1979). *J. Theor. Biol.* **79**, 281–301.

Kurganov, B. I., Klinov, S. V., and Sugrobova, N. P. (1978). *Symp. Biol. Hung.* **21**, 81–104.

Kursanov, A. L. (1940). "Reversible Action of Enzymes in Living Plant Cells. Izd. of Academy of Sciences of USSR, Moscow.

Liou, R.-S., and Anderson, S. (1980). *Biochemistry* **19**, 2684–2688.

Lusis, A. J., Tomino, S., and Paigen, K. (1976). *J. Biol. Chem.* **251**, 7753–7760.

McDaniel, C. F., and Kirtley, M. E. (1975). *Biochem. Biophys. Res. Commun.* **65**, 1196–1200.

McDaniel, C. F., Kirtley, M. E., and Tanner, M. J. A. (1974). *J. Biol. Chem.* **249**, 6478–6485.

McGhee, J. D., and von Hippel, P. H. (1974). *J. Mol. Biol.* **86**, 469–489.

Margreth, A., Muscatello, U., and Anderson-Cedergren, E. (1963). *Exp. Cell. Res.* **32**, 484–509.

Masters, C. J. (1977). *Curr. Top. Cell. Regul.* **12**, 75–105.

Masters, C. J. (1978). *Trends Biochem. Sci.* **3**, 206–208.

Masters, C. J. (1981). *CRC Crit. Rev. Biochem.* **11**, 105–143.

Masters, C. J., Sheedy, R. J., and Winzor, D. J. (1969). *Biochem. J.* **12**, 806–808.

Maurel, P., and Douzou, P. (1978). *In* "Frontiers in Physicochemical Biology" (B. Pullman, ed.), pp. 421–457. Academic Press, New York.

Melnick, R. L., and Hultin, H. O. (1968). *Biochem. Biophys. Res. Commun.* **33**, 863–868.

Monod, J., Wyman, J., and Changeux, J.-P. (1965). *J. Mol. Biol.* **12**, 88–118.

Ogawa, H., Shiraki, H., Matsuda, Y., and Nakagawa, H. (1978). *Eur. J. Biochem.* **85**, 331–337.

Oparin, A. (1934). *Erg. Enzymforsch.* **3**, 57–72.

Poglazov, B. F., Livanova, N. B., and Ostrovskaya, M. V. (1982). *Dokl. Akad. Nauk USSR* **263**, 221–224.

Poltorak, O. M. (1967). *Zh. Fis. Khim.* **41**, 2544–2562.

Poltorak, O. M., and Chukhrai, E. S. (1970). *Vestnik Mosk. Univ., Ser. 2, Khim.* **11**, 133–146.

Poltorak, O. M., and Pryakhin, A. N. (1977). *Vestnik Mosk. Univ., Ser. 2, Khim.* **18**, 251–268.

Poltorak, O. M., Pryakhin, A. N., and Chukhrai, E. S. (1977). *Vestnik Mosk. Univ., Ser. 2, Khim.* **18**, 125–142.

Ratner, J. H., Nitisewojo, P., Hirway, S., and Hultin, H. O. (1974). *Int. J. Biochem.* **5**, 525–533.

Ricard, J., Noat, G., Crasnier, M., and Job, D. (1981). *Biochem. J.* **195**, 357–367.

Rose, J. A., and Warms, J. V. B. (1967). *J. Biol. Chem.* **242**, 1635–1645.

Rothstein, A., Grinstein, S., Ship, S., and Knauf, P. A. (1978). *Trends Biochem. Sci.* **3**, 126–128.

Schrier, S. L. (1967). *Biochim. Biophys. Acta* **135**, 591–598.

Schwarz, G. (1970). *Eur. J. Biochem.* **12**, 442–453.

Shin, B. C., and Carraway, K. L. (1973). *J. Biol. Chem.* **248**, 1436–1444.

Shiraki, H., Ogawa, H., Matsuda, Y., and Nakagawa, H. (1979). *Biochim. Biophys. Acta* **566**, 345–352.

Siekevitz, P. (1959). *Ciba Found. Symp. Regul. Cell Metab.* pp. 17–45.

Sigel, P., and Pette, D. (1969). *J. Histochem. Cytochem.* **17**, 225–237.

Singer, S. J. (1974). *Annu. Rev. Biochem.* **43**, 805–833.

Solti, M., and Friedrich, P. (1976). *Mol. Cell. Biochem.* **10**, 145–152.

Solti, M., Bartha F., Halász, N., Tóth, G., Sirokmán, F., and Friedrich, P. (1982). *J. Biol. Chem.* **256**, 9260–9265.

Southard, J. H., and Hultin, H. O. (1972). *Arch. Biochem. Biophys.* **153**, 468–474.

Steck, T. L. (1974). *J. Cell Biol.* **62**, 1–19.

Steck, T. L. (1978). *J. Supramol. Struct.* **8**, 33–46.

Steitz, T. A., Anderson, W. F., Fletterick, R. J., and Anderson, C. M. (1977). *J. Biol. Chem.* **252**, 4494–4500.

Stewart, M., Morton, D. J., and Clarke, F. M. (1980). *Biochem. J.* **186**, 99–104.

Strapazon, E., and Steck, T. L. (1977). *Biochemistry* **16**, 2966–2971.

Sugrobova, N. P., Eronina, T. B., Chebotareva, N. A., Ostrovskaya, M. V., Livanova, N. B., Kurganov, B. I., and Poglazov, B. F. (1983). *Mol. Biol.* **17**, 430–435.

Tanner, M. J. A., and Gray, S. R. (1971). *Biochem. J.* **125**, 1109–1117.

Tillmann, W., Cordua, A., and Schröter, W. (1975). *Biochim. Biophys. Acta* **382**, 157–171.

Tsai, I.-H., Murthy, S. N. P., and Steck, T. L. (1982). *J. Biol. Chem.* **257**, 1438–1442.

Tuttle, J. P., and Wilson, J. E. (1970). *Biochim. Biophys. Acta* **212**, 185–188.

Von Hippel, P. H., and Schleich, T. (1969). *In* "Structure and Stability of Biological Macromolecules" (S. N. Timasheff and G. D. Fasman, eds.), pp. 251–417. Dekker, New York.

Walsh, T. P., Clarke, F. M., and Masters, C. J. (1977). *Biochem. J.* **165**, 165–167.

Weiss, T. L., Zieske, J. D., and Bernstein, I. A. (1981). *Biochim. Biophys. Acta* **661**, 221–229.

Wilson, J. E. (1968). *J. Biol. Chem.* **243**, 3640–3647.

Wilson, J. E. (1973). *Arch. Biochem. Biophys.* **159**, 543–549.

Wilson, J. E. (1978). *Trends Biochem. Sci.* **3**, 124–125.

Wilson, J. E., Reid, S., and Masters, C. J. (1982). *Arch. Biochem. Biophys.* **215**, 610–620.

Wojtczak, L., and Nałęcz, M. J. (1979). *Eur. J. Biochem.* **94**, 99–107.

Yu, J., and Steck, T. L. (1975). *J. Biol. Chem.* **250**, 9176–9184.

6

Models of Organized Multienzyme Systems: Use in Microenvironmental Characterization and in Practical Application

Nils Siegbahn and Klaus Mosbach

Pure and Applied Biochemistry
Chemical Center, University of Lund
Lund, Sweden

G. Rickey Welch

Department of Biological Sciences
University of New Orleans
New Orleans, Louisiana

ORGANIZED
MULTIENZYME SYSTEMS

271

I. INTRODUCTION

Cell metabolism, by its very nature, is characterized by the action of *sequences of enzymes*, a montage of interdependent chemical reactions connected by coupled reaction–diffusion processes. Despite the apparent air of homogeneity connoted by this picture, the conception of the cell as a bag composed of a bulk aqueous solution uniformly dispersed with enzyme proteins and freely diffusing metabolites has long been abandoned in contemporary cell biology. It is now thought that many (if not most) enzymes of intermediary metabolism operate in structured states, that is, membrane-bound arrays, solid-state aggregates in organelle (e.g., mitochondrion) compartments, or multienzyme complexes *per se* (Srere and Mosbach, 1974; Welch, 1977a). Even the so-called soluble enzymes of the cytosol may function in gellike environments, in association with particulate structures (Clegg, 1984).

A basic issue, of course, concerns the establishment of a qualitative (and, hopefully, quantitative) understanding of cell metabolism characterized by such structural heterogeneity. There are many questions. What is the physicochemical nature of the microenvironments *in situ*? What are the actual concentrations of substrates in these microenvironments? How are labile and/or common metabolic intermediates sequestered? How do diffusional lags affect metabolism? Such questions must be addressed, in order to understand better the efficiency and regulation of cell metabolism.

Any conceptualization of cell metabolism based on kinetic data derived from studies of isolated enzymes is fraught with limitations. Most conventional *in vitro* studies are conducted under highly nonphysiological conditions, for example, in artificial buffer systems, at dilute protein concentration, and/or at unrealistically high substrate—enzyme ratios. Upon isolation, the microenvironment and/or the intrinsic properties of the enzymes may be altered. Indeed, the structural and functional features of many proteins can change merely upon dilution (Minton, 1981).

One approach would be to study enzymes maintained in their natural milieu, either inside the intact cell or in fragments thereof (i.e., isolated organelles or membranes). However, the complexity of these systems makes interpretation difficult. More clear-out results may be obtained with reconstituted multienzyme systems, following prior characterization of the individual enzyme components. This approach has been fruitful in the area of oxidative phosphorylation. However, it is not applicable to the gellike systems for which no exact knowledge of the natural makeup is available.

Another approach, which is the main theme of this chapter, involves the study of model enzyme systems, usually in the immobilized state. Obviously, this approach also has its limitations, because both the structural matrix and the way in which the enzymes are immobilized are usually nonbiological. The virtues of this approach are, however, that these preparations are easily handled, they can be well defined chemically and physically, and they simplify the study of the effects of changes in individual parameters. In contrast, natural systems of enzyme sequences render interpretation of data (e.g., as to the efficiency of the overall process) very difficult, because of the inherent complexity and variables that cannot be controlled. Thus, it is not possible to evaluate accurately the contribution of every parameter that might lead to a change in the catalytic efficiency of enzyme sequences, such as distances between enzymes, diffusional effects, concentrations of intermediates, and membrane (or medium) effects. One might even hope that the immobilization approach will reveal heretofore unknown effects, of biological relevance, that are important in natural systems. In this chapter we explore various functional advantages of organized systems revealed by study of immobilized designs and indicate some of the practical uses in enzyme technology.

II. MODEL SYSTEMS OF ENZYME SEQUENCES

A. General Features of Immobilized Enzyme Systems

We give some concepts and definitions relative to the models of enzyme sequences. A large variety of materials has been used as immobilization supports, mostly solid-phase beads (e.g., agarose, acrylate, glass) but also membranes, fibers, and, in a few cases, water-soluble supports such as dextran. Enzymes are usually immobilized on these matrices (or supports) by one of the following means (Fig. 1): (a) covalent attachment; (b)inclusion within a gel

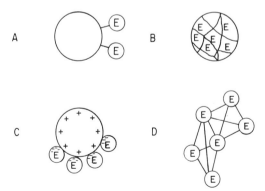

Fig. 1.   Schematic representation of the four major types of immobilized enzyme preparation: (A) covalent binding, (B) entrapment, (C) adsorption, and (D) cross-linking.

lattice, or microencapsulation within semipermeable membrane; (c) adsorption, including hydrophobic and affinity binding; or (d) cross-linking as insoluble aggregates. (For further details and literature references, see Mosbach, 1976, 1983, 1985a; Pye and Wingard, 1974; Pye and Weetall, 1978; Broun et al., 1978; Wingard et al., 1976; Weetall and Royer, 1980; Chibata et al., 1982; Mosbach and Matiasson, 1978).

Upon immobilization, the intrinsic kinetic parameters of an enzyme may or may not be altered. Usually, the specific activity is reduced to some extent. Observed changes caused by factors other than immobilization per se can be divided into effects caused by (a) products of the enzymatic reaction itself and (b) the nature of the matrix microenvironment.

1. Matrix Effects

Enzymes immobilized on charged matrices may, at low ionic strength, be surrounded by a micromilieu characterized by a pH grossly different from that of the bulk solution (Levin et al., 1964). Positively charged matrices generate higher local pH values (Broun et al., 1969; Kay and Lilly, 1970; O'Neill et al., 1971), whereas negatively charged matrices decrease local pH values (Patel et al., 1969; Hornby et al., 1966; see also Hervagault and Thomas, Chapter 8).

Charged matrices may also attract or repel charged substrate or product molecules. Variations in the local substrate concentration result in apparent differences in the K_m values of the immobilized versus the free enzyme. Substrates, inhibitors, and products tend to be distributed between the gel phase where the enzyme is situated and the bulk solution, according to their partition coefficients between the two phases. These coefficients are functions of local conditions, for example, hydrophobicity (Filippusson and Hornby, 1970; Laidler and Sundaram, 1971; Brockman et al., 1973; Johansson and Mosbach, 1974).

2. Diffusional Effects

Diffusion limitations may influence the activity of an immobilized enzyme in many ways. An enzyme-containing gel particle, when stirred in solution, is usually surrounded by a so-called Nernst diffusion layer (or unstirred layer), leading to a concentration gradient from a higher outer substrate concentration in the bulk medium to a lower concentration at the gel surface (Laidler and Sundaram, 1971; Katchalski et al., 1971; Sélégny et al., 1971). The products generated in the enzyme reaction, however, lie in a gradient in the opposite direction.

The Nernst diffusion layer created around a particle in a stirred solution can, from a functional point of view, be treated as a layer of restricted diffusion as compared to the situation in free solution. The thickness of the layer varies with the size of the particle and with the relative movement between the solute and the particulate phase. The thickness of an unstirred layer is usually in the range <400 μm (Wilson and Dietschy, 1974; Lerche, 1976).

The effect of pore diffusion is mainly observed in porous supports (e.g., agarose) and especially when enzymes are entrapped (e.g., in acrylic polymers). The effect on the overall behavior is dependent not only on pore size but also on particle size. Surface-bound enzyme systems operate under negligible pore diffusion influence, whereas particles of entrapped enzymes are markedly influenced. In all systems governed by diffusion restrictions, a clear correlation is observed between the degree of loading with active enzyme molecules and the degree of diffusion restrictions. It has also been shown (Regan et al., 1974) that, within the same batch of an immobilized enzyme preparation, smaller particles offer better "substrate-feeding" conditions than larger particles. This results in lower apparent K_m values for enzymes immobilized on the former. Also, there is likely to be some degree of heterogeneity within an immobilized enzyme preparation with respect to the number of linkages of a covalently bound enzyme molecule as well as in orientation in relation to the matrix (Pye and Chance, 1976). In this context it deserves mentioning that methods have been applied that allow some degree of control of the number of covalent linkages formed between the enzyme and the support (Koch-Schmidt and Mosbach, 1977; Koch-Schmidt et al., 1979; Nilsson and Mosbach, 1981; Nilsson and Larsson, 1983).

The effects of diffusion restrictions on sequentially acting enzyme systems are in a sense contradictory. Following an increase in the loading of enzymes on a support (i.e., a decrease in the distance between the enzymes), the enzymes last in line tend to operate under more favorable conditions; the close proximity between the enzymes favors the kinetics of the next enzyme in line, because of the generation of a favorable microenvironment with respect to intermediate concentration (Mosbach and Mattiasson, 1970; Mattiasson and Mosbach, 1971). The inward diffusion of substrate to the first enzyme, however, because of the diffusion restrictions discussed above, becomes limiting for the system when the "concentration" of this enzyme is higher.

Another factor of great importance in the kinetic behavior of an immobilized enzyme sequence is the activity ratio, that is, the ratio between the different enzyme activities under operational conditions. By altering the activity ratio the steady-state rate can be changed (Mosbach and Mattiasson, 1970), as well as the length of the lag period.

3. Enzyme Concentration and Exclusion Effects

Increasing attention has been given to the fact that highly concentrated enzyme solutions are present in the cell and that this must be considered when interpreting intracellular events. As an example, we cite the case of the glycolytic enzyme phosphofructokinase. It is known from kinetic studies in dilute solutions that this enzyme exhibits allosteric reactions to a broad spectrum of effectors. However, these properties are much less marked when the enzyme is studied *in situ*, that is, at a high protein concentration representative of its natural surroundings (Reeves and Sols, 1973; Karadsheh and Uyeda, 1977). The technique that made such studies possible involved the use of permeabilized cells in which the cell membranes were partially destroyed by toluene treatment. This allowed substrate and product to penetrate freely, whereas macromolecules such as proteins remained inside (Reeves and Sols, 1973; Weitzman, 1973).

With conventional techniques it is difficult to study the effects of high enzyme concentrations in free solution. However, this can be done with immobilized enzyme systems, which are characterized by a bulk solution free of enzyme molecules and a solid phase containing high concentrations of enzymes. Immobilized model systems are characterized by a milieu of high macromolecular content, of protein as well as matrix polymer. Such high concentration of polymer gives rise to exclusion effects and diffusion hindrance and may also create better conditions for interaction between free macromolecules (Laurent, 1971; Ceska, 1971; Backman and Johansson, 1976; Halper and Srere, 1977). To explore the influence of a polymer, soluble enzymes may be studied in media containing varying concentrations of polymers such as polyethylene glycol or dextran. The presence of polymers causes a reduction in free volume available for the solute molecules surrounding the enzymes. This leads to so-called exclusion effects (observed as an apparent increase in concentration of the solute). In addition, at very high polymer concentrations diffusion restrictions may occur. As discussed by Minton (1981), the rate constant k_e for formation of an enzyme–substrate complex may be approximated as

$$k_e \simeq k_e^0 e^{-g\phi}, \tag{1}$$

where k_e^0 is the intrinsic value for "ideal" conditions, ϕ the fraction of solution volume occupied by background molecules, and g a constant. One might expect a number of features (e.g., enzyme stability, reaction rates, equilibria) to vary with ϕ in dense polymeric media (Minton, 1981).

Besides experimental schemes, theoretical models for immobilized systems may also be set up. However, many of the earlier model simulations were based on kinetic values obtained from free enzymes in well-buffered solutions. Hence, in the interpretation of the results, possible contributions of microenvironmental effects were not taken into account. If such factors are considered, then computational models may be a valuable complement to experimental models, or vice versa (see Hervagault and Thomas, Chapter 8).

4. Kinds of Enzyme Arrangement

Evaluation of the influences from each of the microenvironmental parameters for a given system poses extreme complications. It is more feasible to change the models so as to isolate the individual parameters. Thus, surface-bound systems are used when pore diffusion is to be minimized (Matiasson and Mosbach, 1971), and soluble enzyme aggregates are studied to eliminate (or reduce) the influence of the Nernst diffusion layer (Matiasson et al., 1974; Koch-Schmidt et al., 1977; Siegbahn et al., 1985).

Other than the low accessibility found with large substrates (not considered in detail here), the availability of substrate molecules (inward diffusion) for an immobilized enzyme is influenced particularly by diffusion restrictions exerted by the unstirred layer surrounding the matrices and by restricted pore diffusion in the interior of the gels.

Based on their macroscopic properties, the models can be divided into three main groups: (a) particle-bound enzyme systems, (b) enzymes in/on membranes, and (c) soluble enzyme aggregates. The enzyme arrangement in these model systems may be either *random*, which means that the sequentially acting enzymes are separately and randomly distributed within the system, or the enzymes may be aggregated prior to immobilization. Aggregates may be generated either *at random*, with no orientation of the individual enzymes in relation to each other, or *directed*, with active sites, effector sites, etc., arranged in a specific manner.

a. Particle-Bound Enzyme Systems. The systems most studied to date are those with enzyme sequences immobilized on polymer particles. The choice of spherical beads as the support is based on the fact that they are readily available and that most previous studies with immobilized enzymes have been performed using this kind of support. They have the following advantages: (a) a broad spectrum of support materials is available (Mosbach, 1976; Goldstein and Manecke, 1976); (b) a large number of methods for attachment of enzymes to these matrices is known (Mosbach, 1976; Goldstein and Manecke, 1976); (c) a good prior knowledge from less complex systems is at hand; and (d) good assay methods are available for these systems (Mattiasson and Mosbach, 1976). The possible disadvantage is that a heterogeneous size distribution of the support material can give rise to a population of enzyme particles with slightly different properties.

b. Enzymes in/on Membranes. Such designs may be considered a specific form of particulate enzyme systems. They have the advantage that it is relatively easy to prepare well-defined membranes (Thomas and Broun, 1976). A disadvantage is the limited choice of suitable membrane material.

c. Enzyme Aggregates. Cross-linking of proteins can yield precipitates, which can be considered a form of immobilized enzyme (Broun, 1976). In some studies the cross-linking process is terminated prior to precipitation, resulting in high-molecular-weight soluble enzyme aggregates. Enzyme aggregates have the following advantages: (a) the extremely high protein (enzyme) concentration obtainable is similar to that found in natural systems; (b) such systems can be studied with only little interference by a Nernst diffusion layer; and (c) it is easy to achieve close proximity between different enzymes in the sequence. They have the disadvantage that the complexes often contain highly modified enzyme molecules, which may lead to altered enzyme properties.

B. Models of Multienzyme Systems

1. Two-Step Enzyme Systems

Model systems have been constructed for the study of several important questions pertaining to catalysis by enzyme sequences. For example, how does proximity between sequentially acting enzymes affect the efficiency of a system operating under rate-determining concentrations of the intermediate? Also, how is the concentration of intermediate influenced in a milieu characterized by diffusion restrictions such as pore diffusion and the Nernst diffusion layer? Finally, how does an immobilized enzyme sequence respond to inhibition, activation, and so on, by internally generated substances?

A two-step enzyme system was prepared by immobilizing two sequentially acting enzymes on the same particulate matrix (Mosbach and Mattiasson, 1970). The exact distance between these two enzymes and those found in the other enzyme systems to be discussed remains to be established but should yield valuable information on enzyme proximity in relation to overall activity. The system studied was hexokinase and glucose-6-phosphate dehydrogenase:

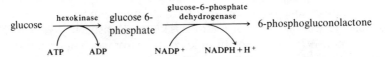

The enzyme activities were assayed using a stirred-batch procedure (Mattiasson and Mosbach, 1976). First, the overall activity in the coupled reaction was assayed; then the two separate enzyme activities were determined, starting with the last enzyme step. A reference system comprising the free enzymes was subsequently prepared by mixing the same number of expressed enzyme units of each enzyme per volume of incubation solution as were measured on the gel. As an additional control, a system consisting of a mixture of the two enzymes immobilized to *separate particles* was studied. It was found that in the initial stage the coimmobilized system was far more efficient, as compared with the soluble enzymes; after a lag phase the two systems operated at identical rates. Figure 2 is a generalized picture of this behavior. Comparisons were made of the enzymes immobilized by different methods. Cyanogen bromide activation of Sepharose gave products with the enzymes bound on the surface of the particles, as well as within the pores of the matrix. Similar results were obtained by entrapping the enzymes in polyacrylamide, or when they were covalently coupled to a copolymer of acrylic acid and acrylamide. In the latter cases, however, the particles were larger and had smaller pores as compared to Sepharose, which resulted in increased diffusion restrictions being exerted on the system. The differences in length of the lag phases between the coimmobilized and the free systems have been interpreted as follows. The product from the first enzyme reaction is present at a higher concentration within the domain of the particle than in the bulk solution, so that in the vicinity of the second enzyme a more favorable concentration of the rate-limiting intermediate substrate, glucose 6-phosphate, is found than in the bulk of the solution. Thus the first enzyme reaction generates, within the microenvironment of the enzyme sequence, a high local concentration of intermediate. This can be due to any of the following: (a) the fact that the product of the first step, glucose 6-phosphate, has a shorter distance to diffuse to the second enzyme in the coimmobilized system than is the case in the soluble system, (b) pore diffusion restrictions, or (c) the presence of an unstirred layer, which impedes diffusion of the intermediates into the bulk solution. The reference systems, the soluble system, and the system with the enzymes immobilized to separate particles all operate in a larger effective volume than the coimmobilized system; hence the buildup of the required intermediate concentration takes a longer time, which is expressed in a longer lag phase.

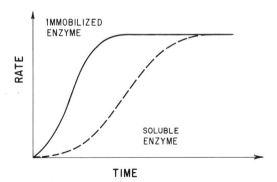

Fig. 2. Comparison of the overall rates of consecutive enzyme reactions catalyzed by coimmobilized systems and by the corresponding enzymes in free solution.

The kinetic behavior of such two-enzyme systems was confirmed independently by theoretical calculations based on membrane-bound enzyme systems (Goldman and Katchalski, 1971; Gondo, 1977).

The effect of varying enzyme activity ratios on the kinetic behavior of immobilized enzyme sequences was investigated (Mosbach and Mattiasson, 1970; Gondo, 1977; Wykes et al., 1975). In one such study (Bouin et al., 1976) the enzymes glucose oxidase and catalase were coimmobilized to silica alumina using the glutaraldehyde coupling procedure:

$$\text{glucose} + O_2 \xrightarrow[\text{oxidase}]{\text{glucose}} H_2O_2 + \text{gluconolactone} \longrightarrow (\text{gluconic acid})$$

with H_2O_2 undergoing, via catalase, $\tfrac{1}{2}O_2 + H_2O$

Reference systems for these coimmobilized, or dual, systems were prepared either with the enzymes coupled separately and then mixed, or with the two enzymes in free solution. Comparison of the efficiencies (in terms of length of lag phase) of dual, mixed, and soluble systems showed clearly that the efficiency of the dual system was always higher than that of the soluble system, which was in turn higher than that of the mixed system, irrespective of the activity ratio between the participating enzymes. [In general, the length of the lag phase decreased as the enzyme activity ratio (E_1/E_2) increased.] Similar effects have also been reported for the system hexokinase–glucose-6-phosphate dehydrogenase (Mosbach and Mattiasson, 1970) and also for alcohol dehydrogenase–lactate dehydrogenase (Wykes et al., 1975).

Two-step enzyme systems in membranes, with a homogeneous enzyme distribution, have also been investigated (Broun et al., 1972). One such sequence studied was as follows:

$$\text{lactose} \xrightarrow{\beta\text{-galactosidase}} \begin{array}{c}\text{glucose}\\ +\\ \text{galactose}\end{array} \xrightarrow{\text{glucose oxidase}} \text{gluconolactone} + H_2O_2$$

The concentration profile for the intermediate, glucose, within the membrane could be maintained much longer than in solution. The advantage of this system is that the concentration of a given metabolite can be maintained locally at a constant value. This system was designed as a model for any insulated compartment where metabolite transformation takes place, rather than to demonstrate effects on the overall activity of an enzyme sequence.

The two-enzyme malate dehydrogenase–citrate synthase system has also been studied (Srere et al., 1973):

$$\text{malate} + NAD^+ \underset{\text{malate dehydrogenase}}{\rightleftharpoons} \text{oxaloacetate} + NADH + H^+$$

$$\text{oxaloacetate} + \text{acetyl CoA} \xrightarrow{\text{citrate synthase}} \text{citrate} + \text{CoA} + H^+$$

Sum: $\text{malate} + NAD^+ + \text{acetyl CoA} = \text{citrate} + NADH + \text{CoA} + 2H^+$

This system involves a thermodynamically unfavorable step, that is, malate dehydrogenase ($K_{eq} = 2.5 \times 10^{-6}$ at pH 8.1). Coimmobilization of this enzyme with citrate synthase (which has a high affinity for oxaloacetate) leads to a microenvironmental situation wherein oxaloacetate is removed from the system, thus producing a higher overall activity during the course of the two-step sequence than that in the free state. The immobilized system was 100% faster in this regard. Apparently, the equilibrium of the malate dehydrogenase reaction is shifted in the immobilized regime (see also Srere, Chapter 1 this volume).

In a subsequent study (Koch-Schmidt *et al.*, 1977), the two enzymes malate dehydrogenase and citrate synthase were chemically cross-linked prior to immobilization (on Sepharose particles), so as to minimize diffusional hindrances. Kinetic measurements revealed that the bienzyme conjugate was more efficient in the initial (lag) phase when it was immobilized than when it was placed in a simple (bulk) aqueous solution. This result indicates that the polymer matrix plays a role in the channeling of intermediate substrate, via impedance of the out-diffusion (see Section III).

Friedman and co-workers have also examined the behavior of immobilized enzymes from the citric acid cycle. In one such study (Erekin and Friedman, 1979), the sequential enzymes fumarase and malate dehydrogenase were immobilized on Sepharose. This system catalyzes the *in vivo* sequence

$$\text{fumarate} \xrightarrow[+\,H_2O]{\text{fumarase}} \text{malate} \xrightarrow[\text{NADH}]{\substack{\text{malate} \\ \text{dehydrogenase}}} \text{oxaloacetate}$$

By varying the ratio of the two enzyme activities in the immobilized state, it was found that the organized scheme is more efficient than the corresponding two-enzyme system in free solution. It should be mentioned that, in order to obtain measurable kinetic data, catalysis here was recorded in the opposite direction to that of the mitochondrial pathway.

Also, Heidepriem *et al.* (1980) studied fumarase–malate dehydrogenase and malate dehydrogenase–citrate synthase bienzyme systems immobilized on porous glass beads. Each of these two immobilized systems gave a rate enhancement of approximately 10-fold over that for the nonassociated (free-solution) case (cf. Srere *et al.*, 1973). The functional role of the heterogeneous (silica) surface in effecting substrate channeling here remains to be elucidated.

Shimizu and Lenhoff (1979) have examined the properties of phosphoglucomutase and glucose-6-phosphate dehydrogenase coimmobilized on *s*-triazine trichloride-activated cellulose. This system catalyzes the sequence

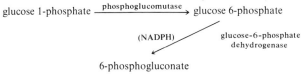

The results of experiments dealing with the effect of dilution on the sequential rate indicated greater efficiency for enzymes coimmobilized on the same surface. Of particular concern is the influence of the Nernst layer. In order for this unstirred layer to provide for compartmentation of the intermediate substrate, the distance between the two sequential enzymes should not be greater than the thickness of that layer. Moreover, efficient operation would require that the second enzyme follow first-order kinetics, so that the intermediate is consumed as fast as it is formed. For unassociated systems free in solution, the only way to "compartmentalize" the intermediate would be to swamp the first enzyme with a high concentration of the second one. Clearly, these problems can be obviated with immobilization in heterogeneous states. The previous authors (Shimizu and Lenhoff, 1979) examined various criteria for determining the channeling efficiency of immobilized two-enzyme systems.

The unique properties of other immobilized two-enzyme systems have been elaborated by Thomas and co-workers (See Hervagault and Thomas, Chapter 8 this volume; see also Mosbach, 1976).

2. Three-Step Enzyme Systems and Beyond

Extension of the previously described scheme to multistep enzyme systems by addition of additional sequentially acting enzymes has made it possible to study the relationships between the length of the lag period and the efficiency of the overall reaction on the one hand, and the number of participating enzymes within the sequence on the other hand.

The overall reaction rates of an immobilized three-step enzyme system (Fig. 3) (Mattiasson and Mosbach, 1971) and that of the corresponding system in free solution were compared. Reaction rates were also measured for

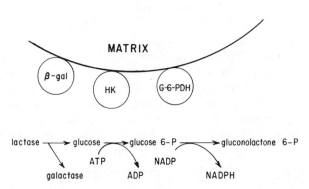

Fig. 3.   Schematic presentation of the matrix-bound three-enzyme system β-galactosidase (β-gal)–hexokinase (HK)–glucose-6-phosphate dehydrogenase (G-6-PDH), with the respective reactants. (See Mattiasson and Mosbach, 1971.)

the last two enzymes in the sequence, both in the immobilized state and in free solution. The efficiency of the initial stage of the overall reaction for the matrix-bound three-step enzyme system was higher than that for the soluble system. A similar trend was observed, although to a lesser extent, when the two-step enzyme systems were compared. The results indicate a cumulative efficiency effect as the number of enzymes participating in the reaction sequence increases. It should be stressed that the enzyme molecules in the sequences studied so far have been randomly distributed in the immobilized phase, probably resulting in a mosaic pattern of enzymes throughout the support.

The enzymes of a cyclic four-step system, the urea cycle, have been coimmobilized (by covalent attachment) on Sepharose (Siegbahn and Mosbach, 1982). The pathway is shown in Fig. 4. The immobilized multienzyme design was found to be more efficient than the corresponding soluble system, in terms of the overall rate. In fact, a true operative cycle was found with the immobilized system. Similar results were obtained by Inada *et al.* (1980), although their system (made by coentrapment in a fibrin membrane matrix) did not allow convenient determination of the individual enzymes participating in the process.

Okamoto *et al.* (1980) coimmobilized the enzymes uricase, allantoinase, allantoicase, and catalase in a fibrin membrane. This system was found to degrade uric acid (a purine derivative) to urea and glyoxylate. Another multienzyme system studied (De Luca and Kricka, 1983) combined as many as 13 different enzyme activities. The coimmobilized system consisted of the 11 enzymes necessary for the conversion of glucose to alcohol, along with two enzymes used to monitor NADH production by light emission.

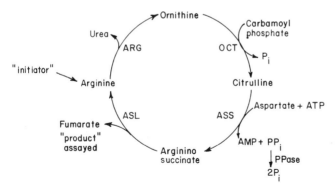

Fig. 4.   Urea cycle: OCT, ornithine carbamoyltransferase; ASS, argininosuccinate synthetase; ASL, argininosuccinate lyase; ARG, arginase; PPase, inorganic pyrophosphatase. (Under artificial conditions the cycle was started with arginine, and fumarate was the product determined. See Siegbahn and Mosbach, 1982.)

Such immobilized "metabolic pathways" may prove useful, not only as models for basic biochemistry, but also in such applied areas as clinical detoxification (Miura et al., 1980, 1981; Mosbach, 1980, 1983).

3. Coenzyme Recycling Systems

The effect of recycling coenzymes, for example, reoxidation of NADH in a two-step enzyme system producing NADH, was investigated by immobilizing lactate dehydrogenase together with malate dehydrogenase and citrate synthase (Srere et al., 1973), as shown in Fig. 5 (see Section II,B,1). Addition of pyruvate to the system resulted in oxidation of the NADH produced in the malate dehydrogenase reaction, with a concomitant conversion of pyruvate to lactate. The efficiency (measured as citrate produced) of the immobilized system compared to that of the free was enhanced in some cases by as much as 400%. Because of the "high" concentration of NADH in the microenvironment, a rapid oxidation of the coenzyme was observed when pyruvate was added to the system. This in turn created a favorable local NAD^+ concentration for malate dehydrogenase, shifting the equilibrium toward high oxaloacetate concentrations. Generally, it appears that kinetic advantages are to be gained when such enzyme systems operate at a limiting coenzyme concentration. The factors responsible for this increased catalytic efficiency are the favorable high NAD^+ concentration and, simultaneously, the low NADH concentration for malate dehydrogenase, created by the action of lactate dehydrogenase.

Recently, a method was developed for site-to-site directed immobilization of enzymes with bis-NAD^+ analogs (Mansson et al., 1983). Lactate dehydrogenase and alcohol dehydrogenase were cross-linked with glutaraldehyde on agarose beads. The cross-linking was performed, while the two enzymes were spatially arranged *with their active sites facing one another*, with the aid of the bis-NAD^+ compound (Figs. 6 and 7). (Subsequent to the immobilization process, the bis-NAD^+ analog was allowed to diffuse out.) This design allows for efficient recycling of the coenzyme NAD^+. By using a third enzyme, lipoamide dehydrogenase, which was also coupled to the same beads and

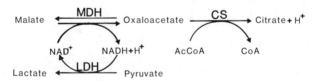

Fig. 5. Schematic representation of the matrix-bound, coenzyme-recycling three-enzyme system malate dehydrogenase (MDH)–citrate synthase (CS)–lactate dehydrogenase (LDH), with the respective reactants. (See Srere et al., 1973.)

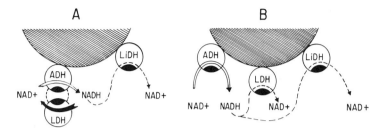

Fig. 6. Recycling (channeling) of NAD(H) in an artificial bienzyme complex. In (A), alcohol dehydrogenase (ADH) and lipoamide dehydrogenase (LiDH) were immobilized directly to tresylchloride-activated Sepharose; lactate dehydrogenase (LDH) was subsequently coupled (via glutaraldehyde) to the bound ADH, using the site-to-site directing agent bis-NAD$^+$. In (B), all three enzymes were simultaneously immobilized directly to the Sepharose support. The LiDH was employed as an enzyme probe (cf. Solti and Friedrich, 1979), to analyze the degree of channeling of the coenzyme. The figure is highly schematic and does not take into account the oligomeric nature of the enzymes. The dark areas indicate the enzyme active sites. (See Mansson *et al.*, 1983.)

$$
\overset{H}{\underset{\underset{NAD}{|}}{\overset{|}{N}}}-CH_2-\overset{O}{\overset{||}{C}}-NH-(CH_2)_6-NH-\overset{O}{\overset{||}{C}}-(CH_2)_4-\overset{O}{\overset{||}{C}}-NH-(CH_2)_6-NH-\overset{O}{\overset{||}{C}}-CH_2-\overset{H}{\underset{\underset{NAD}{|}}{\overset{|}{N}}}^*
$$

Fig. 7. A bis-NAD$^+$ compound used for the orientation of lactate dehydrogenase to alcohol dehydrogenase in Fig. 6. This compound was synthesized by condensing two molecules of N^6-[(6-aminohexyl)carbamoylmethyl]-NAD with adipic acid dichloride. The connection with NAD is through the exocyclic N of adenine. (See Mansson *et al.*, 1983.)

which competes with lactate dehydrogenase for the NADH produced by alcohol dehydrogenase, the effect of site-to-site directed immobilization was studied. In Figure 6A, in which lactate dehydrogenase and alcohol dehydrogenase are covalently oriented, a major portion of the NADH formed by the oxidation of ethanol was recycled efficiently by lactate dehydrogenase. Only a minor portion of NADH was oxidized by the competing lipoamide dehydrogenase.

In Fig. 6B, where all three enzymes had been bound at random to the matrix, by contrast, the proportion of NADH oxidized by lactate dehydrogenase was much smaller. [The enzyme probe scheme used by Solti and Friedrich (1979) bears some semblance to this preparation, although that of the previous authors deals with naturally occurring systems.] (See also Section IV,C.)

4. Other Model Studies

Immobilized multienzyme systems have also been used to model such biological features as active transport (Mosbach, 1976; Mosbach and Mattiasson, 1978; see also Hervagault and Thomas, Chapter 8 this volume).

III. ROLE OF IMMOBILIZED ENZYME SYSTEMS IN THE INTERPRETATION OF MICROENVIRONMENTAL COMPARTMENTATION *IN VIVO*

The biological relevance of model systems to actual *in vivo* situations is, of course, a key concern. In principle, model studies may be designed to test, in a simple yet defined way, the validity of certain hypotheses regarding the nature of enzyme action in the living cell. Effects observed with such models may then be sought in the corresponding biological systems. Conversely, effects observed with biological systems can be "isolated" and further analyzed by designing appropriate model schemes, and new and important effects may be encountered in model studies that were heretofore unknown in the "real" system. In addition, some biological phenomena that were difficult to interpret are readily understood in the light of results obtained with model studies. An example of the latter concerns the puzzling observation that membrane-bound acetylcholinesterase shows a pH dependence different from that of the soluble esterase. The observation was rationalized after results were obtained with enzymes immobilized to synthetic polymers. It was revealed that local pH changes in the microenvironment of the membrane-bound enzyme, resulting from hydrolysis of the substrate, can affect the apparent pH dependence (Silman and Karlin, 1967).

From the model studies described here, together with emerging information on enzyme cytology, we may conclude that enzymes arranged in close proximity in immobilized states are likely to be representative of biological systems. Provided the intermediates are present at rate-limiting concentrations, such systems exhibit shorter transient times and/or higher overall steady-state rates compared to nonclustered systems.

Knowledge from immobilized-enzyme studies has already aided in the understanding of metabolic conditions. For example, the malate dehydrogenase–citrate synthase system discussed in Section II,B,1 has proven valuable in the interpretation of metabolic flux in the citric acid cycle, particularly in relation to compartmentalization of oxaloacetate (see Srere, Chapter 1 this volume).

A similar example is the two-enzyme system aspartate aminotransferase–malate dehydrogenase, which appears to constitute a natural physical complex (Backman and Johansson, 1976). No lag phase was observed in the overall activity of this system when it was studied with stopped-flow techniques. Moreover, there was no equilibration between nascent oxaloacetate and "bulk" oxaloacetate, indicating some kind of compartmentalization of this intermediate (Bryce *et al.*, 1976). This effect is analogous to that in model systems discussed above and in transiently interacting systems (see Friedrich, this volume).

The potential role of exclusion effects in dictating the microenvironmental characteristics of cell metabolism is intriguing. It has been demonstrated that such effects in polymeric media can alter the apparent kinetic constants for enzymes, as well as the stability and associative properties of proteins (Ceska, 1971; Laurent, 1971; Minton, 1981). Exclusion effects were shown to improve the efficiency of a two-step immobilized enzyme system (Mattiasson et al., 1974), and the use of polymeric media in vitro has made it possible to demonstrate a natural interaction between metabolically related enzymes (Backman and Johannsson, 1976; Halper and Srere, 1977). One cannot help but regard such dense polymeric regimes as indicative of microenvironments in vivo (Clegg, 1984; Sitte, 1980; see Srere, Chapter 1 this volume). Interestingly, Friedrich et al. (1977) found the macromolecular associations responsible for formation of a "metabolic channel" to be dependent on the state of dilution.

Microenvironmental compartmentation in vivo may be accomplished by specific physical constraints, as well as by diffusion restrictions. The degree of influence in vivo of such parameters as the Nernst diffusion layer, viscosity barriers, enzyme proximity, and exclusion effects is difficult to evaluate. This is particularly evident for unstirred-layer effects, the existence of which is readily apparent in living cells (Clegg, 1984; Keith, 1979).

In an attempt to evaluate the contribution of factors such as proximity versus diffusion hindrance, bienzyme conjugates of malate dehydrogenase and citrate synthase were prepared by artificial cross-linking (Koch-Schmidt et al., 1977). These aggregates were studied in free solution, as well as immobilized to Sepharose. As indicated in Section II,B,1, the immobilized conjugate was more efficient than that in solution, indicating that diffusional restrictions play an important role. Somewhat surprisingly, however, it was found that the conjugate when in free solution was no more efficient than a reference system comprised of non-cross-linked enzymes free in solution. Therefore, under these conditions, the proximity effect appears to provide little advantage. (However, addition of polyethylene glycol to the system resulted in an advantage of the cross-linked enzyme design over the corresponding free system, most likely due to exclusion effects.) For a globular protein in dilute aqueous solution, the unstirred layer is probably no more than one monolayer of water molecules (Hagler and Moult, 1978). The mean distance between active sites in the artificial bienzyme aggregates is clearly greater than this monolayer thickness. In naturally occurring multienzyme complexes, though, the constituent enzyme molecules are certainly not randomly associated, as evinced by the fact that these systems channel intermediates under in vitro conditions (Welch, 1977a; Gaertner, 1978).

Cross-linked oriented enzyme aggregates, such as the alcohol dehydrogenase–lactate dehydrogenase system discussed in Section II,B,3, (see

Fig. 6), may serve as valuable "macroscopic" models for channeling of metabolite flow in the actual biological systems (Mansson *et al.*, 1983). Variation of the spacer region in the bis-NAD^+ (Fig. 7), used to gauge the site-to-site distance, might allow one to delineate the relative contributions of active-site proximity versus Nernst layer, exclusion effects, etc., in the channeling process.

Welch *et al.* (1983) examined theoretically the role of the microenvironment in enzyme–ligand dissociation processes, as it relates to enzyme action in heterogeneous states. The implications bear upon immobilized systems, as well as upon enzymes operating in dense media *in vivo*. In order to give a molecular representation to the association process $E + L \rightarrow EL$ (where E is enzyme and L could be a *substrate* or a *product* molecule), we write the unitary rate constant k_{+L} as follows:

$$k_{+L} = \frac{3}{2} \frac{D_L V_r}{r_L^2} \left(\frac{1}{2} + \frac{r_0}{\lambda \sqrt{v}} \right)^2 \exp(-E_L/k_B T), \qquad (2)$$

where r_L and D_L are the molecular radius and diffusion coefficient, respectively, of L; r_0 is a linear parameter characteristic of the range of the intermolecular forces between E and L; E_L is the threshold energy required for the ligand L to actually bind to the enzyme; k_B is the Boltzmann constant; T is the absolute temperature; and λ and v are parameters relating to lattice diffusion in liquids (Somogyi and Damjanovich, 1973). A recognition volume V_r is assigned to the binding region (active center) of the enzyme. Within this volume the ligand molecule L can be bound to the enzyme owing to specific intermolecular forces. If L is outside V_r, no specific forces act on it (cf. Hill, 1975).

Now, consider the dissociation of the EL complex. Immediately upon dissociation, the free L molecule possesses an impulsive kinetic energy. It loses this energy by being subject to a viscous drag from the ambient medium. Assuming a Stokes law force as a first approximation, the separation distance S between the enzyme active site and L upon dissociation is determined as

$$S = \frac{m_L v_L}{\alpha \pi \eta R_{EL}}, \qquad (3)$$

where m_L and v_L are the molecular mass and initial (impulsive) velocity of L, R_{EL} the reduced radius for enzyme active site and ligand L, η the medium viscosity, and α a constant (usually taken as 4 or 6). If $S > d$, where d is a critical distance characteristic of the recognition volume, V_r, then the reassociation of E and L has a very low probability. On the other hand, if $S < d$, L loses its kinetic energy within the recognition volume, and the enzyme can rebind it with a probability q [which is contained in the expression of Eq. (2)].

Accordingly, the probability of reassociation of E and L is given by the product $p_E q$, where p_E is the probability that $S \leq d$. Thus, we have an *apparent* dissociation rate constant k'_{-L} defined as

$$k'_{-L} = (1 - p_E q)k_{-L},$$ (4)

where k_{-L} is the intrinsic value (whose explicit expression need not concern us here) (see Somogyi and Damjanovich, 1973). Obviously, p_E depends on the medium viscosity, as well as on the kinetic energy of the nascent ligand molecule L. (This energy is defined by the particular chemical mechanism responsible for dissociation.)

Statistical aspects of the microenvironment led the previous authors (Welch *et al.*, 1983) to the following result:

$$p_E = \int_{\eta_0}^{\infty} f(\eta)\, p_E(\eta)\, d\eta,$$ (5)

where

$$p_E(\eta) = \begin{cases} 0, & \text{if} \quad U > \gamma\eta^2, \\ 1 - \exp\left(-\dfrac{\gamma\eta^2 - U}{k_B T}\right), & \text{if} \quad U \leq \gamma\eta^2. \end{cases}$$ (6)

Here, U is the amount of chemical energy of the dissociation process available for the kinetic energy of the dissociating molecule L, and, the parameter γ is given by

$$\gamma = \frac{\alpha^2 \pi^2 R_{EL}^2 d^2}{2\mu},$$ (7)

where μ is the reduced mass of E and L, and the other symbols are as defined above. The value η_0 in Eq. (5) is $(U/\gamma)^{1/2}$. Finally, $f(\eta)$ is a statistical distribution function for the ambient *microviscosity*.

We note that $\gamma\eta^2$ has units of energy. In fact, it specifies the energy of the "viscosity barrier" tending to keep the ligand L within the "recognition volume" of the enzyme. If the kinetic energy of the nascent L molecule is less than $\gamma\eta^2$, it will remain in the "volume." Calculations (Welch *et al.*, 1983) show that for enzymes (or enzyme complexes) in dilute aqueous solution, where out-diffusion is hindered only by a monolayer of water, the viscosity barrier is virtually nonexistent. Thus, channeling of intermediate substrates under this condition can be effectuated only if the enzyme active sites are oriented in close proximity (or if other physical constraints are operative). Results from the aforementioned model systems lend credence to this picture.

Now, consider the enzyme–ligand process where the setting is a dense, heterogeneous medium (as is the case for immobilized states and for many microenvironments *in vivo*). Dissociating L molecules will experience a wide variety of microviscosities, owing to local heterogeneities. This is especially true for *in vitro* studies using such polymeric media as polyvinylpyrrolidone, polyethylene glycol, or dextran. This is the reason for the use of a microviscosity distribution function, $f(\eta)$, in Eq. (5) [and, therefore, in Eq. (4)]. Generally speaking, there is no exact analytical expression for $f(\eta)$, due to the paucity of information on liquid structure. For purposes of a qualitative calculation, the previous authors (Welch *et al.*, 1983) employed the following form:

$$f(\eta) = (4\eta/H)e^{-2\eta/H}, \tag{8}$$

interrelating microviscosity (η) and the experimentally observed macroviscosity (H). This function has the following characteristics, which are consistent with physical reality:

$$f(\eta = 0) = 0, \tag{9}$$

$$\int_0^\infty f(\eta)\, d\eta = 1, \tag{10}$$

$$\int_0^\infty \eta f(\eta)\, d\eta = H, \tag{11}$$

$$\int_0^\infty (\eta - H)^2 f(\eta)\, d\eta = \frac{1}{2}H^2, \tag{12}$$

The maximum value of $f(\eta)$ here is at $\eta = \frac{1}{2}H$. And, as indicated by Eq. (11), the average value of η is H. Thus, $f(\eta)$ defined by Eq. (8) should be useful in giving at least a qualitative empirical picture of the viscosity effect.

Results of a numerical simulation are shown in Fig. 8. These theoretical findings are quite consistent with experimental data on the effects of viscosity on enzyme action (see Welch *et al.*, 1983, and references cited therein). Generally, the experimental studies show a decrease in K_m (suggesting an increase in affinity for substrate) and/or a decrease in V_{max} (suggesting an increase in affinity for product), with increasing solvent viscosity (see also Welch *et al.*, 1982.)

The effects of viscosity barriers and restricted diffusion, exemplified so well by the model studies with immobilized systems, most likely contributed to the evolutionary advantages of organized multienzyme systems in cell metabolism. For example, such factors may explain, in part, why the density of enzyme molecules is so great on biomembranes (Sitte, 1980; see Srere, Chapter 1 this volume). If metabolically sequential enzymes are arranged in clusters or "patches" (Welch, 1977a; Peterson *et al.*, 1978) on these surfaces, the interenzymatic diffusion distance is greatly reduced. The blanketing effect of

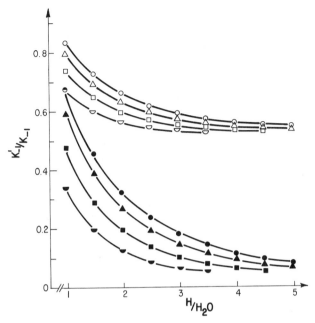

Fig. 8. The change in the value of $k'_{-1}/k_{-1}(=1-qp_E)$ as a function of the mean (macroscopic) viscosity, H (normalized to the mean value for water, $H_{H_2O} = 0.01$ poise). The value of the recombination probability, q, is taken as 0.5 (open symbols) or as 1.0 (closed symbols). The parameter $\gamma H_{H_2O}^2/k_B T$ is taken as 15, while the value of $U/k_B T$ is 5 ($\triangledown$, $\blacktriangledown$), 10 ($\square$, $\blacksquare$), 15 ($\triangle$, $\blacktriangle$), and 20 ($\bigcirc$, $\bullet$). (See Welch, *et al.*, 1983.)

the ordered water (and the electrical double layer) at the membrane–cytoplasm interface would bathe the localized multienzyme systems with high concentrations of intermediate substrates, while the close packing of enzyme molecules would obviate any substantial lateral diffusion (see Adam and Delbrück, 1968). Furthermore, in such structured regimes, electrical effects may play a role in augmenting site-to-site diffusion (Welch, 1977a). As is well known in the theory of reaction kinetics, the effects of a long-range interaction potential, $\psi(r)$, between reacting molecules is to increase the reaction radius (North, 1964). In accordance, Somogyi (1974a,b) showed that the *effective* enzyme–substrate recognition volume V_{eff}, is given by

$$V_{eff} = \int \exp\left[-\frac{\psi(r)}{k_B T}\right] dV_r, \tag{13}$$

where integration is performed over the geometrical structure of the actual recognition volume, V_r. As might be expected, one finds that $K_m \propto V_{eff}^{-1}$ (Somogyi and Damjanovich, 1971).

The ultimate state of enzyme organization is the multienzyme complex, which is characterized by a direct physical interaction of the respective enzyme molecules. In these structures the respective active sites are known to be arranged very closely in space, in some systems generating "composite active sites" (Welch, 1977a). Close juxtaposition (perhaps overlap) of the V_{eff} ranges of the individual enzymes, in conjunction with a high local viscosity in the ambient medium, yields an ideal microenvironment for compartmentation (channeling) of intermediate substrates. A combination of enzyme complex formation and adsorption on a membrane interface (with its high microviscosity) would seem to offer the optimal configuration for biological utilization of the viscosity effect, Nernst layer, etc. (Welch, 1977b).

To conclude this section, we emphasize that, even though it may be difficult (or impossible) to design a model system closely resembling the "real thing," model studies of the type described herein have their value. They bear an important message: Whenever an understanding of metabolic sequences is attempted, one must consider the microenvironment of the participating enzymes, as dictated by the surrounding membrane (matrix), the general cellular milieu itself, and/or the proximity to the next enzyme. Such consideration is a *condicio sine qua non* for the consummation of biochemistry as a holistic science. For additional reading on the topic of microenvironmental compartmentation and metabolic control, see Srere and Mosbach (1974), Friedrich (1974), Masters (1977), Welch (1977a,b, 1984), Srere and Estabrook (1978), and Wombacher (1983).

IV. PRACTICAL APPLICATIONS OF IMMOBILIZED ENZYME SEQUENCES

In this final section we extend the topic of immobilized enzyme sequences to encompass aspects dealing with their present, or potential, practical applications. There is an obvious parallel in the demands for efficiency and regulation of metabolism in a living cell and those required for an industrial process or analytical device where enzymes or whole cells are often utilized as catalysts. Thus, the results obtained (and understanding gained) from studies of a more theoretical nature will also be beneficial to the practical application of immobilized enzymes. In this article we wish to concentrate on examples in which immobilized multistep enzyme systems have been utilized. Previously, we (Mosbach and Mattiasson, 1978) tabulated and reviewed various applied uses of such systems. Here, we shall focus on recent advances in three specific areas. Readers who wish a more general introduction to the topic of "enzyme technology" (enzyme engineering), particularly in areas of medicine and industry-scale processes, are referred to other reviews and books (e.g.,

Mosbach, 1976, 1980, 1983, 1985a; Pye and Wingard, 1974; Pye and Weetall, 1978; Broun *et al.*, 1978; Weetall and Royer, 1980; Chibata *et al.*, 1982; Wingard *et al.*, 1976; Chang, 1977; Maugh, 1984).

A. Analytical Uses

Many immobilized multistep enzyme systems have been employed for analytical purposes. Immobilized preparations are especially suitable for continuous monitoring and for the reuse of expensive enzymes. Moreover, the immobilization technique can arrange the sensor (enzyme) in close proximity to a transducer (measuring device), resulting in a rapid and highly sensitive response.

One such design is the enzyme electrode (Guilbault, 1976). Here, the high selectivity of enzymes is combined with the great sensitivity of electrochemical measurements. In one report (Renneberg *et al.*, 1982), for example, an electrode system with glucose oxidase, peroxidase, and catalase entrapped in gelatin was used for the determination of toxicologically important peroxidase substrates such as bilirubin and aminopyrine. In another case (Verduyn *et al.*, 1983), alcohol oxidase and catalase were coimmobilized in an electrode system designed to measure alcohol in aqueous solution.

Multistep enzyme systems have also been applied to thermal analysis using a device called an enzyme thermistor to amplify the heat response from a primary reaction on a substrate of interest. For example, immobilized glucose oxidase–catalase preparations have proven efficient for determination of glucose (Mattiasson *et al.*, 1976; Danielsson *et al.*, 1977). Such a system, in conjunction with β-glucosidase, has been used for measuring cellobiose formed during degradation of cellulose (Danielsson *et al.*, 1981).

Another analytical technique involves the immobilization of enzymes on the inner surfaces of nylon tubular reactors (Sundaram and Hornby, 1970). Coimmobilization of hexokinase–glucose-6-phosphate dehydrogenase (a system discussed in Section II,B,1) on nylon tubing has led to a method for glucose and ATP analysis (León *et al.*, 1977), and pyruvate kinase–lactate dehydrogenase coimmobilized on such a reactor has been employed for determination of phosphoenolpyruvate, pyruvate, ADP, and NADH (Sundaram, 1978). More recently, a four-step enzyme system was applied to a nylon reactor for creatinine analysis (Ginman *et al.*, 1983). Generally, these immobilized enzyme analytical techniques have been found more efficient than assays in free solution.

Enzyme immobilization has also been applied to bioluminescence assays. Coimmobilization of NAD(P)H:FMN oxidoreductase and luciferase, along with various other coupled enzymes, has provided an assay method for a wide variety of biological substances (Wienhausen and DeLuca, 1982; Wienhausen et al., 1982; Kricka et al., 1983 Roda et al., 1984). Again, one finds these immobilized "enzyme luminometers" to be more efficient (and more stable) than the corresponding soluble systems.

Generally, it can be said from the examples given here that an additional advantage of coimmobilization of sequentially acting enzymes is signal amplification. Similarly, substrate channeling in an immobilized two-step system, hexokinase–glucose-6-phosphate dehydrogenase, has led to enhancement of an enzyme immunoassay (Litman et al., 1980; Ullman et al., 1983).

B. Coenzyme Regeneration

Many enzyme reactions of potential interest in the applied sector require the participation of expensive, dissociable coenzymes (e.g., pyridine nucleotides, adenine nucleotides). Methods for the reuse of these coenzymes, based on their retention and regeneration, have been studied extensively as of late. The actual regeneration can be accomplished by different approaches: chemical, electrochemical, photochemical, and enzymatic (Mosbach, 1978; Wang and King, 1979; Baughn et al., 1978; Furukawa et al., 1980). Of these, use of a recycling enzyme (or cell) is at present the preferred route. Interaction between coenzyme and the enzyme(s) catalyzing the reaction of interest, as well as with the regenerating enzyme, can be allowed to take place in different configurations. Here, we shall indicate some of the more recent techniques (see also Mosbach, 1978).

One design involves a bioreactor mounted with an ultrafiltration membrane allowing flow-through operation. A prerequisite is the binding of the coenzyme to a solid or preferentially water-soluble support (e.g., dextran, polyethylene glycol, polyacrylamide), to increase its molecular mass for retention. The enzyme(s) taking part, however, do not have to be bound to these supports unless favorable proximity or stabilizing effects are sought. Of the various combinations possible, those best studied involve the use of coenzymes coupled to water-soluble supports and with the enzymes kept in solution. In one such system, alanine formation from pyruvate and NH_4^+ with alanine dehydrogenase was studied with dextran-coupled NAD^+ (Davies and Mosbach, 1974) [or polyethyleneimine-coupled NAD^+ (Marconi et al., 1975)] and lactate dehydrogenase as recycling enzyme. The same dextran-coupled NAD^+ preparation, when entrapped by a dialysis membrane surrounding the tip of an electrode together with glutamate dehydrogenase and lactate dehydrogenase, could be put to use in analysis. Thus, with such an enzyme electrode system, glutamate could be determined by the formation of NH_4^+ in an assay mixture containing pyruvate required for the oxidation of the reduced dextran-bound NAD^+ formed (Davies and Mosbach, 1974). A similar system used by other workers, but with polyethylene glycol-bound NAD^+

instead, has employed enzyme couples such as formate dehydrogenase–alanine dehydrogenase and formate dehydrogenase–leucine dehydrogenase for large-scale production of amino acids (Wandrey *et al.*, 1980; Wichmann *et al.*, 1981). Furthermore, the recycling of polyethylene glycol-bound NAD$^+$ by lactate dehydrogenase and alcohol dehydrogenase has also been investigated (Katayama *et al.*, 1983).

There are other variations of this theme. For example, NAD recycling has been investigated in collagen membrane (Morikawa *et al.*, 1978). In this case, alcohol dehydrogenase and lactate dehydrogenase were coimmobilized in collagen, along with high-molecular-weight NAD$^+$–dextran. Also, formate dehydrogenase and malate dehydrogenase were entrapped in polyacrylamide gels, along with a polymerizable NAD$^+$ derivative (Yamazaki *et al.*, 1982). Also, microencapsulated multienzyme systems have been endowed with coenzyme-recycling capabilities (Grunwald and Chang, 1981; Chang *et al.*, 1982; Yu and Chang, 1982). An approach to the regeneration of coenzymes using immobilized hydrogen dehydrogenase (hydrogenase) has been described (Danielsson *et al.*, 1982). The hydrogenase was immobilized on porous glass particles and used in combination with alanine dehydrogenase for alanine formation, while the NADH consumed was regenerated by molecular hydrogen.

Rather specific alternative approaches to the problem of coenzyme regeneration are described in Fig. 9. Approach (A) has been studied with the system alcohol dehydrogenase–NAD$^+$, whereby the enzyme and the spacer-extended NAD$^+$ analog, N^6-[(6-aminohexyl)carbamoylmethyl]-NAD$^+$, are both coupled to the support. In this case, a so-called coupled-substrate assay

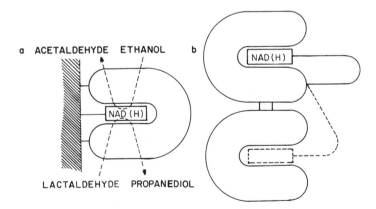

Fig. 9. Schematic representation of two ways (a, b) of "permanently fixing" the coenzyme NAD(H) in the vicinity of the active site of alcohol dehydrogenase. In both cases the NAD$^+$ analog used was N^6-[(6-aminohexyl)carbamoylmethyl]-NAD$^+$. (See Mansson *et al.*, 1978, 1979.)

allows continuous regeneration of the coenzyme while "fixed" in the active site of the enzyme (Gestrelius *et al.*, 1975). In a similar approach, the enzymes alcohol dehydrogenase (Legoy *et al.*, 1980) and myoinositol-1-phosphate synthase (Pittner, 1981) have been coupled simultaneously with NAD^+ to serum albumin, the resulting preparation no longer requiring externally added coenzyme. In alternative (B), the aforementioned NAD^+ analog is coupled directly to the enzyme alcohol dehydrogenase in such a fashion that the coenzyme can reach the active site of the enzyme and is active. In other words, a normally free coenzyme has artificially been turned into a prosthetic group. Preliminary results indicate that this same coenzyme can then swing out to interact with the second regenerating enzyme, which is bound to the first enzyme (Mansson *et al.*, 1978, 1979). Related to this latter approach is the design in Fig. 6, discussed in Section II,B,3. Alternatively, the second enzyme can be replaced by an electrode which regenerates the coenzyme (Torstensson *et al.*, 1980).

C. Immobilized Cells

The epitome of an organized multienzyme system is the living cell itself. Interest in the use of immobilized cells has grown steadily in the last two decades. This is illustrated by the increasing number of publications, starting with the first report on entrapment of cells as indicative of their potential for production purposes (Mosbach and Mosbach, 1966). Thus, whereas in 1973 only seven papers appeared on immobilized cells (compared with 220 on immobilized enzymes), 4 years later more than 50 dealt with immobilized cells (Dunnill, 1980). We discuss this topic only briefly here.

Cells can be immobilized by the four principal procedures applied to enzymes (Fig. 1). Of these, entrapment within a gel network is the most widely employed. Immobilized cells can be manipulated at different levels of viability. Thus, they have found use in the nonviable form, obtained for instance through heating or freezing. Permeabilized cells can be obtained, under slightly less drastic conditions, by brief treatment with organic solvents such as toluene or dimethyl sulfoxide. In its most sophisticated form, with antibiotics such as nystatin, permeabilization involves the formation of small pores in the cell membrane, thereby leaving the entire enzyme package intact. The cell membrane of such cells can subsequently be completely reconstituted and the cells made to propagate. Obviously, the borderline between nonviable and permeabilized cells is flexible. The general advantage gained with these preparations is that inward and outward diffusion of metabolites through the cell membrane, in particular of charged substrates and coenzymes, is greatly facilitated, and many of the enzymes tested are left intact (Felix *et al.*, 1981). Finally, living cells can be used either under stationary or growing conditions.

Immobilized cells find increasing interest for their potential in the following main areas: (a) for the production (including transformation) of useful compounds, (b) in analysis (e.g., microbial electrode sensor, microbe thermistor), (c) for the removal of harmful compounds (e.g, nitrate from drinking water), (d) as microbial fuel cells, (e) when used in an affinity chromatographic fashion for the isolation of specific compounds, and (f) in medicine. We refer the reader to reviews on this topic (e.g., Messing, 1981; Bucke and Wiseman, 1981; Jack and Zajic, 1977; Abbott, 1976; Chibata, 1979; Brodelius, 1978; Linko and Larinkari, 1980; Mattiasson, 1979; Birnbaum *et al.*, 1984; Mosbach, 1982, 1983, 1985b).

V. CONCLUDING REMARKS

From the model studies and examples discussed herein, it is clear that spatial organization of enzyme sequences is advantageous for the execution of cell metabolism, as well as for practical applications. We can look to the future with optimism, as the dialogue between basic and applied enzymologists continues to yield mutually palatable fruits. Results from studies on naturally occurring multienzyme clusters provide useful lessons in the optimal construction of reactors for enzyme technology, while the immobilized systems developed in the applied sector serve as macroscopic models for the elucidation of microenvironmental effects in the biological designs. Although strides are being made, our knowledge of the cellular *milieu intèrieur* is still rather meager. Studies with artificially immobilized enzyme systems, in conjunction with emerging views on cellular infrastructure, will be invaluable to the development of our understanding of enzyme action *in vivo*.

The role of immobilized enzymes in the pure and applied areas is still in its infancy. With increasing appreciation of the heterogeneous nature of the cell interior (e.g., see Clegg, 1984), together with renewed interest in the applied uses of immobilized enzymes (Mosbach, 1980, 1983; Maugh, 1984), the future of enzyme technology and engineering seems bright. Perhaps the merging of solid phase biochemistry with a general understanding of enzyme behavior will pave the way for preparation of artificial biocatalysts that, in many aspects, simulate Nature's own.

REFERENCES

Abbott, B. J. (1976). *Adv. Appl. Microbiol.* **20**, 203.
Adam, G., and Delbruck, M. (1968). *In* "Structural Chemistry and Molecular Biology" (A. Rich and N. Davidson, eds.), p. 198. Freeman, San Francisco, California.
Backman, L., and Johannson, G. (1976). *FEBS Lett.* **65**, 39–43.
Baughn, R. L., Adaltsteinsson, O., and Whitesides, G. M. (1978). *J. Am. Chem. Soc.* **100**, 304.

298   NILS SIEGBAHN et al.

Birnbaum, S., Larsson, P. -O., and Mosbach, K. (1984). In "Solid Phase Biochemistry: Analytical and Synthetic Aspects" (W. H. Scouten, ed.). Wiley, New York.

Bouin, J. C., Atallah, M. T., and Hultin, H. O. (1976). Biochim. Biophys. Acta 438, 23–36.

Brockman, H. L., Law, J. H., and Kezdy, F. J. (1973). J. Biol. Chem. 248, 4965–4970.

Brodelius, P. (1978). Adv. Biochem. Eng. 10, 75.

Broun, G. (1976). In "Methods in Enzymology" (K. Mosbach, ed.), Vol. 44, pp. 263–280. Academic Press, New York.

Broun, G., Sélégny, E., Avrameas, S., and Thomas, D. (1969). Biochim. Biophys. Acta 185, 260–262.

Broun, G., Thomas, D., and Sélégny, E. (1972). J. Membr. Biol. 8, 313–332.

Broun, G., Manecke, G., and Wingard, L. B., eds. (1978). "Enzyme Engineering" (Vol. 4). Plenum, New York.

Bryce, C., Williams, D., John, R., and Fasella, P. (1976). Biochem. J. 153, 571–577.

Bucke, C., and Wiseman, A. (1981). Chem. Ind. p. 234.

Ceska, M. (1971). Experientia 27, 767–768.

Chang, T. M. S., ed. (1977). "Biomedical Applications of Immobilized Enzymes and Proteins," Vols. 1 and 2. Plenum, New York, 1977.

Chang, T. M. S., Yu, Y. T., and Grunwald, J. (1982). In "Enzyme Engineering" (I. Chibata, S. Fukui, and L. B. Wingard, eds.), Vol. 6, pp. 451–455. Plenum, New York.

Chibata , I. (1979). ACS Symp. Ser. p. 187.

Chibata, I., Fukui, S., and Wingard, L. B., eds. (1982). "Enzyme Engineering" Vol. 6, Plenum, New York.

Clegg, J. S. (1984). Am. J. Physiol. 246, R133–R151.

Danielsson, B., Gadd, K., Mattiasson, B., and Mosbach, K. (1977). Clin. Chim. Acta 81, 163–175.

Danielsson, B., Rieke, E., Mattiasson, B., Winquist, F., and Mosbach, K. (1981). Appl. Biochem. Biotech. 6, 207–222.

Danielsson, B., Winquist, F., Malpote, J. Y., and Mosbach, K. (1982). Biotech. Lett. 4, 673–678.

Davies, P., and Mosbach, K. (1974). Biochim. Biophys. Acta 370, 329–338.

De Luca, M., and Kricka, L. J. (1981). Arch. Biochem. Biophys. 226, 285–291.

Dunnill, P. (1980). Philos. Trans. R. Soc. London Ser. B, 290, 409.

Erekin, N., and Friedman, M. E. (1979). J. Solid Phase Biochem. 4, 123–130.

Felix, H. R., Brodelius, P., and Mosback, K. (1981). Anal. Biochem. 116, 462–470.

Filippusson, H., and Hornby, W. E. (1970). Biochem. J. 120, 215–219.

Friedrich, P. (1974). Acta Biochim. Biophys. Acad. Sci. Hung. 9, 159–173.

Friedrich, P., Apró-Kovács, V. A., and Solti, M. (1977). FEBS Lett. 84, 183–186.

Furukawa, S., Sugimoto, Y., Urabe, I., and Okada, H. (1980). Biochimie 62, 629.

Gaertner, F. H. (1978). Trends Biochem. Sci. 3, 63–65.

Gestrelius, S., Mansson, M. O., and Mosbach, K. (1975). Eur. J. Biochem. 57, 529–535.

Ginman, R., Colliss, J. S., and Knox, J. M. (1983). Appl. Biochem. Biotech. 8, 213–226.

Goldman, R., and Katchalski, E. (1971). J. Theor. Biol. 32, 243–257.

Goldstein, L., and Manecke, G. (1976). Appl. Biochem. Bioeng. 1, 23–126.

Gondo, S. (1977). Chem. Eng. J. 13, 153–163.

Grunwald, J., and Chang, T. M. S. (1981). J, Mol. Catal. 11, 83–90.

Guilbault, G. G. (1976). "Handbook of Enzymatic Methods of Analysis." Dekker, New York.

Hagler, A., and Moult, J. (1978). Nature (London) 272, 222–226.

Halper, L., and Srere, P. A. (1977). Arch. Biochem. Biophys. 184, 529–534.

Heidepriem, P. M., Kohl, H. H., and Friedman, M. E. (1980). J. Solid Phase Biochem. 5, 5–9.

Hill, T. L. (1975). Proc. Natl. Acad. Sci. U.S.A. 72, 4918–4922.

Hornby, W. E., Lilly, M. D., and Crook, E. M. (1966). Biochem. J. 98, 420–425.

Inada, Y., Tazawa, Y., Attygalle, A., and Saito, Y. (1980). Biochem. Biophys. Res. Commun. 96, 1586–1591.

Jack, T. R., and Zajic, J. E. (1977). *Adv. Biochem. Eng.* **5,** 143.
Johannsson, A. C., and Mosbach, K. (1974). *Biochim. Biophys. Acta* **370,** 348–353.
Karadsheh, N. S., and Uyeda, K. (1977). *J. Biol. Chem.* **252,** 7418–7420.
Katayama, N., Urabe, I., and Okada, H. (1983). *Eur. J. Biochem.* **132,** 403–409.
Katchalski, E., Silman, I., and Goldman, R. (1971). *Adv. Enzymol.* **34,** 445–536.
Kay, G., and Lilly, M. D. (1970). *Biochim. Biophys. Acta* **198,** 276–285.
Keith, A. D., ed. (1979). "The Aqueous Cytoplasm." Dekker, New York.
Koch-Schmidt, A. C., and Mosbach, K. (1977). *Biochemistry* **16,** 2105–2109.
Koch-Schmidt, A. C., Mattiasson, B., and Mosbach, K. (1977). *Eur. J. Biochem.* **81,** 71–78.
Koch-Schmidt, A. C., Mosbach, K., and Werber, M. M. (1979). *Eur. J. Biochem.* **100,** 213.
Kricka, L. J., Wienhausen, G. K., Hinkley, J. E., and DeLuca, M. (1983). *Anal. Biochem.* **129,** 392–397.
Laidler, K. J., and Sundaram, P. V. (1971). *In* "Chemistry of the Cell Interface" (H. D. Brown, ed.), Part A, pp. 255–296. Academic Press, New York.
Laurent, T. C. (1971). *Eur. J. Biochem.* **21,** 498–506.
Legoy, M. D., Larreta Garde, V., Le Moullec, J. M., Ergan, F., and Thomas, D. (1980). *Biochimie* **62,** 341.
León, L. P., Sansur, M., Snyder, L. R., and Horvath, C. (1977). *Clin. Chem.* **23,** 1556–1562.
Lerche, D. (1976). *J. Membr. Biol.* **27,** 193–205.
Levin, Y., Pecht, M., Goldstein, L., and Katchalski, E. (1964). *Biochemistry* **3,** 1905–1913.
Linko, P., and Larinkari, J., eds. (1980). "Food Process Engineering," Vol. 2. Applied Science Publ. Barking.
Litman, D. J., Hanlon, T. M., and Ullman, E. F. (1980). *Anal. Biochem.* **106,** 223–229.
Mansson, M. -O, Larsson, P.-O., and Mosbach, K. (1978). *Eur. J. Biochem.* **86,** 455–463.
Mansson, M.-O, Larsson, P.-O., and Mosbach, K. (1979). *FEBS Lett.* **98,** 309–313.
Mansson, M.-O., Siegbahn, N., and Mosbach, K. (1983). *Proc. Natl. Acad. Sci. U.S.A.* **80,** 1487–1491.
Marconi, W., Prosperi, G., Giovenco, S., and Morisi, F. (1975). *J. Mol. Catal.* **1,** 111.
Masters, C. J. (1977). *Curr. Top. Cell. Regul.* **12,** 75–105.
Mattiasson, B. (1979). *ACS Symp. Ser.* p. 203.
Mattiasson, B., and Mosbach, K. (1971). *Biochim. Biophys. Acta* **235,** 253–257.
Mattiasson, B., and Mosbach, K. (1976). *In* "Methods in Enzymology" (K. Mosbach, ed.), Vol. 44, pp. 335–353. Academic Press, New York.
Mattiasson, B., Johansson, A. C., and Mosbach, K. (1974). *Eur. J. Biochem.* **46,** 341–349.
Mattiasson, B., Danielsson, B., and Mosbach, K. (1976). *Anal. Lett.* **9,** 217–234.
Maugh, T. H. (1984). *Science* **223,** 474–476.
Messing, R. A. (1981). *Appl. Biochem. Biotech.* **6,** 167.
Minton, A. P. (1981). *Biopolymers* **20,** 2093–2120.
Miura, Y., Urabe, H., Miyamoto, K., and Okazaki, M. (1980). *In* "Enzyme Engineering" (H. H. Weetall and G. P. Royer, eds.), Vol. 5, pp. 247–250. Plenum, New York.
Miura, Y., Urabe, H., Miyamoto, K., and Okazaki, M. (1981). *Artif. Organs* **5,** 72–79.
Mosbach, K., ed. (1976). "Methods in Enzymology," Vol. 44. Academic Press, New York.
Mosbach, K. (1978). *Adv. Enzymol.* **46,** 205–278.
Mosbach, K. (1980). *Trends Biochem. Sci.* **5,** 1–3.
Mosbach, K. (1982). *J. Chem. Tech. Biotechnol.* **32,** 179–188.
Mosbach, K. (1983). *Philos. Trans. R. Soc. London Ser. B* **300,** 355–367.
Mosbach, K. (ed.). (1985a). "Methods in Enzymology." Academic Press, New York. In press.
Mosbach, K. (1985b). *Annu. Rev. Biochem.* In press.
Mosbach, K., and Mattiasson, B. (1970). *Acta Chem. Scand.* **24,** 2093–2100.
Mosbach, K., and Mattiasson, B. (1978). *Curr. Top. Cell. Regul.* **14,** 187–241.
Mosbach, K., and Mosbach, R. (1966). *Acta. Chem. Scand.* **20,** 2807–2810.

Nilsson, K., and Larson, P.-O., (1983). *Anal. Biochem.* **134**, 60–72.

Nilsson, K., and Mosbach, K. (1981). *Biochem. Biophys. Res. Commun.* **102**, 449–457.

North, A. M. (1964). "The Collision Theory of Chemical Reactions in Liquids." Wiley, New York.

Okamoto, H., Tipayang, P., and Inada, Y. (1980). *Biochim. Biophys. Acta* **611**, 35–39.

O'Neill, S. P., Dunnill, P., and Lilly, M. D. (1971). *Biotechnol. Bioeng.* **13**, 337–352.

Patel, A. B., Pennington, S. N., and Brown, H. D. (1969). *Biochim. Biophys. Acta* **178**, 626–629.

Peterson, J. A., O'Keefe, D. H., Werringloer, J., Ebel, R. E., and Estabrook, R. W. (1978). *In* "Microenvironments and Metabolic Compartmentation" (P. A. Srere and R. W. Estabrook, eds.), pp. 433–447. Academic Press, New York.

Pittner, F. (1981). *J. Appl. Biochem. Biotechnol.* **6**, 85.

Pye, E. K., and Chance, B. (1976). *In* "Methods in Enzymology" (K. Mosbach, ed.), Vol. 44, pp. 357–372. Academic Press, New York.

Pye, E. K. and Weetall, H. H. (1978). "Enzyme Engineering," Vol. 3. Plenum, New York.

Pye, E. K., and Wingard, L. B. (1974). "Enzyme Engineering," Vol. 2. Plenum, New York.

Reeves, R. E., and Sols, A. (1973). *Biochem. Biophys. Res. Commun.* **50**, 459–466.

Regan, D. L., Dunnill, P., and Lilly, M. D. (1974). *Biotechnol. Bioeng.* **16**, 333–343.

Renneberg, R., Pfeiffer, D., Scheller, F., and Jänchen, M. (1982). *Anal. Chim. Acta* **134**, 359–364.

Roda, A., Girotti, S., Severino, G., Grigolo, B., Carrea, G., and Borara, R. (1984). *Clin. Chem.* **30**, 206–210.

Sélégny, E., Kernevez, J.-P., Broun, G., and Thomas, D. (1971). *Physiol. Veg.* **9**, 51–63.

Shimizu, S. Y., and Lenhoff, H. M. (1979). *J. Solid Phase Biochem.* **4**, 75–94, 95–107, 109–122.

Siegbahn, N., and Mosbach, K. (1982). *FEBS Lett.* **137**, 6–10.

Siegbahn, N., Månsson, M.-O., and Mosbach, K. (1985). In press.

Silman, I. H., and Karlin, A. (1967). *Proc. Natl. Acad. Sci. U.S.A.* **58**, 1664–1668.

Sitte, P. (1980). *In* "Cell Compartmentation and Metabolic Channeling" (L. Nover, F. Lynen, and K. Mothes, eds.), pp. 17–32. Elsevier, Amsterdam.

Solti, M., and Friedrich, P. (1979). *Eur. J. Biochem.* **95**, 551–559.

Somogyi, B. (1974a). *Acta Biochim. Biophys. Acad. Sci. Hung.* **9**, 175–184.

Somogyi, B. (1974b). *Acta Biochim. Biophys. Acad. Sci. Hung.* **9**, 185–196.

Somogyi, B., and Damjanovich, S. (1971). *Acta Biochim. Biophys. Acad. Sci. Hung.* **6**, 353–364.

Somogyi, B., and Damjanovich, S. (1973). *Acta Biochim. Biophys. Acad. Sci. Hung.* **8**, 153–160.

Srere, P. A., and Estabrook, R. W., eds. (1978). "Microenvironments and Metabolic Compartmentation" Academic Press, New York.

Srere, P. A., and Mosbach, K. (1974), *Annu. Rev. Microbiol.* **28**, 61–83.

Srere, P. A., Mattiasson, B., and Mosbach, K. (1973). *Proc. Natl. Acad. Sci. USA* **70**, 2534–2538.

Sundaram, P. V. (1978). *J. Solid Phase Biochem.* **3**, 185–197.

Sundaram, P. V., and Hornby, W. E. (1970). *FEBS Lett.* **10**, 325–327.

Thomas, D., and Broun, G. (1976). *In* "Methods in Enzymology" (K. Mosbach, ed.), Vol. 44, pp. 901–929. Academic Press, New York.

Torstensson, A., Johansson, G., Mansson, M.-O., Larsson, P.-O., and Mosbach, K. (1980). *Anal. Lett.* **13**, 837–849.

Ullman, E. F., Gibbons, I., Litman, D., Weng, L., and Di Nello, R. (1983). *In* "Immunoenzymatic Techniques" (S. Auremeas *et al.*, eds.), pp. 247–256.. Elsevier, Amsterdam.

Verduyn, C., Van Dijken, J. P., and Scheffers, W. A. (1983). *Biotechnol. Bioeng.* **25**, 1049–1055.

Wandrey, C., Wichmann, R., Bueckmann, A. F., and Kula, M. R. (1980). *In* "Enzyme Engineering" (H. H. Weetall and G. P. Royer, eds.), Vol. 5, pp. 453–456. Plenum, New York.

Wang, S. S., and King, C.-K. (1979). *Adv. Biochem. Eng.* **12**, 119.

Weetall, H. H., and Royer, G. P., eds. (1980). "Enzyme Engineering," Vol. 5. Plenum, New York.

Weitzman, P. D. J. (1973). *FEBS Lett.* **32**, 247–250.

Welch, G. R. (1977a). *Prog. Biophys. Mol. Biol.* **32**, 103–191.

Welch, G. R. (1977b). *J. Theor. Biol.* **68,** 267–291.

Welch, G. R. (1984). *In* "Dynamics of Biochemical Systems" (J. Ricard and A. Cornish-Bowden eds.), pp. 85–101. Plenum, New York.

Welch, G. R., Somogyi, B., and Damjanovich, S. (1982). *Prog. Biophys. Mol. Biol.* **39,** 109–146.

Welch, G. R., Somogyi, B., Matkó, J., and Papp, S. (1983). *J. Theor. Biol.* **100,** 211–238.

Wichmann, R., Wandrey, C., Bueckmann, A. F., and Kula, M. R. (1981). *Biotechnol. Bioeng.* **23,** 2789–2802.

Wienhausen, G., and De Luca, M. (1982). *Anal. Biochem.* **127,** 380–388.

Wienhausen, G. K., Kricka, L. J., Hinkley, J. E., and De Luca, M. (1982). *Appl. Biochem. Biotechnol.* **7,** 463–473.

Wilson, F. A., and Dietschy, J. M. (1974). *Biochim. Biophys. Acta* **363,** 112–126.

Wingard, L. B., Jr., Katchalski-Katzir, E., and Goldstein, L., eds. (1976). "Applied Biochemistry and Bioengineering," Vol. 1, Academic Press, New York.

Wombacher, H. (1983). *Mol. Cell. Biochem.* **56,** 155–164.

Yamazaki, Y., Maeda, H., and Kamibayashi, A. (1982). *Biotechnol. Bioeng.* **24,** 1915–1918.

Yu, Y.-T., and Chang, T. M. S. (1982). *Enzyme Microb. Technol.* **4,** 327–331.

7

Kinetic Analysis of Multienzyme Systems in Homogeneous Solution

Philip W. Kuchel

Department of Biochemistry
University of Sydney
Sydney, New South Wales, Australia

ORGANIZED
MULTIENZYME SYSTEMS

303

I. INTRODUCTION

A. General

This chapter aims (1) to describe the time-dependent behavior of the concentrations of reactants in sequences of enzyme-catalyzed reactions in homogeneous solution and (2) to develop means of estimating parameter values from the experimental data of such systems. The first aim can only be met, in general, by computer-based numerical integration of the nonlinear differential rate equations that describe the flux of reactants through the individual steps of the enzyme sequences. Repeated simulations of a kinetic system may allow the investigator to gain insight into those features of the system which most affect the behavior which is of particular experimental interest. The alternative approach is to obtain an analytical solution to the set of differential equations after making necessary simplifications and approximations; this way the *form* of the solution, usually an explicit function of time, may be amenable to the study of its general mathematical properties. Greater emphasis is given to the latter approach in this chapter, and the mathematics is entirely deterministic. In other words, the individual reaction rate constants have single values for a given set of experimental conditions and are not described by distributions of values as occurs in stochastic models (e.g., Gillespie, 1977).

The second aim involves analyzing the mathematical solutions and the possible application of nonlinear data regression procedures.

Some of the time- and concentration-dependent responses of enzyme systems are not properties of the *isolated* enzymes, but of the "society of enzymes and reactants" in which they are immersed (Kacser and Burns, 1979), just as "the mind" is a product of a neuronal network whose components are not "micro-minds." The stance adopted here is that the complex behavior of large enzyme systems is synthesized from the behavior, as determined by individual study, of isolated components of the system. The *synthetic* model gives, or purports to give, an explanation of the system based on current knowledge of its parts and overall responses. Such models are exemplified by Garfinkel and his colleagues' models of pyruvate metabolism in perfused hearts (Kohn *et al.*, 1977a–c; Garfinkel *et al.*, 1977; Achs *et al.*, 1977; Achs and Garfinkel, 1979, 1982; Garfinkel and Achs, 1979). An alternative approach is to develop a phenomenological model that accounts only for the responses recorded in a coherent set of data; such models are designated *analytic* and in the present context are exemplified by some models of red cell glycolysis and energy metabolism in the cell (Rapoport and Heinrich, 1975; Heinrich and Rapoport, 1975; Heinrich *et al.*, 1977; Reich and Sel'kov, 1981). The essence of this approach is simplicity. However, as will be seen later, it is not always possible to draw a hard distinction between the two approaches.

In summary, this chapter describes rather unique methods of solving the differential rate equations which characterize consecutive enzymatic reactions and gives means of elucidating enzyme–enzyme interactions, slow transitions in enzyme properties induced by changes in the reaction environment, and, on a more practical level, the bases for the optimal design of coupled assays. The central mathematical paradigm specified herein is applied rigorously to systems in homogeneous aqueous solution. Its extension to heterogeneous media (with diffusional resistance) remains for future exploration. (For a review of the latter designs see Engasser and Horvath, 1976.)

B. Basic Concepts

Although the emphasis of this chapter is on enzyme systems, the theory of sequential unimolecular reactions is considered first, since, as will become clear, many descriptions of enzyme systems can be reduced to these simpler cases. Some of the more important general concepts that are used follow.

1. Open and Closed Systems

Systems are described as either *open* or *closed* with respect to mass transfer. The former involves the flow of matter/energy into and possibly out of the system, while in the latter case no exchange takes place with the rest of the universe. Living systems are open while the usual (coupled) enzyme assay, in a test tube, is a closed system. An *adiabatic* system is a closed system in which no energy exchange occurs as well as no mass exchange.

2. Principle of Mass Action

The Principle of Mass Action states that "the rate of a chemical reaction is proportional to the concentrations of the reactants involved in the elementary chemical process." The constant of proportionality is called the *rate constant*, or *unitary rate constant*, to highlight the fact that it applies to an elementary process. A subtlety is the use of *chemical activites* (Moore, 1981, pp. 233–234) and not simply concentrations in rate expressions, but activity coefficients in biological systems are generally taken to be near one; nonideality is observed, however, in some aggregating protein (enzyme) systems (Jeffrey, 1981).

3. Concentration Brackets

Units of concentration are given as mol liter^{-1} and are denoted by square brackets [$\square$] around the symbol for a reactant. Constant volumes of the reaction systems are assumed here, and if necessary it is usually possible to convert rate expressions to account for volume changes should that be needed for realistic modeling (e.g., Hearon, 1949; Kacser and Burns, 1973). In a reaction scheme a reactant concentration is a function of time and other reactant concentrations. Frequently the time dependence is highlighted by the use of a subscript t or (t), e.g., $[S_i]$ denotes the concentration of S_i at any time, $[S_i]_0$ the concentration at zero time in the reaction, and $[S_i]_\infty$ the concentration at infinite time.

4. Molecularity and Order

The *molecularity* of an elementary reaction is the number of molecules involved in each reaction. Usually two molecules collide to give product(s) (molecularity = 2) or a single molecule undergoes fission (or scission; molecularity = 1). The *order* of a reaction is the sum of the powers to which the concentration (or chemical activity) terms are raised in the rate expression. For example, if the reaction rate $= [A]^{1/2}[B]^{1/3}$ the order of the reaction is $\frac{5}{6}$; the order of an enzymatic reaction described by the Michaelis–Menten equation [Michaelis and Menten, 1913; rate $= V_{max}[S]/(K_m + [S])$] is not an integer and ranges from near 0, when $[S] \gg K_m$, to near 1 when $[S] \ll K_m$.

5. Units of Rate Constants

For the sake of brevity, the units of rate constants are not given routinely in this chapter. They obey the dimensional relationship (mol liter^{-1})$^{-(n-1)}s^{-1}$, where n is the order of the reaction. Rate constants are denoted by lower case k, e.g., $k_{\pm i}$, where the sign indicates the direction of the reaction to which it applies, i.e., + for the direction initial reactants → products.

6. Extent of Reaction

This is the fractional progress of a reaction toward completion and is a dimensionless ratio. For example, in a reaction in which the product S_n increases with time the extent of reaction is given by $[S_n]/[S_n]_\infty$; for an initial reactant S_1, whose concentration declines with time, the extent of reaction is given by $([S_1]_0 - [S_1])/[S_1]_0$.

7. Nonlinearity

A mathematical function is *nonlinear* if it contains terms which involve products of variables; these can be either the dependent variables, the reactant concentrations, or the independent variable time. An alternative definition is "the partial derivative of the function with respect to any of the variables is not a constant." While, as will be shown, general analytical procedures exist for linear differential equations, no such *general* strategy exists for nonlinear equations. However, numerical integration can be used to solve either type of differential equation (McCracken and Dorn, 1964; Schied, 1968).

II. UNIMOLECULAR UNIDIRECTIONAL CLOSED SYSTEMS

A. Single Reaction

One of the simplest possible descriptions of a chemical reaction is

$$S_1 \xrightarrow{k_1} S_2. \tag{1}$$

The differential rate equation describing reactant flux is

$$\frac{d[S_1]}{dt} = -k_1[S_1]. \tag{2}$$

The conservation of mass condition is

$$[S]_0 = [S_1] + [S_2]. \tag{3}$$

The initial condition at time zero is given by

$$[S_1]_0 = [S]_0. \tag{4}$$

The solution obtained by using elementary calculus (Ayres, 1952) is

$$[S_1] = [S_1]_0 e^{-k_1 t}. \tag{5}$$

If experimental data, from a chemical kinetic experiment, are thought to be described by Eq. (5), then a plot of $\ln([S_1]/[S_1]_0)$ versus time will yield a line with slope $-k_1$. Alternatively, Eq. (5) can be fitted by nonlinear regression onto the data (Sadler, 1975; Osborne, 1976); this equation is nonlinear in the parameter k_1. The expression for the half-life ($t_{1/2}$), the time when $[S_1] = \frac{1}{2}[S_1]_0$, is given, after simple substitution into Eq. (5), by

$$t_{1/2} = \ln(2)/k_1. \tag{6}$$

B. Two Sequential Reactions

The scheme for two sequential reactions is given by

$$S_1 \xrightarrow{k_1} S_2 \xrightarrow{k_2} S_3. \tag{7}$$

The differential rate equations are

$$\frac{d[S_1]}{dt} = -k_1[S_1], \tag{8}$$

$$\frac{d[S_2]}{dt} = k_1[S_1] - k_2[S_2], \quad \text{and} \tag{9}$$

$$\frac{d[S_3]}{dt} = k_2[S_2]. \tag{10}$$

The conservation of mass condition is

$$[S]_0 = [S_1] + [S_2] + [S_3]. \tag{11}$$

Initial condition: Consider the special situation where only S_1 is present at $t = 0$. Then

$$[S_1]_0 = [S]_0, \quad [S_i]_0 = 0 \quad (i = 2, 3). \tag{12}$$

The solutions of Eqs. (8)–(10) are found in several texts (e.g., Moore, 1981, p. 345; Mellor, 1954, pp. 434–436) and it is important for later sections that they be included here.

Equation (8) and its integral are identical to Eqs. (2) and (5), respectively. The latter is substituted into Eq. (9), which on rearrangement yields

$$\frac{d[S_2]}{dt} + k_2[S_2] = k_1[S_1]_0 e^{-k_1 t}. \tag{13}$$

This first-order differential equation is solved using the integration factor $e^{k_2 t}$ (Ayres, 1952, pp. 12–14; Mellor, 1954, p. 381):

$$[S_2] = e^{-k_2 t} \int_0^t k_1[S_1]_0 e^{-k_1 t} e^{k_2 t} \, dt, \tag{14}$$

$$= k_1[S_1]_0 \left\{ \frac{e^{-k_1 t}}{(k_2 - k_1)} + \frac{e^{-k_2 t}}{(k_1 - k_2)} \right\}. \tag{15}$$

From Eqs. (11)–(13) and (15)

$$[S_3] = [S_1]_0 \left\{ 1 - \frac{k_1 e^{-k_1 t}}{(k_2 - k_1)} - \frac{k_2 e^{-k_2 t}}{(k_1 - k_2)} \right\}. \tag{16}$$

(A general principle that is evident in the two previous solutions is that in a closed system a conservation of mass condition can replace a differential equation in a set of simultaneous equations, thus simplifying the analysis. This also applies for obtaining numerical solutions.)

Equations (5), (15), and (16) will be analyzed further in the context of enzymatic reactions (Sections V and VI).

C. $n - 1$ Sequential Reactions, n Reactants

The scheme is given by

$$S_1 \xrightarrow{k_1} S_2 \xrightarrow{k_2} \cdots \xrightarrow{k_{n-2}} S_{n-1} \xrightarrow{k_{n-1}} S_n. \tag{17}$$

The differential rate equations are

$$\frac{d[S_1]}{dt} = -k_1[S_1] + \quad 0 \quad \cdots + \quad 0 \quad + 0 \quad \cdots + \quad 0 \quad + \quad 0$$

$$\frac{d[S_2]}{dt} = \quad k_1[S_1] \; -k_2[S_2] \cdots + \quad 0 \quad + 0 \quad \cdots + \quad 0 \quad + \quad 0$$

$$\vdots \qquad \vdots \qquad \vdots \qquad \vdots \qquad \vdots \qquad \vdots \qquad \vdots \qquad \vdots \tag{18}$$

$$\frac{d[S_i]}{dt} = \quad 0 \quad + \quad 0 \quad \cdots + k_{i-1}[S_{i-1}] - k_i[S_i] \cdots + \quad 0 \quad + \quad 0$$

$$\vdots \qquad \vdots \qquad \vdots \qquad \vdots \qquad \vdots \qquad \vdots \qquad \vdots$$

$$\frac{d[S_{n-1}]}{dt} = \quad 0 \quad + \quad 0 \quad \cdots + \quad 0 \quad + \quad 0 \quad \cdots + k_{n-2}[S_{n-2}] - k_{n-1}[S_{n-1}]$$

and

$$d[S_n]/dt = k_{n-1}[S_{n-1}]. \tag{19}$$

The conservation of mass condition is

$$[S]_0 = \sum_{i=1}^{n} [S_i]. \tag{20}$$

The solution of Eq. (18) using the method of Section II,B is extremely tedious for $n > 4$. Therefore, a procedure in which Eq. (18) is converted to algebraic equations which are then solved for the dependent variables is introduced (Rodiguin and Rodiguina, 1964; Capellos and Bielski, 1972; Roberts, 1977). The method employs an *integral transform* known as the *Laplace–Carson transform*; its definition and some important properties follow. The application of the method to solve Eq. (18) is taken up again in Section II,E and readers not interested in the mathematical details should skip Section II,D.

D. The Laplace–Carson Transform

1. Definition

Let $F(t)$ be a function of t specified for $t > 0$ (i.e., a time domain function). The Laplace–Carson transform, denoted by $L\{F(t)\}$, is

$$L\{F(t)\} = f(\Psi) = \Psi \int_0^\infty F(t) e^{-\Psi t} dt, \tag{21}$$

where in all the present work Ψ is a real number. Thus the symbol L, which denotes transformation of $F(t)$ into $f(\Psi)$, is called the *Laplace–Carson transform operator*. [The function $f(\Psi)$ exists if there exist real constants $M > 0$ and γ such that for all $t > N$, $|F(t)| < Me^{\gamma t}$ (Spiegel, 1965, p. 2). The function $F(t)$ is said to be a function of exponential order γ as $t \to \infty$, or, briefly, it is of *exponential order*.] This existence condition is satisfied in all the kinetic schemes discussed herein.

2. Properties of the Transform

The properties have been described by Rodiguin and Rodiguina (1964), Capellos and Bielski (1972), and to a lesser extent by Roberts (1977), and the opportunity is taken here to develop some expressions using an approach that is a little more mathematically rigorous than in the earlier works.

a. *Transform of a Constant.* If $F(t) = A$, a constant, then

$$L\{A\} = \Psi \int_0^\infty Ae^{-\Psi t} dt = \Psi \lim_{P \to \infty} \left.\frac{Ae^{-\Psi t}}{-\Psi}\right|_0^P = A. \tag{22}$$

b. *Transform of* $F(t) = e^{-\lambda t}$. This transformation is given by

$$L\{e^{-\lambda t}\} = \Psi \int_0^\infty e^{-\lambda t} e^{-\Psi t} dt,$$

$$= \Psi \left\{ \lim_{P \to \infty} -\frac{1}{(\Psi + \lambda)} e^{-(\Psi + \lambda)t} \Big|_0^P \right\},$$

$$= \frac{\Psi}{(\Psi + \lambda)}. \tag{23}$$

c. Transform of F(t) = t.   The transformation is given by

$$L\{t\} = \Psi \int_0^\infty t e^{-\Psi t} \, dt.$$

Integration by parts (Mellor, 1954, p. 204) gives

$$L\{t\} = \lim_{P \to \infty} \left. -t e^{-\Psi t} \right|_0^P + \int_0^\infty e^{-\Psi t} \, dt,$$

$$= 0 - \left\{ \lim_{P \to \infty} \frac{1}{\Psi} e^{-\Psi t} \Big|_0^P \right\} = \frac{1}{\Psi}. \tag{24}$$

d. Transform of F(t) = t^n.   This transformation is given by

$$L\{t^n\} = \Psi \int_0^\infty t^n e^{-\Psi t} \, dt.$$

Integration by parts as for Eq. (24) yields

$$L\{t^n\} = \lim_{P \to \infty} \left. -t^n e^{-\Psi t} \right|_0^P + n \int_0^\infty t^{n-1} e^{-\Psi t} \, dt,$$

$$= 0 + n \int_0^\infty t^{n-1} e^{-\Psi t} \, dt.$$

Iteration of the process gives

$$n \int_0^\infty t^{n-1} e^{-\Psi t} \, dt = \frac{n(n-1)}{\Psi} \int_0^\infty t^{n-2} e^{-\Psi t} \, dt,$$

$$= \frac{n(n-1)(n-2)}{\Psi^2} \int_0^\infty t^{n-3} e^{-\Psi t} \, dt,$$

$$\vdots$$

$$= \frac{n!}{\Psi^{n-1}} \int_0^\infty e^{-\Psi t} \, dt = \frac{n!}{\Psi^n}. \tag{25}$$

e. Transform of the Sum of Two Functions.   This transformation is equal to the sum of the transforms as follows:

$$L\{F_1(t) + F_2(t)\} = \Psi \int_0^\infty [F_1(t) + F_2(t)] e^{-\Psi t} \, dt.$$

By one of the elementary rules of integration (Spiegel, 1965, p. 12), this becomes

$$= \Psi \int_0^\infty F_1(t)e^{-\Psi t}\,dt + \Psi \int_0^\infty F_2(t)e^{-\Psi t}\,dt. \tag{26}$$

Because of the property expressed in this "theorem," Ψ is said to be a *linear operator* or that it has the *linearity property*.

f. The Key Property for Solving Differential Equations. This property is the transform of the first derivative of a function. Assume that $F(t)$ and its transform are known, then it is required to find the transform of $F'(t)$ (the first derivative with respect to t):

$$L\{F'(t)\} = \Psi \int_0^\infty F'(t)e^{-\Psi t}\,dt.$$

Integration by parts as for Eq. (24) yields

$$L\{F'(t)\} = \Psi\left\{\lim_{P\to\infty} e^{-\Psi t}F(t)\bigg|_0^P\right\} + \Psi\left\{\Psi \int_0^\infty F(t)e^{-\Psi t}\,dt\right\},$$

$$= -\Psi F(0) + \Psi f(\Psi). \tag{27}$$

Thus, the transform of the derivative of a function is equal to the transform of the function multiplied by the transform variable (Ψ), minus the zero time value of the original function (a constant), also multiplied by Ψ.

3. Application to Differential Equations

In solving differential equations, use is made of the previous properties of the transform, especially Eq. (27). Suppose after a series of mathematical manipulations we are left with a function of Ψ; it is then necessary to transform the function back to its original function of t; this then gives the solution of the problem. If the function is one of the elementary expressions given above, then simple inspection yields the original function of t. However, transforms of functions frequently are expressed as ratios of polynomials in Ψ, i.e., rational functions, and these are very important for the theory in subsequent sections. The analysis which gives the original functions of t for four important cases is developed below.

a. Inverse of a Rational Function. Let $f(\Psi)$ be the function:

$$f(\Psi) = \frac{f_1(\Psi)}{f_2(\Psi)} = \frac{a_0\Psi^m + a_1\Psi^{m-1} + \cdots + a_m}{b_0\Psi^n + b_1\Psi^{n-1} + \cdots + b_n}. \tag{28}$$

Assume that any common factors in the numerator and denominator have been canceled and $m \le n$. Decompose $f(\Psi)$ into partial fractions (Spiegel,

1965, pp. 46, 48–51). If all the roots of the polynomials $f_i(\Psi)$ differ, then

$$\frac{f_1(\Psi)}{f_2(\Psi)} = \frac{A_1}{(\Psi - \lambda_1)} + \frac{A_2}{(\Psi - \lambda_2)} + \cdots + \frac{A_n}{(\Psi - \lambda_n)} = \sum_{i=1}^{n} \frac{A_i}{(\Psi - \lambda_i)}, \quad (29)$$

where the λ_i are roots of $f_2(\Psi) = 0$ [i.e., $f_2(\Psi) = \prod_{i=1}^{n} (\Psi - \lambda_i)$]. The decomposition coefficients, A_i, are determined as follows: Multiply Eq. (28) by $(\Psi - \lambda_j)$ to give

$$(\Psi - \lambda_j)\frac{f_1(\Psi)}{f_2(\Psi)} = (\Psi - \lambda_j) \sum_{i=1}^{n} \frac{A_i}{(\Psi - \lambda_i)}. \quad (30)$$

Take the limit $\Psi \to \lambda_j$ of each side of Eq. (30). The limit of the right-hand side is A_j, but on the left the limits of both the numerator and denominator are zero. However, let us define

$$g_1(\Psi) = (\Psi - \lambda_j)f_1(\Psi),$$

then

$$g_1(\lambda_j) = 0, \qquad f_2(\lambda_j) = 0,$$

and by l'Hôpital's rule (Spiegel, 1965, pp. 46–47, 61–62)

$$\lim_{\Psi \to \lambda_j} \frac{g_1(\Psi)}{f_2(\Psi)} = \lim_{\Psi \to \lambda_j} \frac{g_1'(\Psi)}{f_2'(\Psi)}.$$

Differentiation of $g_1(\Psi)$ and $f_2(\Psi)$, and taking the limit $\Psi \to \lambda_j$, give

$$\lim_{\Psi \to \lambda_j} g_1'(\Psi) = \lim_{\Psi \to \lambda_j} [(\Psi - \lambda_j)f_1'(\Psi) + f_1(\Psi)] = f_1(\lambda_j), \quad (31)$$

which is not equal to zero because f_1 and f_2 have distinct roots. Using the product rule of differentiation,

$$\lim_{\Psi \to \lambda_j} f_2'(\Psi) = \lim_{\Psi \to \lambda_j} [(\Psi - \lambda_2)(\Psi - \lambda_3)\cdots(\Psi - \lambda_n)$$
$$+ (\Psi - \lambda_1)(\Psi - \lambda_3)\cdots(\Psi - \lambda_n)$$
$$+ \cdots + (\Psi - \lambda_1)(\Psi - \lambda_2)\cdots(\Psi - \lambda_{j-1})$$
$$\times (\Psi - \lambda_{j+1})\cdots(\Psi - \lambda_n)$$
$$+ \cdots + (\Psi - \lambda_1)(\Psi - \lambda_2)\cdots(\Psi - \lambda_{n-1})].$$

All terms except the product term that does not contain λ_j become zero in the limit; therefore,

$$f_2'(\lambda_j) = (\lambda_j - \lambda_1)(\lambda_j - \lambda_2)\cdots(\lambda_j - \lambda_{j-1})(\lambda_j - \lambda_{j+1})\cdots(\lambda_j - \lambda_n),$$

or, in more concise notation,

$$f'_2(\lambda_j) = \prod_{\substack{i=1 \\ i \neq j}}^{n} (\lambda_j - \lambda_i), \tag{32}$$

which is not equal to zero because the roots are all distinct. Therefore, from Eqs. (30)–(32), a variant of the Heaviside expansion formula (Spiegel, 1965, p. 47) is obtained:

$$A_j = f_1(\lambda_j) \Big/ \prod_{\substack{i=1 \\ i \neq j}}^{n} (\lambda_j - \lambda_i),$$

and

$$f(\Psi) = \frac{f_1(\Psi)}{f_2(\Psi)} = \sum_{j=1}^{n} \left[f_1(\lambda_j) \Big/ \prod_{\substack{i=1 \\ i \neq j}}^{n} (\lambda_j - \lambda_i) \right] \frac{1}{(\Psi - \lambda_j)}. \tag{33}$$

The "original" of function $f(\Psi)$ can now be readily determined since each term of the sum on the right-hand side of Eq. (33) is a constant multiplied by $1/(\Psi - \lambda_j)$; its inverse is obtained as follows. Using Eq. (26) with $F_1(t) + F_2(t) = 1 - e^{-\lambda t}$ as the original function, the transform $g(\Psi)$ is

$$g(\Psi) = \Psi \int_0^\infty (1 - e^{-\lambda t}) e^{-\Psi t} \, dt,$$

$$= \Psi \int_0^\infty e^{-\Psi t} \, dt - \Psi \int_0^\infty e^{-\lambda t} e^{-\Psi t} \, dt.$$

From Eq. (22) with $A = 1$, and Eq. (23),

$$g(\Psi) = 1 - \frac{\Psi}{(\Psi + \lambda)} = \frac{\lambda}{(\Psi + \lambda)}, \tag{34}$$

i.e., the inverse of $1/(\Psi + \lambda)$ is $(1 - e^{-\lambda t})/\lambda$. Therefore, the inverse of $f(\Psi)$ in Eq. (33) is readily determined after taking note of the change of sign of the λ_is between Eq. (35) and Eq. (33):

$$F(t) = \sum_{j=1}^{n} \left[f_1(\lambda_j) \Big/ \lambda_j \prod_{\substack{i=1 \\ i \neq j}}^{n} (\lambda_j - \lambda_i) \right] (e^{\lambda_j t} - 1), \tag{35}$$

which appears in special forms in rows 6–12 in Table I.

b. If $f_2(\Psi)$ has a Root $\lambda = 0$. In accordance with L'Hôpital's rule and with the same mathematical argument used to arrive at Eq. (30) let A_0 be the

TABLE I

TABLE OF FUNCTIONS OF t AND THEIR LAPLACE–CARSON TRANSFORMS

Row	Function of t	Function of Ψ	Conditions
			$a, b, \ldots, \lambda, \lambda_i$ $(i = 1, 2, \ldots, n)$ (all real)
	Particular cases		
1.	A	A	
2.	t	$\dfrac{1}{\Psi}$	
3.	$t^n/n!$	$1/\Psi^n$	n (integer)
4.	$(1/\pm\lambda)(1 - e^{\mp\lambda t})$	$1/(\Psi \pm \lambda)$	$\lambda > 0$
5.	$\dfrac{a}{\lambda} \mp \left\{\dfrac{a \mp \lambda}{\lambda}\right\} e^{\mp\lambda t}$	$\dfrac{\Psi \pm a}{\Psi \pm \lambda}$	$\lambda > 0$
6.	$\dfrac{1}{\lambda_1\lambda_2} - \dfrac{1}{\lambda_1(\lambda_2 - \lambda_1)}e^{-\lambda_1 t} - \dfrac{1}{\lambda_2(\lambda_1 - \lambda_2)}e^{-\lambda_2 t}$	$\dfrac{1}{(\Psi + \lambda_1)(\Psi + \lambda_2)}$	
7.	$\dfrac{1}{(\lambda_2 - \lambda_1)}e^{-\lambda_1 t} + \dfrac{1}{(\lambda_1 - \lambda_2)}e^{-\lambda_2 t}$	$\dfrac{\Psi}{(\Psi + \lambda_1)(\Psi + \lambda_2)}$	
8.	$\dfrac{a}{\lambda_1\lambda_2} - \dfrac{(a - \lambda_1)}{\lambda_1(\lambda_2 - \lambda_1)}e^{-\lambda_1 t} - \dfrac{(a - \lambda_2)}{\lambda_2(\lambda_1 - \lambda_2)}e^{-\lambda_2 t}$	$\dfrac{(\Psi + a)}{(\Psi + \lambda_1)(\Psi + \lambda_2)}$	
9.	$-\dfrac{\lambda_1}{(\lambda_2 - \lambda_1)}e^{-\lambda_1 t} - \dfrac{\lambda_2}{(\lambda_1 - \lambda_2)}e^{-\lambda_2 t}$	$\dfrac{\Psi^2}{(\Psi + \lambda_1)(\Psi + \lambda_2)}$	
	General expressions		
10.	$\dfrac{1}{\prod_{j=1}^{n} \lambda_j} - \sum_{j=1}^{n} \dfrac{1}{\lambda_j \prod_{\substack{k=1 \\ k \neq j}}^{n}(\lambda_k - \lambda_j)}e^{-\lambda_j t}$	$\dfrac{1}{\prod_{j=1}^{n}(\Psi + \lambda_j)}$	
11.	$\dfrac{\prod_{l=1}^{m} a_l}{\prod_{i=1}^{n} \lambda_i} - \sum_{j=1}^{n} \dfrac{\{\prod_{l=1}^{m}(a_l - \lambda_j)\}}{\lambda_j \prod_{\substack{k=1 \\ k \neq j}}^{n}(\lambda_k - \lambda_j)}e^{-\lambda_j t}$	$\dfrac{\prod_{l=1}^{m}(\Psi + a_l)}{\prod_{j=1}^{n}(\Psi + \lambda_j)}$	$m < n$
12.	$\sum_{j=1}^{n} \dfrac{\{\prod_{l=1}^{m}(a_l - \lambda_j)\}}{\prod_{\substack{k=1 \\ k \neq j}}^{n}(\lambda_k - \lambda_j)}e^{-\lambda_j t}$	$\dfrac{\Psi \prod_{l=1}^{m}(\Psi + a_l)}{\prod_{j=1}^{n}(\Psi + \lambda_j)}$	$m < n$
13.	$\dfrac{1}{(n - 1)!}\int_0^t e^{\mp\lambda x}x^{n-1}\,dx$	$\dfrac{1}{(\Psi \pm \lambda)^n}$	$n > 0$
14.	$\dfrac{t^{n-1}}{(n - 1)!}e^{\mp\lambda t}$	$\dfrac{\Psi}{(\Psi \pm \lambda)^n}$	$n > 0$

constant term of $f_1(\Psi)$, then

$$\lim_{\Psi \to 0} \frac{f_1(\Psi)}{f_2(\Psi)} = \frac{f'_1(0)}{f'_2(0)} = \frac{A_0}{\lambda_1 \cdots \lambda_n},$$

and from Eq. 35

$$f(\Psi) = \frac{A_0}{\prod_{i=1}^{n} \lambda_i} + \sum_{j=1}^{n} \frac{f_1(\lambda_j)}{\lambda_j \prod_{\substack{i=1 \\ i \neq j}}^{n} (\lambda_j - \lambda_i)} (e^{\lambda_j t} - 1). \tag{36}$$

c. *Inverse of $f(\Psi) = \Psi f_1(\Psi)/f_2(\Psi)$.* Here f_1 and f_2 are polynomials in Ψ with the degree of f_1 less than that of f_2. We assume that common roots have been canceled. Then, using an argument similar to that used for the development of Eq. (36).

$$f(\Psi) = \sum_{j=1}^{n} \frac{f_1(\lambda_j)}{f'_2(\lambda_j)} \frac{\Psi}{(\Psi - \lambda_j)}.$$

The original function of t is obtained from the sum of the inverses of terms of the form $\Psi/(\Psi - \lambda_j)$, which, according to Eq. (23), yields,

$$f(\Psi) = \sum_{j=1}^{n} \frac{f_1(\lambda_j)}{f'_2(\lambda_j)} e^{\lambda_j t}, \tag{37}$$

and this is given in special forms in rows 7 and 12 in Table I.

d. *Inverse of a Rational Function where the Denominator has Multiple Roots.* In this case let

$$f(\Psi) = f_1(\Psi)/f_2(\Psi),$$

but $f_2(\Psi)$ has, in addition to simple roots, an α-tuple root λ_n:

$$f_2(\Psi) = a_0(\Psi - \lambda_1)(\Psi - \lambda_2) \cdots (\Psi - \lambda_{n-1})(\Psi - \lambda_n)^{\alpha}.$$

The ratio of polynomials is decomposed into partial fractions as before (Section II,D,3,a):

$$\frac{f_1(\Psi)}{f_2(\Psi)} = \frac{A_1}{(\Psi - \lambda_1)} + \frac{A_2}{(\Psi - \lambda_2)} + \cdots + \frac{A_{n-1}}{(\Psi - \lambda_{n-1})} + \frac{{}^1A_n}{(\Psi - \lambda_n)} + \frac{{}^2A_n}{(\Psi - \lambda_n)^2}$$

$$+ \cdots + \frac{{}^{\alpha}A_n}{(\Psi - \lambda_n)^{\alpha}}. \tag{38}$$

In a manner analogous to the development of Eq. (33), multiply both sides of Eq. (38) by $(\Psi - \lambda_j)$:

$$A_j = \frac{f_1(\Psi)}{f_2(\Psi)}(\Psi - \lambda_j) - (\Psi - \lambda_j) \sum_{\substack{i=1 \\ i \neq j}}^{n-1} \frac{A_i}{(\Psi - \lambda_j)} - \cdots - (\Psi - \lambda_j) \sum_{i=1}^{\alpha} \frac{{}^iA_n}{(\Psi - \lambda_n)^i}.$$

Take the limit $\Psi \to \lambda_j$ ($j \neq n$) and the third and fourth terms are zero while the first and second terms yield $A_j = f_1(\lambda_j)/f'_2(\lambda_j)$, as in Eq. (33). Therefore,

$$\frac{f_1(\Psi)}{f_2(\Psi)} = \sum_{j=1}^{n-1} \frac{f_1(\lambda_j)}{f'_2(\lambda_j)} \frac{1}{(\Psi - \lambda_j)} + \frac{{}^1 A_n}{(\Psi - \lambda_n)} + \cdots + \frac{{}^\alpha A_n}{(\Psi - \lambda_n)^\alpha}. \tag{39}$$

Multiplying both sides by $(\Psi - \lambda_n)^\alpha$ gives

$$\frac{f_1(\Psi)}{f_2(\Psi)}(\Psi - \lambda_n)^\alpha = (\Psi - \lambda_n)^\alpha \sum_{j=1}^{n-1} \frac{f_1(\lambda_j)}{f'_2(\lambda_j)} \frac{1}{(\Psi - \lambda_j)} + {}^1 A_n (\Psi - \lambda_n)^{-(\alpha+1)}$$

$$+ \cdots + {}^\alpha A_n.$$

Applying L'Hôpital's rule as before and taking the limit $\Psi \to \lambda_n$ yield

$$^\alpha A_n = \left\{ \frac{f_1(\Psi)}{f_2(\Psi)}(\Psi - \lambda_n)^\alpha \right\}_{\Psi = \lambda_n};$$

that is,

$$^\alpha A_n = \frac{f_1(\lambda_n)}{a_0(\lambda_n - \lambda_1)(\lambda_n - \lambda_2)\cdots(\lambda_n - \lambda_{n-1})}.$$

The other coefficients, $^{\alpha-1}A_n$, $^{\alpha-2}A_n$, etc., are determined by differentiating Eq. (39) with respect to Ψ and then equating Ψ with λ_n. Thus,

$$^{\alpha-1}A_n = \left\{ \frac{d}{d\Psi} \left[\frac{f_1(\Psi)(\Psi - \lambda_n)^\alpha}{f_2(\Psi)} \right] \right\}_{\Psi = \lambda_n}, \tag{40}$$

$$^{\alpha-2}A_n = \frac{1}{2!} \left\{ \frac{d^2}{d\Psi^2} \left[\frac{f_1(\Psi)(\Psi - \lambda_n)^\alpha}{f_2(\Psi)} \right] \right\}_{\Psi = \lambda_n}, \tag{41}$$

and so on to $^1 A_n$. After evaluating all of the decomposition coefficients, $f(\Psi)$ is inverted to its original time-domain function. Each term of the original function is like those of Eq. (35) plus the inverses of the following multiples:

$$\frac{1}{(\Psi - \lambda_n)}, \frac{1}{(\Psi - \lambda_n)^2}, \ldots, \frac{1}{(\Psi - \lambda_n)^\alpha}.$$

The inverse of $1/(\Psi - \lambda_n)^j$ is (see expression 13 in Table I)

$$L^{-1}\left\{ \frac{1}{(\Psi - \lambda_n)^j} \right\} = \frac{1}{(-\lambda_n)^j} - \frac{e^{\lambda_n t}}{(-\lambda_n)^j}\left[1 - \lambda_n t + \frac{(\lambda_n t)^2}{2!} + \cdots + \frac{(-\lambda_n t)^{j-1}}{(j-1)!} \cdots \right]. \tag{42}$$

Thus the full inverse (original) function of time can be derived.

e. *Inverse of* $f(\Psi) = f_1(\Psi)/f_2(\Psi)$, *where* $f_2(\Psi)$ *has Multiple Roots.* The function can be expressed as a sum of partial fractions as in Eq. (38):

$$f(\Psi) = \sum_{j=1}^{n-1} \frac{f_1(\lambda_j)}{f'_2(\lambda_j)} \frac{\Psi}{(\Psi - \lambda_j)} + \frac{{}^1A_n\Psi}{(\Psi - \lambda_n)} + \frac{{}^2A_n(\Psi)}{(\Psi - \lambda_n)^2} + \cdots + \frac{{}^\alpha A_n}{(\Psi - \lambda_n)^\alpha}. \quad (43)$$

The coefficients ${}^iA_n$ are determined as before, and in the final expression it is seen that each term contains the following expression of Ψ:

$$\frac{\Psi}{(\Psi - \lambda_n)^\alpha}, \dots, \frac{\Psi}{(\Psi - \lambda_n)}, \quad \text{and} \quad \frac{\Psi}{(\Psi - \lambda_j)} \quad (j = 1, 2, \dots, n - 1).$$

The inversion of these terms to their original functions of time is carried out according to Eq. (23) to yield the following:

$$\frac{\Psi}{(\Psi - \lambda_j)} \rightarrow e^{\lambda_j t}, \quad \frac{\Psi}{(\Psi - \lambda_n)^2} \rightarrow te^{\lambda_n t}, \quad \dots, \quad \frac{\Psi}{(\Psi - \lambda_n)^\alpha} \rightarrow \frac{t^{\alpha-1}e^{\lambda_n t}}{(\alpha - 1)!}.$$

Therefore, the original function, $F(t)$, is

$$F(t) = L^{-1}\left\{\frac{\Psi f_1(\Psi)}{f_2(\Psi)}\right\} = \sum_{j=1}^{n} \frac{f_1(\lambda_j)}{f'_2(\lambda_j)} e^{\lambda_j t} + {}^1A_n e^{\lambda_n t} + {}^2A_n e^{\lambda_n t}$$

$$+ \cdots + \frac{{}^\alpha A_n e^{\lambda_n t} t^{\alpha-1}}{(\alpha - 1)!}. \quad (44)$$

E. *n* − 1 Sequential Reactions; The Laplace–Carson Solution

The theory of integral transforms is complicated and subtle but the *application* of the method to the solution of Eq. (18), and others like it, is simple, as will be illustrated now.

On repeated use of Eq. (26) and after some rearrangment, Eq. (18) is transformed to

$$\Psi[S_1]_0 = (\Psi + k_1)s_1 + \quad 0 \quad + \quad 0 \quad \cdots + \quad 0 \quad + \quad 0$$

$$0 \quad = \quad -k_1 s_1 \quad + (\Psi + k_2)s_2 + \quad 0 \quad \cdots + \quad 0 \quad + \quad 0$$

$$0 \quad = \quad 0 \quad - \quad k_2 s_2 \quad + (\Psi + k_3)s_3 \cdots + \quad 0 \quad + \quad 0 \quad (45)$$

$$\vdots \quad \quad \vdots \quad \quad \vdots \quad \quad \vdots \quad \quad \vdots \quad \quad \vdots$$

$$0 \quad = \quad 0 \quad + \quad 0 \quad + \quad 0 \quad \cdots - k_{n-2}s_{n-2} + (\Psi + k_{n-1})s_{n-1},$$

where it is assumed that $[S_i]_0 = 0$ $(i = 2, 3, \dots, n)$, and the Laplace–Carson transform of $[S_i]_t$ is given by the corresponding lower case letter. The array of algebraic equations can be written in vector and matrix form:

$$\mathbf{c} = \mathbf{As}, \quad (46)$$

where $\mathbf{c}$ is the vector of constant terms, $\mathbf{s}$ the vector of transformed concentrations, and $\mathbf{A}$ the matrix of coefficients in Eq. (45).

Equation (46) is solved for each s_i using Cramer's rule (Lipschutz, 1968, p. 177) as follows with s_1 as an example:

$$s_1 = \frac{\begin{vmatrix} \Psi[S_1]_0 & 0 & 0 & \cdots & 0 & 0 \\ 0 & (\Psi + k_2) & 0 & \cdots & 0 & 0 \\ 0 & -k_2 & (\Psi + k_3) & \cdots & 0 & 0 \\ \vdots & \vdots & \vdots & \ddots & \vdots & \vdots \\ 0 & 0 & 0 & \cdots & -k_{n-2} & (\Psi + k_{n-1}) \end{vmatrix}}{\begin{vmatrix} (\Psi + k_1) & 0 & 0 & \cdots & 0 & 0 \\ -k_1 & (\Psi + k_2) & 0 & \cdots & 0 & 0 \\ 0 & -k_2 & (\Psi + k_3) & \cdots & 0 & 0 \\ \vdots & \vdots & \vdots & \ddots & \vdots & \vdots \\ 0 & 0 & 0 & \cdots & -k_{n-2} & (\Psi + k_{n-1}) \end{vmatrix}}, \quad (47)$$

where the denominator is the determinant of $\mathbf{A}$. In shorter notation,

$$s_i = \frac{|\mathbf{A}_{i,\mathbf{c}}|}{|\mathbf{A}|},$$

where $|\mathbf{A}_{i,\mathbf{c}}|$ is the determinant of the matrix formed from $\mathbf{A}$ after replacing the ith column by the vector of constant terms. The matrices are both *lower triangular* forms, i.e., the nonzero terms lie below the *leading diagonal* (also called the *trace*). In this case the determinant is simply the product of the terms in the leading diagonal (Lipschutz, 1968, pp. 177–196). Thus,

$$s_1 = \frac{\Psi[S_1]_0(\Psi + k_2)(\Psi + k_3)\cdots(\Psi + k_{n-1})}{(\Psi + k_1)(\Psi + k_2)\cdots(\Psi + k_{n-1})} = \frac{\Psi[S_1]_0}{(\Psi + k_1)}. \quad (48a)$$

The inverse of Eq. (48a) (Table I, row 4) is

$$[S_1] = [S_1]_0 e^{-k_1 t}; \quad (48b)$$

this is as expected from Eq. (5).

Application of Cramer's rule to Eq. (46) to obtain s_2 yields

$$s_2 = \frac{k_1 \Psi[S_1]_0(\Psi + k_3)(\Psi + k_4)\cdots(\Psi + k_{n-1})}{(\Psi + k_1)(\Psi + k_2)\cdots(\Psi + k_{n-1})},$$

$$= \frac{k_1[S_1]_0 \Psi}{(\Psi + k_1)(\Psi + k_2)}.$$

From Table I, row 7, the inverse transform is

$$[S_2] = k_1[S_1]_0\left\{\frac{e^{-k_1t}}{(k_2 - k_1)} + \frac{e^{-k_2t}}{(k_1 - k_2)}\right\}, \tag{49a}$$

which is identical to Eq. (15), as expected.

In the special case where all of the k_i have identical values (k) the expression for s_2 is

$$s_2 = \frac{k[S_1]_0\Psi}{(\Psi + k)^2},$$

which has the inverse, according to row 14 of Table I,

$$[S_2] = k[S_1]_0te^{-kt}. \tag{49b}$$

The general expression, and this is where the transform method is most valuable, is obtained from the inverse of

$$s_j = \frac{k_1k_2k_3\cdots k_{j-1}\Psi[S_1]_0(\Psi + k_{j+1})\cdots(\Psi + k_{n-1})}{(\Psi + k_1)(\Psi + k_2)\cdots(\Psi + k_{j+1})\cdots(\Psi + k_{n-1})},$$

$$= \frac{[S_1]_0\left\{\prod\limits_{i=1}^{j-1} k_i\right\}\Psi}{\prod\limits_{l=1}^{j}(\Psi + k_l)}, \tag{50}$$

which yields, from Table I, row 12, with $m = 0$, the expression for $[S_j]$:

$$[S_j] = [S_1]_0k_1k_2\cdots k_{j-1}\left\{\frac{1}{(k_2 - k_1)(k_3 - k_1)\cdots(k_j - k_1)}e^{-k_1t}\right.$$

$$+ \frac{1}{(k_1 - k_2)(k_3 - k_2)\cdots(k_j - k_2)}e^{-k_2t}$$

$$\left. + \cdots + \frac{1}{(k_1 - k_j)(k_2 - k_j)\cdots(k_{j-1} - k_j)}e^{-k_jt}\right\}. \tag{51a}$$

In other notation, the solution is a sum of j exponentials,

$$[S_j] = [S_1]_0\sum_{i=1}^{j}\frac{1}{\alpha_i}e^{-k_it}, \tag{51b}$$

where the α_i are constants identified with the terms in Eq. (51a), i.e., this is the *general* solution for the intermediate species in a linear sequence of unidirectional first-order reactions in a closed system when $[S_i]_0 = 0$ $(i \neq 1)$.

The special case in which all k_i have the same value (k) has the following expression for $[S_j]$:

$$[S_j] = \frac{k^{j-1}[S_1]_0 t^{j-1} e^{-kt}}{(j-1)!}. \tag{51c}$$

And, by differentiation of this expression and equating it to zero, the time at which the maximum value of $[S_j]$ is attained (t_{max}) is readily seen to be

$$t_{max} = \frac{(j-1)}{k}. \tag{51d}$$

Finally, $d[S_n]/dt = k_{n-1}[S_{n-1}]$. The Laplace–Carson transform of this differential equation is $\Psi s_n = k_{n-1} s_{n-1}$; i.e.,

$$s_n = \frac{k_{n-1} s_{n-1}}{\Psi}. \tag{52}$$

Substitution of Eq. (50), with $j = n - 1$, into Eq. (51) yields

$$s_n = [S_1]_0 \prod_{i=1}^{n-1} k_i \frac{1}{\prod_{l=1}^{n-1}(\Psi + k_{n-1})},$$

and using Table I, row 10, the solution is

$$[S_n] = [S_1]_0 \left\{ 1 - \frac{k_1 \cdots k_{n-1}}{k_1(k_2 - k_1)(k_3 - k_1)\cdots(k_2 - k_{n-1})} e^{-k_1 t} \right.$$
$$- \frac{k_1 \cdots k_{n-1}}{k_2(k_2 - k_1)(k_2 - k_3)\cdots(k_2 - k_{n-1})} e^{-k_2 t}$$
$$\left. - \cdots - \frac{k_1 \cdots k_{n-1}}{k_{n-1}(k_{n-1} - k_1)(k_{n-1} - k_2)\cdots(k_{n-1} - k_{n-2})} e^{-k_{n-1} t} \right\}. \tag{53a}$$

This is the *general* solution for the concentration of final product, $[S_n]$, of a linear sequence of unidirectional reactions in a closed system when $[S_i]_0 = 0$ $(i \neq 1)$; it has the general form

$$[S_n] = [S_1]_0 \sum_{i=1}^{n} \left(1 - \frac{1}{\alpha_i} e^{-k_i t} \right). \tag{53b}$$

Figure 1 demonstrates the relatively complicated behavior of a linear sequence of irreversible unimolecular reactions as described by the scheme in Eq. (17). Because the scheme is irreversible, the time course of $[S_1]$ is the same as for the simple two-reactant system of Eq. (1). Similarly, the expression for $[S_2]$ is given by Eq. (15) [actually it is Eq. (49b) in this special case where all the k_i are equal]. The intermediate compounds $(S_2–S_4)$ display maxima in their time courses then decay to zero concentration while $[S_5]$ follows the sigmoidal

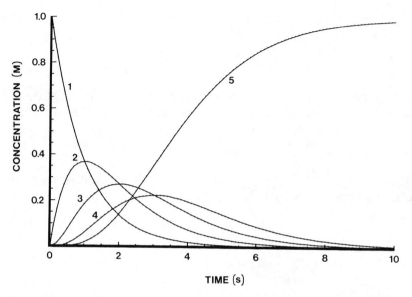

Fig. 1. Evolution of the concentrations of reactants with time in the closed irreversible first-order scheme

$$S_1 \xrightarrow{k_1} S_2 \xrightarrow{k_2} S_3 \xrightarrow{k_3} S_4 \xrightarrow{k_4} S_5,$$

i.e., Eq. (17) with $n = 5$. The parameter values and initial conditions employed in the simulation were $[S_1]_0 = 1$, $[S_i]_0 = 0$ $(i \neq 1)$, $k_i = 1$. The concentrations of reactants are described by the following expressions: $[S_1]$, Eq. (48b); $[S_2]$, Eq. (49); S_3 and S_4, Eq. (51), $j = 3$ and 4, respectively; $[S_5]$, Eq. (53), $n = 5$. The curves are labeled with numbers that are the subscripts of the S_i to which they apply. Although dimensions have been given to the numbers on the axes it is valid to view the simulations as those of a dimensionless scheme and they are therefore more general.

trajectory characteristic of at least two exponentials and as described by Eq. (53).

Figure 2 is similar to the previous one, except that a system of 10 reactants is simulated. Note that in this long irreversible sequence the first four reactants follow time courses that are identical to the five-reactant system; this arises simply because of the irreversible nature of the reactions, i.e., whatever occurs further down the pathway has no effect on earlier reactions. The progress curve of the final reactant, S_{10}, is sigmoidal (like $[S_5]$ in the previous figure) and displays very little "character" that would indicate that it is described by 10 exponential terms [or, more exactly with the present choice of parameters, by terms in t up to t^9 multiplied by an exponential, see Eq. (51c)]; this highlights the enormous problem of parameter determination in real systems that are often far more complex than linear unimolecular irreversible sequences.

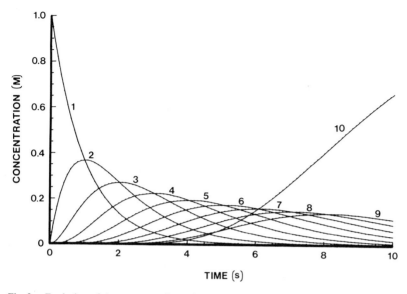

Fig. 2.   Evolution of the concentrations of reactants with time in the closed irreversible first-order scheme

$$S_1 \xrightarrow{k_1} S_2 \xrightarrow{k_2} \cdots \xrightarrow{k_9} S_{10},$$

i.e., Eq. (17) with $n = 10$. The parameter values used in the simulations were the same as in Fig. 1. The time courses of $[S_1]$, $[S_2]$, $[S_3]$, and $[S_4]$ are identical to those in Fig. 1. The curves for $[S_5],\ldots,[S_9]$ are described by Eq. (51) with the appropriate value of j, and $[S_{10}]$ is described by Eq. (53) with $n = 10$. The curves are labeled with numbers which are the subscripts of the S_i to which they apply. The comment about dimensions given in the caption of Fig. 1 also applies here.

III. UNIMOLECULAR BIDIRECTIONAL CLOSED SYSTEMS

A. Single Reaction

The simplest reaction scheme is

$$S_1 \underset{k_{-1}}{\overset{k_1}{\rightleftharpoons}} S_2 \tag{54}$$

and the corresponding differential equations describing reactant changes with time are

$$\frac{d[S_1]}{dt} = -k_1[S_1] + k_{-1}[S_2], \qquad \frac{d[S_2]}{dt} = k_1[S_1] - k_{-1}[S_2]. \tag{55}$$

The conservation condition for this closed system is

$$[S]_0 = [S_1] + [S_2]. \tag{56}$$

Let $[S_1]_0$ and $[S_2]_0$ be nonzero. Use of the "key" property of the Laplace–Carson transform procedure (Section II,D,2,f) leads to the algebraic equations

$$\Psi s_1 - \Psi[S_1]_0 = -k_1 s_1 + k_{-1} s_2, \qquad \Psi s_2 - \Psi[S_2]_0 = k_1 s_1 - k_{-1} s_2. \quad (57)$$

Rearrangement of Eq. (57) yields the nonhomogeneous equations that can be solved using Cramer's rule (Section II,E) to give

$$s_1 = \frac{\Psi[S_1]_0 + k_{-1}([S_1]_0 + [S_2]_0)}{(\Psi + k_1 + k_{-1})},$$

$$s_2 = \frac{\Psi[S_2]_0 + k_1([S_1]_0 + [S_2]_0)}{(\Psi + k_1 + k_{-1})}.$$

The inverse transforms of the s_i are found from Table I, rows 4 and 5:

$$[S_1] = \frac{k_{-1}[S]_0}{(k_1 + k_{-1})}\{1 - \exp[-(k_1 + k_{-1})t]\} + [S_1]_0 \exp[-(k_1 + k_{-1})t].$$

$$(58)$$

Similarly,

$$[S_2] = \frac{k_1[S]_0}{(k_1 + k_{-1})}\{1 - \exp[-(k_1 + k_{-1})t]\} + [S_2]_0 \exp[-(k_1 + k_{-1})t].$$

$$(59)$$

These general expressions are simplified if $[S_2]_0 = 0$ and $[S_1]_0 = [S]_0$. Thus,

$$[S_1] = \frac{[S_1]_0}{(k_1 + k_{-1})}\{k_{-1} + k_1 \exp[-(k_1 + k_{-1})t]\}, \qquad (60)$$

$$[S_2] = \frac{k_1[S_1]_0}{(k_1 + k_{-1})}\{1 - \exp[-(k_1 + k_{-1})t]\}. \qquad (61)$$

The general formulae, Eqs. (58) and (59), and the special case, Eqs. (60) and (61), naturally reduce to the same expression as $t \to \infty$, i.e., when the system reaches equilibrium. Then

$$[S_1]_e = [S_1]_0 k_{-1}/(k_1 + k_{-1}), \qquad (62)$$

$$[S_2]_e = [S_1]_0 k_1/(k_1 + k_{-1}), \qquad (63)$$

where the subscript e signifies the equilibrium condition. Furthermore, it is known that the equilibrium constant for the reaction is

$$K_{eq} = [S_2]_e/[S_1]_e = k_1/k_{-1},$$

which is satisfied by Eqs. (62) and (63).

B. Two Consecutive Reactions

The scheme is

$$S_1 \underset{k_{-1}}{\overset{k_1}{\rightleftharpoons}} S_2 \underset{k_{-2}}{\overset{k_2}{\rightleftharpoons}} S_3 \tag{64}$$

The rearranged transformed differential equations that describe the react-ant fluxes in the scheme are

$$\begin{aligned}
\Psi[S_1]_0 &= (\Psi + k_1)s_1 - \quad k_{-1}s_2 \quad + \quad 0, \\
\Psi[S_2]_0 &= \quad k_1 s_1 \quad - (\Psi + k_{-1} + k_2)s_2 - \quad k_{-2}s_3, \\
\Psi[S_3]_0 &= \quad 0 \quad - \quad k_2 s_2 \quad + (\Psi + k_{-2})s_3.
\end{aligned} \tag{65}$$

For the sake of simplicity consider the particular initial conditions $[S_i]_0 = 0$ $(i \neq 1)$. Using Cramer's rule,

$$s_1 = \frac{\Psi[S_1]_0(\Psi + k_{-1} + k_2)(\Psi + k_{-2})}{\Delta},$$

where Δ is the determinant of the coefficient matrix of Eq. (65):

$$\Delta = \Psi[\Psi^2 + \Psi(k_1 + k_{-1} + k_2 + k_{-2}) + (k_1 k_{-2} + k_1 k_2 + k_{-1}k_{-2})].$$

The roots of Δ, $-\lambda_1$, and $-\lambda_2$, are obtained by solving the quadratic expression to give

$$\lambda_1 = [A + (A^2 - 4B)^{1/2}]/2, \qquad \lambda_2 = [A - (A^2 - 4B)^{1/2}]/2,$$

where $A = k_1 + k_{-1} + k_2 + k_{-2}$, $B = k_1 k_{-2} + k_1 k_2 + k_{-1}k_{-2}$.
Therefore,

$$s_1 = \frac{[S_1]_0(\Psi + k_{-1} + k_2)(\Psi + k_{-2})}{(\Psi + \lambda_1)(\Psi + \lambda_2)}, \tag{66}$$

and the inverse of s_1 is obtained from Table I, row 11:

$$[S_1] = [S_1]_0 \left\{ \frac{(k_{-1} + k_2)k_{-2}}{\lambda_1 \lambda_2} + \frac{(k_{-1} + k_2 - \lambda_1)(k_{-2} - \lambda_1)}{(\lambda_1 - \lambda_2)} e^{-\lambda_1 t} \right.$$

$$\left. + \frac{(k_{-1} + k_2 - \lambda_2)(k_{-2} - \lambda_2)}{(\lambda_2 - \lambda_1)} e^{-\lambda_2 t} \right\}. \tag{67}$$

Now, $\lambda_1 \lambda_2 = B$ and $\lambda_1 + \lambda_2 = A$, so, some simplifications are possible in Eq. (67). Further consideration of Eq. (67) will be left to particular exam-ples of enzyme systems that, after certain approximations, conform to the model in Eq. (64).

A similar analysis to that involved in obtaining an expression for $[S_1]$ gives

$$s_2 = \frac{\Psi[S_1]_0 k_1(\Psi + k_{-2})}{\Delta} = \frac{[S_1]_0 k_1(\Psi + k_{-2})}{(\Psi + \lambda_1)(\Psi + \lambda_2)},$$

and

$$[S_2] = k_1[S_1]_0 \left\{ \frac{k_{-2}}{\lambda_1 \lambda_2} - \frac{(k_{-2} - \lambda_1)}{\lambda_1(\lambda_2 - \lambda_1)} e^{-\lambda_1 t} - \frac{(k_{-2} - \lambda_2)}{\lambda_2(\lambda_1 - \lambda_2)} e^{-\lambda_2 t} \right\}; \quad (68)$$

$$s_3 = \frac{[S_1]_0 k_1 k_2}{\Delta} = \frac{[S_1]_0 k_1 k_2}{(\Psi + \lambda_1)(\Psi + \lambda_2)},$$

and

$$[S_3] = k_1 k_2 [S_1]_0 \left\{ \frac{1}{\lambda_1 \lambda_2} - \frac{1}{\lambda_1(\lambda_2 - \lambda_1)} e^{-\lambda_1 t} - \frac{1}{\lambda_2(\lambda_1 - \lambda_2)} e^{-\lambda_2 t} \right\}. \quad (69)$$

Mathematical expressions to describe the reversible schemes are clearly more difficult to derive than for the irreversible ones. The extension of the

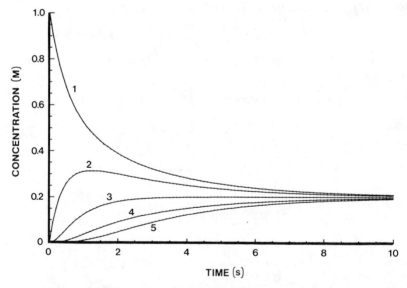

Fig. 3. Reactant concentrations in time in the closed first-order reversible scheme

$$S_1 \underset{k_{-1}}{\overset{k_1}{\rightleftharpoons}} S_2 \underset{k_{-2}}{\overset{k_2}{\rightleftharpoons}} S_3 \underset{k_{-3}}{\overset{k_3}{\rightleftharpoons}} S_4 \underset{k_{-4}}{\overset{k_4}{\rightleftharpoons}} S_5,$$

which is an extension of Eq. (64). The following parameters and initial values were used in the numerical integration of the relevant rate equations: $k_{\pm i} = 1$, $[S_1]_0 = 1$, $[S_i]_0 = 0$ $(i \neq 1)$. The comments made in the caption of Fig. 1 concerning dimensions on axes and labels on curves also apply here.

general analysis to three consecutive reversible reactions involves the solution of a quartic polynomial which is the expression for the determinant of the coefficient matrix of the transformed differential rate equations (Kuchel and Chapman, 1984). Because of its complexity it will not be reproduced here.

The problem of estimating values of the k_i from an experimental system that conforms to Eq. (64) will not be considered here but the general procedure, which uses the expression for the quantities $\lambda_1\lambda_2$ and $(\lambda_1 + \lambda_2)$ that are obtained by graphical analysis of experimental data, is discussed in detail in Section VII,B,2. The kinetic behavior of reversible systems is far more complex than for the irreversible cases, since all reactants, no matter what their position in the pathway, "influence" the concentrations of all others. Figure 3 highlights the difference between the reversible and earlier irreversible schemes shown in Figs. 1 and 2. Attainment of equilibrium of all species is evident as is the fact that the maximum concentration of some reactants is their equilibrium value, while others, S_2 in this case, display "overshoot" of their equilibrium concentration.

IV. COUPLED ENZYME SYSTEMS— POLYNOMIAL SOLUTIONS

A. General

The simplest description of a system of coupled, unidirectional, Michaelis–Menten-type enzymes is as follows:

$$E_1 + S_1 \xrightleftharpoons[k_{-1}]{k_1} E_1S_1 \xrightarrow{k_2} E_1 + S_2$$
$$E_2 + S_2 \xrightleftharpoons[k_{-3}]{k_3} E_2S_2 \xrightarrow{k_4} E_2 + S_3 \tag{70}$$

where the E_i denote enzymes and k_2 and k_4 are the corresponding *turnover numbers* (Laidler and Bunting, 1973, p. 254). A full set of differential (rate) equations written in terms of the individual rate constants may readily be constructed but the analytical integration of this set is possible only in special cases that will be considered later (Section V), or in general by numerical methods.

Figure 4 shows the result of a numerical simulation of the scheme in Eq. (70) using parameter values and concentrations of enzymes and reactants that might be encountered experimentally. There is obvious build up of the product (S_2) of the first enzyme-catalyzed reaction and the appearance of S_3 follows the characteristic sigmoidal path seen in all such systems.

It is possible to proceed with nonnumerical analysis and integrate in general terms, for a small extent of reaction, a set of differential equations that have been written assuming that each of the enzyme–substrate complexes is already in a steady state (Kuchel et al., 1974, 1975). Thus, the rate equations are

$$\frac{d[S_1]}{dt} = -\frac{V_1[S_1]}{K_1 + [S_1]}, \tag{71}$$

$$\frac{d[S_2]}{dt} = \frac{V_1[S_1]}{K_1 + [S_1]} - \frac{V_2[S_2]}{K_2 + [S_2]}, \tag{72}$$

$$\frac{d[S_3]}{dt} = \frac{V_2[S_2]}{K_2 + [S_2]}, \tag{73}$$

where V_1 and V_2 are the maximum velocities, $k_2[E_1]_0$ and $k_4[E_2]_0$, respectively, and K_1 and K_2 are the Michaelis constants, given respectively by $K_1 = (k_{-1} + k_2)/k_1$ and $K_2 = (k_{-3} + k_4)/k_3$. Equations (71) and (72) follow from the Briggs and Haldane (1925) treatment while Eq. (73) follows from the differentiation of the following conservation of mass condition, i.e.,

$$[S]_0 = [S_1] + [S_2] + [S_3]. \tag{74}$$

B. Methods of Solving Equations (71)–(73)

1. Expansion of the Logarithmic Term

Equation (71) is readily integrated by separation of the variables (Haldane, 1930, p. 75) to give

$$V_1 t = ([S_1]_0 - [S_1]) - K_1 \ln([S_1]/[S_1]_0), \tag{75}$$

where it is assumed that $[S_1]_0 = [S]_0$ and $[S_2]_0 = [S_3]_0 = 0$. This equation is physically valid only over the time domain in which the enzyme–substrate complex can be considered to be in an approximate steady state, a point that has been considered by many investigators (Laidler, 1955; Hommes, 1962; Walter and Morales, 1964; Wong, 1965; Walter, 1966).

Note that Eq. (75) expresses $[S_1]$ as an *implicit* function of t [i.e., $t = t([S_1])$] which is not useful in relation to further integration of Eq. (72). In order to obtain $[S_1]$ in Eq. (75) as an *explicit* function of t the logarithmic term is expanded as a power series to obtain

$$V_1 t = ([S_1]_0 - [S_1]) + K_1 \left\{ \left(\frac{[S_1]_0 - [S_1]}{[S_1]_0} \right) - \left(\frac{[S_1]_0 - [S_1]}{[S_1]_0} \right)^2 \frac{1}{2} \right.$$
$$\left. + \cdots + (-1)^{n-1} \left(\frac{[S_1]_0 - [S_1]}{[S_1]_0} \right)^n \frac{1}{n} + \cdots \right\}. \tag{76}$$

Application of the ratio test (Mellor, 1954, p. 272) to the n and $(n + 1)$th terms of the series indicates that the series converges absolutely for all $0 < [S_1] < 2[S_1]_0$, which is nonrestrictive since these conditions are always met experimentally (in a closed system).

Truncation of the series after the squared term gives a quadratic in $[S_1]$. This is solved for $[S_1]$ and after expansion of the square root of the discriminant the following expression is obtained:

$$[S_1] = [S_1]_0 - \frac{V_1[S_1]_0 t}{(K_1 + [S_1]_0)} + \frac{V_1^2 K_1 [S_1]_0}{(K_1 + [S_1]_0)^3} \frac{t^2}{2!}. \tag{77}$$

Retention of higher degree terms necessitates the solution of higher degree polynomials for which a general solution exists up to a quartic (Kline, 1972, pp. 267–270). However, this method of integration fails to yield a simple general solution for other types of rate equations; therefore alternatives discussed in Section IV,B,2 are used.

2. Integration in Series

This method (Ayres, 1952, Chs. 25, 26) assumes that the $[S_i]$ can be expressed as an infinite power series in t that converges for values of t in the neighborhood of $t = 0$, i.e,

$$[S_i] = [S_i]_0 + A_1 t + A_2 t^2 + \cdots + A_n t^n + \cdots. \tag{78}$$

It suffices to illustrate the application of Eq. (78) to $[S_1]$ for which Eq. (71) may be written as

$$V_1[S_1] + K_1[S_1]^{\langle 1 \rangle} + [S_1][S_1]^{\langle 1 \rangle} = 0, \tag{79}$$

where $\langle i \rangle$ denotes the ith derivative with respect to time. Application of Eq. (79) requires evaluations of $[S_1]$, $[S_1]^{\langle 1 \rangle}$, and their product, which follows directly from Eq. (78):

$$[S_1] = [S_1]_0 + A_1 t + A_2 t^2 + \cdots + A_n t^n + \cdots,$$

$$[S_1]^{\langle 1 \rangle} = A_1 + 2A_2 t + 3A_3 t^2 + \cdots + (n + 1)A_{n+1} t^n + \cdots,$$

$$[S_1][S_1]^{\langle 1 \rangle} = A_1[S_1]_0 + (A_1^2 + 2A_2[S_1]_0)t + (3[S_1]_0 A_3 + 3A_1 A_2)t^2$$

$$+ \cdots + \left[\sum_{j=0}^{n} (j + 1)A_{j+1} A_{n-j} \right] t^n + \cdots,$$

where $A_0 = [S_1]_0$.

Substitution of the latter three equations into Eq. (79) with expansion of each term leads to a new power series in t; each coefficient is then equated to zero to yield explicit expressions for $A_1, A_2, \ldots, A_n$. This method leads to

$$
\begin{aligned}
[S_1] = [S_1]_0 &- \frac{V_1[S_1]_0}{(K_1 + [S_1]_0)} t + \frac{V_1^2 K_1 [S_1]_0}{(K_1 + [S_1]_0)^3} \frac{t^2}{2} \\
&+ \frac{V_1^3 K_1 [S_1]_0 (2[S_1]_0 - K_1)}{(K_1 + [S_1]_0)^5} \frac{t^3}{3!} \\
&+ \frac{V_1^4 K_1 [S_1]_0 (K_1^2 - 8K_1[S_1]_0 + 6[S_1]_0^2)}{(K_1 + [S_1]_0)^7} \frac{t^4}{4!} + \cdots.
\end{aligned}
\tag{80}
$$

It is somewhat reassuring to note that the first three terms of Eqs. (77) and (79) are identical. However, the second method is still cumbersome to apply, especially to find expressions for $[S_2]$ and $[S_3]$. When it is recalled that the rate equations [Eqs. (71)–(73)] pertain to only one particular model, the need for an easier method, such as is given below, becomes apparent.

3. Maclaurin Polynomials

Since the concentration of each species S_i in a reaction scheme can be described by a continuous and differentiable function in t (the deterministic stance) it may be expressed as a Taylor series whose coefficients are evaluated at the origin ($t = 0$, $[S_i]_0$, $i = 1, 2, \ldots, n$); i.e., A Maclaurin polynomial (Mellor, 1954, pp. 286–291):

$$
[S_i] = [S_i]_0 + [S_i]_0^{\langle 1 \rangle} t + [S_i]_0^{\langle 2 \rangle} \frac{t^2}{2!} + \cdots + [S_i]_0^{\langle n \rangle} \frac{t^n}{n!} + \cdots.
\tag{81}
$$

Expressions for $[S_i]^{\langle 1 \rangle}$ are simply the given rate equations [Eqs. (71)–(73)]. Higher order derivatives require more labor to obtain them; note that they require successive differentiation of Eqs. (71)–(73) *prior* to the substitution $[S_i] = [S_i]_0$. Equations (71)–(73) may be written as

$$
[S_1]^{\langle 1 \rangle} = -f_1\{[S_1](t)\},
\tag{82}
$$

$$
[S_2]^{\langle 1 \rangle} = f_1\{[S_1](t)\} - f_2\{[S_2](t)\},
\tag{83}
$$

$$
[S_3]^{\langle 1 \rangle} = f_2\{[S_2](t)\},
\tag{84}
$$

where the $[S_i](t)$ are written to emphasize the fact that the $[S_i]$ are functions of t. Also, the equations are *autonomous*, i.e., they do not contain *terms* in t.

By introducing $[S_2]_0$ and the appropriate derivatives, obtained by the chain rule of differentiation, into Eq. (81) we obtain

$$
[S_2] = [S_2]_0 + (f_1 - f_2)t + \left\{ -\frac{df_1}{d[S_1]} f_1 - \frac{df_2}{d[S_2]}(f_1 - f_2) \right\} \frac{t^2}{2!} + \cdots.
\tag{85}
$$

If $[S_2]_0 = 0$, then $(f_2)_0 = 0$ and

$$[S_2]_0 = 0 + (f_1)_0 t + \left\{ -\frac{df_1}{d[S_1]} f_1 - \frac{df_2}{d[S_1]} f_1 \right\} \frac{t^2}{2!} + \cdots. \tag{86}$$

Therefore,

$$\lim_{t \to 0} \left\{ \frac{[S_2]}{t} \right\} = (f_1)_0 = \frac{V_1 K_1}{(K_1 + [S]_0)}. \tag{87}$$

Similar analysis for S_3 is as follows. Note that

$$[S_3] = [S]_0 - [S_1] - [S_2], \tag{88}$$

$$[S_3]^{\langle 1 \rangle} = 0 + f_1 - (f_1 - f_2) = f_2, \tag{89}$$

$$[S_3]^{\langle 2 \rangle} = \frac{df_2}{d[S_2]} (f_1 - f_2), \tag{90}$$

and at $t = 0$, $f_2 = 0$ and $[S_3]_0 = 0$.
Therefore,

$$[S_3] = 0 + 0 + [S_3]^{\langle 2 \rangle} \frac{t^2}{2!} + \cdots. \tag{91}$$

In the present scheme,

$$f_2 = \frac{V_2 [S_2]}{(K_2 + [S_2])}, \quad \text{therefore} \quad \left\{ \frac{df_2}{d[S_2]} \right\}_0 = \frac{V_2}{K_2}, \tag{92}$$

and from Eqs. (91) and (92),

$$[S_3] = \frac{V_2}{K_2} \frac{V_1 [S_1]_0}{(K_1 + [S_1]_0)} \frac{t^2}{2} + \cdots. \tag{93}$$

The expression for $[S_3]$ given in Eq. (93), truncated at the t^2 term, is a parabola in t. Reference to Fig. 4 reveals that the early part of the time course of $[S_3]$ is at least "superficially" parabolic.

Furthermore,

$$\lim_{t^2 \to 0} \left\{ \frac{[S_3]}{t^2} \right\} = \frac{V_2}{K_2} \frac{V_1 [S_1]_0}{(K_1 + [S_1]_0)}. \tag{94}$$

C. Long Sequences of Michaelis–Menten Enzymes

For longer sequences of irreversible reactions like those in Eq. (70) it is readily inferred from the previous section that

$$[S_n] = \frac{df_n}{d[S_n]} f_1 \frac{t^{n-1}}{(n-1)!} + \cdots, \tag{95}$$

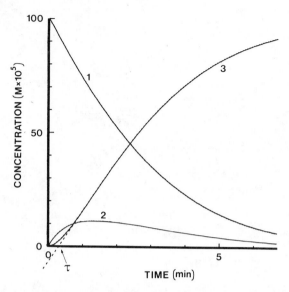

Fig. 4. Evolution of reactant concentrations in a closed coupled enzyme system. The reaction scheme is that of Eq. (70) and the simulation was performed by numerical integration of the appropriate differential equations with the following parameters and initial values: $k_1 = 1 \times 10^7$; $k_{-1} = 1 \times 10^4$; $k_2 = 1 \times 10^3$; $k_3 = 2 \times 10^7$; $k_{-3} = 2 \times 10^4$; $k_4 = 2 \times 10^3$; $[S_1]_0 = 1 \times 10^{-3}$ mol liter^{-1}; $[E_1]_0 = 1 \times 10^{-8}$ mol liter^{-1}; $[E_2]_0 = 2 \times 10^{-8}$ mol liter^{-1}. These parameters give steady-state parameter values as follows: $K_1, K_2 = 1.1 \times 10^{-3}$ mol liter^{-1}; $V_1 = 1 \times 10^{-5}$ mol liter^{-1} s^{-1}; $V_2 = 4 \times 10^{-5}$ mol liter^{-1} s^{-1}. Numerical integration of Eqs. (71)–(73) with these steady-state parameters gives an almost identical simulation result for $t \geq 1$ ms. The extrapolation of the linear part of the progress curve of $[S_3]$ intersects the abscissa at $t = \tau$. An expression for τ is given by Eq. (122a) with $n = 3$.

provided that $[S_i]_0 = 0$ $(i = 2, 3, \ldots, n)$ and f_n is defined as in Eqs. (82)–(84). Thus, in the limit of $t \to 0$,

$$[S_n]/t^{n-1} = (df_n/d[S_n])f_1.$$

In words, the first term of the power series is one with t raised to the power given by one less than the number of chemical species in the sequence. The general expression for an n-member sequence of Michaelis–Menten enzymes is (Kuchel *et al.*, 1974)

$$\lim_{t^{n-1} \to 0} \left\{ \frac{[S_n]}{t^{n-1}} \right\} = \left\{ \prod_{i=2}^{n-1} \frac{V_i}{K_i} \right\} \frac{V_1[S_1]_0}{(K_1 + [S_1]_0)} \frac{1}{(n-1)!}. \tag{96}$$

D. Other Variations

If the f_i of Eqs. (82)–(84) are altered to describe reversible Michaelis–Menten reactions and if $[S_i]_0 = 0$ $(i \neq 1)$, then in the limit $t \to 0$ the initial slope $([S_3]/t^2)$ is still given by Eq. (94). Intuitively, this is not surprising, since in the neighborhood of $t = 0$, the reverse reaction has negligible effect on the time course. Furthermore, since steady states of E_iS_i complexes are assumed in this analysis, as in any other steady-state enzyme kinetic analysis under constant conditions, it is not possible to discern the number of different enzyme–reactant complexes (Roberts, 1977, Ch. 2).

Another "complication" is accumulating-product inhibition (competitive or noncompetitive) of one or both of the enzymes. But this effect is again eliminated by considering only the neighborhood of $t = 0$ as occurs in the analysis of the initial slope of the progress curve defined by Eq. (94).

The presence of the intermediate compound S_2 at $t = 0$ alters the value of the derivative, in the case of the two simple enzymes in Eq. (92), thus modifying Eq. (94); V_2/K_2 is replaced by $V_2 K_2/(K_2 + [S_2]_0)^2$. If S_2 inhibits E_1 in some way, then the rate expression for E_1 must be appropriately modified for the analysis. If, for example, S_2 is a competitive inhibitor, then K_1 in Eq. (94) is replaced simply by $K_1(1 + [S_2]_0/K_I)$ where K_I is the competitive inhibition constant (Laidler and Bunting, 1973, p. 98).

Finally, if f_1 is a rate expression that describes a control enzyme, then Eq. (96) also can be applied (Nichol et al., 1974). In other words, in a coupled assay, as $t \to 0$, the behavior of the system is that of the first enzyme multipled by the dimensionless scaling factor, i.e.,

$$\left\{ \prod_{i=2}^{n-1} \left(\frac{df_i}{d[S_i]} \right)_0 \right\} \frac{t^{n-1}}{(n-1)!}$$

E. Applications

The domain of t over which Eqs. (94) and (96) apply as suitable descriptions of the time course of $[S_3]$ clearly depends on the various parameter values. Figure 5 illustrates the extent of deviation of the approximation from the (true) numerical solution; retention of higher degree terms clearly improves the approximation to a much larger extent of reaction. The t^2 analysis was successfully applied in an investigation of the coupled arginase–urease system in which K_m and V_{max} for arginase were determined from the coupled assay and compared with analysis of arginase alone (Kuchel et al., 1975). The procedure has been suggested as a means for assessing enzyme–enzyme interaction. Deviation of the time course of the coupled reaction from the behavior predicted for the system on the basis of information on each enzyme studied alone could indicate enzyme association (Nichol et al., 1974).

The polynomial-based analysis is useful when rates of the enzymes in a couple cannot be independently varied as can be done with bifunctional

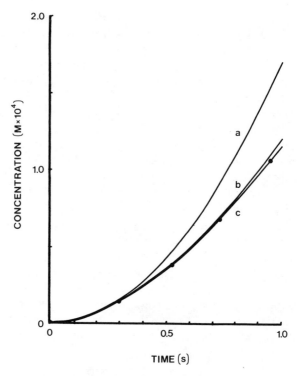

Fig. 5. The increase of final product concentration, $[S_3]$, with time in a sequence of two reactions each catalyzed by a Michaelis–Menten enzyme [Eq. (70)]. Curve c was computed by integrating numerically Eqs. (71)–(73) with the kinetic parameters V_1, $V_2 = 3.3 \times 10^{-3}$ mol liter^{-1} s^{-1}, K_1, $K_2 = 5.165 \times 10^{-3}$ mol liter^{-1} and $[S_1]_0 = 10^{-3}$ mol liter^{-1}. The solid dots were computed by integrating numerically the rate equation corresponding to Eq. (70), which did not presuppose a steady state of the two enzyme–substrate complexes; values of the individual rate constants needed for this calculation were chosen to be consistent with the above-mentioned steady-state parameter values and were $k_1 = 2.0 \times 10^7$ mol liter^{-1} s^{-1}; $k_{-1} = 1.0 \times 10^5$ s^{-1}; $k_2 = 3.3 \times 10^3$ s^{-1}; $[E_1]_0 = 1.0 \times 10^{-6}$ mol liter^{-1}; same corresponding values for E_2. Lines a and b refer to the Maclaurin polynomial solution of Eqs. (71)–(73) with retention of the t^2 and of the t^4 terms, respectively.

enzymes like chorismate mutase-prephenate dehydratase (Koch *et al.*, 1971). However, it is frequently possible to manipulate the ratios of the enzymes to ensure an excess of coupling enzyme(s). This leads to simpler data analysis, as will be considered in Section V. Furthermore, the desire to extract as much information as possible from one time course means that analysis of the whole time course by the methods outlined in Section VIII can, in principle, be used.

V. COUPLED ENZYME SYSTEMS—APPROXIMATIONS

A. General Overview—Two Enzymes

Haldane (1930) considered the problem of integrating Eqs. (71)–(73) and noted that in general they were not integrable, except numerically. However, if K_1 and K_2 are "large" or "small" compared with $[S_1]$ and $[S_2]$, then six cases arise that are amenable to further (semi)quantitative analysis. These cases are summarized in Table II.

With K_1 and K_2 both large the system is quasi-unimolecular and conforms to the scheme in Eq. (7). Its solutions are Eqs. (5), (15), and (16) where the first-order rate constants k_1 and k_2 are replaced by V_1/K_1 and V_2/K_2. Further comments will be made on this system later (Section V,C).

If K_1 and K_2 are both small relative to the substrate concentrations then zero-order kinetics are approached for each enzyme. If $V_1 > V_2$ then $[S_2]$ rises at a rate equal to $V_1 - V_2$ until S_1 is nearly depleted. If $V_1 < V_2$ the rate of the E_1 reaction is limiting so S_1 appears at a rate almost equal to V_1; nevertheless, there must be an initial "lag" phase which will be discussed later.

The cases shown in rows 4 and 5 of Table II are both amenable to much further investigation. When $V_1 > V_2$ the differential equations are

$$d[S_1]/dt = -V_1, \tag{97}$$

$$d[S_2]/dt = V_1 - V_2[S_2]/(K_2 + [S_2]). \tag{98}$$

Separation of variables and integration give

$$\int_0^{[S_2]} \frac{(K_2 + [S_2])d[S_2]}{K_2 V_1 + (V_1 - V_2)[S_2]} = t. \tag{99}$$

After some manipulation (see Section V,C),

$$t = \frac{[S_2]}{(V_1 - V_2)} - \frac{V_2 K_2}{(V_1 - V_2)^2} \ln\left\{1 + \frac{(V_1 - V_2)}{V_1 K_2}[S_2]\right\}. \tag{100}$$

However, this implicit function of $[S_2]$ does not appear to have been of any practical value, especially since a steady state of $[S_2]$ is not attained. The other case, $V_2 > V_1$, has been studied further (Storer and Cornish-Bowden, 1974; Varfolomeev, 1977) although apparently, since the work was not cited, the authors were unaware of Haldane's (1930) earlier solution of the relevant equations. This system is taken up further in Section V,D.

TABLE II

Two Consecutive Michaelis–Menten Enzymes[a]—Approximations That Enable Further Analysis

Row	Relative magnitudes of parameters[b]			Order of each reaction		Solution of differential equations
	K_1	K_2	V_1 V_2	1.	2.	
1	Large	Large	$\gtrsim$	First order	First order	Eqs. (5), (13), (16)
2	Small	Small	$>$	Zero order	Zero order	$[S_1] = [S_1]_0 - V_1 t$; $[S_2] = (V_1 - V_2)t$; $[S_1] = V_2 t$
3	Small	Small	$<$	Zero order	Zero order	$[S_1] = [S_1]_0 - V_1 t$; $[S_3] = V_1 t$
4	Small	Large	$>$	Zero order	First order or Michaelis–Menten	$[S_1] = [S_1]_0 - V_1 t$; Eq. (100)
5	Small	Large	$<$	Zero order	First order or Michaelis–Menten	Eq. (107)
6	Large	Small	$\gtrsim$	First order	Zero order	Eqs. (5), (104)

[a] The rate equations are defined in Eqs. (71)–(73).

[b] The magnitudes of K_1 and K_2 are relative to $[S_1]$ and $[S_2]$, respectively, and the inequality signs indicate the relative magnitudes of V_1 and V_2.

Finally, with the parameter conditions of row 6 of Table II the simplified rate equations are

$$d[S_1]/dt = -(V_1/K_1)[S_1], \tag{101}$$

$$d[S_2]/dt = (V_1/K_1)[S_1] - V_2. \tag{102}$$

With $[S_2]_0$ and $[S_3]_0 = 0$ the solution of Eq. (101) is Eq. (5) with $V_1/K_1 \equiv k_1$, and

$$[S_2] = [S_1]_0\{1 - \exp[-(V_1/K_1)t]\} - V_2 t. \tag{103}$$

Under conditions where the approximations hold, $[S_1]$ declines purely exponentially, and $[S_2]$ reaches a maximum at

$$[S_2]_{max} = [S_1]_0 - \frac{V_2 K_1}{V_1}\left\{1 + \ln\left(\frac{V_1[S_1]_0}{V_2 K_1}\right)\right\}, \tag{104}$$

and $[S_3]$ increases linearly with a time course given by $V_2 t$.

Several important questions, which follow, remain after this semi-quantitative analysis: (1) When can the approximations be applied? (2) Under what conditions can a steady state of $[S_2]$ be expected? (3) What are the optimal conditions for attainment of a steady-state rate of $[S_3]$ production such that the rate reflects that of the first enzyme? These questions will now be addressed.

B. Coupled Enzyme Assays—Zero-Order/First-Order Schemes

1. General

As was mentioned in Section IV, coupled assays are frequently employed when the product of the first reaction is not readily detectable. Thus, *auxiliary* enzymes (Bergmeyer, 1962) are added that result in the formation of products, such as NADH, which are readily and continuously detectable using spectrophotometry. A basic problem in designing the assays is determining the amounts of auxiliary enzymes needed in order that the steady-state rate of product formation is monitored and *not* the rates of the auxiliary enzymes. As was shown in Section IV and Fig. 5, there is an inherent lag in product formation prior to the steady-state rate of product formation. This lag must be analyzed according to the theory in Section IV or, as will be shown here, minimized in order to apply alternative theory. The general strategy with the present analysis is to use excessive amounts of auxiliary enzymes, thus ensuring that $[S_2]$ is always small and less than K_2. Then, the analytical situation summarized in Table II, rows 4 and 5, pertains.

Several investigators (McClure, 1969; Wurster and Hess, 1970; Easterby, 1973; Rudolph *et al.*, 1979) have analyzed the scheme in which E_2 is an irreversible Michaelis–Menten enzyme, i.e.,

$$S_1 \xrightarrow{\ E_1\ } S_2 \xrightarrow{\ E_2\ } S_3. \tag{105}$$

In practice, either S_1 is assumed to saturate E_1 so that $v_1 = V_1$, or only a small fraction of the total amount of S_1 is "used up" over the period in which the rate of production of S_3 is measured. With either of these situations it is possible to calculate the amount of E_2 required to fulfill the requirement of a steady state in $[S_2]$ after a small extent of reaction. The rate equation for $[S_2]$ is

$$d[S_2]/dt = v_1 - (V_2/K_2)[S_2], \tag{106}$$

which has the solution

$$[S_2] = v_1(K_2/V_2)\{1 - \exp[(-V_2/K_2)t]\}. \tag{107}$$

For long times ($t \to \infty$) the exponential term approaches zero and the steady-state concentration of S_2 becomes

$$[S_2]_{ss} = v_1(K_2/V_2). \tag{108}$$

2. Time to Attain a Steady State of $[S_2]$

Since $[E_2] = V_2/k_{2cat}$, where k_{2cat} is the turnover number of E_2, the amount of E_2 required to allow a steady state to be attained in the time t_{ss} is calculated from a rearrangement of Eq. (107):

$$\ln\left(1 - \frac{V_2}{K_2}\frac{[S_2]}{v_1}\right) = -\frac{V_2}{K_2}t. \tag{109}$$

Using Eqs. (108) and (109),

$$\ln\left(1 - \frac{[S_2]}{[S_2]_{ss}}\right) = -\left(\frac{V_2}{K_2}\right)t_{ss},$$

and upon rearrangement,

$$V_2 = -\frac{K_2}{t_{ss}}\ln\left(1 - \frac{[S_2]}{[S_2]_{ss}}\right). \tag{110}$$

Thus, the amount of auxiliary enzyme required to give a maximal velocity V_2 for attainment of, say, 99% of the steady-state rate v_1 in the rate of appearance of S_3 in the time t_{ss} (say 20 s) is

$$V_2 = -\frac{K_2}{20}\ln(1 - 0.99) = K_2(0.23 \text{ mol liter}^{-1}\text{ s}^{-1}).$$

Hence all that is required for this calculation is prior knowledge of K_2, the Michaelis constant of the second enzyme.

3. Upper Limit of Time in the Steady State of [S_2]

Another time, which is called the critical time, t_{crit} (Varfolomeev, 1976), defines an upper limit of the time domain over which the steady state of [S_3] production can reasonably be expected to exist. Suppose, [S_1]$_0 \gg$ [S_3]; under normal experimental conditions a 10-fold excess of [S_1] over [S_3] is sufficient to ensure this, i.e., [S_1]$_0 \gg 10$[S_3].

In the steady state of [S_2], the rate of E_2 is given by the Michaelis–Menten equation with the concentration of S_2 being [S_2]$_{ss}$. Thus,

$$[S_3] \approx \frac{V_2[S_2]_{ss}}{K_2 + [S_2]_{ss}} t = v_{ss}t, \tag{111}$$

where v_{ss} is the steady-state rate of S_3 production. Since [S_1]$_0 \gg$ [S_3] then [S_1]$_0 \gg v_{ss}t$. Therefore,

$$t_{crit} \ll \frac{[S_1]_0}{v_{ss}}. \tag{112}$$

Thus, the time domain in which the system is found, to a reasonable approximation, with a steady state of [S_2] is limited to $t_{ss} < t \ll t_{crit}$ (Varfolomeev, 1976) and for the actual functioning system t_{ss} must be $\ll t_{crit}$; this corresponds to [S_2]$_{ss} \ll$ [S_1]$_0$ (recall that this refers to a *closed* system).

C. Extensions of the Analysis to Long Enzyme Sequences

1. General

The theory of the previous two subsections can be extended to include more than one auxiliary enzyme (Easterby, 1973). As before, the theory is simplified if zero-order kinetics is assumed for the first reaction; this can be approximated in practice if the extent of reaction $S_1 \rightarrow S_2$ is small while allowing attainment of a steady state of all intermediates and a steady state of the rate of production of S_3. Quasi-first-order kinetics of the auxiliary enzymes is achieved if the [S_i] are always $\ll K_i$ ($i \neq 1$ and n).

The experimental attainment of these conditions will be considered further in Section V,E,2.

The reaction scheme to be considered is

$$S_1 \xrightarrow{v_1} S_2 \xrightarrow{V_2/K_2} S_3 \xrightarrow{V_3/K_3} \cdots \xrightarrow{V_{n-1}/K_{n-1}} S_n. \tag{113}$$

First, the mathematical analysis will be developed. The differential rate equations relevant to the scheme in Eq. (113) are

$$d[S_i]/dt = [S_{i-1}]/\tau_{n-1} - [S_i]/\tau_i \quad (i = 2, 3, \ldots, n-1), \quad (114)$$

$$d[S_n]/dt = [S_{n-1}]/\tau_{n-1}, \quad (115)$$

where $\tau_i = K_i/V_i$ and initial conditions are taken to be $[S_i]_0 = 0 \ (i \neq 1)$. The corresponding set of rearranged Laplace–Carson transformed equations is

$$v_1 = \left(\Psi + \frac{1}{\tau_2}\right)s_2 + \quad 0 \quad + \quad 0 \quad \cdots + \quad 0 \quad + \quad 0$$

$$0 = \quad -\frac{1}{\tau_2}s_2 \quad +\left(\Psi + \frac{1}{\tau_3}\right)s_3 + \quad 0 \quad \cdots + \quad 0 \quad + \quad 0$$

$$0 = \quad 0 \quad - \quad \frac{1}{\tau_3}s_3 \quad +\left(\Psi + \frac{1}{\tau_4}\right)s_4 \cdots + \quad 0 \quad + \quad 0 \quad (116)$$

$$\vdots \qquad \vdots \qquad \vdots \qquad \vdots \qquad \vdots \qquad \vdots$$

$$0 = \quad 0 \quad + \quad 0 \quad + \quad 0 \quad \cdots -\frac{1}{\tau_{n-2}}s_{n-2} +\left(\Psi + \frac{1}{\tau_{n-1}}\right)s_{n-1}.$$

As in Eq. (45) the matrix is "lower triangular," so the expression for the determinant (Δ) is simply that of the "trace." Thus,

$$\Delta = \prod_{j=2}^{n-1}\left(\Psi + \frac{1}{\tau_j}\right). \quad (117)$$

By further application of the procedure used in Section II,E,

$$s_i = v_1 \prod_{j=2}^{i-1}\frac{1}{\tau_j}\prod_{k=i+1}^{n-1}\left(\Psi + \frac{1}{\tau_k}\right)\Big/\Delta \quad (i = 2, 3, \ldots, n-1),$$

$$= v_1 \prod_{j=2}^{i-1}\frac{1}{\tau_j}\Big/\prod_{j=2}^{i}\left(\Psi + \frac{1}{\tau_j}\right). \quad (118)$$

The inverse of Eq. (118), obtained from Table I, row 10, is, after some rearrangement [as was obtained, with a minor error, by Easterby (1973), Eq. (9)]

$$[S_i] = v_i\tau_i\left(1 - \sum_{j=2}^{i} A_j \exp[-t/\tau_j]\right), \quad (119)$$

where

$$A_j = \tau_j^{i-1}\prod_{\substack{k=2\\k\neq j}}^{i}\frac{1}{(\tau_j - \tau_k)}.$$

The expression for the progress curve for $[S_n]$ is obtained by substituting Eq. (119) into Eq. (115) and carrying out the straightforward integration to obtain

$$[S_n] = v_1\left(t + \sum_{j=2}^{n-1} B_j e^{-t/\tau_j} - \sum_{j=2}^{n-1} \tau_j\right), \tag{120}$$

where

$$B_j = \tau_j^n \prod_{\substack{k=2 \\ k \neq j}}^{n} \frac{1}{(\tau_j - \tau_k)}.$$

How this expression is used in analyzing data from real enzyme systems will now be considered.

2. Steady State in Enzyme Sequences

As t becomes large the exponential terms in Eqs. (119) and (120) approach zero. So,

$$[S_i]_{ss} = v_1\tau_i \qquad (i = 2, 3, \ldots, n - 1), \tag{121}$$

and

$$[S_n] = v_1\left(t - \sum_{j=2}^{n-1} \tau_j\right). \tag{122}$$

Therefore, a plot of $[S_n]$ versus t yields a straight line segment with an intercept on the time axis at

$$t = \sum_{j=2}^{n-1} \tau_j, \tag{122a}$$

(see Fig. 4), i.e., the so-called *transient time* for the whole sequence is simply the sum of the transient times of the individual steps. This is a general result for mathematically *linear* systems. The ordinate intercept is

$$-v_1 \sum_{j=2}^{n-1} \tau_j,$$

which, from Eq. (121), gives

$$-v_1 \sum_{j=2}^{n-1} \frac{[S_j]_{ss}}{v_1},$$

i.e., the ordinate intercept yields the sum of the steady-state concentrations of the $n - 2$ intermediates.

If any two or more transient times are identical then the previous solutions do not apply. However, these particular cases can be solved readily using the Laplace–Carson theory with Table I and the theory in Sections II,D,3,d and e.

D. Coupled Enzyme Assays—Zero-Order/Michaelis–Menten Schemes

1. General

The scheme to be discussed here is as follows:

$$S_1 \xrightarrow{v_1} S_2 \xrightarrow[E_2]{[V_2[S_2]/(K_2 + [S_2])]} S_3 \tag{123}$$

It was introduced in Section V,A and Table II, rows 4 and 5. Haldane (1930) carried out the early analysis and it was extended by others (Storer and Cornish-Bowden, 1974; Varfolomeev, 1977). It is assumed that S_2 is produced at a constant rate (see Section V,B,4), then

$$\frac{d[S_2]}{dt} = v_1 - \frac{V_2[S_2]}{K_2 + [S_2]}, \tag{124}$$

where v_1 can be specified by a Michaelis–Menten expression or *any* rate function.

Rearrangement of Eq. (124) with $[S_2]_0 = 0$ gives

$$\int_0^{[S_2]} \frac{K_2 + [S_2]}{-(V_2 - v_1)[S_2] + v_1 K_2} d[S_2] = \int_0^t dt. \tag{125}$$

By use of the substitution $y =$ "denominator term" and $d[S_2] = -[1/(V_2 - v_1)] \, dy$ followed by back substitution, the following expression is obtained:

$$t = -\frac{V_2 K_2}{(V_2 - v_1)^2} \ln\left\{ 1 - \frac{(V_2 - v_1)[S_2]}{v_1 K_2} \right\} - \frac{[S_2]}{(V_2 - v_1)}. \tag{126}$$

This is an implicit function of $[S_2]$, so, for a given value of $[S_2]$ the time taken to attain that concentration can be calculated.

2. A Condition for a First-Order Kinetic Approximation

Further analysis of Eq. (126) is valuable because it enables one to define the experimental conditions under which the reaction of E_2 can be considered to behave like a first-order process. First, the logarithmic term must be >0, i.e.,

$$\frac{v_1 K_2 - (V_2 - v_1)[S_2]}{v_1 K_2} > 0, \tag{127}$$

which leads to the inequality $[S_2] < K_2/(V_2/v_1 - 1)$. Thus, the magnitude of $[S_2]$ is bounded above; its value is determined by the Michaelis constant of E_2 and the ratio of v_1 and the maximal velocity of E_2. In Section V,A it was noted that a steady state of $[S_2]$ is only possible if $V_2/v_1 > 1$. If $V_2/v_1 > 2$ then $[S_2]_{ss}$ is less than K_2 (Varfolomeev, 1976); this is an important result which is obtained from Eq. (126). The relationship gives some guide to the conditions under which the first-order approximation of the Michaelis–Menten expression might be valid (viz., the approximation used in Sections V,A and B), and this defines conditions in which the subsequent mathematical analysis may be applied. Thus, if $V_2 \gg v_1$ then $[S_2]_{ss} \ll K_2$. Since $[S_2]_{ss} > [S_2]$ then $[S_i] \ll K_i$ for all $i = 2, 3, \ldots, n - 1$ in a linear sequence. This was discussed in Section V,C,1 specifically in relation to Eqs. (114) and (115).

3. Economization on the Amount of E$_2$ in an Assay System

We discuss here the choice of the concentration of auxiliary enzyme(s) in coupled assays of the type already discussed in this section.

In the steady state $d[S_2]/dt = 0$, so from Eq. (124)

$$[S_2]_{ss} = v_1 K_2 / (V_2 - v_1). \tag{128}$$

Let α be the extent of attainment of the (maximum) steady-state value of $[S_2]$, and as before (Section V,B,2) let $\alpha = 0.99$. Substitution of $[S_2] = \alpha v_1 K_2/(V_2 - v_1)$ into Eq. (126) gives the time taken to attain 0.99 of the steady-state value of $[S_2]$;

$$t_{ss} = -\frac{V_2 K_2}{(V_2 - v_1)^2}\left(\ln(1 - \alpha) + \alpha\frac{v_1}{V_2}\right) \simeq \frac{V_2 K_2}{(V_2 - v_1)^2} 3.6. \tag{129}$$

So, if $V_2 \gg v_1$, then $t_{ss} \approx K_2/V_2$ as was shown for the general system in Section V,B,2.

Storer and Cornish-Bowden (1974) introduced the relationship

$$[S_2] = K_2 v_2/(V_2 - v_2)$$

into Eq. (126), with $v_2 = 0$ at $t = 0$, to obtain

$$t = \frac{V_2 K_2}{(V_2 - v_1)^2} \ln\left\{\frac{v_1(V_2 - v_2)}{V_2(V_1 - v_2)}\right\} - \frac{K_2 v_2}{(V_2 - v_2)(V_2 - v_1)}. \tag{130}$$

This equation is of the form

$$t = \Phi K_2/v_1, \tag{131}$$

where

$$\Phi = \frac{1}{\beta(1 - \gamma)^2} \ln\left\{\frac{(\beta - \gamma)}{\beta(1 - \gamma)}\right\} + \frac{\gamma}{(\beta - \gamma)(1 - \beta)}, \tag{132}$$

with $\beta = V_2/v_1$ and $\gamma = v_2/v_1$. The expression for Φ is dimensionless, as can be seen from Eq. (131), and is a function of the two ratios β and γ. It is therefore possible to calculate Φ values for various values of the ratios; these were tabulated by Storer and Cornish-Bowden (1974). The value of $v_2/v_1(\gamma)$ that is appropriate for use in any given experimental situation depends on the accuracy required; in precise studies v_2/v_1 should be > 0.99.

Note, however, that there is a potential trap in applying this method to estimate the amount of coupling enzyme; if unforeseen inhibition of E_2 occurs due to reactants or buffer components in the assay mixture subsequent data analysis may be invalid. Furthermore, the original model, with (pseudo) zero-order kinetics in the first step, demands that $v_1 \ll V_2$ and that an estimate of K_2 be available.

Despite these potential shortcomings, the theory has been used successfully in the optimal design of a coupled assay—glucokinase coupled to glucose-6-phosphate dehydrogenase (Storer and Cornish-Bowden, 1974).

The theoretical approach cannot be extended to more than two enzymes, for reasons similar to those which barred the further development of solutions in Section IV,B,1. But, the analysis does enable one to estimate an upper limit on the time (t_u) required for the observed rate of product formation to reach a given fraction of v_1. Consider the scheme

$$S_1 \xrightarrow{v_1} S_2 \xrightarrow{v_2} S_3 \xrightarrow{v_3} S_4,$$

with v_1 constant and v_2 and v_3 governed by Michaelis–Menten equations. The time t_1 required for v_2 to reach $0.99v_1$ is obtained from Eq. (129) or (131) and the appropriate values of the ratios β and γ [Eq. (132)] therein. The estimate of t_1 is exactly the same value as for the simple coupled system, which is no surprise since the reactions are considered to be irreversible. If v_2 is considered to be constant, then the time t_2 for v_3 to reach $0.99v_2$ can also be estimated, as was done for t_1. The appropriate replacements of K_3 for K_2, V_3 for V_2, and v_2 for v_1 are required. This calculation overestimates t_2 because v_3 must begin to increase before v_2 reaches $0.99v_1$ so that the time for v_3 to reach $0.99v_1 \times 0.99v_1$ is less than $t_1 + t_2$. Now, the total time t_u cannot be less than the greater of t_1 and t_2, therefore $t_1, t_2 < t_u < t_1 + t_2$. These limits appear to define t_u precisely enough for most practical purposes (Storer and Cornish-Bowden, 1974). Extension of this semiquantitative analysis to higher numbers of auxiliary enzymes follows the same procedure as above and is obviously simple.

E. Two Michaelis–Menten Enzymes in an Open System

1. General

In the previous section the first reaction in the series, with the rate v_1, was taken to be a quasi-zero-order process; this enabled the subsequent mathematical analysis to be carried out. The zero-order requirement can be relaxed since it is really only necessary to assume v_1 to be a constant over a short extent of the total reaction and therefore v_1 can have *any* functional dependence on

$[S_1]$, including the Michaelis–Menten equation. In this section we consider the particular case of v_1 being a Michaelis–Menten-type enzyme.

2. Conditions for Attainment of a Steady State

We now consider parameter value conditions necessary to bring about a steady state in the following system:

$$S_1 \xrightarrow{\text{E}_1} S_2 \xrightarrow{\text{E}_2} S_3,$$

where the two reaction rates are described by Michaelis–Menten expressions (Varfolomeev, 1977); this is the scheme defined by Eqs. (71)–(73) where $[S_1]$ is assumed to be constant and of the value $[S_1]_0$. This arrangement can be approximated experimentally in situations that were discussed before (Section V,B,1). The rate of formation of S_3 is,

$$d[S_3]/dt = V_2[S_2]/(K_2 + [S_2]). \tag{133}$$

Since a steady state of $[S_2]$ is *assumed to have been attained*, $d[S_2]/dt = 0$, and from Eq. (72),

$$\frac{V_1[S_1]_0}{K_1 + [S_1]_0} = \frac{V_2[S_2]_{ss}}{K_2 + [S_2]_{ss}}. \tag{134}$$

On rearrangment,

$$[S_2]_{ss} = \frac{V_1[S_1]_0 K_2}{V_2(K_1 + [S_1]_0) - V_1[S_1]_0}. \tag{135}$$

Since $[S_2]_{ss} > 0$ it follows that the denominator of Eq. (135) must be > 0. Therefore,

$$V_2(K_1 + [S_1]_0) > V_1[S_1]_0 \quad \text{and} \quad V_2 > \frac{V_1[S_1]_0}{K_1 + [S_1]_0}, \tag{136}$$

i.e., $V_2 > v_1$, as was noted in Section V,D,3 for the zero-order/Michaelis–Menten system; the present system is a particular case of that earlier one.

Since the term $[S_1]_0/(K_1 + [S_1]_0)$ is always less than 1 a sufficient condition for establishment of a steady state of $[S_2]$ is $V_1 < V_2$. The value of $[S_2]_{ss}$, as given by Eq. (135), can be expressed another way (Varfolomeev, 1977), i.e.,

$$[S_2]_{ss} = \frac{[S_2]_{max}[S_1]_0}{K_{eff} + [S_1]_0}, \tag{137}$$

where

$$[S_2]_{max} = \frac{K_2}{V_2/V_1 - 1} \quad \text{and} \quad K_{eff} = \frac{K_1}{1 - V_1/V_2}.$$

Therefore, from Eq. (137) it is seen that the value of $[S_2]_{ss}$ is determined by the relative magnitudes of the maximal velocities of E_1 and E_2. And, $[S_2]_{ss}$ has a maximum value when $[S_1]_0 \gg K_{eff}$, i.e., $[S_1]_0 \gg K_1/(1 - V_1/V_2)$. If $V_2 \gg V_1$, $[S_2]$ will always be less than K_2 and the system reverts to that discussed in Section V,B. With $V_2 = 2V_1$, $[S_2]_{max} = K_2$, and with $2V_1 > V_2 > V_1$ the limiting steady-state concentration of S_2 exceeds K_2. Furthermore, $[S_2]_{ss}$ has a maximum value when $[S_1]_0 \gg K_{eff}$, i.e., $[S_1]_0 \gg K_1/(1 - V_1/V_2)$.

F. Two Michaelis–Menten Enzymes with a Steady State of [S₂] and Consumption of S₁

1. General

The analysis here examines the less stringent condition than presented in the previous section, i.e., that $[S_1]$ is *not constant* in the scheme described by Eqs. (71)–(73). The conservation of mass condition is

$$[S]_0 = [S_1] + [S_2] + [S_3]. \tag{138}$$

However, the extent of reaction is assumed to be very small such that

$$[S_1] + [S_2] \gg [S_3]; \tag{139}$$

then Eq. (138) becomes

$$[S]_0 \simeq [S_1] + [S_2]. \tag{140}$$

To reiterate, Eq. (140) is not as stringent a condition as $[S_1] = [S]_0$ as was used in Section V,E. Some interesting information on the early time behavior of the coupled enzymatic reaction can be gleaned by further analysis.

In the time domain encompassing the steady state of $[S_2]$, Eqs. (133) and (134) apply, and substitution of Eq. (140) into Eq. (134) yields

$$\frac{V_1([S]_0 - [S_2]_{ss})}{K_1 + ([S]_0 - [S_2]_{ss})} = \frac{V_2[S_2]_{ss}}{K_2 + [S_2]_{ss}}. \tag{141}$$

This equation can be rearranged to give a quadratic in $[S_2]_{ss}$ with the following roots:

$$[S_2]_{ss} = \frac{1}{2}\left\{[S]_0 + \frac{V_1 K_2 + V_2 K_1}{V_2 - V_1} \right.$$
$$\left. \pm \sqrt{\left[[S]_0 + \frac{V_1 K_2 + V_2 K_1}{V_2 - V_1}\right]^2 - \frac{4[S]_0 V_1 K_2}{V_2 - V_1}}\right\}. \tag{142}$$

Clearly, the relationship holds only if $V_1 \neq V_2$. Furthermore, earlier analysis

(Section V,D) showed that the condition $V_2 > V_1 > v_1$ must hold for attainment of a steady state of $[S_2]$, and this enables a decision to be made on which sign before the irrational term applies in this physical system; it must be the negative sign, because if it were positive $[S_2]_{ss}$ would be $> [S]_0$ which is impossible in the closed system. Therefore, from Eq. (142) the dependence of $[S_2]_{ss}$ on the ratio V_1/V_2 can be plotted (Varfolomeev, 1976).

2. Determination of Enzyme Parameters

From Eq. (142) it is readily shown that if $V_1 \ll V_2$ and $[S]_0 \gg K_1$ and K_2, then

$$[S_2]_{ss} \simeq \frac{[S]_0 V_1 K_2/(V_2 - V_1)}{[S]_0 + \{(V_1 K_2 + V_2 K_1)/(V_2 - V_1)\}}. \tag{143}$$

Therefore, the relationship of the steady-state concentration of S_2 to $[S]_0$ has the form of a rectangular hyperbola, with the expressions for $[S_2]_{max}$ and K_{eff} as defined in Eq. (137). The behavior of $[S_2]_{ss}$ is like that of a Michaelis–Menten enzyme; at high values of $[S]_0$, $[S_2]$ attains the level $[S_2]_{max}$; at low values of $[S]_0$, $[S_2]_{ss}$ shows a linear dependence on $[S]_0$.

Furthermore, if in a coupled enzyme system the ratio V_1/V_2 is independently varied and the effect on the value of $[S_2]_{ss}$ is analyzed, then using Eq. (143) the ratio K_1/K_2 can be computed (Varfolomeev, 1976).

G. Sequences of Michaelis–Menten Enzymes in a Steady State

1. General

One means of circumventing the problem, outlined in Section IV,A, of the intractability of integrating the differential equations of the n-enzyme system [of which the set of Eqs. (71)–(73) is the two-enzyme example] is to consider another approximation, i.e., that the system has already attained a steady state. Thence, all derivatives in the set of rate equations are set to zero. In the sequence with v_1 a constant value,

$$S_1 \xrightarrow{v_1} S_2 \longrightarrow \ldots \longrightarrow S_n, \tag{144}$$

the conservation of mass condition is

$$v_1 t = \sum_{i=2}^{n} [S_i], \tag{145}$$

or

$$[S_n] = v_1 t - \sum_{i=2}^{n-1} [S_i].$$

In the steady state,

$$[S_n] = v_1\left(t - \sum_{i=2}^{n-1} \frac{[S_i]_{ss}}{v_1}\right), \tag{146}$$

which is reminiscent of Eq. (120) for $t \to \infty$. The point of intersection of the asymptote drawn to the progress curve, $[S_n]$ versus t, intersects the time axis at

$$\tau = \sum_{2}^{n-1} \frac{[S_i]_{ss}}{v_1} = \sum_{i=2}^{n-1} \tau_i, \tag{147}$$

where $\tau_i = [S_i]_{ss}/v_1$.

In the particular case where the rate of each reaction is described by a Michaelis–Menten expression,

$$\frac{d[S_i]}{dt} = 0 = v_1 - \frac{V_i[S_i]_{ss}}{K_i + [S_1]_{ss}} \qquad (i = 2, 3, \ldots, n - 1). \tag{148}$$

Hence,

$$[S_i]_{ss} = v_1 K_i/(V_i - v_1), \tag{149}$$

so from Eq. (147),

$$\tau_i = K_i/(V_i - v_1). \tag{150}$$

Therefore, the total transient time is (Easterby, 1981)

$$\tau = \sum_{i=2}^{n-1} \frac{K_i}{(V_i - v_1)}. \tag{151}$$

[A check is to make $V_i \gg v_1$, and the expression reverts, as expected, to that of Eq. (120) for $t \to \infty$.]

2. Effect of Inhibitors on the Sequence

If one (or more) of the reactants, say S_j, in the sequence described by Eq. (144) is a competitive inhibitor of E_i then the corresponding K_i must be modified appropriately (see Section IV,D); an apparent value "K_i" = $K_i(1 + [S_j]_{ss}/K_1)$ replaces K_i in Eq. (151).

Analogously, a noncompetitive inhibitor of E_i modifies the apparent V_i which then appears in Eq. (151) (Easterby, 1981).

3. Profile of Concentrations

If it is assumed that the sum of the concentrations of those intermediates $(S_2, S_3, \ldots, S_{n-1})$ that are in a steady state can be determined $([S]_T)$ (all may have a common moiety which can be detected spectrophotometrically), and if $[S_i]_{ss} \ll K_i$ $(i = 2, 3, \ldots, n - 1)$ then it is possible to predict the concentration of each species from a knowledge of the V_is and K_is. For the five-reactant sequence,

$$[S]_T = [S_2]_{ss} + [S_3]_{ss} + [S_4]_{ss},$$

$$0 = (V_2/K_2)[S_2]_{ss} - (V_3/K_3)[S_3]_{ss} + 0, \tag{152}$$

$$0 = 0 + (V_3/K_3)[S_3]_{ss} - (V_4/K_4)[S_4]_{ss}.$$

Using Cramer's rule to solve for $[S_i]$, with $[S_4]$ as an example,

$$[S_4]_{ss} = \begin{vmatrix} 1 & 1 & [S]_T \\ \dfrac{V_2}{K_2} & -\dfrac{V_3}{K_3} & 0 \\ 0 & \dfrac{V_3}{K_3} & 0 \end{vmatrix} \div \begin{vmatrix} 1 & 1 & 1 \\ \dfrac{V_2}{K_2} & -\dfrac{V_3}{K_3} & 0 \\ 0 & \dfrac{V_3}{K_3} & -\dfrac{V_4}{K_4} \end{vmatrix}. \tag{153}$$

Equation (153) reduces to

$$[S_4]_{ss} = [S]_T \frac{V_2}{K_2}\frac{V_3}{K_3} \Big/ \left(\frac{V_3}{K_3}\frac{V_4}{K_4} + \frac{V_2}{K_2}\frac{V_3}{K_3} + \frac{V_2}{K_2}\frac{V_4}{K_4} \right). \tag{154}$$

In addition, the rate expression for $[S_5]$ follows directly from Eq. (154), thus

$$\frac{d[S_5]}{dt} = \frac{V_4}{K_4}[S_4]_{ss} = \frac{[S]_T}{K_2/V_2 + K_3/V_3 + K_4/V_4}. \tag{155}$$

By induction, the general solution for the concentration of the $(n - 1)$th species in a long sequence is

$$[S_{n-1}]_{ss} = [S]_T \Big/ \left(1 + \frac{V_{n-1}}{K_{n-1}} \sum_{i=2}^{n-2} \frac{K_i}{V_i} \right) = [S]_T \Big/ \left(1 + \frac{1}{\tau_{n-1}} \sum_{i=2}^{n-1} \tau_i \right), \tag{156}$$

where τ_i is defined as K_i/V_i, as for Eq. (115), and it follows that

$$\left(\frac{d[S_n]}{dt} \right)_{ss} = [S]_T \Big/ \sum_{i=2}^{n-1} \frac{K_i}{V_i}. \tag{157}$$

The other $[S_i]$ can be found readily, as above, thus defining the reactant concentration profile and its dependence on all of the enzyme kinetic parameters. The inverse problem of obtaining information on V_i and K_i from the experimental determination of the $[S_i]$ profile seems a "tall order" for large enzyme sequences, except for the fact that Eq. (121) holds. So, if values of v_i and $[S_i]_{ss}$ are known from an experiment the ratios $\tau_i = K_i/V_i$ can be found directly using Eq. (121).

A numerical simulation of the more complicated reversible system is presented in Fig. 6. The mathematical analysis is not discussed here because the labor involved in deriving the expressions does not appear to warrant the

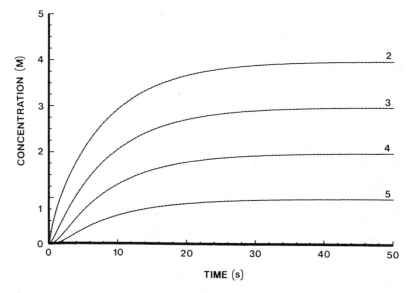

Fig. 6. Time course for the open scheme

$$\longrightarrow S_1 \underset{\text{[constant]}}{\overset{k_1}{\longrightarrow}} S_2 \underset{k_{-2}}{\overset{k_2}{\rightleftharpoons}} S_3 \underset{k_{-3}}{\overset{k_3}{\rightleftharpoons}} S_4 \underset{k_{-4}}{\overset{k_4}{\rightleftharpoons}} S_5 \overset{k_5}{\longrightarrow}$$

The parameter values for the numerical simulation were $k_{\pm i} = 1$, $[S_1] = 1$. Although the axes have been labeled with dimensions of mol liter^{-1} (which may seem biologically unrealistic) and s, the results could have been presented equally well as those of a dimensionless scheme. The curves have been labeled with a number that corresponds to the subscript of the S_i to which it applies.

effort at this stage. However, it is easy to see that the results shown in Fig. 6 are distinctly different from those predicted analytically for the corresponding *irreversible* system described by Eq. (156).

VI. OTHER FEATURES OF COUPLED ENZYME SYSTEMS

A. Transitions between Steady States in Linear Enzyme Sequences

1. General

From a purely mathematical viewpoint, one way of making tractable the integration of a set of simultaneous nonlinear ordinary differential equations, such as arises from the scheme in Eq. (70), is to assume that equilibrium or a steady state has been attained. It is then possible to consider, in a quantitative manner, the time evolution of concentrations of reactants after small perturbations away from the stationary state. Experimentally, it is usually a simple matter to arrange for a closed system to attain *equilibrium* within a reasonable time. On the other hand, conditions for a *steady state* can be more difficult to attain but they can involve an automatic titration apparatus, slow release of initial reactant from a "depot", and uptake of product by sequestration (e.g., onto an ion exchange resin) or evaporation (as occurs with gaseous products). Perturbation of the system can be brought about by "jumps" in temperature, pressure, pH, ionic strength, or the addition of one or more reactants and inhibitors; in other words, short "pulses" of matter or energy are introduced into the system.

The mathematical analysis is very effectively exploited in temperature relaxation kinetic studies (Eigen and De Maeyer, 1963; Schwarz, 1968; Schwarz and Engel, 1972; Roberts, 1977, Ch. 7). The general analytical procedure will now be developed for enzyme sequences.

2. Perturbation of a Two-Enzyme System in a Steady State

Consider two enzymes in a scheme the same as Eq. (123). The concentration of S_2 at any time can be written as

$$[S_2](t) = [S_2]_{ss} + \Delta[S_2](t), \tag{158}$$

where $[S_2]_{ss}$ is the steady-state concentration of S_2, and $\Delta[S_2](t)$ is that component of $[S_2]$ that varies with time; the t in parentheses stresses the time dependence of $[S_2]$. If the change in the steady-state concentration is small, i.e.,

$$\Delta[S_2] \ll [S_2]_{ss}, \tag{159}$$

then

$$d[S_2]/dt = v_1 - V_2[S_2]/(K_2 + [S_2]). \tag{160}$$

This can be simplified by noting that

$$d[S_2]/dt = d\Delta[S_2]/dt, \tag{161}$$

and the Taylor series expansion of the hyperbolic term in Eq. (160) is carried out as follows (see Section IV,B,3). Define f_2 as

$$f_2([S_2]) = V_2[S_2]/(K_2 + [S_2]). \tag{162}$$

Then the Taylor series expansion of f_2 is

$$f_2([S_2]) = f_2([S_2]_{ss}) + \left(\frac{df_2}{d[S_2]}\right)_{ss} \Delta[S_2] + \cdots, \tag{163}$$

$$= \frac{V_2[S_2]_{ss}}{K_2 + [S_2]_{ss}} + \frac{V_2 K_2}{(K_2 + [S_2]_{ss})^2} \Delta[S_2] + \cdots. \tag{164}$$

Substitute Eq. (164) into Eq. (160) and (161) noting that the first term is equal to v_1, and then

$$\frac{d[S_2]}{dt} = -\frac{V_2 K_2}{(K_2 + [S_2]_{ss})^2} \Delta[S_2]. \tag{165}$$

The solution of Eq. (165) is

$$[S_2] = \Delta[S_2]_0 \exp\left[\frac{-V_2 K_2}{(K_2 + [S_2]_{ss})^2} t\right], \tag{166}$$

where $\Delta[S_2]_0$ is the *amplitude* of the perturbation and is experimentally determined as $|[S_2]_{ss_1} - [S_2]_{ss_2}|$, i.e., the difference between the concentration of S_2 in each steady state. Thus, the kinetics of the transition from one steady state to another are characterized by a *relaxation time* which by definition is the reciprocal of the first-order rate constant,

$$V_2 K_2/(K_2 + [S_2]_{ss})^2 \qquad \text{(dimensions: s}^{-1}\text{)}. \tag{167}$$

If $[S_2]_{ss} \ll K_2$, then the expression for the relaxation time, τ, becomes K_2/V_2.

If the absolute value of $[S_2]_{ss}$ is indeterminable, but only a measurement that is proportional to $|[S_2]_{ss_1} - [S_2]_{ss_2}|$, then V_2 and K_2 may still be found as follows. From Section V,G,1 and Eq. (148).

$$[S_2]_{ss} = v_1 K_2/(V_2 - V_1). \tag{168}$$

Substitution of this into Eq. (167) yields the rate of transition $1/\tau$:

$$1/\tau = (V_2 - v_1)^2/V_2 K_2. \tag{169}$$

Plots of $\sqrt{1/\tau}$ versus v_1 give an intercept on the v_1 axis of $-V_2/\sqrt{V_2 K_2}$ and a slope of $-\sqrt{1/V_2 K_2}$ which, coupled with the estimate of V_2, yields K_2.

$$\frac{d\Delta[S_2]}{dt} = -\frac{V_2 K_2}{(K_2 + [S_2]_{ss})^2} \Delta[S_2] + \qquad 0 \qquad \cdots + \qquad 0$$

$$\frac{d\Delta[S_3]}{dt} = \frac{V_2 K_2}{(K_2 + [S_2]_{ss})^2} \Delta[S_2] - \frac{V_3 K_3}{(K_3 + [S_3]_{ss})^2} \Delta[S_3] \cdots + \qquad 0$$

$$\vdots \qquad\qquad \vdots \qquad\qquad \vdots \qquad\qquad \vdots$$

$$\frac{d\Delta[S_i]}{dt} = \qquad 0 \qquad + \qquad 0 \qquad \cdots + \frac{V_{i-1} K_{i-1}}{(K_{i-1} + [S_{i-1}]_{ss})^2} \Delta[S_{i-1}]_{ss}$$

$$\vdots \qquad\qquad \vdots \qquad\qquad \vdots \qquad\qquad \vdots$$

$$\frac{d\Delta[S_{n-1}]}{dt} = \qquad 0 \qquad + \qquad 0 \qquad \cdots + \qquad 0$$

SCHEME 1

3. Perturbation of a Multienzyme System in a Steady State

Consider the scheme of unidirectional Michaelis–Menten enzymes in which the first reaction step is one of constant rate v_1, i.e.,

$$S_1 \xrightarrow{v_1} S_2 \xrightarrow{E_1} \cdots \xrightarrow{E_{n-1}} S_n. \tag{170}$$

A small perturbation of v_1 will affect the steady-state concentrations of all reactants and also the rate of accumulation of $[S_n]$ (Varfolomeev, 1977). In a manner analogous to that in Section VI,A,2,

$$[S_2] = [S_2]_{ss} + \Delta[S_2],$$
$$\vdots$$
$$[S_i] = [S_i]_{ss} + \Delta[S_i],$$
$$\vdots$$
$$[S_{n-1}] = [S_{n-1}]_{ss} + \Delta[S_{n-1}]. \tag{171}$$

The displacement from the steady-state concentration is assumed to be very small, i.e.,

$$\Delta[S_i] \ll [S_i]_{ss} \qquad (i = 2,3,\ldots,n-1).$$

The kinetics of the transition from one steady state to another are then described by a set of ordinary *linear* differential equations, in $\Delta[S_i]$, with constant coefficients, as shown in Scheme 1.

With the substitution $k_i = V_{i+1} K_{i+1}/(K_{i+1} + [S_{i+1}]_{ss})^2$ and $[S_i] = \Delta[S_{i+1}]$ the set of differential equations is seen to be identical to Eq. (18). Solution would therefore be given by Eqs. (51)–(53), except for the important

$$
\begin{array}{cccccc}
+ & 0 & \cdots+ & 0 & + & 0 \\[2mm]
+ & 0 & + & 0 & + & 0 \\[2mm]
\vdots & \vdots & \vdots & & \vdots & \\[2mm]
-\dfrac{V_i K_i}{(K_i + [S_i]_{ss})^2}\Delta[S_i]_{ss}\cdots+ & & 0 & & 0 & \\[2mm]
\vdots & & \vdots & & \vdots & \\[2mm]
+ & 0 & \cdots+\dfrac{V_{n-2}K_{n-2}}{(K_{n-2}+[S_{n-2}]_{ss})^2}\Delta[S_{n-2}]_{ss} & - \dfrac{V_{n-1}K_{n-1}}{(K_{n-1}+[S_{n-1}]_{ss})^2}\Delta[S_{n-1}]_{ss}
\end{array}
\tag{172}
$$

modification that $\Delta[S_i]_0$ is *not* zero. Therefore, further analysis using the theory in Section II,D is required; however, because of its complexity and since no new concepts emerge from the analysis, it will not be given here. Suffice it to say that the mathematical *form* of the transition functions is

$$\Delta[S_i] = \sum_{j=2}^{n-1} C_{ij} e^{-t/k_j}. \tag{173}$$

In principle, it is possible to determine the number of intermediates in a sequence of reactions from analyzing the number of exponentials in the relaxation profile (Eigen and De Maeyer, 1963; Schwarz and Engel, 1972; Schwarz, 1968; Varfolomeev, 1977; Hearon, 1981). However, it is rarely possible to extract more than three exponentials from any complicated time course. On the other hand, if any of the components of the reaction are individually identifiable, for example, if they have peculiar fluorescence characteristics, then it may be possible to monitor progress of the changes in a single species. This will greatly assist the analysis of the number of transient species and the determination of the *individual* relaxation times.

4. Stability of Steady States

In the previous analysis it was assumed that after a perturbation the system would evolve to another steady state "near" the original one. The final state of a closed system is always stable; after a perturbation, no matter how large it may be, the system always returns to the equilibrium state (Hearon, 1953). However, in open systems this is not always the case; the system then is said to be in an *unstable* equilibrium (or in an unstable stationary state). The solution of the differential rate equations in these systems contains exponential terms with exponents that are not real numbers or have positive real components. Thus, a range of possibly complicated behaviors can result. We will not discuss these matters here but good discussions can be found in Reich and Sel'kov (1981), Heinrich et al. (1967), and Hearon (1953).

5. Optimization of Enzyme Concentrations in Linear Sequences—A Theoretical Biological Model

The following model and analysis is obviously simplistic but it serves to raise an interesting notion that must be of importance in the energy economy of living cells, i.e., the energy cost of producing enzymes, in relation to their catalytic activity. The analysis is very much in the spirit of some earlier work by Waley (1964) and recently in a slightly different context by Cleland (1979).

Consider the scheme in Eq. (170). We ask what the concentration of each E_i would be in order to minimize the "energy cost" to the cell in producing enzymes in order that the transition time of the system from one steady state to another should be a given value τ.

Since each peptide bond that is synthesized results in the hydrolysis of approximately four high-energy phosphate bonds (ATP) the cost to the cell of producing an enzyme is proportional to the molecular weight of the protein per active (catalytic) site. Now, $V_i = k_{cat_i}[E_i]$, where $[E_i]$ is taken to be the *total* concentration of the enzyme E_i and k_{cat_i} is the turnover number. For the purpose of the present exercise it is assumed that all the k_{cat_i} have the same value, i.e.,

$$V_i/V_j = [E_i]/[E_j]. \tag{174}$$

The relaxation time for the traversal from one steady state to another, τ, is given by the generalization of Eq. (169) and Section VI,A,3 as

$$\tau = \sum_{i=2}^{n-1} \frac{V_i K_i}{(V_i - v_1)^2}. \tag{175}$$

The cost, to the cell, of each enzyme E_i is proportional to the molecular weight of the enzyme. Therefore, the total cost of the whole enzyme system is proportional to

$$C = M_2 V_2 + M_3 V_3 + \cdots + M_{n-1} V_{n-1}. \tag{176}$$

Thus, we wish to find the relative values of the V_is (given their respective M_is) that minimize the total cost, C, of catalyzing a transition to a new steady state with the rate constant $1/\tau$. This is a problem with a conditional minimum and is best solved using *Lagrange's method of undetermined multipliers* (Mellor, 1954, pp. 300–301). Partial differentiation of Eq. (175), with respect to the V_i, yields the following expression which is equal to zero since τ is a constant:

$$\sum_{i=2}^{n-1} \frac{K_i(V_i + v_1)}{(V_i - v_1)^3} \partial V_i = 0. \tag{177}$$

Partial differentiation of Eq. (176) with respect to V_i gives

$$\sum_{i=2}^{n-1} M_i \partial V_i = 0. \tag{178}$$

Multiplication of Eq. (178) by the undetermined multiplier λ produces

$$\sum_{i=2}^{n-1} \left(\frac{K_i(V_i + v_1)}{(V_i - v_1)^3} + \lambda M_i \right) \partial V_i = 0. \tag{179}$$

But λ is arbitrary and can be *chosen* so that

$$\frac{K_2(V_2 + v_1)}{(V_2 - v_1)^3} + \lambda M_2 = 0. \tag{180}$$

But, since V_i, V_{i+1} ($i = 3, 4, \ldots, n - 1$) are independent we also have

$$\frac{K_i(V_i + v_1)}{(V_i - v_1)^3} + \lambda M_i = 0. \tag{181}$$

Therefore,

$$\lambda = -\frac{M_i(V_i - v_1)^3}{K_i(V_i + v_1)} = -\frac{M_j(V_j - v_1)^3}{K_j(V_j + v_1)}, \tag{182}$$

and any given V_i can be found as a function of any given V_j by solution of the cubic polynomial which emerges from the appropriate rearrangement of Eq. (182). We will not develop this solution any further here, but choose to simplify the analysis by assuming first-order kinetics of each reaction, i.e., $V_i \gg v_1$. Hence,

$$\lambda = -\frac{M_i V_i^2}{K_i} = -\frac{M_j V_j^2}{K_j}, \tag{183}$$

that is,

$$V_i = V_j \left(\frac{M_j K_i}{M_i K_j}\right)^{1/2}, \tag{184}$$

which is the general solution.

If we let $j = 2$ and $V_i \gg v_1$, then from Eq. (175)

$$\tau = \frac{K_2}{V_2} + \sum_{i=3}^{n-1} \frac{K_i}{V_2(M_2 K_i/M_i K_2)^{1/2}}. \tag{185}$$

It follows that

$$V_2 = \frac{K_2}{\tau} \left\{ 1 + \sum_{i=3}^{n-1} \left(\frac{M_i K_i}{M_2 K_2}\right)^{1/2} \right\}, \tag{186}$$

$$V_3 = \frac{K_3}{\tau} \left\{ 1 + \sum_{\substack{i=2 \\ i \neq 3}}^{n-1} \left(\frac{M_i K_i}{M_3 K_3}\right)^{1/2} \right\}, \tag{187}$$

and in general,

$$V_l = \frac{K_l}{\tau} \left\{ 1 + \sum_{\substack{i=2 \\ i \neq l}}^{n-1} \left(\frac{M_i K_i}{M_l K_l}\right)^{1/2} \right\}. \tag{188}$$

Thus we have an expression which gives the optimal values of the V_is, hence the amount of protein required to be synthesized to allow a steady-state transition with a specified transition time τ. It should be noted that intracellular conditions may place upper limits on the enzyme concentrations, such that alternative means (e.g., spatial organization) may have been devised in the evolution of some metabolic processes (Welch, 1977; Welch and Keleti, 1981; Keleti and Welch, 1984). The problem considered here has been presented by Cleland (1979), although not developed using the Lagrange multiplier theory, for the optimization of coupled enzyme assays where the cost of the auxiliary enzymes is an important consideration. In this case the M_i terms refer to the actual monetary cost of the enzymes and the aim is to reduce the total cost (C) for an acceptable value of τ, the latter being chosen using the theory of Sections V,B,2 and V,D,3.

B. A Special Case

In Section IV,B,1 the analytical solution of the Michaelis–Menten expression for a closed system was presented [Eq. (75)]. Consider now an enzyme with competitive inhibition by the product of the reaction. The appropriate rate equation is (Laidler and Bunting, 1973, p. 92)

$$\frac{d[S_1]}{dt} = -\frac{V_1[S_1]}{K_1\{1 + [([S_1]_0 - [S_1])/K_I]\} + [S_1]}. \tag{189}$$

The conservation of mass expression has been used in the denominator to give $[S_2]$ in terms of $[S_1]$. In the special situation that $K_I = K_1$, Eq. (189) becomes,

$$\frac{d[S_1]}{dt} = -\frac{V_1[S_1]}{K_1 + [S_1]_0}. \tag{190}$$

This is a first-order rate equation since the value of the coefficient of $[S_1]$ is constant. An example of an enzyme with a rate equation like that of Eq. (189) is arginase. The K_I for ornithine in the beef liver arginase reaction (3 mM) is not too dissimilar to that of the K_m of arginine (5 mM) at physiological pH (Kuchel et al., 1975).

The solution of Eq. (190) is, of course,

$$[S_1] = [S_1]_0 \exp\left[-\left(\frac{V_1}{K_1 + [S]_0}\right)t\right]. \tag{191}$$

This illustrates the important point that a mechanistically complex series of reactions can yield a very simple progress curve (Haldane, 1930). Pedagogically, the expression is useful since it is a simple closed solution of the Michaelis–Menten equation and simulations of time courses of reactions can be performed readily using a pocket calculator.

It is possible to build on Eq. (190) and consider a special case of a coupled enzymatic reaction:

$$S_1 \underset{S_3}{\overset{E_1}{\rightleftharpoons}} S_2 \overset{E_2}{\longrightarrow} S_4 \tag{192}$$

The differential rate equations relevant to Eq. (192) are

$$\frac{d[S_1]}{dt} = -\frac{V_1[S_1]}{K_1(1 + [S_3]/K_{I_1}) + [S_1]},$$

$$\frac{d[S_2]}{dt} = -\frac{d[S_1]}{dt} - \frac{V_2[S_2]}{K_2(1 + [S_4]/K_{I_2}) + [S_2]}. \tag{193}$$

Conservation of mass conditions are

$$[S_1]_0 = [S_1] + [S_2] + [S_4], \tag{194}$$

$$[S_1]_0 = [S_1] + [S_3]. \tag{195}$$

If $K_{I_1} = K_1$ and $K_{I_2} = K_2$, then the differential equations simplify to

$$\frac{d[S_1]}{dt} = -\left\{\frac{V_1}{K_1 + [S_1]_0}\right\}[S_1], \tag{196}$$

$$\frac{d[S_2]}{dt} = \left\{\frac{V_1}{K_1 + [S_1]_0}\right\}[S_1] - \frac{V_2[S_2]}{(K_2 + [S_1]_0 - [S_1])}. \tag{197}$$

The solution of Eq. (196) is Eq. (191) and this is substituted into Eq. (197) to give

$$\frac{d[S_2]}{dt} = \left\{\frac{V_1}{K_1 + [S]_0}\right\}\exp\left[-\left\{\frac{V_1}{K_1 + [S_1]_0}\right\}t\right]$$

$$- \frac{V_2[S_2]}{\{K_2 + [S_1]_0 - [S_1]_0\exp\{-[V_1/(K_1 + [S_1]_0)]t\}\}}. \tag{198}$$

Let $b = -V_1/(K_1 + [S_1]_0)$, $c = V_2$, $g = (K_2 + [S_1]_0)$, $x = \exp(bt)$, $y = [S_2]$, and $f = [S_1]_0$, then Eq. (198) becomes

$$y = \left(\frac{x}{g - fx}\right)^{-c/bd}\int -\left(\frac{x}{g - fx}\right)^{c/bd} dx + K;$$

let $-c/bg = 1$, then

$$y = -\left(\frac{x}{g - fx}\right)(g \ln x - fx) + K. \tag{199}$$

This was only readily integrated if c/bg was an integer; we chose the value -1 for this ratio, i.e., $V_2 = V_1(K_2 + [S_1]_0)/(K_1 + [S_1]_0)$. Therefore,

$$[S_2] = \frac{1}{(K_2 + [S_1]_0)\exp\left(\frac{V_1}{K_1 + [S_1]_0}t\right) - [S_1]_0}\left\{\frac{V_1(K_2 + [S_1]_0)}{K_1 + [S_1]_0}\right.$$

$$\times t\exp\left(\frac{V_1}{K_1 + [S_1]_0}t\right) + [S_1]_0\bigg\} - \frac{[S_1]_0}{K_2}. \tag{200}$$

This is a cumbersome but nonetheless *closed* solution to a, albeit very particular, coupled enzyme system. It is likely to be of little value experimentally, but it is of some mathematical interest.

VII. COUPLED ENZYMES IN THE PRE-STEADY STATE

A. General

Steady-state kinetic analysis under a fixed set of environmental conditions (pH, ionic strength, temperature) for coupled enzyme assays can yield only estimates of V_{max} and K_m and steady-state isomerization constants (e.g., Roberts, 1977). Only by an analysis of the transient phase of each enzymatic reaction can the unitary rate constants be evaluated. Recent interesting work on the thermal-variation method for the determination of values of rate constants in a simple mechanism has appeared (Lin *et al.*, 1982) but we assume, in this chapter, that constant temperature applies throughout a reaction time course. Analytical solutions of the rate equations that describe even the simplest enzyme couple [Eq. (70)] can be obtained only under two extreme conditions of parameter values, (1) when the initial substrate concentration is much greater than the enzyme concentration, and (2) when the enzyme concentration is much greater than the initial substrate concentration (Kuchel and Roberts, 1974). These conditions allow the species of highest concentration to be considered constant, thus linearizing the previously nonlinear differential rate equations. In the first case (1) the analytical solution describes a transient phase followed by a steady-state rate of product formation. In the second case (2) no steady state of enzyme–substrate complex eixsts at any stage of the reaction. Furthermore, the reaction must be followed with specialized equipment capable of detecting very low reactant concentrations.

B. Two Single Intermediate Enzymes with Low Enzyme Levels

1. Solution of Differential Equations

We develop here the analysis first presented by Kuchel and Roberts (1974). Consider the scheme in Eq. (70). In the transient phase (i.e., before attainment of a steady state of $[E_iS_i]$) the amount of product S_2 is of the same order as that of E_2. If $[E_2]$ is sufficiently high, then during the transient period (and possibly even longer) $[E_2S_2]$ will be much smaller than free $[E_2]$. Under this condition $[E_2]$ can be regarded as constant and equal to $[E_2]_0$, i.e., $[E_2] \gg [E_2S_2]$ and therefore $[E_2] \simeq [E_2]_0$. In addition, it is assumed that $[S_1]_0 \gg [E_1]_0$, so $[S_1] \simeq [S_1]_0$ in the time domain under study. Therefore, in the reaction scheme, with the above-mentioned approximations, there are five time-dependent variables. The rearranged Laplace–Carson transformed differential equations that describe the kinetics of the system are

$$\Psi[E_1]_0 = (\Psi + k_1[S_1]_0)e_1 - \quad (k_{-1} + k_2)e_1s_1 \quad + \quad\quad 0 \quad\quad + \quad\quad 0$$

$$0 \quad = \quad -k_1[S_1]_0e_1 \quad + (\Psi + k_{-1} + k_2)e_1s_1 + \quad\quad 0 \quad\quad + \quad\quad 0 \quad\quad (201)$$

$$0 \quad = \quad\quad 0 \quad\quad - \quad\quad k_2e_1s_1 \quad\quad + (\Psi + k_2[E_2]_0)s_2 - \quad\quad k_{-3}e_2s_2$$

$$0 \quad = \quad\quad 0 \quad\quad + \quad\quad 0 \quad\quad - \quad\quad k_3[E_2]_0s_2 \quad + (\Psi + k_{-3} + k_4)e_2s_2$$

and $\Psi s_3 = k_4 e_2 s_2$.

The solution of Eq. (201) for each of the four variables is obtained using Cramer's rule (Section II). Thus,

$$e_1 = \frac{(\Psi + k_{-1} + k_2)[E_1]_0}{(\Psi + \lambda_1)}, \tag{202}$$

where

$$\lambda_1 = k_1[S_1]_0 + k_{-1} + k_2, \tag{203}$$

$$e_1s_1 = \frac{k_1[E_1]_0[S_1]_0}{(\Psi + \lambda_1)}, \tag{204}$$

$$s_2 = \frac{k_1k_2[E_1]_0[S_1]_0(\Psi + k_{-3} + k_4)}{(\Psi + \lambda_1)(\Psi + \lambda_2)(\Psi + \lambda_3)}, \tag{205}$$

$$e_2s_2 = \frac{k_1k_2k_3[E_1]_0[S_1]_0[E_2]_0}{(\Psi + \lambda_1)(\Psi + \lambda_2)(\Psi + \lambda_3)}, \tag{206}$$

$$s_3 = \frac{k_1k_2k_3k_4[E_1]_0[S_1]_0[E_2]_0}{\Psi(\Psi + \lambda_1)(\Psi + \lambda_2)(\Psi + \lambda_3)}, \tag{207}$$

and

$$\lambda_2 = [A + (A^2 - 4B)^{1/2}]/2, \tag{208}$$

$$\lambda_3 = [A - (A^2 - 4B)^{1/2}]/2, \tag{209}$$

$$A = k_3[E_2]_0 + k_{-3} + k_4, \tag{210}$$

$$B = k_3k_4[E_2]_0. \tag{211}$$

The time evolution of the various concentrations is obtained by inverting the functions, which are standard forms, using Table I. The concentration of final product, $[S_3]$, as a function of time is

$$[S_3] = \frac{\Pi_{i=1}^4 k_i[E_1]_0[E_2]_0[S_1]_0 t}{\lambda_1\lambda_2\lambda_3} - P_1(1 - e^{-\lambda_1 t})$$

$$- P_2(1 - e^{-\lambda_2 t}) - P_3(1 - e^{-\lambda_3 t}), \tag{212}$$

where

$$P_i = \prod_{i=1}^{4} k_i [E_1]_0 [E_2]_0 [S_1]_0 \bigg/ \lambda_i^2 \prod_{\substack{j=1 \\ j \neq i}}^{3} (\lambda_j - \lambda_i). \tag{213}$$

Substitution of the λ_i into the first term of Eq. (212) gives

$$[S_3] = v_1 t - \sum_{i=1}^{3} P_i (1 - e^{-\lambda_i t}), \tag{214}$$

where

$$v_1 = \frac{k_2 [E_1]_0 [S_1]_0}{[S_1]_0 + K_1}, \tag{215}$$

and

$$K_1 = (k_{-1} + k_2)/k_1. \tag{216}$$

Each constant P_i can be expressed in terms of only v_1 and λ_i as follows:

$$P_i = v_1 \prod_{i=1}^{3} \lambda_i \bigg/ \lambda_i^2 \prod_{\substack{j=1 \\ j \neq i}}^{3} (\lambda_j - \lambda_i). \tag{217}$$

Hence,

$$[S_3] = v_1 \left\{ t - \sum_{i=1}^{3} \frac{(1 - e^{-\lambda_i t})}{\lambda_i^2} \left[\prod_{j=1}^{3} \lambda_j \bigg/ \prod_{\substack{j=1 \\ j \neq i}}^{3} (\lambda_j - \lambda_i) \right] \right\}, \tag{218}$$

and thus the equation is fully determined by four constants, namely, $v_1, \lambda_1, \lambda_2,$ and λ_3. If $[E_2]k_4 > 2[E_1]k_2$, then, according to Section V, $[S_2]$ will always be less than K_2 for all times even into the steady-state region of the second enzymatic reaction. Under these conditions E_2 is very "under-saturated," so free E_2 is greatly in excess of the complex E_2S_2 and Eq. (218) will remain valid for remarkably long times. At larger values of t the exponential terms vanish (note that the λ_i are all positive) and Eq. (218) becomes

$$[S_3] = v_1 t - \sum_{i=1}^{3} P_i. \tag{219}$$

Therefore, a plot of $[S_3]$ versus t approaches the linear asymptote which has a slope v_1 (Fig. 7), i.e., the slope is equal to the rate of the first reaction.

The backward extrapolation of the line intersects the t axis at the value τ (Fig. 7) and the expression for τ in terms of the parameters in the mechanism is

$$\tau = \sum_{i=1}^{3} \frac{P_i}{v_1}. \tag{220}$$

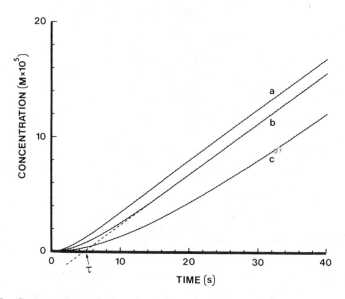

Fig. 7. Concentration of final product in the pre- and intra-steady-state regions of a two-enzyme reaction with enzyme concentrations low compared with the first substrate, and with differing concentrations of the coupling enzyme. The curves are described by Eq. (218), with the following parameter values and initial conditions: $k_1 = 10^7$; $k_{-1} = 10^4$; $k_2 = 10^3$; $k_3 = 2 \times 10^7$; $k_{-3} = 2 \times 10^4$; $k_4 = 2 \times 10^3$; $[S_1]_0 = 10^{-3}$ mol liter^{-1}; $[E_1]_0 = 10^{-8}$ mol liter^{-1}. The various concentrations of E_2 were a, 2×10^{-7} mol liter^{-1}; b, 1×10^{-7} mol liter^{-1}; c, 4×10^{-8} mol liter^{-1}. The slopes of the dotted line on curve b and the linear portion of curve a give the steady-state velocity v_1 [given by Eq. (215) in terms of the unitary rate constants], and the intercept on the abscissa yields the value of τ, for which the analytical expression is given by Eq. (220).

The intersection on the $[S_3]$ axis occurs at

$$[S_3] = -\sum_{i=1}^{3} P_i = -v_1\tau = \beta. \tag{221}$$

Inspection of Eqs. (208) and (209) reveals that, in general, $\lambda_2 \gg \lambda_3$ and it follows that

$$\lambda_3 \simeq \frac{\lambda_2\lambda_3}{(\lambda_2 + \lambda_3)} = \frac{k_3 k_4 [E_2]_0}{k_3 [E_2]_0 + k_{-3} + k_4} = \frac{V_2}{[E_2]_0 + K_2} \simeq \frac{V_2}{K_2}, \tag{222}$$

where V_2 and K_2 are the steady-state parameters of E_2. If $\lambda_1 \gg \lambda_3$, then $P_3 \gg P_1$ and P_2 and the intercept, of the previously mentioned asymptote, on the $[S_3]$ axis approximates to $-P_3$, but

$$P_3 = \frac{v_1\lambda_1\lambda_2\lambda_3}{\lambda_3^2(\lambda_1 - \lambda_3)(\lambda_2 - \lambda_3)} \simeq \frac{v_1}{\lambda_3} = \frac{v_1 K_2}{V_2}. \tag{223}$$

Thus, the expression for the intercept, $-v_1 K_2/V_2$, agrees with that given by the analysis which began with the assumption of steady states in $E_1 S_1$ and $E_2 S_2$ in Sections V,B and C, [see Eq. (122)] (Kuchel and Roberts, 1974; Easterby, 1973; Hess and Wurster, 1970). This is a reassuring validation of the consistency of both methods of analysis and is a powerful victory for the present analysis which is obviously more general than the earlier one. Figure 7 shows simulations for particular sets of realistic parameters in the coupled enzyme system described in this section. It was shown (Kuchel and Roberts, 1974) that the coincidence of numerical and analytical solutions is excellent, with some deviation (a few percent in the time courses shown) being evident when E_2 is lowest.

2. Experimental Evaluation of k_is

Extrapolation of the linear asymptote to the progress curve of $[S_3]$ back to intersect the ordinate gives the value of β which is defined by Eq. (221). Furthermore, v_1 is the steady-state rate of formation of $[S_3]$ [Eq. (219)], i.e., the slope of the linear asymptote measured from the data. So,

$$[S_3] - v_1 t - \beta = \sum_{i=1}^{3} P_i e^{-\lambda_i t}. \tag{224}$$

Therefore, a plot of $\ln\{[S_3] - v_1 t - \beta\}$ versus t *may* have three (almost) linear segments corresponding to the regions where the exponentials containing λ_1, λ_2, and λ_3 dominate the value of the function. Over long times, Eq. (224) reduces to a single exponential, thus enabling estimates to be made of the small valued λ and its coefficients from the slope and ordinate intercept of the log plot. Let these "slow parameters" be λ_j and P_j. Substitution of the values of $P_j e^{-\lambda_j t}$ from the left-hand side of Eq. (224) for each time in the data set is followed by the taking of logarithms; plotting the values of $\ln([S_3] - v_1 t - \beta - P_j e^{-\lambda_j t})$ versus t thus enables the next λ and P to be determined. This process is called the "peeling procedure" since exponentials are "peeled off" the log curve one at a time. The analysis may not always be straightforward because, (1) one of the λ_i may be too large (relaxation too fast) to be followed experimentally, or (2) if two of the λ_i are very similar in value only one exponential will be evaluated, thus giving a false impression of the complexity of the underlying system.

The peeling procedure forms the basis of some computer-based curve fitting procedures (Mancini and Pilo, 1970) but, as it stands, the main disadvantage of the method as presented here is the lack of estimates of the accuracy of the fitted parameter values. Therefore, the previous analysis should be viewed as a means of obtaining initial estimates of the parameters prior to *nonlinear* regression (Osborne, 1976) of the analytical function onto the data. An even more general approach to fitting data from kinetic systems is given in Section VIII.

3. The Second Enzyme

In a coupled enzyme assay the second enzyme may be amenable to separate individual study, i.e., λ_2 and λ_3 may be determined in an independent kinetic study (steady state and pre-steady state) of E_2 with S_2. The pre-steady-state study can be carried out with $[S_2] \gg [E_2]$ or $[S_2] \ll [E_2]$ (Hijazi and Laidler, 1972). The latter case is that which has most bearing on the present problem, so it will be analyzed. The transform of $[S_3]$ is obtained from the system of differential equations for E_2 and is also simply derived from Eqs. (207)–(211);

$$s_3 = \frac{k_3 k_4 [E_2]_0 [S_2]_0}{(\Psi + \lambda_2)(\Psi + \lambda_3)}. \tag{225}$$

The original function is

$$[S_3] = \frac{k_3 k_4 [E_2]_0 [S_2]_0}{\lambda_2 \lambda_3} - P_1 e^{-\lambda_2 t} - P_2 e^{-\lambda_3 t}, \tag{226}$$

where

$$P_1 = -P_2 = \frac{k_3 k_4 [E_2]_0 [S_2]_0}{\lambda_2 (\lambda_3 - \lambda_2)}, \tag{227}$$

and λ_2 and λ_3 are given by the same expressions as in Eqs. (208) and (209). Note also that the first term of Eq. (226) is simply $[S_2]_0$, since $\lambda_2 \lambda_3 = B = k_3 k_4 [E_2]_0$. Hence the following equation describes $[S_3]$ as a function of time in terms of $[S_2]_0$ and the two λs:

$$[S_3] = [S_2]_0 - \frac{\lambda_3}{(\lambda_3 - \lambda_2)} e^{-\lambda_2 t} - \frac{\lambda_2}{(\lambda_2 - \lambda_3)} e^{-\lambda_3 t}. \tag{228}$$

The exponential peeling procedure (Section VII,B,2) can be used to obtain estimates of λ_2, λ_3, and $\lambda_3/(\lambda_3 - \lambda_2)$ and $\lambda_2/(\lambda_3 - \lambda_2)$. From Eq. (228),

$$\ln\{[S_2]_0 - [S_3]\} = \ln\{P_1 e^{-\lambda_2 t} + P_2 e^{-\lambda_3 t}\}, \tag{229}$$

and at relatively large values of t one expontial term remains; if $\lambda_2 \gg \lambda_3$ then the $e^{-\lambda_3 t}$ term remains. So, for large values of t, Eq. (229) can be usefully approximated to

$$\ln\{[S_2]_0 - [S_3]\} \simeq \ln P_3 - \lambda_3 t. \tag{230}$$

Therefore, the asymptote to the plot of the left-hand side of Eq. (229) versus t

has a slope $-\lambda_3$ and ordinate intercept $\ln P_3$. With P_3 and λ_3 now known a plot is constructed of the left-hand side of the following equation versus t;

$$\ln\{[S_2]_0 - [S_3] - P_3 e^{-\lambda_3 t}\} = \ln P_2 - \lambda_2 t. \tag{231}$$

The slope of such a plot is $-\lambda_2$ and the ordinate intercept is $\ln P_2$. From Eqs. (208)–(211), $\lambda_2 + \lambda_3 = A = k_3[E_2]_0 + k_{-3} + k_4$ and $\lambda_2\lambda_3 = B = k_3 k_4[E_2]_0$. A series of experiments may be conducted at various concentrations of $[E_2]_0$. So, a plot of $\lambda_2\lambda_3$ versus $[E_2]_0$ gives a slope of $k_3 k_4$. A plot of $(\lambda_2 + \lambda_3)$ versus $[E_2]_0$ gives a slope of k_3 and an ordinate intercept of $(k_{-3} + k_4)$. Therefore, from the slopes of the two previous plots the values of k_3 and k_4 can be determined, and with a knowledge of k_4 and the intercept of the second plot, the value of k_{-3} can be found. An additional check on k_4 comes from the steady-state parameter $V_2 = k_4[E_2]_0$ and a check on parameter consistency comes from $K_2 = k_{-3} + k_4/k_3$. If $\lambda_2 \gg \lambda_3$, then from Eq. (222) a plot of $1/\lambda_3$ versus $1/[E_2]_0$ has a slope of K_2/k_4 and an ordinate intercept of $1/k_4$.

In conclusion, since λ_2 and λ_3 can be determined from an independent study of $[E_2]$ this leaves only λ_1 unknown in Eq. (218); λ_1 may be further analyzed to yield values of k_1, k_{-1} and k_2: by doing experiments with varying $[S_1]_0$, it is seen from Eq. (203) that a plot of λ_1 versus $[S_1]_0$ gives the value of k_1 from the slope and the value of $(k_{-1} + k_2)$ from the ordinate intercept. From steady-state kinetics, $V_1 = k_2[E_1]_0$, which enables an estimate to be made of k_2, thus further enabling k_{-1} to be evaluated from the ordinate intercept of the previous plot. As before, consistency of the unitary rate constants with K_1 is used as a "cross-check."

C. Two Double Intermediate Enzymes with Low Enzyme Levels

The single enzyme counterpart of this system was analyzed by Maguire et al. (1974) and Kassera and Laidler (1970). The following analysis is that of Kuchel and Roberts (1974). The scheme of reactions is

$$
\begin{aligned}
E_1 + S_1 &\underset{k_{-1}}{\overset{k_1}{\rightleftharpoons}} E_1 S_1 \xrightarrow{k_2} E_1 S_2 \xrightarrow{k_3} E_1 + S_2 \\
E_2 + S_2 &\underset{k_{-4}}{\overset{k_4}{\rightleftharpoons}} E_2 S_2 \xrightarrow{k_5} E_2 S_3 \xrightarrow{k_6} E_2 + S_3
\end{aligned}
\tag{232}
$$

The set of differential rate equations that describe reactant fluxes in this system gives a closed analytical solution if the approximation methods of Section VII,B are used. Naturally, the number of exponential terms in the expressions for the reactant concentration is increased. For example, the Laplace–Carson transform of $[S_3]$ is

$$s_3 = \prod_{i=1}^{6} k_i[E_1]_0[E_2]_0[S_1]_0 \bigg/ \Psi \prod_{i=1}^{5} (\Psi + \lambda_i), \tag{233}$$

where $\lambda_1 = [A + (A^2 - 4B)^{1/2}]/2,$ (234)

$\lambda_2 = [A - (A^2 - 4B)^{1/2}]/2,$ (235)

$A = k_1[S_1]_0 + k_{-1} + k_2 + k_3,$ (236)

$B = k_1(k_2 + k_3)([S_1]_0 + K_1),$ (237)

$K_1 = k_3(k_{-1} + k_2)/[k_1(k_2 + k_3)],$ (238)

$\lambda_3 = [C + (C^2 - 4D)^{1/2}]/2,$ (239)

$\lambda_4 = [C - (C^2 - 4D)^{1/2}]/2,$ (240)

$C = k_4[E_2]_0 + k_{-4} + k_5,$ (241)

$D = k_4 k_5 [E_2]_0,$ (242)

$\lambda_5 = k_6.$ (243)

Since $\lambda_1 \lambda_2 = k_1(k_2 + k_3)([S_1]_0 + K_1)$ and $\lambda_3 \lambda_4 = \lambda_4 \lambda_5 [E_2]_0$, the original function, obtained using Table I, is

$$[S_3] = v_1 \left[t - \sum_{i=1}^{5} \frac{(1 - e^{-\lambda_i t})}{\lambda_i^2} \left(\prod_{i=1}^{5} \lambda_i \bigg/ \prod_{\substack{j=1 \\ j \neq i}}^{5} (\lambda_j - \lambda_i) \right) \right].$$ (244)

The progress curve of $[S_3]$ is similar to that of the simpler scheme [Eq. (218)] but with two additional exponential terms due to the extra species $E_1 S_2$ and $E_2 S_3$. Furthermore, in order to extract unique values of the k_is it is even more important, than in the previous case, that the second enzyme should be studied alone.

D. Hysteretic Enzymes—A Pre-Steady-State Model

The slow reversible transformation of the activity of an enzyme, even over the time scale normally encountered in steady-state enzyme kinetics experiments and in response to changes in the chemical environment, has been described as *enzyme hysteresis* or *enzyme memory* (Frieden, 1970; Hatfield *et al.*, 1970; Ray and Hatfield, 1970); for a while the enzyme continues to behave as it would in the "old" environment even though conditions have changed, and it only slowly changes to its "new" form. In purely abstract mathematical terms there is nothing really new about the models needed to describe this phenomenon. This phenomenon results, simply, from large differences in the magnitudes of groups of rate constants that characterize the enzyme system. Threonine deaminase (Hatfield *et al.*, 1970) and wheat germ hexokinase (Ricard *et al.*, 1977; Buc *et al.*, 1977) are enzymes which show a slow acceleration of the steady-state progress curves of product formation in certain buffer conditions.

A simple scheme that describes the *hysteretic* or *enzyme memory* phenomenon is now analyzed in a manner that considers also the pre-steady-state time domain (Roberts, 1977). Consider a single enzyme E that (slowly) isomerizes to a second form F, which, for the sake of simplicity, is taken to be totally inactive; the transition can be thought of as occurring in response to the addition of the enzyme solution to a reaction medium:

$$F \underset{k_3}{\overset{k_{-3}}{\rightleftharpoons}} E + S_1 \underset{k_{-1}}{\overset{k_1}{\rightleftharpoons}} ES_1 \overset{k_2}{\longrightarrow} E + S_2 \tag{245}$$

The initial conditions at $t = 0$ are $[F] = [F]_0$, $[E] = [E]_0$, $[S_2]_0 = 0$. The main conservation of mass condition is $[E]_T = [E]_0 + [F]_0 = [E] + [F] + [ES_1]$. Also, we will assume that the substrate concentration, $[S_1]_0$, is much greater than $[E]_0$ such that $[S_1] \simeq [S_1]_0$ at all times of interest. The set of simultaneous Laplace–Carson transformed differential equations that correspond to Eq. (245) and the specified initial conditions are

$$\Psi[E]_0 = (\Psi + k_1[S_1]_0 + k_3)e - k_{-3}f - (k_2 + k_{-1})es_1,$$

$$\Psi[F]_0 = -k_3e + (\Psi + k_{-3})f + 0,$$

$$0 = -k_1[S_1]_0e + 0 + (\Psi - k_{-1} + k_2)es_1, \tag{246}$$

and

$$v_1 = \frac{d[S_2]}{dt} = k_2[ES_1]. \tag{247}$$

The solution of Eq. (246) for es_1, using the methods of Sections II and III, is

$$es_1 = \frac{\Psi k_1 k_{-3}[S_1]_0[F]_0}{\Psi(\Psi + \lambda_1)(\Psi + \lambda_2)} + \frac{\Psi(\Psi + k_{-3})k_1[E]_0[S_1]_0}{\Psi(\Psi + \lambda_1)(\Psi + \lambda_2)}, \tag{248}$$

where

$$\lambda_1 = [A + (A^2 - 4B)^{1/2}]/2, \tag{249}$$

$$\lambda_2 = [A - (A^2 - 4B)^{1/2}]/2, \tag{250}$$

$$A = k_1[S_1]_0 + k_{-1} + k_2 + k_3 + k_{-3}, \tag{251}$$

$$B = (k_3 + k_{-3})(k_{-1} + k_2) + k_1 k_{-3}[S_1]_0. \tag{252}$$

The inverse of Eq. (248), using Table I, is

$$[ES_1] = k_1 k_{-3}[S_1]_0[F]_0 \left\{ \frac{1}{\lambda_1 \lambda_2} - \frac{e^{-\lambda_1 t}}{\lambda_1(\lambda_2 - \lambda_1)} - \frac{e^{-\lambda_2 t}}{\lambda_2(\lambda_1 - \lambda_2)} \right\}$$

$$+ k_1[E]_0[S_1]_0 \left\{ \frac{k_{-3}}{\lambda_1 \lambda_2} - \frac{(k_{-3} - \lambda_1)e^{-\lambda_1 t}}{\lambda_1(\lambda_2 - \lambda_1)} - \frac{(k_{-3} - \lambda_2)e^{-\lambda_2 t}}{\lambda_2(\lambda_1 - \lambda_2)} \right\}. \tag{253}$$

From Eq. (247) and the conservation condition for $[E]_T$,

$$v_1 = k_1 k_2 k_{-3} [S_1]_0 [E]_T \left\{ \frac{1}{\lambda_1 \lambda_2} - \frac{e^{-\lambda_1 t}}{\lambda_1 (\lambda_2 - \lambda_1)} - \frac{e^{-\lambda_2 t}}{\lambda_2 (\lambda_1 - \lambda_2)} \right\}$$
$$+ k_1 k_2 [S_1]_0 [E]_0 \left\{ \frac{e^{-\lambda_1 t}}{(\lambda_2 - \lambda_1)} + \frac{e^{-\lambda_2 t}}{(\lambda_1 - \lambda_2)} \right\}. \tag{254}$$

For large values of t, Eq. (254) becomes

$$v_1 = v_f = k_1 k_2 k_{-3} [S_1]_0 [E]_T / \lambda_1 \lambda_2, \tag{255}$$

where the subscript f refers to the rate in the "final state" of the reaction. Now, from Eqs. (249)–(252), $\lambda_1 \lambda_2 = B$, so Eq. (255) becomes

$$v_f = \frac{V_1 [S_1]_0}{K_1 (1 + k_3/k_{-3}) + [S_1]_0}, \tag{256}$$

with $V_1 = k_2 [E]_T$ and $K_1 = (k_{-1} + k_2)/k_1$ as usual (Section IV).

Consider now the very short time domain where a steady state of $[ES_1]$ is established but the extent of substrate conversion to product is still very small. In general, the inequality $\lambda_1 \gg \lambda_2$ can be verified by inspection of Eqs. (249) and (250), so Eq. (254) is approximated by

$$v_1 = k_1 k_2 k_{-3} [S_1]_0 [E]_T \left\{ \frac{1}{\lambda_1 \lambda_2} - \frac{e^{-\lambda_2 t}}{\lambda_1 \lambda_2} \right\}$$
$$+ \frac{k_1 k_2 [S_1]_0 [E]_0 e^{-\lambda_2 t}}{\lambda_1}, \tag{257}$$

i.e., even at very short times, because of the large values of λ_1, $e^{-\lambda_1 t}$ has a small value. Now, redefine a new "zero time" that coincides with the time domain when the $e^{-\lambda_1 t}$ terms are very small. Set t to zero, and Eq. (257) becomes,

$$v_0 = \frac{k_1 k_2 [S_1]_0 [E]_0}{\lambda_1}. \tag{258}$$

Thus Eq. (257) can be written as

$$v_t = (v_0 - v_f) e^{-\lambda_2 t} + v_f, \tag{259}$$

which is the expression derived by Hatfield et al. (1970). The expression for $[S_2]$ at any time is most simply derived by integration of Eq. (247):

$$[S_2] = k_1 k_2 k_{-3} [S_1]_0 [E]_T \left\{ \frac{t}{\lambda_1 \lambda_2} - \frac{1 - e^{-\lambda_1 t}}{\lambda_1^2 (\lambda_1 - \lambda_2)} - \frac{1 - e^{-\lambda_2 t}}{\lambda_2^2 (\lambda_1 - \lambda_2)} \right\}$$
$$+ k_1 k_2 [S_1]_0 [E]_0 \left\{ \frac{1 - e^{-\lambda_1 t}}{\lambda_1 (\lambda_2 - \lambda_1)} + \frac{1 - e^{-\lambda_2 t}}{\lambda_2 (\lambda_1 - \lambda_2)} \right\}. \tag{260}$$

If we consider, as before, that $\lambda_1 \gg \lambda_2$, then using Eqs. (256) and (258), the following expression for $[S_2]$ is obtained:

$$[S_2] = v_f t + [(v_0 - v_f)/\lambda_2](1 - e^{-\lambda_2 t}). \tag{261}$$

The λ_2 in Eq. (261) could be treated as a phenomenological rate constant, but here we have ascribed some mechanistic meaning to it. The progress of $[S_2]$ with time displays an initial "lag" with a final steady-state rate of v_f. A linear asymptote to the final steady-state time course of $[S_2]$ intersects the time axis at $(v_0 - v_f)/\lambda_2$. If $[E]_0 = 0$, then $v_0 = 0$ and the rate of appearance of $[S_2]$ increases with time and the previous intercept yields directly an estimate of the value of λ_2; if $[F]_0 = 0$, then the rate declines with time.

The hysteretic behavior of the scheme of Eq. (245) is shown in Fig. 8. The two time courses were obtained with (b) and without (a) reversibility of the F to E "slow" isomerization. Inspection of the list of parameter values used for the simulation reveals the large differences (five orders of magnitude) between the "slow" isomerization rate constants and the slowest of the reaction-pathway constants, k_2. These large differences in values of rate constants must exist to give rise to the hysteresis phenomenon.

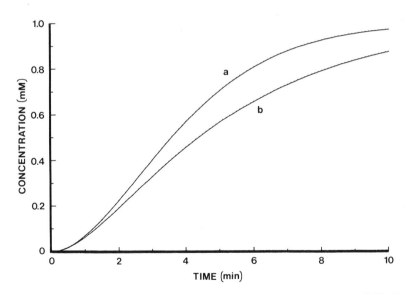

Fig. 8. Enzyme hysteresis, or "enzyme memory," according to the scheme in Eq. (245) with a, irreversible formation of E from F; and b, reversible formation of E from F. The following parameter and initial values were used in the simulation: $k_1 = 1 \times 10^7$, $k_{-1} = 1 \times 10^4$, $k_2 = 1 \times 10^3$, $k_{-3} = 1 \times 10^{-2}$; for curve a, $k_3 = 0$, and for curve b, $k_3 = 1 \times 10^{-2}$.

E. Enzyme–Enzyme Interaction

1. General

It was shown in Section III,B,3 that the Maclaurin polynomial procedure could be used to develop general expressions that relate the early time course of a coupled enzymatic reaction to the mechanisms of the individual enzymes. Indeed, if competitive inhibition of the reaction by the products occurs as these compounds accumulate, the fact that the mathematical solution applies at $t = 0$ means that the behavior of the system at longer times must be described using other mathematical means. Nevertheless, using the procedures already described in this section it is possible to show that if two enzymes, catalyzing consecutive reactions that alone are described by the simple Michaelis–Menten equation, interact heterogeneously, then the overall sequence behaves as if it were composed of control enzymes (Nichol *et al.*, 1974). This situation also pertains with ambiquitous enzymes which associate with membranes in response to binding a metabolite; the best examples are the binding of aldolase to myofibrils in response to changes in ionic strength (Masters, 1977, 1981) and the binding of hexokinase to brain mitochondria which is sensitive to ATP, phosphate, and glucose 6-phosphate levels. The bound hexokinase is less susceptible to inhibition by glucose 6-phosphate than is the free enzyme (Wilson, 1968, 1980). In effect, the other enzymes, proteins, or membrane components are acting as "effector molecules," a role often thought of as the exclusive prerogative of low-molecular-weight species (see Friedrich, Chapter 3, and Kurganov, Chapter 5, this volume).

2. A Simple System

The simplest coupled enzyme scheme which involves enzyme–enzyme interaction is

$$
\begin{array}{c}
\mathrm{E_1 + S_1} \underset{k_{-1}}{\overset{k_1}{\rightleftarrows}} \mathrm{E_1 S_1} \xrightarrow{k_2} \mathrm{E_1 + S_2} \\
\mathrm{E_1 E_2} \overset{k_5}{\underset{k_{-5}}{\rightleftarrows}} \\
\mathrm{E_2 + S_2} \underset{k_{-3}}{\overset{k_3}{\rightleftarrows}} \mathrm{E_2 S_2} \xrightarrow{k_4} \mathrm{E_2 + S_3}
\end{array}
\tag{262}
$$

Several possible ranges of parameter values exist for this basic scheme, but for simplicity, it will be assumed that the catalytic activity of E_2 is unchanged in the complex state while only uncomplexed E_1 is catalytically active. The basic set of differential rate equations can be written as before with account taken of terms involving k_5 and k_{-5}. As in Section VII,B, the important assumption that $[S_1]$ and $[E_2]$ are constant renders the differential equations linear and therefore amenable to Laplace–Carson transformation; specifically, $[S_1] \simeq [S_1]_0$ and $[E_2] \simeq [E_2]_0$ for all times under consideration.

Solution of the algebraic transformed equations and subsequent inversion and other algebraic manipulation (Section II) yield

$$[S_3] = \frac{V_1[S_1]_0 t}{K_1(1 + [E_2]_0 k_5/k_{-5}) + [S_1]_0} + \sum_{i=1}^{4}\left(\prod_{j=1}^{4} k_j\right)[E_1]_0[E_2]_0[S_1]_0$$

$$\times \left\{\left[(\lambda_i - k_{-5})\middle/ \lambda_i^2 \prod_{\substack{j=1 \\ j \neq i}}^{4}(\lambda_j - \lambda_i)\right](1 - e^{-\lambda_i t})\right\}. \tag{263}$$

The expression for the λ_is is not given but it can be seen from Eq. (263) that at large values of t the second term approaches zero and a plot of $[S_3]$ versus t becomes linear with a slope smaller than v_1; this is due to the $(1 + [E_2]_0 k_5/k_{-5})$ term of K_1 in the denominator. In other words, E_2 by its association with E_1 acts as a competitive inhibitor of E_1 with an inhibition constant k_{-5}/k_5. Furthermore, the steady-state rate equation is of the form expected for this type of inhibition (Laidler and Bunting, 1973, p. 92).

Therefore, Eq. (263) may provide a basis for experimentally detecting associations between enzymes when both may be studied separately as well as in combination. Deviation of product formation from the response predicted with no interaction *may* be interpreted in terms of the present theory which in turn may enable the determination of rate constants after a data fitting exercise.

Certainly with the particular model chosen here, the heterogeneous enzyme–enzyme interaction may be detected at high protein concentrations using a range of mass migration techniques (Nichol and Winzor, 1972). However, enzyme–enzyme interactions may be mediated by effector molecules in the reaction mixture as the reaction proceeds, thus making the present kinetic experiments a feasible approach to detecting interaction. Furthermore, interactions at protein concentrations much less than those detectable by most mass migration methods can theoretically be detected by the present enzyme–kinetic procedure.

F. Coupled Single-Intermediate Enzymatic Reactions at High Enzyme Concentrations

The reaction scheme of Eq. (70) can also be studied under conditions of high enzyme concentration. This should facilitate the concentration-dependent association of enzymes and then increase the likelihood of detecting an enzyme–enzyme interaction which may lead to deviation of the reaction time course from its predicted (no interaction) path. The basic idea expressed above was the motivation for stopped-flow kinetic analyses carried out on several coupled enzyme systems by Hess's group (Hess and Wurster, 1970; Barwell and Hess, 1970; Wurster and Hess, 1970). The analysis that is developed here, however, is that of Kuchel and Roberts (1974).

Assume that $[E_1]_0 \gg [S_1]_0$ and $[E_2]_0 \gg [S_2]$ for all times under study. The transformed differential equations of Eq. (201) are used here except that the one for e is replaced by one for s. Therefore,

$$\Psi s_1 - \Psi[S_1]_0 = -k_1[E_1]_0 s_1 + k_{-1}e_1 s_1, \tag{264}$$

and the equation for es_1 is changed to

$$\Psi es_1 = k_1[E_1]_0 s_1 - (k_{-1} + k_2)e_1 s_1. \tag{265}$$

The remaining equations of Eq. (201) apply here. The expression for s_3 is

$$s_3 = \prod_{i=1}^{4} k_i \left([E_1]_0[E_2]_0[S_1]_0 \middle/ \prod_{j=1}^{4} (\Psi + \lambda_j) \right), \tag{266}$$

where λ_1 and λ_2 have the same *form* as in Eqs. (203) and (208) with

$$A = k_1[E_1]_0 + k_{-1} + k_2, \tag{267}$$

$$B = k_1 k_2[E_1]_0; \tag{268}$$

λ_3 and λ_4 have the *same forms* as in Eqs. (208) and (209), respectively.

Equation (266) is inverted using Table I, and using the fact that $\lambda_1\lambda_2 = k_1 k_2[E_1]_0$ and $\lambda_3\lambda_4 = k_3 k_4[E_2]_0$, the following expression for $[S_3]$ is obtained:

$$[S_3] = [S_1]_0 \left\{ 1 - \sum_{i=1}^{4} e^{-\lambda_i t} \prod_{\substack{j=1 \\ j \neq i}}^{4} \frac{\lambda_j}{(\lambda_j - \lambda_i)} \right\}. \tag{269}$$

Thus, it can be seen from Eq. (269) that there is no steady-state rate of product formation. The concentration of S_3 ($[S_3]$) rises in a complicated (four) exponential manner to completion at the final concentration $[S_1]_0$. This behavior can be seen in the simulations in Fig. 9.

In general, $\lambda_1 \gg \lambda_2$ and $\lambda_3 \gg \lambda_4$, then

$$\lambda_2 \simeq \frac{\lambda_1\lambda_2}{(\lambda_1 + \lambda_2)} = \frac{V_1}{[E_1]_0 + K_1}, \qquad \lambda_4 \simeq \frac{\lambda_3\lambda_4}{(\lambda_3 + \lambda_4)} \simeq \frac{V_2}{[E_2]_0 + K_2}. \tag{270}$$

Application of the (exponential) peeling procedure (Section VII,B,2) will, under numerically favorable conditions, give values for each λ_i. Hence plots of $1/\lambda_2$ versus $1/[E_1]_0$ and $1/\lambda_4$ versus $1/[E_2]_0$ allow values of k_2, k_4, and the Michaelis constants K_1 and K_2 to be determined.

The simulated time courses, for a set of physically realistic parameters, in the coupled system considered in this subsection are shown in Fig. 9. For the parameters chosen, the analytical and numerical integration solutions are virtually identical (Kuchel and Roberts, 1974); this serves to highlight the rather surprising "robustness" of the present analytical solutions.

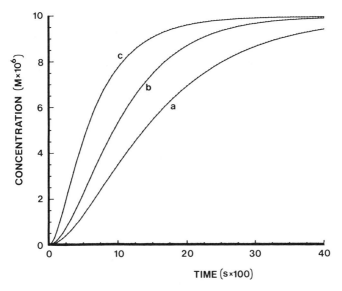

Fig. 9. Pre-steady-state product formation in a coupled enzyme system with high enzyme concentration compared with substrate, and with varying concentrations of E_1. The curves are described by Eq. (269) and the following parameter values and initial conditions were applied: $k_1 = 10^7$; $k_{-1} = 10^4$; $k_2 = 10^3$; $k_3 = 2 \times 10^7$; $k_{-3} = 2 \times 10^4$; $k_4 = 2 \times 10^3$; $[S_1]_0 = 10^{-5}$ mol liter^{-1}; $[E_2]_0 = 10^{-5}$ mol liter^{-1}. The various concentrations of E_1 were a, 1×10^{-5} mol liter^{-1}; b, 2×10^{-5} mol liter^{-1}; c, 8×10^{-5} mol liter^{-1}.

VIII. PARAMETER ESTIMATION IN SYSTEMS OF DIFFERENTIAL EQUATIONS

A. General

This section deals with the general problem of determining parameter values and estimates of their reliability for any system of differential equations, be they linear or nonlinear. The particular cases of enzyme and chemical kinetic schemes described by nonlinear differential equations are those discussed in several previous sections and also include virtually any model of a metabolic pathway. In many of the earlier examples of the time-dependent behavior of chemical reaction schemes, analytical solutions of the differential equations were obtained. However, in many physically realistic schemes, explicit expressions for each species as a function of time cannot be found. We must therefore resort to numerical integration in order to simulate the time evolution of the system. There are now numerous examples of complex numerical models of metabolism in the literature (e.g., Section I: Kohn *et al.*, 1977a–c; Achs *et al.*, 1977, Garfinkel *et al.*, 1977; Kuchel *et al.*, 1977).

Therefore, the determination of rate parameters in many realistic numerical models will involve numerical integration.

B. The General Mathematical Approach

Mathematically stated, the problem is as follows (Rosenbrock and Storey, 1966). A set of differential equations is given by

$$\frac{d[S_i]}{dt} = f_i([S], k, t) \qquad (i = 1, 2, \ldots, n), \qquad (271)$$

where k is a p-vector of rate constants, and the S_i are reactants which include enzyme–reactant complexes and final products, $[S]$ is the n-vector of these species, and t is time. It is assumed that k has the "true" value k^*, but the individual values of the elements in the vector remain to be determined. In most practical kinetic systems the concentration (or chemical activities) of all species cannot be monitored simultaneously; instead, combinations of concentrations may be monitored or, maybe, only the concentration of the final product of the reaction sequence can be monitored. Thus, observations $(u_j)_r$ (combinations of species, or a subset of $\{S_i\}$) are made such that

$$(u_j)_r = g_j([S]_{t_r}, \zeta_r) \, (j = 1, 2, \ldots, m; r = 1, 2, \ldots, R; R \le n), \qquad (272)$$

where ζ_r is an R-vector of random variables affecting the measurement of u_j at time t_r. It is assumed also that the mathematical forms of f_i and g_j are known. For example, in many cases g_j is simply $[S_i]$, or if a combination of two species is measured, g_j may be given by $[S_i] + [S_k]$ with $i \ne k$.

The parameter determination problem is to obtain an *estimate* (k) of the vector of "true" parameter values (k^*) from the R observations of u_j. Note that if the derivatives (rates of change of reactant concentrations) of $[S_i]$ could be measured directly the problem would be much simpler; corresponding values of f_i and g_i would exist and the estimation of k^* would not involve the solution of differential equations. In the present context, the derivatives usually cannot be measured with sufficient accuracy to permit the direct approach.

Equation (271) can be integrated numerically using a range of different algorithms (McCracken and Dorn, 1964; Schied, 1968) that are frequently available in computer "libraries." Of special value in solving arrays of stiff differential equations (i.e., where "stiff" refers to the large differences which exist between the eigenvalues describing the dynamics of the system; these are the λ_is in the previous example) is the algorithm of Gear (1968); this in fact was the algorithm used for the numerical simulations shown in the figures herein. The integration is carried out using an initial estimate of the k_i $(i = 1, 2, \ldots, p)$; it is for this step that an approximate analytical solution of the system under study may be valuable. At each time t_r for which data $(u)_r$ are available, the difference (*residual*) between the data and the simulation is calculated. Then for the R time values the function $F(\mathbf{k})$, which is the sum of the squares of the residuals, is evaluated:

$$F(\mathbf{k}) = \sum_{r=1}^{R} \sum_{j=1}^{m} \{(u_j)_r - w_j g_j([\mathbf{S}], \zeta)\}^2, \qquad (273)$$

where w_i is a weighting factor which incorporates information, introduced by the investigator, on the reliability of each datum, i.e., equal weights implies equal reliability of each datum.

The function value $F(\mathbf{k})$ is minimized using standard numerical techniques designed for locating minima of functions of multiple variables in multidimensional space; the space here is the p-dimensional space of parameters. Among the minimization procedures are the *simplex search* (Nelder and Mead, 1965), modified Marquardt procedure (Sadler, 1975; Osborne, 1976), and the older "*hill climbing*" methods such as the Newton–Raphson method (Schied, 1968; Rosenbrock and Storey, 1966). When the minimum value of $F(\mathbf{k})$ has been located by the computer algorithm the parameter set $\mathbf{k}_{min}$ is called the *best fit*. This is a valuable start, but an estimate of the reliability of the "*best fit*" parameters is usually required.

C. Reliability of Parameter Estimates

Intuitively, the most direct means of assessing standard deviations of parameters in fits of linear or nonlinear systems of differential equations is the "Monte Carlo" method (Hammersley and Handscomb, 1965). A "best fit" data set is produced by simulating the system under study [Eq. (271)] using the "best fit" parameters. Then "numerical noise" is added to the synthetic data set. This can be done using a Gaussian random number generator, available in most computer libraries on larger instruments, in which the investigator specifies the mean (the datum) and standard deviation of the distribution. The standard deviation can be estimated from other experiments by repeated assays of one concentration of reactant in a specified domain of concentrations. The random numbers generated for each datum are added to form a set of "synthetically noisy" data. A large number of sets are produced; about 30 is usually sufficient (Rosenbrock and Storey, 1966). Then the minimum of the function $F(\mathbf{k})$ [Eq. (273)] is obtained using the "best fit" parameters as a starting point. Because the starting point will be numerically near the minimum, the 30-fold repetition of the fitting procedure is not tolerably time consuming (Chandler et al., 1972). Therefore, 30 sets of parameter estimates are obtained, from which it is a simple matter to obtain means (which should be the best fit parameter values) plus standard deviations of each estimate.

Other less computer-time-consuming methods have been developed for estimating parameter reliabilities (Rosenbrock and Storey, 1966). They rely on linear approximations to $F(k)$ [Eq. (273)] in the region of the minimum. The function $F(k)$ is in fact nonlinear in many systems so the Monte Carlo method is usefully applied in initial studies as a check on the linear approximation (Chandler et al., 1972).

In conclusion, a general method, which is described above, exists for the estimation of parameter values and their reliability in sets of differential equations. The equations are integrated numerically, or use can be made of analytical solutions developed using the methods of this chapter. The foregoing analysis encompasses completely, as a subset, the methods of progress curve analysis recently applied in steady-state enzyme kinetics (Duggleby and Morrison, 1977, 1978; Orsi and Tipton, 1979; Duggleby, 1983).

IX. CONCLUSION

This chapter is concerned with sequences of enzymatic reactions and analytical (mathematical) expressions that describe their kinetic behavior in homogeneous solution. Basic definitions necessary for the mathematical description of the kinetic systems are given and this is followed by a consideration of simple first-order uni- and bidirectional chemical schemes. The development of analytical expressions to describe the kinetics of sequences of first-order reactions is shown to be a difficult process, but the procedure is expedited by the introduction and use of the Laplace–Carson operator method. The mathematical properties of this linear operator are explored in some detail and a table of important general transformations is presented. The application to systems with diffusional resistance remains to be studied (cf. Engasser and Horvath, 1976).

In order to solve the differential equations that describe even the simplest realistic coupled enzyme system it is shown that approximations are necessary. First, solutions of the differential equations are developed for a small time domain near the origin. This entails integration of the Michaelis–Menten equation and a subsequent approximation of the solution by a truncated series. Then, integration in series is used, followed by an approach using Maclaurin polynomials to obtain series solutions that give reasonable approximations to the relevant time courses for small extents of reaction.

Second, other approximations are considered and these are applied to the models prior to mathematical analysis rather than *in* the analysis, as is the case for the polynomial solutions. Several cases that were originally discussed by Haldane (1930) are developed. The analyses are used to obtain expressions for the transient time in a linear sequence of enzymatic reactions. Furthermore, the amount of coupling enzyme required in a coupled assay, to ensure that a steady-state rate of product formation is within a specified fraction of the rate of substrate depletion, is calculated using one of the formulae. Third, "open" sequences of enzymes are studied from a theoretical viewpoint and expressions are obtained for the concentrations of intermediate species in the sequences. Fourth, perturbation theory, as applied to steady-state enzyme sequences, is developed to obtain expressions for the transition time between steady states. A simple but pedagogically rewarding theory of the energy cost of transitions between steady states is developed.

Attention then turns from steady-state enzyme kinetic schemes to those dealing with the pre-steady-state phase of coupled enzymatic reactions. The theory that is developed relies heavily on the Laplace–Carson operator method. Several different experimental arrangements that include coupled reactions with high enzyme concentrations relative to substrates are discussed in great mathematical detail. The notion that enzyme kinetic studies of coupled reactions could be used to investigate heterogeneous enzyme–enzyme association is introduced and some relevant theory is developed. Also, the concept of enzyme memory or hysteresis is introduced and a model system is analyzed in the pre- and intra-steady-state regions of the reaction and a very general result concerning the mechanisms of hysteresis emerges. The pre-steady-state analysis is entirely consistent with earlier steady-state models and is therefore more valuable since the kinetic expressions cover a much longer time domain.

Finally, the general kinetic problem of the determination of the values of parameters and their accuracy in models that describe real experimental systems, and that are arrays of (non) linear differential equations, is introduced. This theory encompasses the analysis used for fitting theoretical progress curves to real data in steady-state enzyme kinetics.

ACKNOWLEDGMENTS

The work was supported in part by a grant from the Australian National Health and Medical Research Council. The assistance of the following people has been invaluable: Mr. B. T. Bulliman with computing; Ms. Sue Williams with word processing; Dr. D. V. Roberts for many illuminating discussions on enzyme kinetics and unrelated matters, and Prof. L. W. Nichol who set me on the research path which winds through the present topic.

REFERENCES

Achs, M. J., and Garfinkel, D. (1979). *Am. J. Physiol.* **236,** R21–R30.
Achs, M. J., and Garfinkel, D. (1982). *Am. J. Physiol.* **242,** R533–R544.
Achs, M. J., Kohn, M. C., and Garfinkel, D. (1977). *Am. J. Physiol.* **232,** R174–R180.
Ayres, F., Jr. (1952). "Schaum's Outline of Theory and Problems of Differential Equations." McGraw-Hill, New York.
Barwell, C. J., and Hess, B. (1970). *Hoppe-Seyler's Z. Physiol. Chem.* **351,** 1531–1536.
Bergmeyer, H. U., ed. (1962). "Methods of Enzymatic Analysis." pp. 3–13. Boehringer, London.
Briggs, G. E., and Haldane, J. B. S. (1925). *Biochem. J.* **19,** 338–339.
Buc, J., Ricard, J., and Meunier, J. C. (1977). *Eur. J. Biochem.* **80,** 593–601.
Capellos, C., and Bielski, B. H. J. (1972). "Kinetic Systems: Mathematical Description of Chemical Kinetics in Solution." Wiley, New York.
Chandler, J. P., Hill, D. E., and Spivey, H. O. (1972). *Comp. Biomed. Res.* **5,** 515–534.
Cleland, W. W. (1979). *Anal. Biochem.* **99,** 142–145.
Duggleby, R. G. (1983). *Biochim. Biophys. Acta* **744,** 249–259.
Duggleby, R. G., and Morrison, J. F. (1977). *Biochim. Biophys. Acta* **481,** 297–312.
Duggleby, R. G., and Morrison, J. F. (1978). *Biochim. Biophys. Acta* **526,** 398–409.
Easterby, J. S. (1973). *Biochim. Biophys. Acta* **293,** 552–558.
Easterby, J. S. (1981). *Biochem. J.* **199,** 155–161.
Eigen, M., and DeMaeyer, L. (1963). In "Technique of Organic Chemistry" (S. L. Friess, E. S. Lewis, and A. Weissberger, eds.), Vol. VIII/(II), pp. 895–1054. Wiley, New York.
Engasser, J. M., and Horvath, C. (1976). *Appl. Biochem. Bioeng.* **1,** 127–220.
Frieden, C. (1970). *J. Biol. Chem.* **254,** 5788–5799.
Garfinkel, D., and Achs, M. J. (1979). *Am. J. Physiol.* **236,** R31–R39.
Garfinkel, D., Kohn, M. C., and Achs, M. J. (1977). *Am. J. Physiol.* **232,** R181–R186.
Gear, C. W. (1968). *Proc. IFIP Congr. Edinburgh* **I,** 187–193.
Gillespie, D. T. (1977). *J. Phys. Chem.* **81,** 2340–2361.
Haldane, J. B. S. (1930). "Enzymes." Longmans, London.
Hammersley, J. M., and Handscomb, D. C. (1965). "Monte Carlo Methods." Methuen, London.
Hatfield, G. W., Ray, W. J., and Umbarger, H. E. (1970). *J. Biol. Chem.* **245,** 1748–1753.
Hearon, J. Z. (1949). *Bull. Math. Biophys.* **11,** 83–95.
Hearon, J. Z. (1953). *Bull. Math. Biophys.* **15,** 121–141.
Hearon, J. Z. (1981). *Math. Biosci.* **56,** 129–140.
Heinrich, R., and Rapoport, T. A. (1975). Biosystems **7,** 130–136.
Heinrich, R., Rapoport, S. M., and Rapoport, T. A. (1977). *Prog. Biophys. Mol. Biol.* **32,** 1–82.
Hess, B., and Wurster, B. (1970). *FEBS Lett.* **9,** 73–77.
Hijazi, N. H., and Laidler, K. J. (1972). *Can. J. Chem.* **50,** 1440–1442.
Hommes, F. A. (1962). *Arch. Biochem. Biophys.* **96,** 28–31.
Jeffrey, P. D. (1981). In "Protein-Protein Interactions" (C. Frieden and L. W. Nichol, eds.), pp. 213–256. Wiley, New York.
Kacser, H., and Burns, J. A. (1973). In "Rate Control of Biological Processes" (D. D. Davies, ed.), pp. 65–104. Cambridge Univ. Press, London and New York.
Kacser, H., and Burns, J. A. (1979). *Biochem. Soc. Trans.* **7,** 1149–1161.
Kassera, H. P., and Laidler, K. J. (1970). *Can. J. Chem.* **48,** 1793–1802.
Keleti, T., and Welch, G. R. (1984). *Biochem. J.* **223,** 299–303.
Kline, M. (1972). "Mathematical Thought from Ancient to Modern Times." Oxford Univ. Press, London and New York.
Koch, G. L., Shaw, D. C., and Gibson, F. (1971). *Biochim. Biophys. Acta* **229,** 795–804.
Kohn, M. C., Achs, M. J., and Garfinkel, D. (1977a). *Am. J. Physiol.* **232,** R153–R158.

Kohn, M. C., Achs, M. J., and Garfinkel, D. (1977b). *Am. J. Physiol.* **232**, R159–R166.

Kohn, M. C., Achs, M. J., and Garfinkel, D. (1977c). *Am. J. Physiol.* **232**, R167–R173.

Kuchel, P. W., and Chapman, B. E. (1984). *J. Theor. Biol.* **105**, 569–589.

Kuchel, P. W., and Roberts, D. V. (1974). *Biochim. Biophys. Acta* **364**, 181–192.

Kuchel, P. W., Nichol, L. W., and Jeffrey, P. D. (1974). *J. Theor. Biol.* **48**, 39–49.

Kuchel, P. W., Nichol, L. W., and Jeffrey, P. D. (1975). *J. Biol. Chem.* **250**, 8222–8227.

Kuchel, P. W., Roberts, D. V., and Nichol, L. W. (1977). *Aust. J. Exp. Biol. Med. Sci.* **55**, 309–326.

Laidler, K. J. (1955). *Can. J. Chem.* **33**, 1614–1624.

Laidler, K. J., and Bunting, P. S. (1973). "The Chemical Kinetics of Enzyme Action." Clarendon, Oxford.

Lin, S. X., Chou, K. C., and Wong, J. T. F. (1982). *Biochem. J.* **207**, 179–181.

Lipschutz, S. (1968). "Schaum's Outline of Theory and Problems of Linear Algebra." McGraw-Hill, New York.

McClure, W. R. (1969). *Biochemistry*, **8**, 2782–2786.

McCracken, D. D., and Dorn, W. S. (1964). "Numerical Methods and Fortran Programming." Wiley, New York.

Maguire, R. J., Hijazi, N., and Laidler, K. J. (1974). *Biochim. Biophys. Acta* **341**, 1–14.

Mancini, P., and Pilo, A. (1970). *Comput. Biomed. Res.* **3**, 1–14.

Masters, C. J. (1977). *Curr. Top. Cell. Regul.* **12**, 75–105.

Masters, C. J. (1981). *CRC Crit. Rev. Biochem.* **11**, 105–143.

Mellor, J. W. (1954). "Higher Mathematics for Students of Chemistry and Physics" (4th Ed.). Dover, New York.

Michaelis, L., and Menten, M. L. (1913). *Biochem. Z.* **49**, 333–369.

Moore, W. J. (1981). "Physical Chemistry" (5th Ed.). Longman, London.

Nelder, J. A., and Mead, R. A. (1965). *Computer J.* **7**, 308–313.

Nichol, L. W., Kuchel, P. W., and Jeffrey, P. D. (1974). *Biophys. Chem.* **2**, 354–358.

Nichol, L. W., and Winzor, D. J. (1972). "Migration of Interacting Systems." Clarendon, Oxford.

Orsi, B. A., and Tipton, K. F. (1979). *In* "Methods in Enzymology" (D. L. Purich, ed.), Vol. 63A, pp. 159–183. Academic Press, New York.

Osborne, M. R. (1976). *J. Aust. Math. Soc. B* **19**, 343–357.

Rapoport, T. A., and Heinrich, R. (1975). *Biosystems* **7**, 120–129.

Ray, W. J., and Hatfield, G. W. (1970). *J. Biol. Chem.* **245**, 1753–1754.

Reich, J. G., and Sel'kov, E. E. (1981). "Energy Metabolism of the Cell A Theoretical Treatise." Academic Press, New York.

Ricard, J., Buc, J., and Meunier, J. C. (1977). *Eur. J. Biochem.* **80**, 581–592.

Roberts, D. V. (1977). "Enzyme Kinetics." Cambridge Univ. Press, London and New York.

Rodiguin, N. M., and Rodiguina, E. N. (1964). "Consecutive Chemical Reactions." Van Nostrand-Reinhold, Princeton, New Jersey.

Rosenbrock, H. H., and Storey, C. (1966). "Computational Techniques for Chemical Engineers." Pergamon, Oxford.

Rudolph, F. B., Baugher, B. W., and Beissner, R. S. (1979). *In* "Methods in Enzymology" (D. L. Purich, ed.), Vol. 63A, pp. 22–43. Academic Press, New York.

Sadler, D. R. (1975). "Numerical Methods for Non-Linear Regression." Univ. of Queensland Press, St. Lucia.

Schied, F. (1968). "Schaum's Outline of Theory and Problems of Numerical Analysis." McGraw-Hill, New York.

Schwarz, G. (1968). *Rev. Mod. Phys.* **40**, 206–218.

Schwarz, G., and Engel, J. (1972). *Angew. Chem. (Int. Ed.)* **11**, 568–575.

Spiegel, M. R. (1965). "Schaum's Outline of Theory and Problems of Laplace Transforms." McGraw-Hill, New York.

Storer, A. C., and Cornish-Bowden, A. (1974). *Biochem. J.* **141,** 205–209.
Varfolomeev, S. D. (1976). *Mol. Biol.* **11,** 564–581 (Engl. trans. pp. 430–443).
Varfolomeev, S. D. (1977). *Mol. Biol.* **11,** 790–880 (Engl. trans. pp. 612–620).
Waley, S. G. (1964). *Biochem. J.* **91,** 514–517.
Walter, C. (1966). *J. Theor. Biol.* **11,** 181–206.
Walter, C., and Morales, M. F. (1964). *J. Biol. Chem.* **239,** 1277–1283.
Welch, G. R. (1977). *J. Theor. Biol.* **68,** 267–291.
Welch, G. R., and Keleti, T. (1981). *J. Theor. Biol.* **93,** 701–735.
Wilson, J. E. (1968). *J. Biol. Chem.* **243,** 3640–3647.
Wilson, J. E. (1980). *Curr. Top. Cell. Regul.* **16,** 1–44.
Wong, J. T. Z. (1965). *J. Am. Chem. Soc.* **87,** 1788–1793.
Wurster, B., and Hess, B. (1970). *Hoppe-Seyler's Z. Physiol. Chem.* **351,** 1537–1544.

8

Theoretical and Experimental Studies on the Behavior of Immobilized Multienzyme Systems

J. F. Hervagault and D. Thomas

Laboratoire de Technologie Enzymatique
E.R.A. n° 338 du C.N.R.S.
Université de Compiègne, Compiegne, France

I. INTRODUCTION

The living cell must be regarded as a spatial organization—highly hierarchical, structured, integrated, and compartmentalized. Within that cell, the majority of enzymes are fixed, reversibly or irreversibly, onto membrane structures. They can thus either become spatially organized into sequential systems or play a role as oriented carriers. The appearance of integrated multienzyme systems constitutes a higher level of compartmentalization.

In order for the cell to perform its various functions, enzymes are amalgamated into distinct metabolic pathways. Thus, regulation may operate at the level of (1) each enzymatic entity, (2) the different pathways, and (3) the segregation between these pathways. This brief glimpse shows the outstanding importance of some ideas that must be considered in order to reach a better knowledge of cellular metabolism: (1) Mass transfer phenomena, convection movements, and diffusional hindrance; (2) influence of the microenvironment, e.g., physical or ionic interactions and the existence of local concentrations; and (3) compartmentalization at the supramolecular level.

Cori (1956) indicated long ago that compartmentation offers a convenient means of regulating metabolism. Numerous pieces of evidence, microscopic as well as kinetic and analytic, have confirmed the hypothesis that highly structured organization within the cell is essential to metabolic functions.

By applying the kinetics of elementary collisions, Pollard (1963) concluded that, within cells larger than 1 μm, the diffusional term becomes a limiting step for the accomplishment of metabolic processes, and consequently, a submicroscopic order is required. Similarly, the works of Hübscher et al. (1971), Hess and Boiteux (1972), and also Srere (1972), dealing with the glycolytic enzymes, all demonstrated this necessary organization.

Kempner and Miller (1968), by centrifugation studies on a unicellular alga, Euglena gracilis, have shown that within that cell, no enzyme is found free or unlinked. Lloyd Davies et al. (1972) calculated that 60% of rat liver cell enzymes are linked to structures.

At the cellular level, the existence of organelles (nuclei, mitochondria, chloroplasts, lysosomes, peroxisomes, vacuoles, etc.) separated from each other by membranes can be considered the simplest and most obvious compartmentalization. In the absence of organelles, a functional segregation is effected by the presence of stable and isolable enzymatic complexes. It is even likely that compartmentalization exists in the absence of both organelles and/or complexes. If so, the compartment may be either a region with weak protein–protein interactions or a microenvironment due to unstirred layers (see articles by Friedrich and Kurganov in this volume). A detailed review with numerous references, dealing with the different aspects of metabolic compartmentation, has been published by Srere and Mosbach (1974).

From an experimental point of view, the in situ accessibility of parameters ruling the behavior of these microenvironments is proving limited indeed. In order partly to overcome these difficulties, enzymologists have attempted to reproduce artificially cellular conditions in such a way that diffusional constraints and local concentrations can be accurately taken into account (see Siegbahn et al., this volume). At that level, the so-called heterogeneous enzymology is certainly an important step toward a better understanding of underlying laws governing metabolic pathways. In particular, the feasibility of manufacturing membranous supports having well-defined geometrical and physicochemical properties facilitates mathematical modeling which takes into account the coupling between metabolites, diffusion and/or compartmentation, and enzymatic reactions.

The methods used to fix enzymes on insoluble supports have been described in numerous publications (Silman and Katchalski, 1966; Gryskiewicz, 1971;

Chibata, 1978). The method developed in our laboratory (Broun et al., 1973) consists of a co-cross-linking between an inert protein, such as albumin, gelatin, or hemoglobin, and the enzyme protein with the help of a bifunctional agent, viz. glutaraldehyde. This method allows a homogeneous distribution of active sites throughout the thickness of the membrane. By the mere fact of their membranous shape, our supports can be used for kinetic studies with either symmetrical or asymmetrical boundary concentrations. Moreover, compartmentalized experimental setups can be designed.

Preliminary studies by Prigogine (1955), Kedem and Katchalski (1963), and De Simone and Caplan (1969) used the laws of thermodynamics of irreversible processes to develop mathematical models for coupling diffusion and reaction.

Later, other researchers approached the problem by kinetic treatments (Kasche et al., 1971; Blaedel et al., 1972; Takeski and Laidler, 1973; Engasser and Horvath, 1976). In these models, the experimental supports consisted of either membranes or spherical particles. They were also analyzed by applying the steady-state hypothesis of either zero- or first-order kinetics.

Our approach is different and more general (Kernevez, 1972; Kernevez and Thomas, 1975). We assume that (1) the evolution of the system can be studied, (2) any reaction can be modeled, and (3) diffusion phenomena occurring inside the membrane can be taken into account. This theoretical process allows the mathematical treatment to reflect accurately the reality of the phenomena observed with experimental models.

Most works performed in the field of immobilized enzymes have dealt with simple enzyme systems. By comparison, few studies have focussed on the behavior of multienzyme systems (or multistep enzyme systems), although these are of prime importance in the operation of the cellular network.

Mitchell and Moyle (1958) anticipated that locating two enzymes catalyzing two consecutive reactions in close proximity would affect the behavior of these enzymes. A transport phenomenon cannot arise within a homogeneous structure. Thus, in order to become significant at a macroscopic level, the translocation phenomenon resulting from microscopic heterogeneities requires the contribution of a second enzyme that will create a modification in the product of the first enzymatic reaction. In that case, heterogeneity can be considered as a coupling between different enzyme molecules linked in such a way that diffusion of the product of the first reaction is faster than its leakage into the surrounding medium. A possible approach to studying such weak enzyme–enzyme interactions is obviously the use of enzymes immobilized together in/on a solid matrix.

Brown et al. (1968), Wilson et al. (1968), and Vasileva et al. (1969) published the first experimental results dealing with multistep enzyme systems fixed on insoluble supports. More significant results in that field are from Mosbach's group, who fixed the various enzymes by covalent bridges onto supports of sepharose or polyacrylamide (Mosbach and Mosbach, 1967; Mosbach, 1970). Mosbach and co-workers have studied numerous sequential multistep enzyme systems containing either two (Mosbach and Mattiasson, 1970) or three (Mattiasson and Mosbach, 1971; Srere et al., 1973) enzymes. Detailed accounts of these studies are provided in the article by Siegbahn et al. in this volume.

Explanations of the experimental data given by the authors were purely qualitative: in the case of an immobilized system the product of the first reaction is available at higher concentrations for the second one compared to the same system in solution. This behavior is due mainly to the proximity of active sites, which creates a microenvironment (compartmentalization of the intermediates) and avoids leaks of metabolites into the outside medium by diffusion and/or dilution.

The first in vitro studies and numerical computations, made on a stable multienzyme complex (isolated from Neurospora crassa and consisting of five consecutive enzymes from the biosynthetic pathway of aromatic amino acids) provided confirmation of the former observations (Welch and Gaertner, 1975): transient times for the global reaction were 15 times lower in the case of the complex compared to an hypothetical nonaggregated system. It was concluded from these data that the aggregated multienzyme system compartmentalizes intermediate substrates during the course of the overall reaction. It was also suggested that, in addition to segregating intermediates of competing pathways, reduction of the transient time may be an important consequence of the containment of intermediates within a physically associated enzyme sequence.

Other studies have also been carried out in the field of immobilized multienzyme systems. Particularly notable are those of Bouin et al. (1976) on a cyclic glucose oxidase–catalase system and those of Shimizu and Lenhoff (1979) on the phosphoglucomutase–glucose-6-phosphate dehydrogenase system.

II. GENERAL THEORETICAL TREATMENT OF DIFFUSION REACTIONS

As far as we are aware, the first theoretical analysis of a two-enzyme system carrying out two consecutive reactions and taking into account diffusional hindrance was done by Goldman and Katchalski (1971). It seems necessary to remind the reader briefly of that study, on the one hand because it is close to our own considerations and on the other hand to emphasize the main differences between their approach and ours.

The system considered (see Fig. 1) consists of an impermeable membrane, impregnated on its surface with two uniformly distributed enzymes (E_1 and

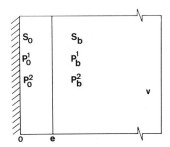

Fig. 1. An impermeable membrane impregnated on its surface with two uniformly distributed enzymes carrying out the sequence

$$S \xrightarrow{E_1} P^1 \xrightarrow{E_2} P^2$$

in contact with a solution of volume v. The unstirred layer at the membrane–solution interface is of thickness e. Subscript 0 denotes concentration at $x = 0$; b as subscript denotes concentration in the bulk of the solution at $x > e$. (After Goldman and Katchalski, 1971.)

E_2), in contact with a solution of a finite volume v.

$$S \xrightarrow{E_1} P^1 \xrightarrow{E_2} P^2$$

The bulk of the solution is stirred to assure a homogeneous distribution of the metabolites. The volume confined within the two boundaries $x = 0$ and $x = e$ is assumed to represent an unstirred layer.

In addition, the following boundary conditions were chosen in the kinetic analysis:

1. The activity of the first enzyme is independent of the second.
2. The enzymatic reactions follow first-order kinetics.
3. The concentration of the substrate of enzyme 1 is constant throughout the enzymatic reaction.
4. The concentrations of the products of both reactions are time dependent.
5. The system is of finite volume.

Goldman and Katchalski showed that the concentrations of the products P_b^1 and P_b^2 increased linearly with time, at least for the first minutes of the reaction. For systems consisting of enzymes of high activities, the reactions were diffusion controlled; the activity of E_1 was limited by the rate of diffusion of substrate from the bulk of the solution and the activity of E_2 approached that exhibited by E_1. The analysis of a corresponding homogeneous system revealed the appearance of an initial lag period for the production of P_b^2. The length of that lag period was a function of the catalytic as well as the physical parameters of the system.

The rate of production of the end product in the first stages of the reaction was markedly higher in the immobilized system than that predicted for a corresponding homogeneous system. These results agree with Mosbach's and Welch's observations, particularly.

Our approach to the study of immobilized multienzyme systems is both theoretical and experimental. This approach presupposes that the model should account as accurately as possible for the experimental facts. As a consequence, our starting assumptions are less restrictive than those of Goldman and Katchalski (1971).

In our supports, enzymes are distributed homogeneously throughout the thickness of the membrane: diffusion of the various metabolites across this active layer must be considered. Enzymatic reaction terms can either be Michaelian or allosteric (effector dependent or not) as well as pH or potential dependent. Internal regulation can take place in the enzymatic sequence. Finally, the behavior of the system must be studied either in a transient regime or in an evolutionary regime (succession of quasi-steady states).

The following theoretical development deals with the simplest case of two sequential enzymes immobilized within the same membrane:

$$S \xrightarrow{E_1} P \xrightarrow{E_2} Q$$

A. Equations Describing Homogeneous Solutions

$$S \xrightarrow{E_1} P \xrightarrow{E_2} Q$$

In solution, [S], [P], and [Q] are functions of time, enzyme concentrations, and initial concentrations of S, P, and Q. Equations governing their evolution are given by

$$\frac{d[S]}{dt} = -v_1, \qquad \frac{d[P]}{dt} = v_1 - v_2, \qquad \text{and} \qquad \frac{d[Q]}{dt} = v_2 \qquad (1)$$

where
$$v_1 = \frac{V_{MS}[S]}{K_{MS} + [S]} \qquad \text{and} \qquad v_2 = \frac{V_{MP}[P]}{K_{MP} + [P]}$$

Initial conditions are given by $[S](0) = S_0$; $[P](0) = P_0$ and $[Q](0) = Q_0$. Equations (1) can be solved by using the numerical method of Runge–Kutta (1964) (see also Kuchel, this volume).

B. Equations Describing Membranous Systems

1. General Case

The Michaelis–Menten relation is, by definition, valid only for a homogeneous and isotropic medium. It cannot be applied to describe overall kinetics of a system inside a membrane, where substrate and product concentrations differ from their outside concentrations and their value is also different at each point of the thickness of the membrane. Nevertheless, it is possible to consider an elementary volume small enough for concentrations to be considered as uniform. Thus, in each elementary volume, the Michaelian formalism remains valid.

Let us consider a membrane with a thickness e, separating two compartments 1 and 2 with substrate concentrations A_1 and A_2, respectively. The reference axis is perpendicular to the membrane surface and $x = 0$ is chosen for the interface between membrane and compartment 1 (Fig. 2). Obviously, A will be a function of both time and space. The change in substrate concentration will be controlled by substrate diffusion and enzymatic reaction acting simultaneously (Kernevez and Thomas, 1975; Kernevez, 1980).

$$\frac{\partial A}{\partial t} = \left(\frac{\partial A}{\partial t}\right)_{\text{diffusion}} + \left(\frac{\partial A}{\partial t}\right)_{\text{reaction}} \tag{2}$$

If the diffusion term is expressed by Fick's second law, Eq. (2) becomes

$$\frac{\partial A}{\partial t} = D_A\frac{\partial^2 A}{\partial x^2} - V_M f(A), \qquad f(A) = A/(K_A + A) \tag{3}$$

where D_A is the diffusion coefficient for A, which is assumed to be independent of concentration and space, and V_M is the maximum rate of the reaction. The resulting Eq. (3) is a nonlinear partial differential equation. A similar equation can be written for the product. It is convenient to rewrite Eq. (3) by using new units linked to the system, viz. space unit, the thickness e of the membrane; time unit, the characteristic time for diffusion Θ proportional to the time lag of the membrane: $\Theta = e^2/D_A$; concentration unit: the Michaelis constant K_A.

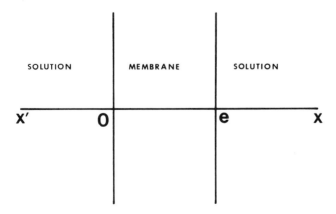

Fig. 2. Schematic representation of a membrane.

Thus, x/e is replaced by x and t/Θ by t, so that x and t become dimensionless, and Eq. (3) becomes

$$\frac{\partial a}{\partial t} = \frac{\partial^2 a}{\partial x^2} - \sigma f(a), \qquad \text{where} \qquad f(a) = a/(1 + a), \tag{4}$$

$$a = A/K_A \qquad \text{and} \qquad \sigma = V_M \Theta/K_A = (V_M/K_A)(e^2/D_A)$$

A similar equation can be written for the product:

$$\frac{\partial p}{\partial t} = \frac{D_P}{D_A} \frac{\partial^2 p}{\partial x^2} + \sigma f(a), \tag{5}$$

where $p = P/K_A$ and D_P is the diffusion coefficient for P.

The parameter σ, which is the ratio between the two characteristic times for diffusion and reaction (K_A/V_M), defines by itself the global behavior of the system and accounts for the importance of diffusional limitations: as σ increases, the diffusional limitations within the membrane become more important (see Engasser and Horvath, 1976).

2. Two-Enzyme Systems

In a transient regime, Eq. (1) must be rewritten (in normalized units)

$$\frac{\partial s}{\partial t} = \frac{\partial^2 s}{\partial x^2} - W_1$$

$$\frac{\partial p}{\partial t} = \alpha \frac{\partial^2 p}{\partial x^2} + W_1 - W_2, \qquad \alpha = D_P/D_S \tag{6}$$

$$\frac{\partial q}{\partial t} = \beta \frac{\partial^2 q}{\partial x^2} + W_2, \qquad \beta = D_Q/D_S$$

where $\qquad W_1 = \sigma_1[s/(1 + s)], \qquad\qquad \sigma_1 = V_{MS}e^2/K_{MS}D_S,$

$\qquad\qquad W_2 = \sigma_2 p/[(K_{MP}/K_{MS}) + p], \qquad \sigma_2 = V_{MP}e^2/K_{MS}D_S$

Let Eqs. (6) describe the evolution of the system during a short stage of filling up of the membrane. Initial conditions are then given by Eq. (7).

$$s(x, 0) = p(x, 0) = q(x, 0) = 0 \tag{7}$$

That is to say, at time zero, the membrane is empty of any substrate, and boundary conditions are fixed (Dirichlet's type).

$$s(0, t) = s_0, \qquad\qquad s(1, t) = s_1$$

$$p(0, t) = p_0, \qquad\qquad p(1, t) = p_1 \tag{8}$$

$$q(0, t) = q_0 = 0, \qquad q(1, t) = q_1 = 0$$

Figure 3 illustrates the S, P, and Q concentration evolution for a bienzyme system in solution (Fig. 3a) and immobilized within an immersed membrane (Fig. 3b).

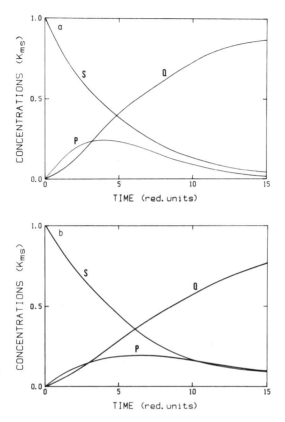

Fig. 3. S, P, and Q concentration evolution for a simple bienzyme system in solution (a) and immobilized inside a membrane (b). At time zero the bulk of the solution contains only substrate ([S] = K_{MS}), and the membrane is empty of any metabolite. The time unit is e^2/D_S (characteristic time for diffusion), and the concentration unit is K_{MS}.

If the system is studied for a longer interval of time, such as a few hours, boundary concentrations will vary and a more accurate description of the system is given by

$$\frac{\partial^2 s}{\partial x^2} - W_1 = 0,$$

$$\alpha \frac{\partial^2 p}{\partial x^2} + W_1 - W_2 = 0, \qquad (9)$$

$$\beta \frac{\partial^2 q}{\partial x^2} + W_2 = 0,$$

where α, β, W_1, and W_2 are as in Eq. (6).

S, p, and q are still functions of time and space but the time derivative has disappeared, thus defining a quasi-steady state. Initial and boundary conditions become, respectively, (for s)

$$s_0(0) = a \quad \text{and} \quad s_1(0) = b \tag{10}$$

$$\frac{ds_0}{dt} - \chi\frac{\partial s}{\partial x}(0, t) = 0 \quad \text{and} \quad \frac{ds_1}{dt} + \chi\frac{\partial s}{\partial x}(1, t) = 0 \tag{11}$$

where $\chi = \Omega e/V$ = volume of the membrane/volume of the bulk compartment.

Note that in the case of an immersed membrane, entering and leaving fluxes from both faces of the membrane contribute to the variation of concentration in the outside compartment. It is thus fitting to replace the boundary conditions [Eq. (11)] by a single equation in which $s_0(t) = s_1(t) = \alpha(t)$.

$$\frac{d\alpha}{dt} + 2\chi\frac{\partial s}{\partial x}(1, t) = 0, \quad \alpha(0) = \alpha_0 \tag{12}$$

and

$$\chi = \text{membrane volume/solution volume}$$

C. Solution of Equations

The equations described above can be solved numerically by a finite-difference implicit method.

1. Transient Regime

Let us divide the space interval $(0, 1)$ into J equal intervals with length Δx, and the time interval $(0, T)$ into N equal intervals with length Δt such that $\Delta x = 1/J$ and $\Delta t = T/N$.

The approximation of s $(j\Delta x, n\Delta t)$ is denoted by s_j^n. We get, in the (x, t) plane, a grid in the rectangle $(0 < x < 1, 0 < t < T)$ made of lines parallel to Ox and Ot with abcissae $x = j\Delta x$ $(0 < j < J)$ and ordinates $t = n\Delta t$ $(0 < n < N)$ as shown in Fig. 4a.

Let us describe briefly the principle of this implicit method to solve equations giving s [Eqs. (6)–(8)]: The values of s_j^n for $n = 0$, $j = 0$, and $j = J$ are known (see Fig. 4b). Suppose that the s_j^n values $(1 < j < J - 1)$ are known. Thus, in order to obtain the s_j^{n+1} values, we have at our disposal the following equations, which are the "discretization" of Eq. 6 (for s).

$$\frac{s_j^{n+1} - s_j^n}{\Delta t} - \frac{s_{j+1}^{n+1} + s_{j-1}^{n+1} - 2s_j^{n+1}}{(\Delta x)^2} + \sigma\frac{s_j^{n+1}}{1 + s_j^{n+1}} = 0 \tag{13}$$

where $1 < j < J - 1$.

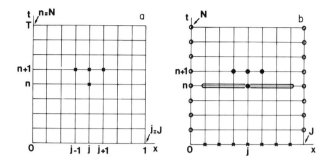

Fig. 4. The finite difference implicit method: (a) symbols used for time and space intervals, (b) principle of the method for solving equations.

We thus get a nonlinear system with $J - 1$ equations whose solution will give $s_1^{n+1}, s_2^{n+1}, \cdots s_{j-1}^{n+1}$.

Profiles shown in Fig. 5 illustrate the evolution of metabolite concentrations as calculated numerically by the method described above. For an explanation of the system of equations and parameters used, see Section III,A,3.

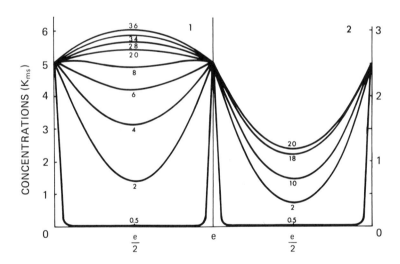

Fig. 5. Transient regime evolution of concentration profiles as calculated numerically. Time in seconds is indicated on each curve. These computations deals with the bienzyme system described in Section III,A,3. Profiles (1) and (2) refer to the evolution of uric acid (S) and xanthine (I) respectively. Parameter values were as follows: $i_0 = i_1 = 10^{-7}$ mol/cm^3; $s_0 = s_1 = 2 \times 10^{-7}$ mol/cm^3; $V_{MI} = 6 \times 10^{-5}$ mol/cm^3/h; $V_{MS} = 1.5 \times 10^{-5}$ mol/cm^3/h; $K_{MI} = 2 \times 10^{-8}$ mol/cm^3; $K_{MS} = 4 \times 10^{-8}$ mol/cm^3; $K_1 = 2 \times 10^{-8}$ mol/cm^3; $D_1 = D_S = 2.5 \times 10^{-3}$ cm^2/h and $e = 5 \times 10^{-3}$ cm ($\sigma_1 = 15$; $\sigma_2 = 3.75$).

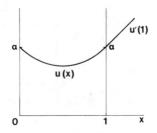

Fig. 6.   Definition of the function $\alpha \rightarrow f(\alpha)$

2. Evolving Regime

Suppose the active membrane is immersed. Thus, boundary conditions are expressed by Eq. (12), and this equation can be rewritten as

$$\frac{d\alpha}{dt} = f(\alpha), \qquad \alpha(0) = \alpha_0, \tag{14}$$

where the function $\alpha \rightarrow f(\alpha)$ is defined as follows (see Fig. 6).

$$-\frac{\partial^2 u}{\partial x^2} + \frac{u}{1 + u} = 0,$$

$$u(0) = u(1) = \alpha, \tag{15}$$

$$f(\alpha) = -2\chi \frac{du}{dx}(1).$$

Eqs. (14) and (15) are solved by the Runge–Kutta and a finite-difference method, respectively.

III. EXPERIMENTAL STUDIES

A. Regulation of Sequential Multistep Enzyme Systems

1. A Simple Bienzyme System and Profile Visualization

The immobilization of enzyme systems, in general, gives rise to diffusional limitations on supplies of substrate (S) at the level of the enzyme active sites embedded within the structure. The main consequence of these mass transfer constraints is the establishment of metabolite concentration profiles across the membrane thickness. The kinetic behavior of multienzyme systems, in particular, is directly linked to the existence of local concentrations. Although the concentration profiles can be easily calculated from the diffusion-reaction equations (see Section II), little attention has been paid to their experimental visualization.

Graves (1973) used a double-beam laser spectrophotometer to demonstrate existence of local pH values in an immobilized monoenzyme film. Sernetz and Puchinger (1976) have done similar work based on fluorescence measurements with a single bead bearing immobilized enzymes. As the protein films we use exhibit a very regular structure, they are easy to treat for electron microscopy. Using EM, Barbotin and Thomas (1974) have described a visualization of local concentration profiles in our artificial monoenzyme membrane; the authors used a horseradish peroxidase 3,3′-diaminobenzidine (DAB) system described by Graham and Karnovsky (1966). This study was extended to the case of a sequential bienzyme system in which glucose oxidase and horseradish peroxidase (HRP) were distributed within the structure (Malpiece *et al.*, 1980a,b).

$$\text{Glucose} + O_2 \xrightarrow{\text{GO}} H_2O_2 + \text{gluconolactone}$$

$$1.6\,H_2O_2 + \text{DAB} \xrightarrow{\text{HRP}} \text{Poly(DAB)} + 1.6\,H_2O$$

The oxidation of DAB results in a highly osmiophilic polymer, which permits high-contrast electron microscopic observations with short incubation periods.

The glucose oxidase catalytic activity is consistent with a ping-pong mechanism. With regard to peroxidase, kinetic results are in agreement with a random bi-bi mechanism (with quasi-equilibrium). Using the following symbols, *viz.* $S_1 = [\text{Glu}]$; $S_2 = [O_2]$; $S_3 = [H_2O_2]$; $S_4 = [\text{DAB}]$, $S_5 = [\text{Poly(DAB)}]$, and D_i, referring to the diffusion coefficient for species i, the mass balance equations ruling the system are given by Eqs. (16).

$$\frac{\partial S_1}{\partial t} - D_1 \frac{\partial^2 S_1}{\partial x^2} + V_1 = 0$$

$$\frac{\partial S_2}{\partial t} - D_2 \frac{\partial^2 S_2}{\partial x^2} + V_1 = 0$$

$$\frac{\partial S_3}{\partial t} - D_3 \frac{\partial^2 S_3}{\partial x^2} - V_1 + V_2 = 0 \tag{16}$$

$$\frac{\partial S_4}{\partial t} - D_4 \frac{\partial^2 S_4}{\partial x^2} + (V_2/1.6) = 0$$

$$\frac{\partial S_5}{\partial t} - (V_2/1.6) = 0$$

where $S_i = $ constant ($1 \leq i \leq 4$) at wall and $S_i = 0$ ($1 \leq i \leq 5$) at time zero.

Introducing the dimensionless parameters

$$s_i = S_i/K_{MS_3}, \qquad t = tD_3/e^2, \qquad \text{and} \qquad x = x/e$$

with the reaction term for glucose oxidase given by

$$V_1 = V_{M_1}/(1 + \lambda_1/s_1 + \lambda_2/s_2)$$

and the reaction term for peroxidase given by

$$V_2 = V_{M2}/(1 + 1/s_3)(1 + \lambda_4/s_4)$$

where $\lambda_1 = K_{MS_1}/K_{MS_3}$, $\lambda_2 = K_{MS_2}/K_{MS_3}$, and $\lambda_4 = K_{MS_4}/K_{MS_3}$, Eqs. (16) become

$$\frac{\partial s_1}{\partial t} - d_1\frac{\partial^2 s_1}{\partial x^2} + \sigma_1\frac{1}{1 + \lambda_1/s_1 + \lambda_2/s_2} = 0$$

$$\frac{\partial s_2}{\partial t} - d_2\frac{\partial^2 s_2}{\partial x^2} + \sigma_1\frac{1}{1 + \lambda_1/s_1 + \lambda_2/s_2} = 0$$

$$\frac{\partial s_3}{\partial t} - \frac{\partial^2 s_3}{\partial x^2} - \sigma_1\frac{1}{1 + \lambda_1/s_1 + \lambda_2/s_2} + \frac{1}{(1 + 1/s_3)(1 + \lambda_4/s_4)} = 0 \qquad (17)$$

$$\frac{\partial s_4}{\partial t} - d_4\frac{\partial^2 s_4}{\partial x^2} + \frac{1}{1.6}\sigma_2\frac{1}{(1 + 1/s_3)(1 + \lambda_4/s_4)} = 0$$

$$\frac{\partial s_5}{\partial t} - \sigma_2\frac{1}{(1 + 1/s_3)(1 + \lambda_4/s_4)} = 0$$

where $\sigma_1 = e^2 V_{M1}/D_3 K_{MS_3}$, $\sigma_2 = e^2 V_{M2}/D_3 K_{MS_3}$, and $d_i = D_i/D_3$ $(1 \le i \le 4)$.

The poly(DAB) concentration profiles within the protein membrane, as visualized by electron microscopy, show the spatial location of the second enzymatic activity (peroxidase). The absence of any polymeric product is obvious in the case of a membrane bearing no peroxidase activity (Fig. 7a). On the other hand, in the presence of the enzymatic activity, the poly(DAB) precipitate is easily observed (Fig. 7b): a nondroplet amorphous precipitate is visible on both edges of the membrane, indicating a symmetrical distribution of the peroxidase activity through the thickness.

The behavior of a sequential bienzyme system in solution is ruled by the relative values of both enzyme activities, and the lowest activity is the limiting step. When immobilized, a well-defined and constant rate between both activities can give rise to very different behavior depending on the diffusional limitation effects.

Many profiles can be obtained depending on the parameter values, and three important groups can be derived from them. One group is composed of

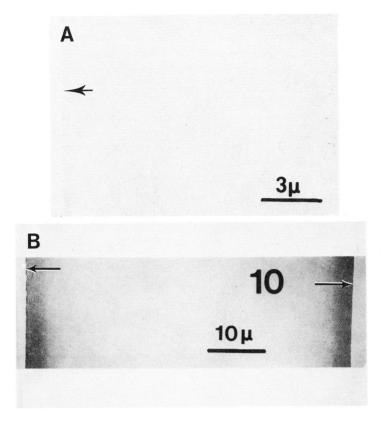

Fig. 7. Immobilized glucose oxidase–horseradish peroxidase system: thin transverse section of a membrane: (a) membrane (× 3500) without enzymatic activity. No precipitate can be seen, although this membrane was incubated in 0.11 M glucose, 2.65 × 10^{-4} M oxygen and 10^{-3} M DAB. The arrow points to the edge of the membrane. (b) active membrane (× 1700). Note in both cases the regularity of the structure.

flat profiles obtained when external H_2O_2 is added: when a bienzyme system is immobilized, the concentration profile of the second activity cannot be demonstrated if the intermediate product (H_2O_2, here) is absent in the bulk solution. Homogeneity and uniformity of the peroxidase activity are a result of adequate H_2O_2 concentrations added to the surrounding medium. This H_2O_2 concentration was calculated to make V_1 close to V_2 at each point inside the membrane.

Two other groups exemplify obvious diffusional limitations. In one case, the boundaries of the membrane are highly contrasted, especially when the external DAB concentration is low and/or when the peroxidase activity is high enough to increase the diffusional constraints. Indeed, the bienzyme system is controlled by the diffusion across the membrane of one of the substrates of the first enzyme (O_2), and as a consequence the peroxidase will be active mainly near the edges. In the other case, the polymer accumulates in the middle of the membrane when the glucose oxidase activity is high with respect to the peroxidase activity. The poly(DAB) precipitation is a direct visualization of the H_2O_2 profile, provided that the external DAB is sufficient. Although the results obtained with this system are of prime importance for histochemical purposes, it is also noteworthy that they are of general interest and can be applied to any immobilized multienzyme system. The study of this model system leads to the main conclusion that there is no geometrical similarity between the enzyme distribution patterns and the insoluble product distribution. In other words, due to the existence of local concentrations engendered by diffusional phenomena, no overlap between active-site locations and activities occurs.

2. Feedback Control

Section III,A,1 deals with a simple sequential bienzyme system without inhibition and/or activation effect. In the same way, model systems developed by Mosbach's group do not take into account internal regulation. Nevertheless, one of the more important reasons for the existence of multienzyme systems might be the necessity to develop more efficient means to control metabolic events.

The following three bienzyme systems illustrate the regulatory effects of internally generated (feedback control including reversibility), as well as externally added, substrates or products on enzymes participating in the sequences. Feedback control is certainly the more common type of regulation encountered within metabolic pathways. As an illustration and to deepen the kinetic analysis, we have chosen a model bienzyme system with feedback inhibition. β-Glucosidase (EC 3.2.1.21) hydrolyzes β-glucosides, and glucose oxidase (EC 1.1.3.4.) oxidizes glucose into gluconolactone, which inhibits the first enzyme, glucosidase (Lecoq *et al.*, 1975). The whole system can be written as

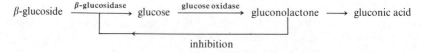

The choosen β-glucoside was salicin, which gives two products: glucose and saligenin. Glucose is not an inhibitor of the first enzyme, and saligenin is neither an inhibitor nor an activator of either enzyme and is not transformed by the second enzyme. β-Glucosidase catalyzes a reaction following first-order kinetics; glucose oxidase catalyzes a Michaelian reaction. Experiments and numerical simulations were performed with both enzymes homogeneously distributed in solution, with both enzymes in a simple membrane (M_{1+2} system), or with each enzyme in a separate membrane ($M_1 + M_2$ system). Their kinetic behaviors were analyzed in order to observe the relative effects of feedback inhibition in solution or in a structured system. Let S, P, Q, and I denote, respectively, the salicin, glucose, gluconolactone, and gluconic acid concentrations; Ω, the solution–membrane interface area; v, the bulk solution volume; k, the β-glucosidase V_M:K_M ratio; K_i the β-glucosidase–lactone inhibition constant; and D_i, the diffusion coefficient for substance i. Values of V_M and K_M are given for glucose oxidase. Thus, the time dependence of concentrations at each membrane point are given by the conservation of mass equation.[1]

One-membrane system (M_{1+2}):

$$\frac{\partial S}{\partial t} = D_S \frac{\partial^2 S}{\partial x^2} - k \frac{1}{1 + I/K_i} S$$

$$\frac{\partial P}{\partial t} = D_P \frac{\partial^2 P}{\partial x^2} + \frac{1}{1 + I/K_i} S - V_M \frac{P}{K_M + P}$$

$$\frac{\partial I}{\partial t} = D_I \frac{\partial^2 I}{\partial x^2} + V_M \frac{P}{K_M + P} \tag{18}$$

$$\frac{\partial Q}{\partial t} = D_Q \frac{\partial^2 Q}{\partial x^2}$$

Boundary conditions are given by $\partial i/\partial t = J_i \Omega/v$, where ($\partial i/\partial t$) is the variation of concentration of substance i in the bulk solution and J_i is the flux of i through the solution–membrane interface.

The enzyme behavior in solution is ruled by the same equations, without the diffusion terms $D_i(\partial^2 i/\partial x^2)$.

Two-membrane system ($M_1 + M_2$), membrane bearing the first enzyme (M_1):

$$\frac{\partial S}{\partial t} = D_S \frac{\partial^2 S}{\partial x^2} - k \frac{1}{1 + I/K_i} S$$

$$\frac{\partial P}{\partial t} = D_P \frac{\partial^2 P}{\partial x^2} + k \frac{1}{1 + I/K_i} S \tag{19}$$

$$\frac{\partial I}{\partial t} = D_i \frac{\partial^2 I}{\partial x^2}, \qquad \frac{\partial Q}{\partial t} = D_Q \frac{\partial^2 Q}{\partial x^2}$$

[1] Experiments were carried out at pH 5.25, and the spontaneous lactone hydrolysis is negligible at that pH.

Membrane bearing the second enzyme (M_2):

$$\frac{\partial S}{\partial t} = D_S \frac{\partial^2 S}{\partial x^2}$$

$$\frac{\partial P}{\partial t} = D_P \frac{\partial^2 P}{\partial x^2} - V_M \frac{P}{K_M + P} \qquad (20)$$

$$\frac{\partial I}{\partial t} = D_I \frac{\partial^2 I}{\partial x^2} + V_M \frac{P}{K_M + P}$$

$$\frac{\partial Q}{\partial t} = D_Q \frac{\partial^2 Q}{\partial x^2}$$

Boundary conditions are: $(\partial i/\partial t)_{\text{solution}} = -\dfrac{J_{i1}\Omega_1}{v} - \dfrac{J_{i2}\Omega_2}{v}$

In Fig. 8, the concentration of saligenin in the bulk solution is plotted as a function of time for the three systems.

Let us point out that the inhibition of β-galactosidase is most effective when both enzymes are coimmobilized within the same membrane, and least effective when they are in two different membranes. Although this curve, dealing with raw concentration data, gives some kinetic information, the use of the ratio of the instantaneous velocity of the first enzyme (v_t) to its velocity at time zero without inhibitor (v_0) is more amenable to mathematical description.

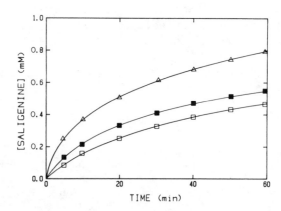

Fig. 8.   Bienzyme system with feedback. Saligenine concentration as a function of time for the β-glucosidase–glucose oxidase system in solution (■); in a one-membrane (M_{1+2}) system (□); and in a two-membrane ($M_1 + M_2$) system (△). At time zero, enzymes or enzyme membranes are immersed in a stirred solution of salicin.

The value of the ratio $v_t{:}v_0$ is plotted versus time in Fig. 9a. With this better defined kinetic information, the observation about the relative efficiencies of the feedback effect obtained in these three systems remains valid. The results of computer simulations using the leading parameters of these systems are given in Fig. 9b.

In the two-membrane system ($M_1 + M_2$), β-glucosidase is less inhibited. This observation can be explained by the production of the inhibitor (lactone) outside the second membrane. This is in agreement with previous results (Thomas *et al.*, 1974) showing that a membrane-bearing enzyme is less sensitive to an external inhibitor than is the same enzyme in solution.

In the one-membrane system, the inhibition is more effective than in solution because the inhibitor appears *in situ*. After a few seconds, the lactone concentration should be very low when produced in solution but is already quite high locally when produced inside the membrane. This appears quite clearly from the calculated evolution of inhibitor concentration profiles (see Fig. 10). The effect of the production rate of lactone inside the membrane is shown by the variation of β-galactosidase velocity as a function of different enzyme ratios (Fig. 11).

The membrane enzyme activity depends on the local inhibitor concentration and not on its concentration in the bulk solution. The $v_t{:}v_0$ ratio curve

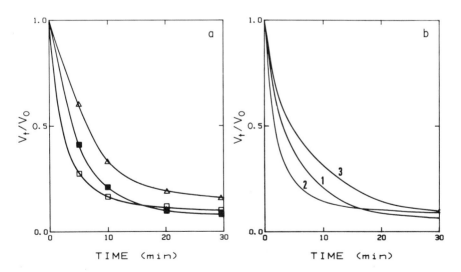

Fig. 9. (a) Experimental $v_t : v_0$ ratio (v_t, instantaneous reaction rate; v_0, initial reaction rate) as a function of time for the bienzyme system in solution ($\blacksquare$), in one membrane ($\square$), and in two membranes ($\triangle$). (b) Computed simulations of the $v_t : v_0$ ratio as a function of time for the bienzyme in solution (1), in one membrane (2), and in two membranes (3). All parameter values are as in (a).

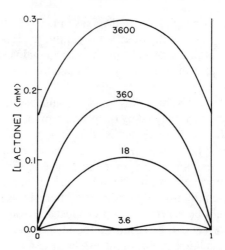

Fig. 10. Instantaneous lactone concentration profiles inside the β-glucosidase–glucose oxidase membrane (M_{1+2}) during the system evolution. Time in seconds is indicated on each curve. The abcissa represents the thickness of the membrane, x/e.

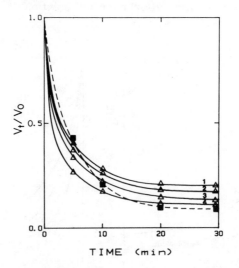

Fig. 11. Experimental $v_t : v_0$ ratio as a function of time for the bienzyme in solution (■) and in one membrane ($\triangle$) for different amounts of glucose oxidase activity (mol/cm^3/h) 1, 0.05×10^{-3}; 2, 0.1×10^{-3}; 3, 0.55×10^{-3}; 4, 1.04×10^{-3}.

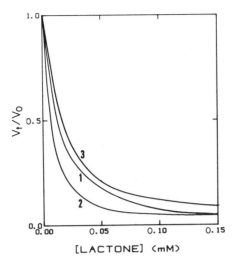

Fig. 12. Calculated $v_t : v_0$ ratio as a function of lactone concentration for the bienzyme in solution (1), in one membrane (2), and in two membranes (3).

for the one-membrane (M_{1+2}) system crosses the solution curve after a certain delay (about 15 min) for experimental data as well as for computed results. Lactone production may be lower in the membrane than in solution, reversing the effect observed during the first few minutes. Alternatively, the curve of $v_t : v_0$ ratio versus lactone concentration for the one-membrane systems does not cross the solution curve (Fig. 12).

3. Inhibition

The second system to be discussed deals with a bienzymatic sequence that in solution exhibits an obvious kinetic incompatibility: xanthine oxidase is the first enzyme and uricase is the second. Xanthine, the first substrate, is a competitive inhibitor of uricase (Hervagault *et al.*, 1975).

$$\text{Xanthine} \xrightarrow{\text{xanthine oxidase}} \text{uric acid} \xrightarrow{\text{uricase}} \text{allantoin}$$
$$(+O_2) \qquad\qquad (+H_2O_2) \qquad\qquad (+CO_2)$$

The reaction can be represented schematically as

$$I \xrightarrow{E_1} S \xrightarrow{E_2} P$$

Experiments were performed with the enzyme molecules in solution and immobilized within the same membrane.

Denoting V_{M1} and V_{M2} as maximum enzyme activity, K_{M1} and K_{M2} as Michaelis constants for E_1 and E_2, and K_i as the inhibition constant, the kinetic equations for I, S, and P in the case of the free enzyme system are

$$\frac{dI}{dt} = -V_{M1}\frac{I}{K_{M1} + I},$$

$$\frac{dS}{dt} = V_{M1}\frac{I}{K_{M1} + I} - V_{M2}\frac{S}{S + K_{M2}(1 + I/K_i)}, \qquad (21)$$

$$\frac{dP}{dt} = V_{M2}\frac{S}{S + K_{M2}(1 + I/K_i)}$$

and if dimensionless parameters of thickness (e), e^2/D_I, and K_{M2} are used as space, time, and concentration units, respectively, the continuity equations for I, S, and P inside the membrane can be written as

$$\frac{\partial i}{\partial t} = \frac{\partial^2 i}{\partial x^2} - \sigma_1\frac{i}{i + K_{M1}/K_{M2}}$$

$$\frac{\partial s}{\partial t} = \frac{\partial^2 s}{\partial x^2} + \sigma_1\frac{i}{i + K_{M1}/K_{M2}} - \sigma_2\frac{s}{s + 1 + (K_{M2}/K_i)i} \qquad (22)$$

$$\frac{\partial p}{\partial t} = \frac{\partial^2 p}{\partial x^2} + (D_s/D_p)\sigma_2\frac{s}{s + 1 + (K_{M2}/K_i)i}$$

where $\sigma_1 = (V_1 e^2)/(K_{M2}D_I)$ and $\sigma_2 = (V_2 e^2)/(K_{M2}D_S)$

Boundary conditions are as in Eq. system (5) (immersed membrane). The effect of xanthine on the kinetic behavior of uricase was studied first in the absence of xanthine oxidase. Experiments were performed with the enzyme in solution and immobilized. The resulting activity is given in Fig. 13.

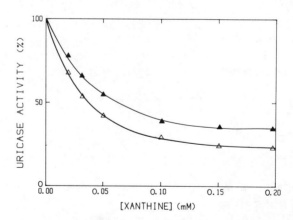

Fig. 13.   Uricase activity as a function of xanthine concentration for the enzyme–membrane system (▲) and for the free enzyme system (△). Uricase activity is expressed as a percentage of the activity in the absence of xanthine.

Uricase activity is less sensitive to inhibition inside the membrane. As already mentioned, this result can be explained by the rules governing heterogeneous enzyme kinetics: the lower the diffusion and the quicker the reaction rate, the higher the diffusional limitations. Any modification of the reaction rate gives rise to an effect on the diffusional limitations. Therefore, the effect of the inhibitor is compensated by the decrease in the diffusion constraints. The effect of xanthine on the behavior of uricase is then studied in the presence of xanthine oxidase inside the membrane. It is noteworthy that the instantaneous uricase activity in solution depends only on the substrate and inhibitor concentrations and not on the uricase concentration. Inside the membrane, the second enzyme, viz. xanthine oxidase is able to strongly modulate the local substrate concentration (uric acid) on the level of uricase. The enzyme–membrane activity will depend not only on the substrate concentration in the bulk solution but also—and mainly—on the local internal concentrations.

The uricase reaction rate was studied as a function of xanthine concentrations in the bulk solution, with a constant uric acid concentration. This study was performed with membranes bearing different xanthine oxidase activities and the same uricase activity. The results are given in Fig. 14.

When the xanthine oxidase concentration is increased, the apparent inhibition effect on uricase is decreased. For a given xanthine oxidase activity (e.g., 15×10^{-3} I.U./cm^2), the uricase reaction does not seem dependent on the xanthine concentration any more. Higher xanthine oxidase activity values lead to a "negative" inhibition effect: the uricase activity is higher with

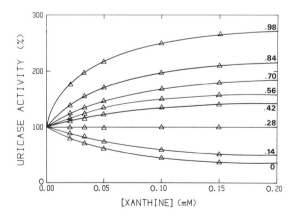

Fig. 14. Uricase activity as a function of xanthine concentration for enzyme membranes bearing different xanthine oxidase activities, expressed as a fraction of maximal activity on each curve. The reaction rates are expressed as a percentage of the activity in the absence of xanthine. Uricase activities without xanthine are the same for any membrane.

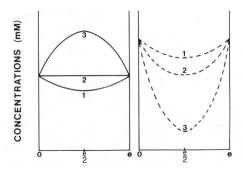

Fig. 15. Calculated concentration profiles (1, 2, and 3) inside the membrane for uric acid (——) and xanthine (--). For increasing values of xanthine oxidase activity, the concentrations in each point inside the membrane are expressed as a fraction of the concentration at the boundary in the solution. Parameter values are the same as in Fig. 5(Section II,C,1) except for V_{MI}: 2×10^{-5} mol/cm^3/h, 4×10^{-5} mol/cm^3/h, and 1×10^{-4} mol/cm^3/h, for profiles 1, 2, and 3, respectively.

inhibitor than without it. The same relationship (not shown here) was computed numerically from Eq. (22) and from the basic parameter values of the system. The numerical results calculated by taking into account the diffusion reaction phenomena are in good agreement with the experimental results. The best way to discuss these results is by introducing the concentration profiles (see Fig. 15).

In the presence of xanthine oxidase inside the membrane, xanthine molecules are locally transformed into uric acid molecules. In the membrane structure, xanthine concentration is lower and uric acid concentration is higher than in the bulk solution. Hence the relative concentration values of the two metabolites are deeply modified. Owing to its competitive nature, the inhibition is decreased by this phenomenon. When the xanthine oxidase activity is great enough, the uric acid concentration in the membrane is higher than outside in the bulk solution and the global uricase activity is higher than in the absence of xanthine.

It was of interest to study the time-dependent behavior of such a bienzyme system. During the first 2 h, the membrane system produced three times more allantoin molecules than the free enzyme system.

4. Reversibility

The two systems described in Sections III,A,1 and III,A,2 dealt with sequential reactions that were thermodynamically favorable. It is of interest to ask what would be the effect if one step in the sequence were unfavorable in the direction of the sequence to be studied. This aspect was examined by Srere *et al.* (1973) in studying the sequence malate dehydrogenase–citrate synthetase (plus lactic dehydrogenase to recycle NAD^+ via pyruvate). In that sequence, the malate dehydrogenase reaction favors the production of malate (and NAD^+). Thus, the system would soon reach equilibrium if no conversion of oxaloacetate to citrate occurred. Coimmobilization leads to a decrease in the mean distance between the different enzymes, diffusional hindrances of the products, and a steeper concentration gradient of oxaloacetate between malate dehydrogenase and citrate synthetase molecules, and hence a higher mass transfer. All this leads to higher citrate synthetase activity. The observed effect can also be expressed as an apparent shift in the equilibrium constant for malate dehydrogenase. The bienzyme system we describe here (Le Moullec and Thomas, 1977) deals with the phosphoglucoisomerase (PGI)–glucose-6-phosphate dehydrogenase (Glu-6-PD) system, coimmobilized within the same membrane:

$$\text{Fru 6-P} \xrightleftharpoons{\text{PGI}} \text{Glu 6-P} \xrightarrow[\text{NADP}^+]{\text{Glu-6-PD}} \text{Glu-6-P lactone} + \text{NADPH} + \text{H}^+$$

Experimentally, we can introduce or withhold NADP in order to test the active membrane as a mono- or bienzyme system.

In an immobilized reversible monoenzyme system, after a few seconds the substrate concentration is smaller inside the membrane than in the bulk solution, the product concentration is locally higher, and the ratio between substrate and product concentrations is close to equilibrium. Due to the diffusional limitations there is a stronger feedback effect on the velocity of the enzymatic reaction by the product accumulation inside the membrane. If there is a quantitative difference between the immobilized and the soluble system in the time until equilibrium is reached, nevertheless, and whatever the system is, the equilibrium constant remains the same.

Experiments carried out in the presence or absence of $NADP^+$ gave the results as presented in Fig. 16. There is experimental evidence for a shift of the apparent value of the equilibrium constant in the presence of a second enzyme activity. The variation of the apparent value R of this constant was studied as a function of the $NADP^+$ concentration (Fig. 17). These R values are here noted (R_{eq}) and represent the ratio Glu 6-P/(Glu 6-P + Fru-6-P) giving a velocity for the isomerase reaction equal to zero.

The results obtained with this system show that the introduction of diffusional constraints leads to modifications not only at the level of enzyme kinetic parameters but also at the level of apparent thermodynamic parameters such as the equilibrium constant. A more theoretical explanation of the observed experimental results is possible by considering an oversimplified model: diffusion and reaction are separated in a fictitious microdialysis bag.[2]

---

[2] It is important to note that numerical results obtained with such a model are qualitatively the same as those obtained with a distributed model, as described in Section II.

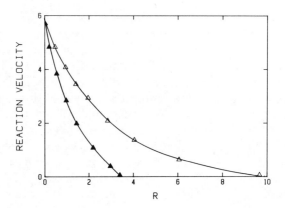

Fig. 16.   Immobilized glucose-6-phosphate isomerase reaction velocity (IU/cm^2 membrane) as a function of the ratio R under zero-order kinetics for $NADP^+$ ($\triangle$) and without it ($\blacktriangle$).

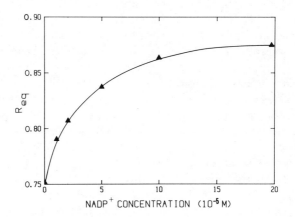

Fig. 17.   Value of the shift of the apparent R_{eq} ratio as a function of the $NADP^+$ concentration in the external solutions.

Under steady-state conditions there is a balance between the reaction rate and the diffusion from or to the bulk solution. The contribution of diffusion is approximated by Fick's first law, and it is always possible to write each reaction rate as $- V_{Mi} f(S, P)$, with $f(S, P)$ any dimensionless function ranging between zero and 1.

Consider the system

$$S \xrightarrow{E_1(V_{M1})} P \xrightarrow{E_2(V_{M2})} Q$$

Setting the thickness of the bulk reservoir (containing the enzymes) equal to the thickness (e) of the diffusional wall (membrane without enzyme) yields

$$(D_s/e^2)(S_0 - S) = V_{M1}f(S, P) \tag{23}$$

$$(D_p/e^2)(P_0 - P) = -V_{M1}f(S, P) + V_{M2} \tag{24}$$

The concentration unit is $(S_0 + P_0)$ with S_0 and P_0 referring to the metabolite concentrations in the external medium.

Let us define

$$\sigma_i = \frac{V_{Mi}e^2}{D(S_0 + P_0)}$$

with the (realistic) assumption that $D_S = D_P$. Equations (23) and (24) then become

$$S_0 - S = \sigma_1 f(S, P), \tag{25}$$

$$P_0 - P = \sigma_1 f(S, P) + \sigma_2, \tag{26}$$

and

$$S = S_0 + P_0 - P - \sigma_2. \tag{27}$$

When the velocity of the reversible reaction is zero, $f(S, P) = 0$ and

$$S - \alpha P = 0 \tag{28}$$

with the equilibrium ratio $\alpha = (S/P)_{eq}$. Equations (27) and (28) give

$$P = 1 - \sigma_2/1 + \alpha \tag{29}$$

and from Eqs. (26) and (29)

$$(P_0)_{eq} = [1/(1 + \alpha)](1 + \sigma_2\alpha). \tag{30}$$

That is to say,

$$R_{eq} = R_{eq}^*(1 + \sigma_2\alpha), \tag{31}$$

with R_{eq} referring to the apparent equilibrium R value for the enzyme membrane and R_{eq}^* referring to the equilibrium R value for the free enzyme in solution.

B. Vectorial Behavior

The studies dealing with immobilized enzyme systems are mainly performed under symmetrical boundary conditions. The various systems presented so far deal with this aspect. Nevertheless, it is obvious that within biological systems, enzymes and/or multi-enzyme systems often work between compartments with different metabolite concentrations on both sides. Membranous supports allow a study of enzyme behavior under asymmetrical boundary conditions in a well-defined context.

Asymmetrical conditions are able to generate vectorial behavior. The vectorial catalysis reflects two kinds of phenomena: (1) The asymmetry in the function created by a gradient between the boundaries of a membrane homogeneous in structure; and (2) the asymmetry in the structure of the membrane itself. Examples will illustrate each of these two aspects.

1. Asymmetrical Behavior (Functional Asymmetry)

The xanthine oxidase–uricase bienzyme system previously described (Section III,A,3) was used to illustrate this first aspect (Hervagault et al., 1976). Nevertheless, in the present case, the membrane—with a homogeneous repartition of both enzyme molecules—is not immersed, but separates two compartments 1 and 2. Continuity equations for I, S, and P are the same as Eq. (22) except for boundary conditions. Here, indeed, they are not constrained, so that

$$\frac{\partial i}{\partial t} + \chi \frac{\partial i}{\partial v} = 0 \quad \text{where} \quad \partial/\partial v = \begin{cases} -\partial/\partial x & \text{for} & x = 0 \\ +\partial/\partial x & \text{for} & x = e \end{cases} \tag{32}$$

with $\chi = \Omega e / V$.

Initial conditions for I (xanthine) are

$$i(0,0) = 0, \qquad i(e,0) = I_0 / K_{M2} \tag{33}$$

and initial conditions for S (uric acid) are

$$s(0,0) = s(e,0) = S_0 / K_{M2} \tag{34}$$

Due to the enzymatic reaction, substrate molecules are entering the membrane on both sides. The flux is leaving compartment 2, that is, [I] > 0. There is a quantitative difference from classic experiments dealing with an immersed membrane, but the behavior is qualitatively similar; substrate molecules are consumed on both sides of the membrane. With a homogeneous bienzyme membrane, a qualitatively different behavior is observed (Fig. 18). Under the same boundary conditions for I and S as described above, urate molecules are leaving and entering the membrane by compartments 1 and 2, respectively. The urate concentration is decreasing in the first compartment and increasing in the second one. In this way, a concentration gradient is generated between the membrane boundaries. The entering and leaving fluxes, measured experimentally, are given as a function of time in Fig. 19.

The behavior of the mono- and bienzyme systems is explained by the membrane concentration profiles as calculated numerically. Substrate con-

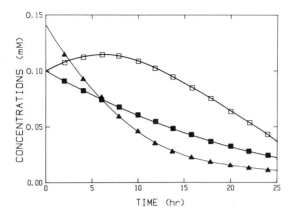

Fig. 18. Time-dependent evolution of uric acid concentrations for the bienzyme membrane (xanthine oxidase–uricase) between compartments 1 ($\square$) and 2 ($\blacksquare$). Xanthine concentration is given for compartment 2 ($\blacktriangle$) (the concentration in compartment 1 remains equal to zero). Initial concentrations for uric acid are symmetrical and equal to 10^{-7} mol/cm^3. Initial xanthine concentration is zero and 1.4×10^{-7} mol/cm^3 in compartments 1 and 2, respectively.

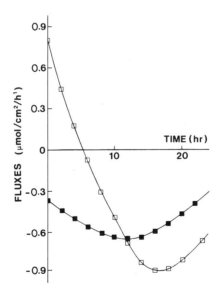

Fig. 19. Measured fluxes as a function of time for uric acid in compartments 1 ($\square$) and 2 ($\blacksquare$) for the bienzyme membrane system during the experiment shown in Fig. 18.

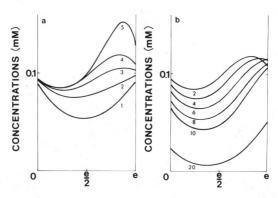

Fig. 20. (a) Concentration profiles of uric acid inside the membrane under stationary state conditions for constant V_{M2} values (uricase activity): 1, 0; 2, 0.40 × 10^{-4}; 3, 0.8 × 10^{-4}; 4, 0.14 × 10^{-3}; 5, 0.5 × 10^{-3} mol/cm^3/h. For profiles 3, 4, and 5, a pseudo-active transport effect is observed: $S_1 < S_2$, and uric acid molecules are entering and leaving compartments 1 and 2, respectively. (b) Concentration profiles of uric acid inside the membrane during evolution (2, 4, 6, 8, 10, and 20 h) for system starting under pseudo-active transport conditions.

centrations are given in Fig. 20a as a function of the $a + b$ abcissa in the membrane. If xanthine oxidase activities are high enough, fluxes at both membrane boundaries are of the same sign: substrate molecules enter at one side and leave at the other one. The global behavior is explained physicochemically by the existence of two domains in the membrane with concentrations lower and higher, respectively, than the concentration in the bulk solution. Figure 20b shows the evolution of substrate profiles as a function of time when the system is starting under pseudo-active transport conditions. The homogeneous bienzyme membrane gives rise to a global vectorial effect, with asymmetrical boundary conditions. In other words, the bienzyme membrane that is homogeneous in structure becomes heterogeneous in function.

2. Active Transport Effect (Structural Asymmetry)

It is possible to construct a multienzyme model having an asymmetrical distribution of active sites in the membrane. Such a system has been studied by Broun *et al.* (1972). A structured, multilayer bienzyme membrane was composed of two active protein layers adjacent to one another, sandwiched between two selective films. The enzyme layers carried, respectively, hexokinase and phosphatase cross-linked with albumin; both were impregnated with ATP and covered on their external surfaces by selective films permeable to glucose but impermeable to glucose 6-phosphate (Fig. 21).

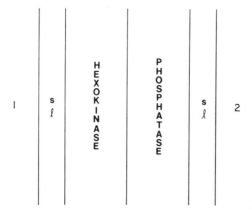

Fig. 21. Diagram of a double-layered membrane covered on its external side by two selective films. The system gives an active transport effect for glucose with consumption of ATP. Sl, selective layers.

In this asymmetrical membrane, glucose is temporarily phosphorylated:

$$\text{Glucose} + \text{ATP} \xrightarrow{\text{hexokinase (E}_1\text{)}} \text{glucose 6-phosphate} + \text{ADP}$$

$$\text{Glucose 6-phosphate} \xrightarrow[\text{(E}_2\text{)}]{\text{phosphatase}} \text{glucose} + \text{inorganic phosphate}$$

Since the sum of the two reactions is $\text{ATP} \rightarrow \text{ADP} + \text{inorganic phosphate}$, the system behaves chemically as a simple ATPase.

The asymmetrical membrane exemplifies a spatially and metabolically sequential enzyme array. When it is placed between glucose solutions, glucose entering the system is transiently phosphorylated and the system behaves chemically as a simple ATPase. In the hexokinase layer the glucose is phosphorylated to glucose 6-phosphate, which diffuses into the phosphatase layer, where it is promptly dephosphorylated.

In the first layer ($0 < x < \frac{1}{2}$), the enzyme E_1, $S \xrightarrow{E_1} P$ is immobilized with the reaction rate

$$F(x, s, p) = \sigma s \left(1 + \frac{K_M}{K_p} p + s \right)^{-1}, \qquad 0 < x < \tfrac{1}{2}. \tag{35}$$

In the second one ($\frac{1}{2} < x < 1$), E_2 is immobilized, $P \xrightarrow{E_2} S$. The reaction rate is

$$F(x, s, p) = -\sigma' p (K'_M / K_M)^{-1}, \qquad \tfrac{1}{2} < x < 1, \tag{36}$$

where $\sigma' = \dfrac{V_{M'}}{V_{M'}} \dfrac{x^2}{D_s}$ and $V_{M'}$ and $K_{M'}$ refer to E_2.

The equations governing the substrate and product concentrations are

$$\frac{\partial s}{\partial t} - \frac{\partial^2 s}{\partial x^2} + F(x, s, p) = 0$$

$$\frac{\partial p}{\partial t} - \frac{D_p}{D_s} \frac{\partial^2 p}{\partial x^2} - F(x, s, p) = 0. \tag{37}$$

The boundary conditions for p are of the Neumann type

$$\frac{\partial p}{\partial x} = 0 \quad \text{for} \quad x = 0 \quad \text{and} \quad x = 1.$$

Boundary conditions for s are of the Dirichlet type for $x = 0$ and of mobile boundary conditions for $x = 1$ are

$$s(0, t) = \alpha \quad \text{and} \begin{cases} s(1, 0) = \beta \\ \dfrac{\partial s}{\partial t} + \dfrac{\partial s}{\partial x} = 0 \quad \text{for } x = 1. \end{cases} \tag{38}$$

Initial conditions are

$$s(x, 0) = p(x, 0) = 0. \tag{39}$$

Examination of the glucose and the glucose 6-phosphate concentration profiles illustrated in Fig. 22 shows that this membrane can act as a glucose pump. The time evolution of the glucose concentration in the receptor

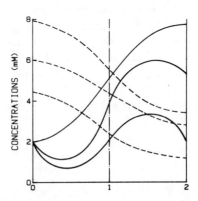

Fig. 22. "Active transport" of glucose hexokinase–phosphatase system. The membrane separates two compartments; the glucose concentration [S_1] within the first (donor) compartment can be considered constant, the volume being far larger than that of the second (receptor) compartment, in which the concentration [S_2] changes. Glucose concentration profiles (——) and glucose 6-phosphate concentration profiles (---). The lowest profiles represent initial conditions [S_1] = [S_2] evolving over a period of time; the uppermost profiles represent stationary state conditions (net fluxes, zero).

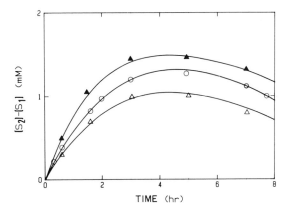

Fig. 23. Evolution of glucose concentration on the receptor side (S_2) in the case where the donor compartment concentration is constant (S_1). These experimental results are obtained with three initial concentration values: $S_1 = 2.8 \times 10^{-3}\ M$ ($\triangle$); $S_1 = 5.6 \times 10^{-3}\ M$ ($\bigcirc$), and $S_1 = 1.12 \times 10^{-2}\ M$ ($\blacktriangle$).

compartment is shown in Fig. 23. That in the donor compartment remains constant. The receptor concentration increases regularly at first; it then attains a plateau, which can be explained in terms of the concentration profiles. However, the concentration falls again once the initial ATP charge of the membrane is depleted. The pumping effect may be reduced by the addition of fructose, which competes with glucose for the hexokinase, as shown in Fig. 24. The possibility of chemodiffusional coupling in this and similar systems is clearly dictated by the global analog of the Curie principle.

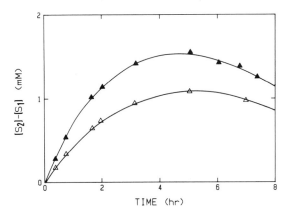

Fig. 24. Evolution of the glucose concentration in the receptor compartment in the presence of fructose ($2.8 \times 10^{-3}\ M$) showing a competitive phenomenon. ($\triangle$), With fructose; ($\blacktriangle$), without fructose. Experimental conditions are the same as in Fig. 23 ($S_1 = 1.12 \times 10^{-2}\ M$).

C. A Dissipative Structure Linked to an Acid–Base Metabolism

The system to be described next does not deal with a consecutive set of reactions, insofar as the two enzymatic steps are apparently independent. Nevertheless, this example illustrates the potential of artificial enzyme membranes. Clearly, the effect of diffusional hindrances, in addition to the close spatial location of enzymatic entities, can give rise to completely new behavior whose appearance is almost impossible in an homogeneous medium.

An artificial membrane bearing two different enzymes (glucose oxidase and urease) in a spatially homogeneous fashion is produced by using the method previously described. The glucose oxidase reaction increases the pH, and the urease reaction decreases the pH. The pH activity profiles show an autocatalytic effect for glucose oxidase in the range of pH values greater than the optimum; for the urease, the autocatalytic effect appears in pH values smaller than the optimum (Fig. 25). When the two enzymes are mixed together, the global pH variation is zero for one well-defined pH value.

The active membrane separates two compartments, and it is possible to get this pH value throughout the system, in the presence of the two substrates, by the transient use of a buffer. The pH values outside are controlled and H^+ fluxes measured by pH-stat systems. After small asymmetrical perturbations of the pH values at the boundaries (0.05), an inhomogeneous pH distribution arises spontaneously inside the membrane. The initial perturbations are amplified, and the pH values in the compartments tend to evolve in opposite directions. The H^+ fluxes entering and leaving the membrane can be determined by pH-stat measurements. If the boundary pH values are not maintained constant by a pH-stat, the system evolves to a new stationary state characterized by a pH gradient of 2 pH units across the membrane. It seems

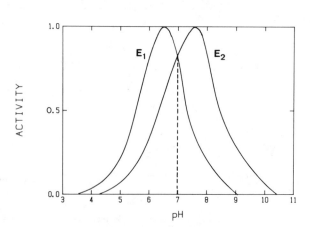

Fig. 25. Enzymatic activities of glucose oxidase (E_1) and urease (R_2) as a function of pH. Activities are given as a fraction of the maximal activity, for both enzymes.

possible to explain this behavior by the diffusion reaction equations. Paradoxically, this experimental system is similar to the absolutely fictitious system described by Turing (1952). In this system, the simplest case consists of two cells connected by a permeable wall. In each cell, a chemical reaction takes place in which one substance is converted into another one by an autocatalytic process. With a well-defined homogeneous distribution of matter, the concentrations do not change with time. If, however, a fluctuation of a few percent occurs, a rather important change will take place, as in the experimental system previously presented.

In both systems, a positive fluctuation in the concentration in one compartment does not result in a negative value of the derivatives with time, but in a positive one. In these systems the initial homogeneous steady state is unstable, and the system will develop spontaneously toward a nonhomogeneous concentration distribution (Thomas, Goldbeter, and Lefever, unpublished data).

IV. CONCLUSION AND PERSPECTIVES

Our aim was to study the modifications of the kinetic behavior in sequential enzyme systems after their immobilization within artificial membrane structures. With the help of some examples we have demonstrated that the introduction of diffusional constraints—existence of local concentrations—can deeply modulate this behavior. Such an approach led us to bring out some general laws governing "heterogeneous enzymology."

Thus, in classical homogeneous kinetics, the influence of (competitive) inhibitors upon the enzymatic reaction depends only on the value of the K_M/K_i ratio. The existence of both heterogeneity and local concentrations makes this inhibitory effect more or less efficient depending on whether the inhibitor appears inside (endogenous) or outside (exogenous) the support. That is, the structure can act as either an amplifier or a buffer with regards to an enzyme effector. Moreover, this notion of inhibition can lose significance under particular conditions, leading to an apparent activation of the system. Thermodynamic parameters of a reaction, such as the equilibrium constant, may also be shifted. This behavior, quantitatively different from that observed in solutions, though linked to the diffusion of metabolites and the proximity of catalytic sites, may appear for numerous enzyme species randomly distributed inside the structure, and with symmetrical boundary conditions. However, it is also possible to integrate sequential activities in a determined spatial arrangement: asymmetry can be achieved by either applying a gradient of an effector, (for example, an inhibitor, across the membrane), or placing layers side by side, each one containing an enzyme of the sequence. Under such conditions of functional asymmetry, phenomena intrinsically linked to the structures, such as active or pseudo-active transport, come into sight. In other words, scalar chemical phenomena may give rise to vectorial catalysis.

All the observed phenomena are modeled at both quantitative and qualitative levels simply by taking into account the coupling between enzymatic reactions involved in the sequences and diffusion of the various metabolites.

Thus the global behavior of any multienzyme system is, in our model, completely defined by a set of partial differential equations complete with adequate initial and boundary conditions. An interactive approach between the experimental observation of the phenomena studied and the elaboration of mathematical models accounting for them appear essential. The transposition of a biochemical situation within a well-defined context into its mathematical formulation requires a precise knowledge of the various parameter values. In our systems, the study of the diffusion reaction coupling takes into account two sets of parameters: kinetics (e.g., V_M, K_M, K_I, etc.) and physics (diffusion coefficients, support dimensions, etc.). The first parameters are measured by classical methods, and the membranous shape of our supports makes the determination of the second ones easier. In addition, this geometry allows us to develop a simple analytical expression of the diffusion phenomena.

The numerical simulations having confirmed the observations, the model allows a second step to guide the experimenter's investigations. A study of the influence of the parameters can lead to the appearance of either particular situations or unexpected behavior. Consequently, the model has not only an heuristic meaning in itself but can help in the discovery of unexpected potentialities. If the experiment supports the model, the latter can in turn promote the experiment.

Nevertheless, the multienzyme systems studied here are a preliminary stage in the attempts to acquire a better knowledge of phenomena likely to be involved *in vivo*. Indeed, for the sake of getting reliable bases for further work, our approach dealt with restricted conditions and situations: (1) the structure of the solid phase was macroscopically homogeneous, (2) one particular dimension of space was favored, and (3) sequences were limited to two simple enzymatic reactions.

Obviously, further investigations are required in three directions:

1. The study of heterogeneous membranes, such as biphasic structures (proteins–lipids), is of prime importance. The effect of the physicochemical properties of the support and, more precisely, the influence of fixed and/or mobile charges can lead to very exciting kinetic modulations (Gestrelius *et al.*, 1972). Situations in which diffusion coefficients are not constants (functions of concentration and space) have not been considered in our model.

2. Although the experimental approach to multistep sequences with more

than two or three enzymes is full of difficulties and drawbacks (Mosbach, 1972), much information can be obtained with such systems: efficiency of cross-linked regulations, channeling effects in linear or branched sequences, etc. Multistep enzyme systems can be realistically regarded as spatial organizations in two or three dimensions. Preliminary works (Milan, 1975) have been carried out on regulatory loops at both metabolic and spatial (two-dimensional) levels.

3.   Numerous enzymes exhibiting sophisticated regulatory properties are involved in any pathway: the best-known examples concern allosteric enzymes and substrate-inhibited enzymes, which give rise to complex behaviors such as multistability and/or spatiotemporal structures. When coupled, still more refined phenomena are likely to occur (El Rifai et al., 1980). Classical enzymology, as it has been treated until recently, does not permit explanations of some phenomena and cannot justify direct extrapolation from its results to models accounting for reality in vivo: enzymes are isolated from their natural environment and studies are carried out in homogeneous and isotropic media.

Our approach does not claim to explain cellular reality, but simply attempts to get closer to it by introducing new factors that must necessarily play a fundamental role in the organization of metabolic pathways. The use of artificial membranes allows easy access to a high level of understanding and potential applications of enzymatic systems. Yet, this step is made possible only by knowledge acquired in the fields of fundamental enzymology and metabolic networks. The contribution of diffusion phenomena to enzymatic reactions plays a great part in the estimation of effects that mass transfer, local concentrations, microenvironments, and compartmentalization may have on the kinetic behaviors and on the regulation mechanisms of enzymatic systems.

REFERENCES

Barbotin, J. N., and Thomas, D. (1974). J. Histochem. Cytochem. **22,** 1048–1059.

Blaedel, W. J., Kissel, T. R., and Boguslaski, R. C. (1972). Anal. Chem. **44,** 2030–2037.

Bouin, J. C., Atallah, M. T., and Hultin, H. O. (1976). Biochim. Biophys. Acta **438,** 23–36.

Broun, G., Thomas, D., and Selegny, E. (1972). J. Membr. Biol. **8,** 313–332.

Broun, G., Thomas, D., Gellf, G., Domurado, D., Berjonneau, A. M., and Guillon, C. (1973). Biotechnol. Bioeng. **15,** 259–375.

Brown, H. D., Patel, A. B., and Chattopadhyay, S. K. (1968). J. Chromatogr. **35,** 101–103.

Chibata, I. (1978). "Immobilized Enzymes." Halsted, New York.

Cori, C. F. (1956) In "Currents in Biochemical Research" (D. E. Green, ed.), pp. 198–214. Wiley (Interscience), New York.

De Simone, J. A., and Caplan, S. R. (1969). Symp. Interphase Mass Transf. Biol. Syst., New Orleans.

El-Rifai, M. A., Elnashaie, S. S., and Aboulfath, H. (1980). J. Solid Phase Biochem. **5,** 235–243.

Engasser, J. M., and Horvath, C. (1976). Appl. Biochem. Bioeng. **1,** 127–220.

Gestrelius, S., Mattiasson, B., and Mosbach, K. (1972). Biochim. Biophys. Acta **276,** 339–343.

Goldman, R., and Katchalski, E. (1971). J. Theor. Biol. **32,** 243–257.

Graham, M. C., and Karnovsky, M. J. (1966). J. Histochem. Cytochem. **14,** 291.

Graves, D. J. (1973). *In* "Enzyme Engineering" (E. K. Pye and L. B. Wingard, Jr., eds.), Vol. 2, pp. 253–258. Plenum, New York.

Gryskiewicz, J. (1971). *Folia Biol.* **19**, 119–150.

Hervagault, J. F., Joly, G., and Thomas, D. (1975). *Eur. J. Biochem.* **51**, 19–23.

Hervagault, J. F., Duban, M. C., Kernevez, J. P., and Thomas, D. (1976). *J. Solid Phase Biochem.* **1**, 81–89.

Hess, B., and Boiteux, A. (1972). *In* "Protein-Protein Interactions" (R. Jaenicke and E. Helmreich, eds.), pp. 271–297. Springer-Verlag, Berlin and New York.

Hubscher, G., Mayer, R. J., and Hansen, H. J. M. (1971). *Bioenergetics* **2**, 115–118.

Kasche, V., Lundquist, H., Bergman, R., and Axen, R. (1971). *Biochem. Biophys. Res. Commun.* **45**, 615–621.

Kedem, O., and Katchalsky, A. (1963). *Trans. Faraday Soc.* **59**, 1918.

Kempner, E. S., and Miller, J. H. (1968). *Exp. Cell Res.* **51**, 141–156.

Kernevez, J. P. (1972). Thesis, Paris (A. O. 7246).

Kernevez, J. P. (1980). *In* "Enzyme Mathematics" (J. L. Lions, G. Papanicolaou, and R. T. Rockfellar, eds.). North-Holland Publ., Amsterdam.

Kernevez, J. P., and Thomas, D. (1975). *Appl. Math. Optim.* **1**, 222–285.

Lecoq, D., Hervagault, J. F., Broun, G., Joly G., Kernevez, J. P., and Thomas, D. (1975). *J. Biol. Chem.* **250**, 5496–5500.

Le Moullec, J. M., and Thomas, D. (1977). *J. Biol. Chem.* **252**, 2611–2614.

Lloyd-Davies, KOA., Michell, R. H., and Coleman, R. (1972). *Biochem. J.* **127**, 357–368.

Malpiece, Y., Sharan, M., Barbotin, J. N., Personne, P., and Thomas, D. (1980a). *J. Biol. Chem.* **255**, 6883–6890.

Malpiece, Y., Sharan, M., Barbotin, J. N., Personne, P., and Thomas, D. (1980b). *J. Histochem. Cytochem.* **28**, 961–968.

Mattiasson, B., and Mosbach, K. (1971). *Biochim. Biophys. Acta* **235**, 253–257.

Milan, C. (1975). Dissertation, Compiegne.

Mitchell, P., and Moyle, J. (1958). *Nature* (London) **182**, 372.

Mosbach, K. (1970). *Acta Chem. Scand.* **24**, 2084–2097.

Mosbach, K. (1972). *Biotechnol. Bioeng. Symp.* N **3**, 189–194.

Mosbach, K., and Mattiasson, B. (1970). *Acta Chem. Scand.* **24**, 2098–2100.

Mosbach, K., and Mosbach, R. (1967). *Acta Chem. Scand.* **20**, 2807–2810.

Pollard, E. (1963). *J. Theor. Biol.* **4**, 98–112.

Prigogine, I. (1955). "Introduction to the Thermodynamics of Irreversible Process." Wiley (Interscience), New York.

Runge-Kutta (1964). *In* "Elements of Numerical Analysis" (P. Henrici ed.). Wiley, New York.

Sernetz, M., and Puchinger, H. (1976). *In* "Methods in Enzymology" (K. Mosbach, ed.), Vol. 44, pp. 373–379.

Shimizu, S. Y., and Lenhoff, H. M. (1979). *J. Solid Phase Biochem.* **4**, 75–122.

Silman, I. H., and Katchalski, E. (1966). *Annu. Rev. Biochem.* **35**, 873–908.

Srere, P. A. (1972). *In* "Energy Metabolism and the Regulation of Metabolic Processes in Mitochondria" (M. A. Mehlman and R. W. Hanson, eds.). Academic Press, New York.

Srere, P. A., and Mosbach, K. (1974). *Annu. Rev. Microbiol.* **28**, 61–83.

Srere, P. A., Mattiasson, B., and Mosbach, K. (1973). *Proc. Natl. Acad. Sci. U.S.A.* **70**, 2534–2538.

Takeshi, K., and Laidler, K. J. (1973). *Biochim. Biophys. Acta* **302**, 1–12.

Thomas, D., Bourdillon, C., Broun, G., and Kernevez, J. P. (1974). *Biochemistry* **13**, 2995.

Turing, A. M. (1952). *Philos. Trans. R. Soc. London Ser. B* **237**, 37.

Vasileva, N. V., Balaevskaya, T. O., Gogilashvili, L. Z., and Serebrovskaya, K. B. (1969). *Biokhimiya* **34**, 641 (in Russian).

Welch, G. R., and Gaertner, F. H. (1975). *Proc. Natl. Acad. Sci. U.S.A.* **72**, 4218–4222.

Wilson, R. J. H., Kay, G., and Lilly, M. D. (1968). *Biochem. J.* **108**, 845–853.

9

Long-Range Energy Continua and the Coordination of Multienzyme Sequences *in Vivo*

G. Rickey Welch

Department of Biological Sciences
University of New Orleans
New Orleans, Louisiana

Michael N. Berry

Department of Clinical Biochemistry
School of Medicine
Flinders University of South Australia
Bedford Park, South Australia, Australia

ORGANIZED
MULTIENZYME SYSTEMS

I. INTRODUCTION

In preceding chapters we have viewed the kinetic/catalytic properties of multienzyme systems, as observed in various states of organization and in artificially designed as well as naturally occurring configurations. One cannot help but be impressed with the unique functional features (e.g., substrate–product *channeling*, reduction of diffusional transit time, coordinated allosteric effects) engendered by such structured regimes. There is a profound teleonomic sense that organization is the rule, rather than the exception, in reflecting enzyme character *in vivo*. A perusal of the literature today reveals evidence for enzyme organization in virtually all major metabolic pathways. The oganizational mode may entail formation of protein–protein complexes and/or adsorption to cytological substructures. With the notion of "locational specificity" as part and parcel of enzyme action, enzymology has become a "cytosociological" science (Welch and Keleti, 1981).

In vitro studies with isolated multienzyme complexes and with immobilized systems have, indeed, given us many clues as to the optimization of intermediary metabolism in the living cell. Importantly, we come to realize that there is more to cell metabolism than simply a superposition (or montage) of *individual* enzymes. There is a kind of "wholeness," or "connectedness," which exceeds the reductionistic picture. A "holonomic" perspective (Bohm, 1980) leads one to look more deeply into the nature of molecular processivity within the confines of these organized states, wherein the concern is not so much with the flow of matter as it is with the flow of energy (*viz.*, free energy). At this juncture, one must meld enzymology with emergent principles of cellular bioenergetics. It is now apparent that particulate structures of the living cell are permeated by such energy modes as mobile protonic states ("proticity") and strong electric fields. The organization of multienzyme systems at cytosol–particulate interfaces juxtaposes them to such long-range energy continua. This arrangement seems beyond fortuity. In fact, there are growing indications that certain types of enzymes are designed to "plug into" these energy modes.

One commonly thinks of *energy* as the "capacity to perform work." Hence, the question arises as to the possible role(s) of the "energy continua" in the molecular processes of organized enzyme systems. There are two aspects for consideration: substrate–product translocation and chemical reaction dynamics. Intermediary metabolism, regarded as multienzyme sequences, involves basically these two kinds of process. In the present chapter, we wish to discuss the manner by which long-range energy modes may permeate the superstructure of organized enzyme systems and affect localized molecular events therein. We begin by looking at the conservation and transduction of *chemical* free energy within enzyme aggregates.

It must be emphasized at the outset that empirical evidence in support of the suggested role of these external energy continua is, at present, somewhat scanty and circumstantial. Our treatment, then, is based heavily on theoretical modeling. Yet, this picture is logically consistent with (and, in fact, interrelates) developing ideas in cytology and bioenergetics, as well as with the properties of protein molecules and, therefore, is gaining increasing support from cell biologists, as well as from enzymologists and biochemists. The "holonomic view" of enzymology reveals that chemical processes within the microenvironments of the cellular *milieu intèrieur* may have rather bizarre qualities compared to the counterpart processes *in vitro*.

II. TRANSDUCTION AND CONSERVATION OF CHEMICAL FREE ENERGY WITHIN MULTIENZYME AGGREGATES

Before discussing external energy modes, it is instructive to probe the ways in which chemical free energy is handled within the confinement of organized enzyme systems. Our analysis is based on an elegant notion, the *free energy complementarity principle*, propounded by Lumry and co-workers (Lumry and Biltonen, 1969; Lumry, 1971). This principle was developed originally from consideration of the thermodynamic aspects of protein conformation relative to physiological function. As regards enzyme reactions, the principle holds that the free energy changes associated with motion of the system along the functional conformation coordinate tend to complement the free energy changes of the bound chemical subsystem (substrate) undergoing chemical reaction. The net effect of this complementation, as the combined protein and chemical subsystems move along the reaction coordinate, is to approach an average free energy profile such that the up-and-down swings are smoothed out (see Fig. 1). We would agree with the previous authors (Lumry and Biltonen, 1969), in that *evolution toward complementarity* in protein reactions is "the most refined form of adaptation . . . the major form of evolution today." It applies to the single enzyme reaction, as well as to the operation of multienzyme complexes. In order to relate the latter systems, we will extend Lumry's principle not only to features of the protein conformation *per se* but also to characters specified by the spatial contiguity of the interacting enzymes.

The unique functional properties of organized multienzyme systems, as observed *in vitro*, generally fall into two categories: (i) The clustering of the component protein moieties (either among themselves or with a membraneous matrix) may produce entities that have intrinsic catalytic properties unlike those of the separate proteins, that is, the physical association may stabilize

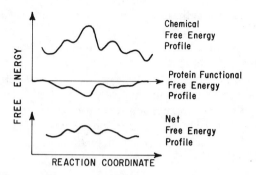

Fig. 1. Simplified example of the flattening of the net free energy profile through complementarity developed between the chemical parts of the protein-supported process and the conformational contribution from the functional conformation process of the protein. (Adapted from Lumry and Biltonen, 1969).

and/or enhance the *overall* activity of the enzyme sequence. (ii) The assembly into an organized cluster of enzymes may alter the efficiency of the overall process, even if the intrinsic catalytic activities of the components are not changed upon association; advantages here result simply from the proximal juxtaposition of the constituent active sites within the organized system (Welch, 1977a; Keleti, 1978). The thermodynamic/kinetic aspects of these organizational properties have been discussed in detail elsewhere (Keleti, 1975, 1978; Welch and Keleti, 1981; Welch, 1977a,b; Welch and DeMoss, 1978). Specifically, studies on various physicochemical characteristics of enzyme action suggest that Nature has a repertoire of "cytosociological factors" (Welch and Keleti, 1981) that can be brought to bear on the free energy profiles of the component reactions in organized states. We briefly describe some of these below.

(i)   *Structural effects* of Laidler and Bunting (1973). It is found that many enzymes undergo a reversible conformational change during the course of the reaction process. (Perhaps a slightly unfolded state makes the active site more available to the substrate.) In multienzyme aggregates the individual proteins might be stabilized in optimally "open" configurations, obviating some postcatalytic refolding.

(ii)   *Protein-configurational fluctuation* (e.g., Karush, 1950). Many individual proteins, particularly components of interacting systems, exist in solution as an equilibrium mixture of a number of configurations of approximately equal energy (Duffy, 1971; Shnoll and Chetverikova, 1975; Straub and Szabolcsi, 1964). In many of these cases (involving enzyme reactions), it has been observed that only one configuration of a given protein is optimal for catalytic function. Consequently, one must associate with the transition state of the ES complex not only a requisite energetic fluctuation, but also a specific configurational fluctuation in the protein. Notably, it is found that formation of multienzyme complexes and membrane-bound enzyme arrays can "freeze" each component protein into a single (optimal) form (Welch, 1977a).

   (iii)   *Chemical activation* (Rabinovitch and Flowers, 1964; Carrington and Garvin, 1969). Essentially every type of elementary chemical reaction yields products, initially, with a nonequilibrium energy distribution (e.g., excited internal vibrational states). For homogeneous systems in solution, the fate of such excited states is rapid relaxation (e.g., via collisional deactivation). However, in physically associated multienzyme systems, a portion of this energy released by the chemical subsystem may be retained within the protein superstructure for specific utilization in subsequent catalytic events (Lumry and Biltonen, 1969; Likhtenshtein, 1966; Wyman, 1975; Hill, 1977; Welch, 1977a; Blumenfeld, 1983).

   (iv)   *Coupling of entropy–enthalpy phase relations.* *Phase relation* refers to the relationship between the enthalpy and entropy changes accompanying, for example, conformational transitions in proteins. Lumry and Biltonen (1969) have emphasized the import of the phase relation in the subtle conformational changes which occur in virtually all enzyme reactions. (Indeed, this is the very basis of their complementarity principle.) This thermodynamic relation determines the free energy minima along the reaction coordinate and, thus, the most stable conformational states. It also determines the magnitude of the transition-state thermodynamic parameters specific for conformational changes. Lumry and Biltonen (1969) proposed that many enzyme proteins, by virtue of a variable phase relation, serve as free energy reservoirs (or buffers) during the course of the catalyzed reaction. They extended this idea to multienzyme complexes, using the term *supercomplementarity* to indicate a complementarity adjustment producing a chemical–mechanical balance of free energy spread through a system of enzymes and not confined to individual subunits. Accordingly, physical organization of related enzymes offers an important evolutionary unification, allowing greater energetic coupling (through the protein medium) of connected chemical processes. This is a natural extension of factor (iii) above.

   (v)   *The entatic state* (Vallee and Williams, 1968). Certain types of enzymes can be energetically poised for catalytic action in the absence of substrate. In such cases, the geometry of the active site can generate "internal activation," due to conformational stress. A substrate molecule entering such a domain would find itself under attack by unusually active groups. Within the confines of a multienzyme complex, there is a greater potential for component enzymes to be stabilized in "entatic" conformations (Welch, 1977a). For example, the active sites may be so arranged as to receive nascent (compartmentalized), intermediate substrates in strained or distorted forms resembling partially the transition state (Andrews et al., 1973; Ovadi and Keleti, 1978).

(vi)   *Electrostatic contributions* (Laidler and Bunting, 1973). Consider the enzyme and substrate as ions. For reactions in solution, the free energy for formation of an activated complex from two ions will contain a part due to the free energy change associated with the electrostatic forces between the two reactants as they are brought together. For the formation of an enzyme–substrate (ES) complex in solution, the *net* charge of the protein will affect the overall interaction potential. Hence, in general, the electrostatic interaction term will not be negative (rate enhancing). Within the confinement of an organized multienzyme system, a nascent intermediate–substrate molecule might not "see" the overall charge of the protein (or matrix structure), but only that in the vicinity of the active site. Also, the value of the dielectric constant might be significantly lower (than for water) in structured regimes. Thus, the translocation of substrates may be enhanced significantly (Welch, 1977a).

(vii)   *Other diffusivity contributions.* Consider the diffusion-controlled case, in which the rate of association of enzyme (E) and substrate (S) is limited by the diffusion process. Then, we can write the rate constant (k_d) for ES formation generally as

$$k_d = \Omega D_{ES} r_c, \tag{1}$$

where Ω is a geometrical factor relating to the orientation restrictions governing interaction of E and S; D_{ES} is a composite diffusion coefficient, equal to $D_E + D_S$ (for all practical purposes, $D_{ES} = D_S$); and r_c is a "capture radius" characteristic of ES formation (Eigen, 1974). Clearly, the evolution of cellular infrastructure can lead to the creation of specialized microenvironments that tailor Eq. (1) to the needs of particular enzymatic processes (Coleman, 1973; Srere and Mosbach, 1974; Srere and Estabrook, 1978; Masters, 1977; Welch et al., 1983). Two obvious means are alteration in microviscosity and distortion of the local diffusion field by specific electrostatic forces [e.g., according to factor (vi) above] (Welch, 1977a; Somogyi, 1974).

(viii)   *Other viscosity effects.* By virtue of an inverse relationship between diffusion and viscosity coefficients, factor (vii) shows how the medium viscosity affects the transport of substrate to the enzyme active site. Moreover, the viscosity of the ambient medium can influence the catalytic constant as well, by determining the rate of transfer of momentum from solvent to the protein body, which transfer relates directly to the "energization" of the protein for the rate process (Damjanovich and Somogyi, 1971; Gavish, 1978; DeBrunner and Frauenfelder, 1982; Welch et al., 1982; Somogyi et al., 1984). Accordingly, the role of intracellular microenvironmental viscosity (e.g., in a heterogeneous milieu) in regulating enzyme activity has been emphasized (Somogyi et al., 1984; Somogyi and Damjanovich, 1975; Beece et al., 1980; Welch et al., 1982).

The list of factors above is not intended to be exhaustive. Rather, it simply emphasizes some of the more readily discernible "sociological" influences on enzyme action *in vivo*. The familiar effects of *allosteric interactions* are included under the umbrella of this list [e.g., in factor (ii)]. In fact, the discovery of allosterism in enzyme function was perhaps the first clue as to the importance of "enzyme sociology" in the living cell, a notion to which Monod *et al.* (1965) alluded in their classic article! Some of these physicochemical factors reduce the free energy of the transition-state barriers along the reaction profile of a multienzyme sequence, whereas others elevate the free energy valleys (defined by the intermediate substrate–product states). The result is a smoothing of the overall flow process.

Thus, we see an orderly evolutionary progression. Alteration in the complementarity features of the conformational dynamics yields improvement in the kinetics of the single enzyme. But evolution can go only so far in optimizing intermediary metabolism by acting on *individual* enzymes. For instance, the protein structure itself cannot increase the rate at which the substrate diffuses to the enzyme. Organization of enzyme sequences in (on) membranous compartments solves this problem (particularly in larger eukaryotic cells), and formation of multienzyme complexes provides for a higher unification and coordination via dynamic coupling of component enzymes according to the supercomplementarity principle.

Lumry's free energy complementarity leads us to view enzyme reactions at a deeper level. Through a fluid and variable interaction with the bound chemical subsystem, *the protein steps from the role of a mere catalyst into that of a reactant*, and this concept is critically important for our understanding of the coupling of enzyme reactions to external energy continua *in vivo*.

III. ENZYMES: BIOCHEMICAL ELECTRODES AND PROTODES

Let us consider the reactive properties of the protein molecule, which suggest that enzymes are suited to function as field-effect electronic/protonic elements in the execution of chemical reactions.

A. Structural–Functional Bases of Enzyme Action

Throughout the 100-year history of enzymology (Gutfreund, 1976), the major emphasis has been on the *active center* of the enzyme molecule. We have used the transition-state free energy change, $\Delta G^{\ddagger}$, as a "window" into the workings of this reaction site, decomposing catalytic processes into entropic ($\Delta S^{\ddagger}$) and enthalpic ($\Delta H^{\ddagger}$) contributions. Indeed, all of the customary sources of catalytic power can, via $\Delta S^{\ddagger}$ or $\Delta H^{\ddagger}$, be formulated in the guise of *transition-state stabilization* (Fersht, 1985).

From the enthalpic ($\Delta H^{\ddagger}$) side, one finds in the enzymology literature a number of energetic factors that lead to decrease in $\Delta G^{\ddagger}$. Generally, these fall into one of the three categories: (i) catalysis by *rack mechanisms* (whereby the bound, ground-state substrate is strained or distorted), (ii) *general acid–base catalysis* (involving proton transfer to or from the transition state), and (iii) *electrophilic/nucleophilic catalysis* (usually involving covalent stabilization of the transition state). Underlying these various catalytic modes is the importance of *local electric fields*, in the microenvironments of enzyme active centers. It has been suggested that the geometry of active-center dipole groups is engineered in evolution to provide optimal "solvation" (electrostatic stabilizaton) of polar transition states (Warshel, 1978).

What of the role(s) of the protein matrix in active-center events? Considering the relatively small volume occupied by the actual reaction site, one might ask: Why are enzymes so big? It is now apparent that the protein as a whole is designed to provide a specific solvent medium for a given chemical reaction (Welch *et al.*, 1982), wherein the combined chemical and protein subsystems engage in a fluid and variable exchange of free energy, *facilitating the entrance of the bound chemical system into its transition state*. Accordingly, we are led to picture the protein matrix as an intermediary, a "deterministic" mediator, between a localized chemical reaction coordinate and the ambient medium. A crucial point here is that, while the local active-site configuration defines the physicochemical nature of the substrate–product transition state(s), the protein molecule determines the rate at which this state(s) is reached.

If the enzyme is viewed as a macromolecular free energy transducer, an immediate question concerns which kinds of useful energy are transduced. It is quite clear that mechanical (i.e., kinetic/potential) forms are involved (Welch *et al.*, 1982). Considering the predominance of activated protonic/electronic states during catalytic events at the active center, it is natural to ponder a role of the protein molecule as a *protical/electrical transducer*. There is an increasing indication, from both experimental and theoretical fronts, that such transduction does, indeed, occur in *individual* proteins and that it may be involved in catalytic processes (Wang, 1968; Somogyi *et al.*, 1984; Welch and Berry, 1983). Interest in this type of energy transduction is heightened further when we relate the cytological juxtaposition of organized enzyme systems and sites of protonic/electronic sources (see Section IV). It is quite plausible that, in many cases, the enzyme molecule connects the active center not only mechanically to the surroundings, but, also protically/electrically.

Let us look at elements of protein structure which suggest such energy transduction design. The obvious possibilities are regions of local secondary structure, viz., α-helix and β-structure. First, we consider the α-helix. The α-helix possesses a marked dipole moment, originating in the alignment of the dipole moments of the individual peptide units (Hol *et al.*, 1978). The helix dipole generates a considerable electric field, whose strength increases with helix length, particularly in a medium (viz., the interior of a globular protein or a membraneous phase) with low dielectric constant.

Hol *et al.* (1978) suggested that the electric field of the α-helix is "a significant factor which must be included in the discussion of the properties of proteins." The previous authors proposed three possible functional roles of this field in enzyme activity: (i) binding of charged substrates or coenzymes, (ii) long-range attraction and/or orientation of charged substrates, and (iii) contribution to catalytic events. Several widely diverse types of enzymes are known from structural studies to bind negatively charged compounds at the N-terminal region of an α-helix. Moreover, Hol *et al.* (1978) specify a number of different kinds of enzymes (e.g., NAD dehydrogenases), whose active sites are known to be located at the N-terminus of an α-helix. Moieties at such helix termini appear to function in proton-transfer networks, which activate (deprotonate) catalytic nucleophiles, and/or in stabilization of negatively charged (e.g., oxyanion) transition states.

Other interesting roles of α-helical elements have been suggested. Types of α-helices may constitute the framework through which proton transport occurs in the well-known, membraneous proton pumps (Van Duijnen and Thole, 1981; Krimm and Dwivedi, 1982; Dunker, 1982). Also, the α-helix seems appropriately designed for vectorial conduction of coherent vibrational excitations (see Section III,B).

Regular β-structures in some enzymes (e.g., serine proteases) most likely contribute to local fields involved in substrate-binding and/or catalytic events. It appears that β-structures provide an additional way of integrating the substrate into the protonic "circuitry" of the protein matrix (Metzler, 1979).

Another structural array that may be particularly important in proton semiconduction (see Section IV) is a hydrogen-bonded network (Fig. 2)

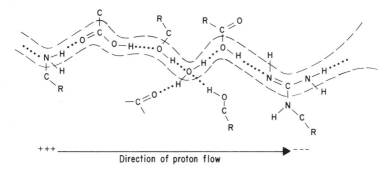

Fig. 2. A series of hydrogen bonds (·····) serving as a proton-conducting pathway within a cytoplasmic or intramembrane protein system. (See Nagle and Tristram-Nagle, 1983).

formed by the interdigitated R groups of parallel α-helix segments *or* parallel β-structures (Nagle and Tristram-Nagle, 1983). Obviously, such a network could function efficiently as a proton conductor only within a relatively hydrophobic environment (viz., a membrane phase *or* inside a globular protein). As shown by Nagle *et al.* (1980), this design can be a *modus operandi* for proton pumps, protochemical reactions, and even proton-driven mechanical work (e.g., cyclic conformational changes, generated by migrating "faults" that follow proton "hopping," see Section IV).

Long-range, mobile protonic states are finding increasing relevance to many cellular processes, stemming from the original pioneering suggestion as to their role in electron-transfer phosphorylation (Mitchell, 1979). This kind of energy continuum is emerging as a unifying theme in cell metabolism. Its importance demands that we begin looking at enzyme structure–function with new perspectives (see Section IV). Relating to basic enzyme action, as expressed by Wang (1968), such protonic states may represent "a broad catalytic principle which transcends the idiosyncracies of individual enzymes." The three enthalpic catalytic factors discussed above may all be subservient to this "principle." That is to say, *facilitated proton transfer* in the enzyme–substrate complex plays a key role in most enzyme catalyses. [This is obvious in the case of acid–base catalysis. It is more subtly involved in rack and covalent mechanisms. For example, nucleophilic (e.g., hydroxyl, sulfhydryl) groups at the active center usually require "activation" by proton shuttles or charge–relay networks.]

A similar notion has been advanced by Metzler (1979) in a review on tautomerism in enzyme catalysis. As noted by the previous author, "the rapid equilibration of isomers in which one or more hydrogen atoms change positions is a prevalent phenomenon among biochemical substances." He proposed that enzyme molecules have evolved generally to conjugate with such protonic/electronic chemical configurations and that tautomeric effects in enzyme–substrate complexes probably play basic roles in catalysis and regulation. As discussed by Metzler (1979), proteins contain many tautomeric groups, such as the peptide–amide groups and many of the amino acid sidechains, as well as bound coenzymes. The nonpolar nature of the protein interior favors the formation of extended hydrogen-bond networks (and ensuing charge mobility) among these various groups. Charge transfer through these chains is likely to be fundamentally important to enzyme action. In particular, addition or removal of a proton at one end of a chain of such hydrogen-bonded groups will produce a local field effect at a distance (and vice versa), via the continuum of the network (see also Banacky, 1981).

Superimposed on these designs may be long-range electronic semiconduction states, which may be important in some enzymes, especially in organized states *in vivo* (Lewis, 1979). Various modes of electronic (and electron–phonon) coupling between enzyme and substrate have been offered as theoretical possibilities (Volkenstein, 1981; Conrad, 1979). Most discussions of the role of electronic semiconduction (and tunneling) in enzymes have focused on the membrane-associated components of electron-transport phosphorylation.

B. Protein Dynamics

As indicated in Section III,A above, there has been an increasing awareness in recent years of the role of the protein matrix as a "mechanical mediator" between the enzyme active center and the thermal properties of the bulk medium. Naturally, the question arises as to the interaction between excited conformational (e.g., vibrational) states and excited protonic/electronic states in the enzyme molecule. This consideration is particularly important when one notes that virtually all enzymatic mechanisms involve some kind of mechanical force acting through electrostatic interactions generated across chemical bonds in the substrate.

There have emerged various theoretical models (Welch *et al.*, 1982; Somogyi *et al.*, 1984; Welch, 1985), which propose that spatiotemporal ordering of the fluctuational behavior of the protein molecules serves an integral role in enzyme catalysis. A notable principle, from all of the models, is the unity of the enzyme molecule and the ambient medium. We find that biological evolution has apparently come to grips with the random field in the medium by designing a macromolecular structure that is capable of "borrowing" and collimating that random energy source in an anisotropic fashion.

Notwithstanding, there is something deceptive and misleading about this developing picture. Accepting the premise that most (if not all) enzymes of intermediary metabolism operate in organized regimes *in vivo*, there is a certain degree of artificiality in our construction of general theoretical models that attempt to "unify" the dynamic protein molecule with a bulk aqueous medium. We contend that by probing the above theoretical models more deeply, we discover a hint (from the fluctuational behavior of the protein in bulk solution) as to the mode of operation in the organized states.

External and internal hydrogen bonding plays an integral role in the dynamic tertiary (and quaternary) structure of globular proteins (Lumry and Rosenberg, 1975; Ikegami, 1977). Following tenets of certain of the dynamic enzyme models (Somogyi *et al.*, 1984), one sees that gaps ("faults," "defects") in the internal bonding arrangement might elevate *locally* the free energy of the system, electrostatically and mechanically, as well as protonically (see Section III,A). Catalytic functions would, then, depend on precise internal fault states (Lumry and Rosenberg, 1975; Ikegami, 1977). Such faults (or defects) arise, in a protein dissolved in aqueous solution, in conjunction with binding relaxation of bound water, fluctuating proton-transfer processes and charge-density fluctuations at the surface, etc. [Inside the protein, these faults can migrate, for example, by proton-hopping (Nagle and Tristram-Nagle, 1983). The energy required for a single "hop" is about kT (i.e., thermal energy).] Thus, isolated in bulk solution, the internal defect pattern of a given protein is subject to a random generator from the solute–solvent system. The catalytic turnover numbers for isolated enzymes *in vitro* reflect this random field.

Consideration of the roles of mobile protons in enzyme structure, function, and evolution has led to the postulation that externally derived high-energy protons may function in the modulation of enzymatic events in organized states *in vivo* (Nagle and Morowitz, 1978; Nagle and Tristram-Nagle, 1983; Dunker, 1982; Berry, 1981; Welch and Berry, 1983; see also Kell and Westerhoff, this volume). When "hooked up," via hydrogen bond networks, to the energy continuum of a proton electrochemical potential difference (e.g., at a membrane interface), the conformational–dynamical aspects of enzyme action may be governed by a proton flow. The mechanics of this proton conformational interaction (Volkenstein, 1981) has been elaborated by Nagle *et al.* (1980). Moreover, the localized proton flow might serve a stoichiometric role as a source/sink of protons for protochemical events at active centers, as well as functioning in substrate translocation (see Section IV).

Hydrogen bond structures in proteins may also couple to other energy continua in organized states. Another likely source would be the electric fields at the surfaces of cytological particulates. A mode of operation here has been expounded by Fröhlich (1975, 1980). In fact, some 15 years ago Fröhlich (1968, 1969) conjectured that coherent excitations of proteins should play an important role in their biological activity. Over the years model calculations have supported this claim, and some experimental evidence has been found (Fröhlich and Kremer, 1983). Specifically, this notion maintains that the activities of localized enzymes couple to electric fields (and other energy sources) via excitation of metastable, highly polar states in the proteins. The modality involves longitudinal electric dipolar oscillations of hydrogen bond units, which are modeled as active *phonons* (i.e., quasi-particle lattice vibrations). Such systems can store external in specific modes (and subsequently do work with it), via a phonon condensation phenomenon analogous to the condensation of a Bose–Einstein gas, if the energy is pumped in above a critical rate. It was proposed that this modality may be involved in driving the conformational dynamics of organized enzyme systems, as well as substrate translocation processes.

Coherent hydrogen bond phonons (e.g., in α-helical regions) can propagate under some conditions in the form of dispersionless wavepackets termed *solitons*. This has been suggested as a transduction device in the operation of proton-motivated "molecular engines," ATP-linked muscle contraction, *inter alia* (Dunker, 1982; Davydov, 1981; Bilz *et al.*, 1981; DelGuidice *et al.*, 1982; Scott, 1981).

Again, we note the possible relevance of mobile electronic states, here in relation to protein dynamics. Interaction of electronic and nuclear degrees of freedom is well known in solid-state physics. An example is electron–phonon coupling in crystal lattice dynamics. The roles of such modes in enzyme action have been discussed by others (Volkenstein, 1981; Conrad, 1979; Caserta and Cervigni, 1974). The idea of long-range electronic semiconduction was proposed long ago by Szent-Györgyi (1941) as an integrative principle in cellular processes. Generally, this notion has not been taken seriously (at least for large supramolecular distances) because of the observed nature of the conduction band in isolated proteins, perceived problems in insulating the flow, etc., though one should note that in organized states *in vivo* such problems may be obviated (Lewis, 1979) (see Section IV,C,1,a).

IV. AN ELECTROCHEMICAL INTERPRETATION OF METABOLISM

Building on considerations of the last two sections, we offer some novel ideas on the manner by which the action of enzyme systems of intermediary metabolism may actually be geared to these localized energy continua. We base our approach on an electrochemical hypothesis proposed by Berry (1981). Of course, the electrochemical character of electron-transport phosphorylation is well known (see Kell and Westerhoff, Chapter 2, this volume). Apart from the prescient suggestions of a few individuals (Szent-Györgyi, 1941; Del Duca and Fuscoe, 1965; Bockris and Strinivasan, 1967; Fröhlich, 1968, 1969), however, the general importance of electrochemical processes in intermediary metabolism has received little attention until comparatively recently. There is now a gradual recognition that the principles of solution chemistry do not provide a satisfactory basis for explaining many phenomena associated with the living state (Welch, 1984). Concomitantly, the importance of vectorial electrochemical reactions in living cells has been increasingly emphasized (Bockris, 1969, 1980; Mitchell, 1979; Cope, 1980). In general, however, descriptions of electrochemical phenomena have been restricted to processes associated with cellular membranes recognized as possessing electrodic properties (see Kell and Westerhoff, this volume). Events in the cytoplasmic compartment have been excluded from consideration on the assumption that reactions occurring in this milieu would be scalar in nature, involving soluble chemical components interacting simply by diffusion and random collision. Doubt is cast on the validity of this assumption with the emergence of new ideas concerning cellular infrastructure, enzyme organization, and mechanisms of vectorial charge transfer. Actually, one finds that electrochemical events occur within all compartments of the cell and are essential for the establishment and maintenance of what we recognize as the "living state." An electrochemical approach, preliminary though it may be in present form, leads to a deeper understanding of metabolism and its regulation.

A. Structural Analogies between Living Cells and Electrochemical Devices

Most biochemists have long since discarded the notion that cells consist of a membranous sac containing a soup full of enzymes and substrates. The recognition that cells contain discrete organelles, such as nuclei, mitochondria, peroxisomes, endoplasmic reticulum, Golgi apparatus, and lysosomes, is virtually universal. Nevertheless, in examining reports dealing with the intermediary metabolism of these units, the impression is often gained that the organelles themselves are still regarded as membranous sacs with soluble contents and that outside the organelles, enzymes and substrates are thought to interact entirely by processes involving free diffusion and random collision.

How realistic is this view? Let us consider the properties of the hyaloplasmic space, the so-called cytosol of the cell. There is much evidence that a substantial quantity of the cellular water assumes a structured form (Clegg, 1984; Drost-Hansen and Clegg, 1979; Keith, 1979). Moreover, the penetrating studies of the electron microscopists have now revealed that surrounding all cellular organelles and pervading the whole cytoplasm is a delicate fine structure: the cytoskeleton, comprising microtubules, microfilaments, and a variety of other structural elements, such as intermediate filaments (Lazarides, 1980; Fisher and Phillips, 1979). In motile cells the cytoskeleton is particularly extensive, and it seems likely that an architecture of similar nature exists in all eukaryotic cells.

Recently, a further level of organization has been recognized. Morphological investigations have convincingly demonstrated an ubiquitous proteinaceous ultrastructure, *a microtrabecular lattice*, which pervades the cytoplasmic compartment of living cells (Wolosewick and Porter, 1979; Porter and Tucker, 1981). It is linked to the cytoskeleton and embraces the cellular organelles. This interconnection of cellular organelles is readily indicated, for example, in damaged (Berry and Simpson, 1962) or intact (Berry and Friend, 1969) isolated liver cells, where the mitochondria show no Brownian motion, even when displaced to one pole of the cell by high speed centrifugation ($100,000 \times g$) of hepatocyte suspensions (M. N. Berry, unpublished observations). In K. R. Porter's view (personal communication), all catalytic activity of the so-called soluble cytoplasm may be carried out by enzymes adsorbed to the microtrabecular lattice or to intracellular membranes.

Support for this surface view of cell metabolism has come from centrifugation studies on whole cells (Kempner and Miller, 1968; Zalokar, 1960), fractionation of cell fragments (Coleman, 1973; Masters, 1977, 1981), and extraction of the microtrabecular lattice (Schliwa et al., 1981). Additional evidence is seen from studies with enzyme systems localized within organelles (see Srere, this volume).

Experimental and theoretical calculations (Sitte, 1980; Srere, 1981) of protein concentrations in association with cytomembranes and organelles indicate high, crystallike densities of protein molecules in(on) particulate structures of the cell. There is a remarkable homology in the surface area-to-volume ratio for *all* membranous cytological substructures. Such considerations led Sitte (1980) to propose that all cytomembranous elements have evolved in a common fashion to function as effective protein collectors in the operation of cell metabolism.

Thus, the living cell can be pictured as a two-phase system: *a solid-state* phase, within and adjacent to which the enzyme-catalyzed reactions of intermediary metabolism take place; and *a bulk aqueous phase*, which functions largely in such subservient roles as thermal buffering, and distribution of common substrates, regulatory substances, and salt ions. This bulk phase ramifies throughout the cell in intimate relationship to the components of the solid-state phase and can be likened to an intracellular circulatory system. In other words, the cytoplasm can be regarded as having the structure and composition of a gel and, hence, represents a system with an enormous surface/volume ratio (Clegg, 1984; Loud, 1968; Blouin et al., 1977). This design is highly appropriate for an electrochemical system, the solid-state phase representing a multielectrode array, and the aqueous phase the electrolyte (Bockris, 1980). It is at the surface between the two phases that electrical (protical) processes would be initiated; the greater the surface area of the interface, the greater the possible current flow.

B. Examples of Electrochemical Phenomena in Living Cells

There is now general agreement that the synthesis of ATP in the transducing membranes of mitochondria, chloroplasts, and bacteria is an electrochemical process (Mitchell, 1979; Nicholls, 1982; Williams, 1978; Racker, 1980; Bockris and Tunuli, 1979), although the exact molecular mechanisms remain to be elucidated. While some modification (see Kell and Westerhoff, this volume) is required on Mitchell's original "chemiosmotic hypothesis," which argues that a proton electrochemical potential difference across a transducing membrane can promote the synthesis of ATP, there is a consensus that electrical energy is part and parcel of the driving force. A vectorial electric field permits the phosphorylation of ATP, a process that, in the absence of the energy-dependent charge separation induced by a redox reaction or by light, would be highly unfavorable thermodynamically. In fact, recent preliminary reports describe the synthesis of ATP by chloroplasts and by submitochondrial particles in response to an applied electric field (Teissie et al., 1981; Avron, 1982).

Electrochemical phenomena are by no means confined to reactions involving the synthesis of ATP within transducing membranes. In recent years, electrical changes within cells have been observed in processes as diverse as differentiation (Jaffe, 1981) and exocrine secretion (Williams, 1981). It is now appreciated that so-called nonexcitable cells respond to certain stimuli, often hormonal in nature, with a transient alteration in the polarity of their plasma membrane (Zierler and Rogus, 1981). This event invariably precedes any measurable chemical change. Indeed, for both the catecholamines and for insulin a case has been made that an alteration of membrane polarity is an initial and essential step in their cellular action (Zierler and Rogus, 1981).

In addition to these events at the plasma membrane, the electrogenic energy-dependent nature of the translocation of many types of ions or molecules across intracellular membranes has been widely recognized (Wilson, 1978; Wiskich, 1977). Moreover, there is now compelling evidence for cytoplasmic electrochemical processes. Studies indicate that intracellular ionic currents can steer protein molecules to their ultimate destination within Cecropia oocytes by a process of ultramicroelectrophoresis (Woodruff and Telfer, 1980). Moreover, steady cytoplasmic ion currents are found to play a role in the growth of fucoid eggs (Robinson and Jaffe, 1975), the regeneration of the stumps of amputated amphibian limbs (Jaffe, 1981), and the healing of mammalian epidermal wounds (Jaffe, 1981). They can also be readily demonstrated at several stages of the life cycle of the water mold Blastocladiella emersonii (Stump et al., 1980; see also Pohl, 1983). Electrical fields are likely to be involved in a host of as yet unexplored biological processes at the subcellular and supracellular levels (see Fröhlich and Kremer, 1983).

C. Mechanisms of Charge Conduction

1. Charge Transfer by Semiconduction

a. Conduction of Electrons.   The electron affinity of O_2 or other acceptors overcomes the ionization potential of H atoms derived from metabolic fuel, resulting in the separation of protons and electrons, which pass down their electrochemical potential gradients via separate paths. The transfer of electrons between carriers embedded within a membrane is a well-recognized phenomenon. Nevertheless, there is still considerable uncertainty as to the mechanism of electronic charge transfer. Although doubts have been raised concerning the feasibility of electron conduction between (and within) proteins, some evidence has now accrued that such semiconduction does indeed take place (Ladik, 1976; Lewis, 1979; Pethig, 1979; Pethig and Szent-Györgyi, 1980). Proteins existing in the solid-state phase of the cell as supramolecular complexes may manifest electron transfer between the individual moieties by semiconduction, rather than by diffusion and random collision of dissociable prosthetic groups. Although a bound prosthetic group within the redox protein may well be the primary site of interaction with substrate, electron flow from the prosthetic group through the body of the protein may subsequently occur (Cope, 1980; Pethig, 1979). Likewise, charge transfer between contiguous proteins may not involve prosthetic groups at the immediate site at which electron transfer between the macromolecules takes place.

At present there is little information available concerning these processes, which can be regarded essentially as mechanisms involving chemical radicals (Pryor, 1976). As electron flow occurs, there is created a series of radical intermediates (compounds containing an unpaired electron), the process being analogous to that known to occur, for example, in the mitochondrial, chloroplast, and endoplasmic reticulum electron transport systems (Nohl and Hegner, 1978; Loach and Hales, 1976; White and Coon, 1980). Generally, free radicals are considered to be highly reactive species (Pryor, 1976). However, it seems likely that the radical product formed within the solid-state phase may move only a very short distance before donating its unpaired electron to the next component of the metabolic sequence. Moreover, the enzyme–coenzyme radical, created in the redox reaction, may well be relatively stable by virtue of the opportunity for charge delocalization within the protein molecule (Henriksen et al., 1976).

While there may still be some argument as to the importance of electron semiconduction in nonmembranous areas of the cell, there is general agreement that some kind of electron conduction within membranes plays a fundamental role in the living state, ultimately providing the primary driving force for endergonic metabolism. Even so, details of these processes remain obscure, in particular the molecular mechanisms relating to energy transduction. The theories of Mitchell (1979) have led to the generally held view that electric fields established as a consequence of electron flow within and between cellular macromolecules generate a protonic electrochemical gradient permitting the flow of proton current (proticity) within the cell.

b. Conduction of Protons.   The conduction of protons within proteins has been experimentally established (Hille, 1976; Metzler, 1979; Kraut, 1977; Wang, 1968), and a number of plausible mechanisms have been proposed. One attractive theory (Nagle and Morowitz, 1978; Nagle and Tristram-Nagle,

1983), as mentioned in Section III, envisages protons as flowing in an ordered and vectorial manner through a series of hydrogen bond chains within proteins, a process analogous to the flow of protons in ice (Onsager, 1969) (Fig. 2). It seems possible that several polypeptides could interact to form an extended series of hydrogen bond chains permitting charge transfer between individual macromolecules (Nagle and Morowitz, 1978). The ice-like nature of the ordered hydration shell surrounding contiguous cytoplasmic proteins would not be anticipated to provide a barrier to proton conduction. Indeed, structured water itself might well provide hydrogen bond chains for the vectorial conduction of protons along macromolecular or membranous surfaces (Clegg, 1984; Kell, 1979).

It is likely, therefore, that living systems possess mechanisms for vectorial conduction of not only the electrons, but also the protons which originate from various metabolic reactions. Important sources and sinks for protons are the numerous redox reactions that occur within the living cell. These generate or consume protons, depending on the direction of the reaction. An example is the dehydrogenase-type reaction, for example, lactate dehydrogenase:

$$\text{lactate} + \text{NAD}^+ \rightleftharpoons \text{pyruvate} + \text{NADH} + \text{H}^+ \tag{1}$$

Other major sources of protons are the reactions involving cleavage of the phosphate anhyride bonds of ATP (or other high-energy compounds). For example, the phosphorylation of glucose generates a proton in the kinase-type reaction, for example, hexokinase:

$$\text{glucose} + \text{ATP}^{4-} \rightleftharpoons \text{glucose 6-phosphate}^{2-} + \text{ADP}^{3-} + \text{H}^+ \tag{2}$$

as do other phosphorylation reactions involving the cleavage of ATP. Likewise, the splitting of ATP, catalyzed by Na^+, K^+-ATPase, generates a proton:

$$\text{ATP}^{4-} + \text{H}_2\text{O} \rightleftharpoons \text{ADP}^{3-} + \text{P}^{2-} + \text{H}^+ \tag{3}$$

Conversely, the central reaction of oxidative phosphorylation [reaction (3) in reverse] consumes a proton, as do other ATP-synthesizing reactions, e.g., pyruvate kinase:

$$\text{phosphoenolpyruvate}^{3-} + \text{ADP}^{3-} + \text{H}^+ \rightleftharpoons \text{pyruvate}^- + \text{ATP}^{4-} \tag{4}$$

"Current" can flow only if the generated protons are insulated from the bulk aqueous phase. Evidence from *in vitro* enzymological studies provides an indication of how this may be achieved. In kinase-type reactions, in which the mechanism of catalysis has been explored, it has been found that reactions take place deep within the active site, a cleft from which bulk water is excluded (Anderson *et al.*, 1979). Even *in vitro*, protons are not released immediately into the medium on ATP splitting (Gutfreund and Trentham, 1975). Thus, it can be predicted that *in vivo* they do not instantaneously dissipate their energy through hydration in the bulk aqueous phase, but rather flow vectorially for a finite period within the kinase and possibly along contiguous enzymes of the associated metabolic pathway, before entering the bulk aqueous phase of the cell. Similar considerations should apply to proton release or consumption in NAD(P)-linked reactions (Gutfreund and Trentham, 1975) and others (Wang, 1968). In these types of processes, the enzyme participates not only as a catalyst of the chemical reaction but also as a mediator of charge (energy) transfer. In fact, vectorial charge relay is a not unusual feature of enzyme catalysis (Hille, 1976; Metzler, 1979).

It seems probable that proton current, generated by redox reactions and by ATP cleavage, flows through intracellular "circuits," specific for the various metabolic functions. Proton-conducting pathways along the planes of energy-transducing membranes have been proposed (see Kell and Westerhoff, Chapter 2, this volume). The idea of a "protoneural network" of proton transfer proteins is particularly attractive. Insulation of the proton flow along hydrogen bond chains from protein to protein might be facilitated by the ambience of structured water at the surface of the particulates. Continuity would be ensured, though, by direct protein–protein interaction, involving quaternary-type [or quinary-type (McConkey, 1982)] contacts. As is well known, such interactions involve hydrogen bonding across an interface that is highly hydrophobic in nature (Metzler, 1979; Hopfinger, 1977). Such interconnection may be important in allowing proton flow to extend into the bulk aqueous phase via the microtrabecular lattice. (A similar role of the lattice in electron semiconduction was suggested by Lewis, 1979).

Through the medium of this proticity, the cell accomplishes a variety of mechanical, osmotic, thermal, and electrical kinds of work (Skulachev, 1980). The role of ATP and similar compounds in these functions is related not merely to their possession of high-energy bonds, but also to their capability of serving as stable, diffusible sources of proton current. The presence of equilibrating mechanisms, such as the adenylate kinase system (Atkinson, 1977), ensures that local ATP depletion at any point in the cell is rapidly buffered.

2. Charge Transfer in the Bulk Aqueous Phase

The bulk aqueous phase represents much of the mass of the cell. The question arises as to whether redox reactions *in vivo* take place in this phase. *In vitro*, the generally accepted mechanism for redox reactions in an aqueous medium involves *hydride ion transfer* (Schellenberg, 1970). This term is used to denote the passage of one proton and two electrons from substrate to coenzyme, with an additional proton being released to the medium. The reduced coenzyme can then transfer the electrons and proton to an acceptor molecule, an additional proton being taken up from the medium.

The present state of our knowledge is too limited to come to firm conclusions about the importance of redox reactions in the bulk aqueous phase. Recently, evidence for the existence of specific channels conducting reducing equivalents between cellular compartments was obtained (Berry, 1980; Berry et al., 1985); and this would argue against the presence of freely diffusible coenzyme charge carriers. Nevertheless, there is a possibility that charge transfer by processes involving diffusion of soluble components does take place in localized regions of the cell. If this is the case, there must be mechanisms for the coordination of solid-state semiconduction with bulk aqueous phase charge transfer. This seems a profitable field for further investigation.

The bulk aqueous phase also acts as a store for key metabolites such as ATP. It may also serve as an energy store over and above its content of nucleotides. The cellular membranes enclose various compartments (Mollenhauer and Morre, 1978), and it seems likely that most (if not all) of these membranes maintain ion and proton gradients across them (Wilson, 1978; Wiskich, 1977; Reeves and Reames, 1981; Heinz et al., 1981; Rosier et al., 1981). These gradients, thus, represent free energy stores that can be utilized for cellular work. It should be emphasized, however, that these gradients are established and maintained as a consequence of redox reactions in the solid-state phase of the cell. They are produced as a result of energy flow and do not reflect thermodynamic equilibrium.

D. Coordination of Proton and Metabolic Fluxes

It is also necessary to address the question as to how the rate of charge transfer is coordinated with metabolic flux. This could be achieved readily if the proton flow along a series of enzymes forming a metabolic pathway were coordinated both in time and space with the movement of (anionic) reaction products between the active centers of the corresponding enzymes (Fig. 3). Under such conditions, the flow of protons through contiguous enzymes would be anticipated to bring about a synchronization of catalysis and substrate–product translocation along the sequence. For such a scheme to be operative in multienzyme systems, there must be rather close juxtaposition of the enzyme components. The issue here is one of channeling of intermediate substrates. The most expedient means of achieving this would involve a precise processing of one molecule of the pathway substrate at a time along a sequentially acting multienzyme complex. Such a compartmentation view is consistent with accumulating experimental measurements in vivo, indicating

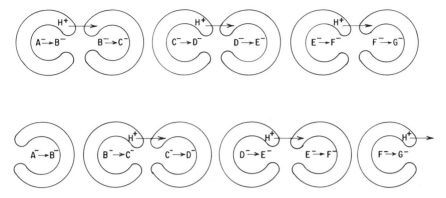

Fig. 3. A schematic representation of coordination of proton flow and metabolic flux along a metabolic pathway. Proton flow between contiguous enzymes brings about conformational changes that facilitate passage of metabolic products between enzymes of the pathway.

that substrate–enzyme molar ratios are often around unity (or, in some cases, less than unity).[1]

The view that many of the multienzyme systems, organized in the solid-phase regions *in vivo*, actually form protein–protein complexes (generating two-dimensional "crystalline arrays", see Section IV, A), lends credibility to the role of local energy continua (e.g., a mobile protonic state) in facilitating not only enzyme catalytic events but also the processing of intermediates. Thus, we consider that proton flow acts to drive the conformational changes occurring during enzyme function in the manner of a cyclical engine (see Fig. 3). The most expedient operation of a multienzyme array would entail *cyclical matching* of sequential enzymes, such that one enzyme in the aggregate is at the beginning of a conformational cycle at just the moment when the preceding enzyme–product complex is at the end of a cycle. Optimal conditions for efficiency in energy flow, as well as in metabolite channeling, would be obtained, provided that the translocation process is coordinated with proton flow within the enzyme ensemble. When these enzymes are isolated in bulk solution *in vitro*, any mobile protons required for catalysis would be seen as extracted from (or donated to) the solvent via the protein matrix. Studying a

---

[1] The alternative to this extreme holistic view of intermediary metabolism is the reductionistic idea that the steady-state pool levels of intermediate substrates are at the whim of K_m/V_{max} ratios of the *individual* (noninteracting) enzymes (Welch, 1977b). Although one might suppose the existence of physical (e.g., viscosity) barriers on the out-diffusion of intermediate substrates in localized microenvironments (Welch *et al.*, 1983), it is difficult to imagine how such a metabolic process could operate without inefficiencies (e.g., leakage, side reactions) in the absence of rather tight coupling.

dehydrogenase-type enzyme, Bennett *et al.* (1982), for example, revealed a stoichiometric proton release from an (as yet undetermined) ionizable group on the protein, triggered by a substrate-induced conformational change.

This translocation modality would demand energetic matching of the organized enzymes, according to the protoneural hypothesis (see Kell and Westerhoff, Chapter 2, this volume). In addition, it would require a degree of complementarity between a given enzyme and the enzyme–product complex of the preceding metabolic step (as opposed to the free product molecule). This notion was proposed by Friedrich, as part of his dynamic compartmentation model (see Friedrich, Chapter 3, this volume). Experimental support for this model has come, for example, from studies of glycolytic enzymes (Weber and Bernhard, 1982).

E. Influence of Cellular Electric Fields on Metabolic Processes

The likelihood that electrical fields are generated and sustained in living cells offers the opportunity for an alternative electrochemical approach to the understanding of metabolic systems. The fact that a large portion of the cell contains an aqueous medium of high dielectric is not seen as a bar to this viewpoint, since evidence suggests that the majority of cellular reactions occur in close association with membranes or with the microtrabecular lattice and could be insulated from the bulk aqueous phase. It is now evident that redox and ion-translocating systems occur within all major membranes of the cell (Wilson, 1978; Wiskich, 1977; Crane, 1977; Kimmich, 1981; Heinz *et al.*, 1975; Low and Crane, 1978) and that the formation and maintenance of electric fields may be a general feature of cellular membrane function. These fields, as determined by measurement of membrane potentials (Rottenberg, 1979; Bashford and Smith, 1979; Hoek *et al.*, 1980) (which may in many instances give understimates of localized field strength) can be remarkably large, on the order of 10^5 to 10^6 V/cm. The domain of the field may extend beyond the confines of the membrane to the adjacent cytoplasmic region.

As we saw in Section IV,C,1,b above, kinases and dehydrogenases are capable of generating or consuming proton charge. When adsorbed to intracellular particulates, therefore, they are potentially sensitive to the influence of local electric fields. Since these fields are vectors, it can be anticipated that they will affect the vectorial movement of charge (e.g., protons) in domains of the organized enzymes. (In this regard, α-helical elements may be particularly important; see Section III,A.) Hence, the energy contained in an electric field could be harnessed, for example, to drive a reaction involving ATP synthesis or cleavage, as first argued by Mitchell's chemiosmotic hypothesis for mitochondrial oxidative phosphorylation

(Mitchell, 1979). There seems no *a priori* reason why this concept cannot be applied to reactions associated with all intracellular membranes and the cytoplasmic interface thereto. The possibility of a vectorial electrical component to a variety of localized enzyme reactions needs serious consideration.

In addition, Fröhlich's theory (see Section III,B) suggests a means by which the protein matrix itself can couple dynamically to local electric fields. Through long-range Coulombic interactions involving dipole matching between protein and substrate, the Fröhlich model also suggests a manner of site-to-site metabolite translocation. It is quite possible that the mobile proton mode is inextricably associated with other transduction modes (e.g., that proposed by Fröhlich). The possibility of such a unification remains to be explored (see Dunker, 1982).

V. LONG-RANGE ENERGY CONTINUA AND METABOLIC REGULATION

In the preceding section, we argued the case that such entities as a mobile protonic state and localized electric fields could represent power sources of intermediary metabolism. Naturally, this leads to a holonomic view of the notion of *metabolic regulation*. Intimately related is the question of free energy dissipation (and conservation) in metabolism, as this is the focal point in the standard paradigm for identifying regulatory steps (Newsholme and Start, 1973; Herman *et al.*, 1980). From biochemistry textbooks, we are taught that the ability of living systems to carry out endergonic processes is attributable to the special properties of a class of molecules containing high-energy bonds, the paragon of which is ATP. Thus, ATP is regarded as a "functional coupling agent" (Atkinson, 1977) between catabolic and anabolic processes. The belief that the great majority of cellular reactions utilizing high-energy phosphate bonds are scalar in nature, involving diffusion and random collision of metabolites and coenzymes, leads to the pervading assumption that the direction and rate of metabolic flow are determined by the ratios of the adenine nucleotides (and in some instances inorganic phosphate), in conjunction with the pathway mass-action ratios. It is commonly held that the flux through each pathway is controlled specifically by the activities of allosteric enzymes positioned at key rate-limiting, irreversible steps. This notion is supported by measurements indicating that mass-action ratios for the putative regulatory reactions differ significantly from the corresponding equilibrium constants.

The foregoing surface view of intermediary metabolism, along with the ambience of long-range energy continua, casts a spectre of mismeasure on the usual assessment of metabolic regulation. Fractionation studies (Masters,

1977, 1981; Bessman and Geiger, 1980; Herman *et al.*, 1980) have revealed that many of the enzymes associated with cellular membranes are kinases and dehydrogenases, a number of which have been deemed *regulatory*. In calculating the nonequilibrium character (free energy change) for such enzyme reactions, the "free" proton component in the stoichiometry is usually ignored, by assuming that it equilibrates (by hydration) in a buffered bulk phase. A substantial portion of the free energy loss arising from ATP cleavage, for example, is associated with the liberation of the proton (Banks and Vernon, 1978). The parallel flow of protons through hydrogen bond chains in multienzyme aggregates during the catalytic sequence represents an "energized" state (Nagle and Tristram-Nagle, 1983), relaxing cyclically as protons pass (along with reaction products) to the contiguous enzymes (or to the proton-conducting superstructure). Accordingly, some of the free energy associated with the individual substrate transformations is conserved and transduced within the enzyme assemblage. The proton motive force permeating the organized systems in the living cell may, in principle, poise a given enzyme reaction at any distance from equilibrium. It follows that the determination of the free energy changes of individual enzyme reactions from their behavior isolated in aqueous solution may have little relevance to conditions extant *in vivo*.

Moreover, the free energy character of an enzyme reaction may be influenced by local electric fields in membrane-associated systems. This could be exerted by the effect of the electric field on binding/release of substrate–product molecules (Fröhlich, 1975, 1980), or on the proton activity in the protein matrix (Berry *et al.*, 1983). Metabolic regulation *in vivo* might, therefore, involve a complex and variable interplay of electro(proto)chemical factors, mass-action ratios of pathway substrates, and ligand-induced allosteric actions. It is of interest to note, in this regard, that the allosteric properties of enzymes sometimes differ when studied in free solution versus in membrane-adsorbed states (see Kurganov, Chapter 5, this volume).

VI. CONCLUDING REMARKS

The details of biological electrochemical mechanisms remain obscure, and it is not feasible to ratiocinate thereupon without indulging to a substantial degree in speculation and conjecture. Nonetheless, the fundamental features of the theoretical approaches in Sections IV and V flow logically and consistently from a merger of enzymology, bioenergetics, and cell biology. Most notable is the congruity with the structural–functional properties of proteins, principles of enzyme organization, recent anatomical knowledge of intracellular matrices, and the properties of energy-transducing particulate structures in the cell. The traditional reductionistic manner of extracting and studying individual enzymes *in vitro* is biased against any long-range, energy-transduction modality. Removal of this bias is paving the way for empirical approaches which recognize the synergistic nature of living systems.

The time seems well overdue for initiation of multidisciplinary approaches to the understanding of enzyme action and metabolic regulation in organized states *in vivo*. Theoretical constructs and analogical reasoning from various areas are required, as we pry into the functional meshwork of the intact metabolic machinery. As the "surface view" of intermediary metabolism reaches fruition, we appreciate the import of studies from artificially immobilized enzyme systems. These "macroscopic" models have provided useful clues as to the nature of enzyme action in heterogeneous states. Now, with the emergence of "microscopic" pictures of naturally occurring multienzyme clusters, basic enzymology finds itself in a position to reciprocate in the realm of technological applications; for these biological systems epitomize such concepts as "efficiency" and "economy," which may well transcend to the level of human industry.

REFERENCES

Anderson, C. M., Zucker, F. H., and Steitz, T. A. (1979). *Science* **204**, 375–380.

Andrews, P. R., Smith, G. D., and Young, I. G. (1973). *Biochemistry* **12**, 3492–3498.

Atkinson, D. E. (1977). "Cellular Energy Metabolism and Its Regulation." Academic Press, New York.

Avron, M. (1982). *Int. Congr. Biochem., 12th* p. 26 (Abstr.)

Banacky, P. (1981). *Biophys. Chem.* **13**, 39–47.

Banks, B. E. C., and Vernon, C. A. (1978). *Trends Biochem. Sci.* **3**, 156–158.

Bashford, C. L., and Smith, J. C. (1979). "Methods in Enzymology" (S. Fleischer and L. Packer, eds.), Vol. 55, pp. 569–586. Academic Press, New York.

Beece, D., Eisenstein, L., Frauenfelder, H., Good, D., Marden, M. C., Reinisch, L., Reynolds, A. H., Sorenson, L. B., and Yue, K. T. (1980). *Biochemistry* **19**, 5147–5157.

Bennett, A. F., Buckley, P. D., and Blackwell, L. F. (1982). *Biochemistry* **21**, 4407–4413.

Berry, M. N. (1980). *FEBS Lett.* **117** (Suppl.), K106–K120.

Berry, M. N. (1981). *FEBS Lett.* **134**, 133–138.

Berry, M. N., and Friend, D. S. (1969). *J. Cell Biol.* **43**, 506–520.

Berry, M. N., and Simpson, F. O. (1962). *J. Cell. Biol.* **15**, 9–17.

Berry, M. N., Grivell, A. R., and Wallace, P. G. (1983). *Pharmacol. Biochem. Behav.* **18** (Suppl. 1), 201–207.

Berry, M. N., Grivell, A. R., and Wallace, P. G. (1985). *In* "Comprehensive Treatise on Electrochemistry" (S. Srinivasan and Yu. A. Bhizmadzhev, eds.), Vol. 10. *Bioelectrochemistry*. Plenum, New York, in press.

Bessman, S. P., and Geiger, P. J. (1980). *Curr. Top. Cell. Regul.* **16**, 55–86.

Bilz, H., Buttner, H., and Fröhlich, H. (1981). *Z. Naturforsch.* **36B**, 208–212.

Blouin, A., Bolender, R. P., and Weibel, E. R. (1977). *J. Cell. Biol.* **72**, 441–455.

Blumenfeld, L. A. (1983). "Physics of Bioenergetic Processes." Springer-Verlag, Berlin and New York.

Bockris, J.O'M. (1969). *Nature (London)*, **224**, 775–777.

Bockris, J.O'M. (1980). *In* "Bioelectrochemistry" (H. Keyzer and F. Gutmann, eds.), pp. 5–17. Plenum, New York.

Bockris, J.O'M., and Strinivasan, S. (1967). *Nature (London)* **215**, 197.

Bockris, J.O'M., and Tunuli, M. S. (1979). *J. Electroanal. Chem.* **100**, 7–12.

Bohm, D. (1980). Wholeness and the Implicate Order." Routledge & Kegan Paul, Boston.

Carrington, T., and Garvin, D. (1969). *In* "Comprehensive Chemical Kinetics" (C. H. Bamford and C. F. H. Tipper, eds.), Vol. 3, pp. 107–181. Elsevier, Amsterdam.

Caserta, G., and Cervigni, T. (1974). *Proc. Natl. Acad. Sci. U.S.A.* **71**, 4421–4424.

Clegg, J. S. (1984). *Am. J. Physiol.* **246**, R133–R151.

Coleman, R. (1973). *Biochim. Biophys. Acta* **300**, 1–30.

Conrad, M. (1979). *J. Theor. Biol.*, **79**, 137–156.

Cope, F. W. (1980). *In* "Bioelectrochemistry" (H. Keyzer and F. Gutman, eds.), pp. 297–329. Plenum, New York.

Crane, R. K. (1977). *Rev. Physiol. Biochem. Pharmacol.* **78**, 99–159.

Damjanovich, S., and Somogyi, B. (1971). *Proc. Eur. Biophys. Congr., 1st* **6**, 133.

Davydov, A. S. (1981). *Physica* **3D**, 1–22.

DeBrunner, P. G., and Frauenfelder, H. (1982). *Annu. Rev. Phys. Chem.* **33**, 283–299.

Del Duca, M. G., and Fuscoe, J. M. (1965). *Int. Sci. Technol.* **3**, 56–67.

Del Guidice, E., Doglia, S., and Milani, M. (1982). *Phys. Scr.* **26**, 232–238.

Drost-Hansen, W., and Clegg, J. eds. (1979). "Cell-Associated Water." Academic Press, New York.

Duffy, P. (1971). *Biochim. Biophys. Acta* **244**, 606–612, 613–617.

Dunker, A. K. (1982). *J. Theor. Biol.* **97**, 95–127.

Eigen, M. (1974). *In* "Quantum Statistical Mechanics in the Natural Sciences" (B. Kursunoglu, S. L. Mintz, and S. M. Widmayer, eds.), pp. 37–61. Plenum Press, New York.

Fersht, A. (1985). "Enzyme Structure and Mechanism," 2nd Ed. Freeman, San Francisco, California.

Fisher, M. M., and Phillips, M. J. (1979). *In* "Progress in Liver Diseases" (H. Popper and F. Schaffner, eds.), Vol. 6, pp. 105–121. Grune & Stratton, New York.

Fisher, M. M., and Phillips, M. J. (1979). *In* "Progress in Liver Diseases" (H. Popper and F. Schaffner, eds.), Vol. 6, pp. 105–121. Grune Stratton, New York.

Fröhlich, H. (1968). *Int. J. Quant. Chem.* **2**, 641–649.

Fröhlich, H. (1969). *In* "Theoretical Physics and Biology" (M. Marois, ed.), pp. 13–22. Elsevier, Amsterdam.

Fröhlich, H. (1975). *Proc. Natl. Acad. Sci. U.S.A.* **72**, 4211–4215.

Fröhlich, H. (1980). *Adv. Electron. Electron Phys.* **53**, 85–152.

Fröhlich, H., and Kremer, F., eds. (1983). "Coherent Excitations in Biological Systems" Springer-Verlag, Berlin and New York.

Gavish, B. (1978). *Biophys. Struct. Mech.* **4**, 37–52.

Gutfreund, H. (1976). *FEBS Lett.* **62** (Suppl.), E1-E2.

Gutfreund, H., and Trentham, D. R. (1975). *In* "Energy Transformation in Biological Systems" (G. E. W. Wolstenholme and E. W. Fitzsimons, eds.), pp. 69–86. Elsevier, Amsterdam.

Heinz, E., Geck, P., and Pietrzyk, C. (1975). *Ann. N. Y. Acad. Sci.* **264**, 428–441.

Heinz, A., Sachs, G., and Schafer, J. A. (1981). *J. Membr. Biol.* **61**, 143–153.

Henriksen, T., Melo, T. B., and Saxebol, G. (1976). *In* "Free Radicals in Biology" (W. A. Pryor, ed.), Vol. 2, pp. 213–256. Academic Press, New York.

Herman, R. H., Cohn, R. M., and McNamara, P. D., eds. (1980). "Principles of Metabolic Regulation in Mammalian Systems." Plenum, New York.

Hill, T. L. (1977). "Free Energy Transduction in Biology" Academic Press, New York.

Hille, B. (1976). *Annu. Rev. Physiol.* **38**, 139–152.

Hoek, J. B., Nicholls, D. G., and Williamson, J. R. (1980). *J. Biol. Chem.* **255**, 1458–1464.

Hol, W. G. J., van Duijnen, P. T., and Berendsen, H. J. C. (1978). *Nature (London)* **273**, 443–446.

Hopfinger, A. J. (1977). "Intermolecular Interactions and Biological Organization" Wiley, New York.

Ikegami, A. (1977). *Biophys. Chem.* **6**, 117–130.

Jaffee, L. F. (1981). *Fed. Proc., Fed. Am. Soc. Exp. Biol.* **40**, 125–127.

Karush, F. (1950). *J. Am. Chem. Soc.* **72**, 2705–2713.

Keith, A. D., ed. (1979). "The Aqueous Cytoplasm" Dekker, New York.

Keleti, T. (1975). *Proc. FEBS Meet., 9th* **32**, 3–27.

Keleti, T. (1978). *Symp. Biol. Hung.* **21**, 107–124.

Kell, D. B. (1979). *Biochim. Biophys. Acta* **549**, 55–99.

Kempner, E. S., and Miller, J. H. (1968). *Exp. Cell. Res.* **51**, 150–156.

Kimmich, G. A. (1981). *Fed. Proc., Fed. Am. Soc. Exp. Biol.* **40**, 2474–2479.

Kraut, J. (1977). *Annu. Rev. Biochem.* **46**, 331–358.

Krimm, S., and Dwivedi, A. M. (1982). *Science* **216**, 407–408.

Ladik, J. (1976). *Int. J. Quant. Chem. Quant. Biol. Symp.* **3**, 237–246.

Laidler, K. J., and Bunting, P. S. (1973). "The Chemical Kinetics of Enzyme Action (2nd Ed.). Oxford Univ. Press, London and New York.

Lazarides, E. (1980). *Nature (London)* **283**, 249–256.

Lewis, J. (1979). *Ciba Found. Symp.* **67**, 65–82.

Likhtenshtein, G. I. (1966). *Biofizika* **11**, 24–32.

Loach, P. A., and Hales, B. J. (1976). *In* "Free Radicals in Biology" (W. A. Pryor, ed.), Vol. 1, pp. 199–237. Academic Press, New York.

Loud, A. V. (1968). *J. Cell. Biol.* **37**, 27–46.

Low, H., and Crane, F. L. (1978). *Biochim. Biophys. Acta* **515**, 141–161.

Lumry, R. (1971). *In* "Electron and Coupled Energy Transfer in Biological Systems" (T. King and M. Klingenberg, eds.), pp. 1–116. Dekker, New York.

Lumry, R., and Biltonen, R. (1969). *In* "Structure and Stability of Biological Macromolecules (S. N. Timasheff and G. D. Fasman, eds.), pp. 65–212. Dekker, New York.

Lumry, R., and Rosenberg, A. (1975). *Colloq. Int. C.N.R.S.* No. 246, pp. 53–62.

McConkey, E. H. (1982). *Proc. Natl. Acad. Sci., U.S.A.* **79**, 3236–3240.

Masters, C. J. (1977). *Curr. Top. Cell. Regul.* **12**, 75–105.

Masters, C. J. (1981). *CRC Crit. Rev. Biochem.* **11**, 105–143.

Metzler, D. E. (1979). *Adv. Enzymol.* **50**, 1–40.

Mitchell, P. (1979). *Eur. J. Biochem.* **95**, 1–20.

Mollenhauer, H. H., and Morre, D. J. (1978). *Subcell. Biochem.* **5**, 327–359.

Monod, J., Wyman, J., and Changeux, J.-P. (1965). *J, Mol. Biol.* **12**, 88–118.

Nagle, J. F., and Morowitz, H. J. (1978). *Proc. Natl. Acad. Sci. U.S.A.* **75**, 298–302.

Nagle, J. F., and Tristram-Nagle, S. (1983). *J. Membr. Biol.* **74**, 1–14.

Nagle, J. F., Mille, M., and Morowitz, H. J. (1980). *J. Chem. Phys.* **72**, 3959–3971.

Newsholme, E. A., and Start, C. (1973). "Regulation in Metabolism." Wiley, New York.

Nicholls, D. G. (1982). "Bioenergetics." Academic Press, New York.

Nohl, H. and Hegner, J. (1978). *Eur. J. Biochem.* **82**, 563–567.

Onsager, L. (1969). *Science* **166**, 1359–1364.

Ovádi, J., and Keleti, T. (1978). *Eur. J. Biochem.* **85**, 157–161.

Pethig, R. (1979). "Dielectric and Electronic Properties of Biological Materials." Wiley, New York.

Pethig, R., and Szent-Györgyi, A. (1980). *In* "Bioelectrochemistry" (H. Keyzer and F. Gutmann, eds.), pp. 227–265. Plenum, New York.

Pohl, H. A. (1983). *In* "Coherent Excitations in Biological Systems" (H. Fröhlich and F. Kremer, eds.), pp. 199–210. Springer-Verlag, Berlin and New York.

Porter, K. R., and Tucker, J. B. (1981). *Sci. Am.* **244**, 40–51.

Pryor, W. A. (1976). *In* "Free Radicals in Biology" (W. A. Pryor, ed.), Vol. 1, pp. 1–49. Academic Press, New York.

Rabinovitch, B. S., and Flowers, M. C. (1964). *Q. Rev. Chem. Soc.* **18**, 122–167.

Racker, E. (1980). *Fed. Proc., Fed. Am. Soc. Exp. Biol.* **39**, 210–215.

Reeves, J. P., and Reames, T. (1981). *J. Biol. Chem.* **256**, 6047–6053.

Robinson, K. R., and Jaffe, L. F. (1975). *Science* **187**, 70–72.

Rosier, R. N., Tucker, D. A., Meerdink, S., Jain, I., and Gunter, T. E. (1981). *Arch. Biochem. Biophys.* **210**, 549–564.

Rottenberg, H. (1979). *In* "Methods in Enzymology" (S. Fleischer and L. Packer, eds.), Vol. 55, pp. 547–569. Academic Press, New York.

Schellenberg, K. A. (1970). *In* "Pyridine Nucleotide-Dependent Dehydrogenases" (H. Sund, ed.), pp. 15–29. Springer-Verlag, Berlin and New York.

Schliwa, M., van Blerkom, J., and Porter, K. R. (1981). *Proc. Natl. Acad. Sci. U.S.A.* **78**, 4329–4333.

Scott, A. C. (1981). *In* "Nonlinear Phenomena in Physics and Biology" (R. N. Enns, B. L. Jones, R. M. Miura and S. S. Rangnekar, eds.), pp. 7–82. Plenum, New York.

Shnoll, S. E., and Chetverikova, E. P. (1975). *Biochim. Biophys. Acta* **403**, 89–97.

Sitte, P. (1980). *In* "Cell Compartmentation and Metabolic Channeling" (L. Nover, F. Lynen, and K. Mothes, eds.), pp. 17–32. Elsevier, Amsterdam.

Skulachev, V. P. (1980). *Can. J. Biochem.* **58**, 161–175.

Somogyi, B. (1974). *Acta Biochim. Biophys. Acad. Sci. Hung.* **9**, 175–184, 185–196.

Somogyi, B., and Damjanovich, S. (1975). *J. Theor. Biol.* **48**, 393–401.

Somogyi, B., Welch, G. R., and Damjanovich, S. (1984). *Biochim. Biophys. Acta* (*Rev. Bioenerg.*) **768**, 81–112.

Srere, P. (1981). *Trends Biochem. Sci.* **6**, 4–7.

Srere, P. A., and Estabrook, R. W., eds. (1978). "Microenvironments and Metabolic Compartmentation." Academic Press, New York.

Srere, P. A., and Mosbach, K. (1974). *Annu. Rev. Microbiol.* **28**, 61–83.

Straub, F. B., and Szabolcsi, G. (1964). *In* "Molekularnaya Biologiya," p. 182. Izd. Nauka, Moscow.

Stump, R. F., Robinson, K. R., Harold, R. L., and Harold, F. M. (1980). *Proc. Natl. Acad. Sci. U.S.A.* **77**, 6673–6677.

Szent-Györgyi, A. (1941). *Nature* (*London*) **148**, 157–159.

Teissie, J., Knox, B. E., Tsong, T. Y., and Wehrle, J. (1981). *Proc. Natl. Acad. Sci. U.S.A.* **78**, 7473–7477.

Vallee, B. L., and Williams, R. J. P. (1968). *Proc. Natl. Acad. Sci. U.S.A.* **59**, 498–505.

Van Duijnen, P. T., and Thole, B. T. (1981). *Chem. Phys. Lett.* **83**, 129–133.

Volkenstein, M. K. (1981). *J. Theor. Biol.* **89**, 45–51.

Wang, J. H. (1968). *Science* **161**, 328–334.

Warshel, A. (1978). *Proc. Natl. Acad. Sci. U.S.A.* **75**, 5250–5254.

Weber, J. P., and Bernhard, S. A. (1982). *Biochemistry* **21**, 4189–4194.

Welch, G. R. (1977a). *Prog. Biophys. Mol. Biol.* **32**, 103–191.

Welch, G. R. (1977b). *J. Theor. Biol.* **68**, 267–291.

Welch, G. R. (1984). *In* "Dynamics of Biochemical Systems" (A. Cornish-Bowden and J. Ricard, eds.), pp. 85–101, Plenum, New York.

Welch, G. R., ed. (1985). "The Fluctuating Enzyme." Wiley, New York, in press.

Welch, G. R. and Berry, M. N. (1983). *In* "Coherent Excitations in Biological Systems" (H. Fröhlich and F. Kremer, eds.), pp. 95–116. Springer-Verlag, Berlin and New York.

Welch, G. R., and DeMoss, J. A. (1978). *In* "Microenvironments and Metabolic Compartmentation" (P. A. Srere and R. W. Estabrook, eds.), pp. 323–344. Academic Press, New York.

Welch, G. R., and Keleti, T. (1981). *J. Theor. Biol.* **93**, 701–735.

Welch, G. R., Somogyi, B., and Damjanovich, S. (1982). *Prog. Biophys. Mol. Biol.* **39,** 109–146.
Welch, G. R., Somogyi, B., Matkó, J., and Papp, S. (1983). *J. Theor. Biol.* **100,** 211–238.
White, R. E., and Coon, M. J. (1980). *Annu. Rev. Biochem.* **49,** 315–356.
Williams, J. A. (1981). *Fed. Proc. Fed. Am. Soc. Exp. Biol.* **40,** 128–134.
Williams, R. J. P. (1978). *FEBS Lett.* **85,** 9–19.
Wilson, D. B. (1978). *Annu. Rev. Biochem.* **47,** 933–965.
Wiskich, J. T. (1977). *Annu. Rev. Plant Physiol.* **28,** 45–69.
Wolosewick, J. J., and Porter, K. R. (1979). *J. Cell. Biol.* **82,** 114–139.
Woodruff, R. I., and Telfer, W. H. (1980). *Nature (London)* **286,** 84–86.
Wyman, J. (1975). *Proc. Natl. Acad. Sci. U.S.A.* **72,** 3983–3987.
Zalokar, M. (1960). *Exp. Cell Res.* **19,** 114–132.
Zierler, K., and Rogus, E. M. (1981). *Fed. Proc., Fed. Am. Soc. Exp. Biol.* **40,** 121–124.

Index